U0627248

W. Greiner

QUANTUM MECHANICS

SPECIAL CHAPTERS

Springer-Verlag Berlin Heidelberg GmbH

Greiner
Quantum Mechanics
An Introduction 3rd Edition

Greiner
Quantum Mechanics
Special Chapters

Greiner · Müller
Quantum Mechanics
Symmetries 2nd Edition

Greiner
Relativistic Quantum Mechanics
Wave Equations 2nd Edition

Greiner · Reinhardt
Field Quantization

Greiner · Reinhardt
Quantum Electrodynamics
2nd Edition

Greiner · Schäfer
Quantum Chromodynamics

Greiner · Maruhn
Nuclear Models

Greiner · Müller
Gauge Theory of Weak Interactions
2nd Edition

Greiner
Mechanics I
(in preparation)

Greiner
Mechanics II
(in preparation)

Greiner
Electrodynamics
(in preparation)

Greiner · Neise · Stöcker
Thermodynamics and Statistical Mechanics

Walter Greiner

QUANTUM MECHANICS

SPECIAL CHAPTERS

With a Foreword by
D. A. Bromley

With 120 Figures,
75 Worked Examples and Problems

Springer

Professor Dr. Walter Greiner

Institut für Theoretische Physik der
Johann Wolfgang Goethe-Universität Frankfurt
Postfach 11 19 32
D-60054 Frankfurt am Main
Germany

Street address:

Robert-Mayer-Strasse 8–10
D-60325 Frankfurt am Main
Germany

email: greiner@th.physik.uni-frankfurt.de

Title of the original German edition: *Theoretische Physik,* Ein Lehr- und Übungsbuch,
Band 4a: Quantentheorie, Spezielle Kapitel, 3. Aufl., © Verlag Harri Deutsch, Thun 1989

1st Edition 1998, 2nd Printing 2001

ISBN 978-3-540-60073-2

Library of Congress Cataloging-in-Publication Data.

Greiner, Walter, 1935 – [Quantenmechanik, English] Quantum mechanics. Special chapters / Walter Greiner; with a foreword by D. A. Bromley, p. cm. Includes bibliographical references and index
ISBN 978-3-540-60073-2 ISBN 978-3-642-58847-1 (eBook)
DOI 10.1007/978-3-642-58847-1

1. Quantum theory, 2. Electrodynamics, 3. Quantum field theory, 4. Mathematical physics. I. Greiner, Walter, 1935 – Theoretische Physik, English, Band 4a. II. Title. QC174.12.G74513 1998 530.12-dc21 97-24126

Foreword to Earlier Series Editions

More than a generation of German-speaking students around the world have worked their way to an understanding and appreciation of the power and beauty of modern theoretical physics – with mathematics, the most fundamental of sciences – using Walter Greiner's textbooks as their guide.

The idea of developing a coherent, complete presentation of an entire field of science in a series of closely related textbooks is not a new one. Many older physicists remember with real pleasure their sense of adventure and discovery as they worked their ways through the classic series by Sommerfeld, by Planck and by Landau and Lifshitz. From the students' viewpoint, there are a great many obvious advantages to be gained through use of consistent notation, logical ordering of topics and coherence of presentation; beyond this, the complete coverage of the science provides a unique opportunity for the author to convey his personal enthusiasm and love for his subject.

The present five-volume set, *Theoretical Physics*, is in fact only that part of the complete set of textbooks developed by Greiner and his students that presents the quantum theory. I have long urged him to make the remaining volumes on classical mechanics and dynamics, on electromagnetism, on nuclear and particle physics, and on special topics available to an English-speaking audience as well, and we can hope for these companion volumes covering all of theoretical physics some time in the future.

What makes Greiner's volumes of particular value to the student and professor alike is their completeness. Greiner avoids the all too common "it follows that ..." which conceals several pages of mathematical manipulation and confounds the student. He does not hesitate to include experimental data to illuminate or illustrate a theoretical point and these data, like the theoretical content, have been kept up to date and topical through frequent revision and expansion of the lecture notes upon which these volumes are based.

Moreover, Greiner greatly increases the value of his presentation by including something like one hundred completely worked examples in each volume. Nothing is of greater importance to the student than seeing, in detail, how the theoretical concepts and tools under study are applied to actual problems of interest to a working physicist. And, finally, Greiner adds brief biographical sketches to each chapter covering the people responsible for the development of the theoretical ideas and/or the experimental data presented. It was Auguste Comte (1798–1857) in his *Positive Philosophy* who noted, "To understand a science it is necessary to know its history". This is all too often forgotten in modern physics teaching and the bridges that Greiner builds to the pioneering figures of our science upon whose work we build are welcome ones.

Greiner's lectures, which underlie these volumes, are internationally noted for their clarity, their completeness and for the effort that he has devoted to making physics an integral whole; his enthusiasm for his science is contagious and shines through almost every page.

These volumes represent only a part of a unique and Herculean effort to make all of theoretical physics accessible to the interested student. Beyond that, they are of enormous value to the professional physicist and to all others working with quantum phenomena. Again and again the reader will find that, after dipping into a particular volume to review a specific topic, he will end up browsing, caught up by often fascinating new insights and developments with which he had not previously been familiar.

Having used a number of Greiner's volumes in their original German in my teaching and research at Yale, I welcome these new and revised English translations and would recommend them enthusiastically to anyone searching for a coherent overview of physics.

Yale University
New Haven, CT, USA
1989

D. Allan Bromley
Henry Ford II Professor of Physics

Preface

Theoretical physics has become a many-faceted science. For the young student it is difficult enough to cope with the overwhelming amount of new scientific material that has to be learned, let alone obtain an overview of the entire field, which ranges from mechanics through electrodynamics, quantum mechanics, field theory, nuclear and heavy-ion science, statistical mechanics, thermodynamics, and solid-state theory to elementary-particle physics. And this knowledge should be acquired in just 8–10 semesters, during which, in addition, a Diploma (Masters) thesis has to be worked on and examinations prepared for. All this can be achieved only if the academic teachers help to introduce the student to the new disciplines as early on as possible, in order to create interest and excitement that in turn set free essential new energy.

At the Johann Wolfgang Goethe University in Frankfurt am Main we therefore confront the student with theoretical physics immediately, in the first semester. Theoretical Mechanics I and II, Electrodynamics, and Quantum Mechanics I – An Introduction are the basic courses during the first two years. These lectures are supplemented with many mathematical explanations and much support material. After the fourth semester of studies, graduate work begins, and Quantum Mechanics II – Symmetries, Statistical Mechanics and Thermodynamics, Relativistic Quantum Mechanics, Quantum Electrodynamics, the Gauge Theory of Weak Interactions, and Quantum Chromodynamics are obligatory. Apart from these, a number of supplementary courses on special topics are offered, such as Hydrodynamics, Classical Field Theory, Special and General Relativity, Many-Body Theories, Nuclear Models, Models of Elementary Particles, and Solid-State Theory.

This volume of lectures provides an important supplement on the subject of *Quantum Mechanics*. These *Special Chapters* are in the form of overviews on various subjects in modern theoretical physics. The book is devised for students in their fifth semester who are still trying to decide on an area of research to follow, whether they would like to focus on experiments or on theory later on.

The observation by Planck and Einstein that a classical field theory – electrodynamics – had to be augmented by corpuscular and nondeterministic aspects stood at the cradle of quantum theory. At around 1930 it was recognized that not only the radiation field with photons but also matter fields, e.g. electrons, can be described by the same procedure of second quantization.

Within this formalism, matter is represented by operator-valued fields that are subject to certain (anti-)commutation relations. In this way one arrives at a theory describing systems of several particles (field quanta) which in particular provides a very natural way to formulate the creation and annihilation of particles. Quantum field theory has become the language of modern theoretical physics. It is used in particle and high-energy physics, but also the description of many-body systems encountered in solid-state, plasma, nuclear, and atomic physics make use of the methods of quantum field theory.

We use second quantization (creation and annihilation operators for particles and modes) extensively. The lectures begin with the quantization of the electromagnetic fields. As well as the state vectors with a well-defined (sharp) number of photons, the coherent (Glauber) states are discussed, followed by absorption and emission processes, the lifetime of exited states, the width of spectral lines, the self-energy problem, photon scattering, and Cherenkov radiation. In between it seemed fit to elucidate on the Aharanov–Bohm and Casimir effects. Many applications are hidden in Exercises and Examples (e.g. two-photon decay, the Compton effect, photon spectra of black bodies).

Fermi and Bose statistics and their relationship with the way of quantization (commutators, anticommutators) are discussed in the third chapter. Here also, tripple commutators leading to para-Bose and para-Fermi statistics are reflected upon. After describing quantum fields with interaction (Chap. 4) we address renormalization problems, not in full (as done in the lectures on quantum electrodynamics and on field quantization), but in a rather elementary way such that the student gets a feeling for the problems, their difficulties, and their solution.

In Chaps. 6 to 9 the methods of quantum field theory are applied to topics in solid-state and plasma physics: quantum gases, superfluidity, pair correlations (Hanbury-Brown–Twiss effect and Cooper pairs), plasmons and phonons, and the quasiparticle concept give an impression of the flavor of these fields. The following chapters are devoted to the structure of atoms and molecules, containing many fascinating subjects (Hartree, Hartree–Fock, Thomas–Fermi methods, the periodic system of elements, the Born–Oppenheimer approach, various types of elementary molecules, oriented orbitals, hybridization, etc.).

Finally we present an elementary exhibition of Feynman path integrals. The method of quantization using path integrals, which essentially is equivalent to the canonical formalism, has gained increasing popularity over the years. Apart from their elegance and formal appeal, path-integral quantization and the related functional techniques are particulary well suited to the implementation of conditions of constraint, which is necessary for the treatment of gauge fields. Nowadays any student of physics should at least know where and how the canonical and the path-integral formalisms are connected.

Like all other lectures, these special chapters are presented together with the necessary mathematical tools. Many detailed examples and worked-out problems are included in order to further illuminate the material.

It is clear from what we have said so far that these lectures are meant to give an elementary (but not naive) overview of special subjects a student may hear about in colloquia and seminars. The lectures may help to furnish better orientation in the vast field of interesting modern physics.

We have profitted a lot from excellent text books, such as

E.G. Harris: *A Pedestrian Approach to Quantum Field Theory* (Wiley, New York 1972),
G. Baym: *Lectures on Quantum Mechanics* (W.A. Benjamin, Reading, MA 1974),
L.D. Landau, E.M. Lifshitz: *Quantum Mechanics* (Pergamon, Oxford 1977),

which have guided us to some extend in devising certain chapters, examples, and exercises. We recommend them for additional reading. The biographical notes on outstanding physicists and mathematicians were taken from the Brockhaus Lexikon.

This book is not intended to provide an exhaustive introduction to all aspects of quantum mechanics. Our main goal has been to present an elementary introduction to the methods of field quantization and their applications in many-body physics as well as to special aspects of atomic and nuclear physics. We hope to attain this goal by presenting the subjects in considerable detail, explaining the mathematical tools in a rather informal way, and by including a large number of examples and worked exercises.

We would like to express our gratitude to Drs. J. Reinhardt, G. Plunien, and S. Schramm for their help in preparing some exercises and examples and in proofreading the German edition of the text. For the preparation of the English edition we enjoyed the help of Priv. Doz. Dr. Martin Greiner. Once again we are pleased to acknowledge the agreeable collaboration with Dr. H.J. Kölsch and his team at Springer-Verlag, Heidelberg. The English manuscript was copy edited by Dr. Victoria Wicks.

Frankfurt am Main, *Walter Greiner*
August 1997

Contents

Contents of Examples and Exercises

1. Quantum Theory of Free Electromagnetic Fields

From the lectures on classical electrodynamics we know Maxwell's equations as the basic equations describing all classical electromagnetic phenomena. To account for quantum effects the Maxwell equations have to be quantized. We are thus led to quantum electrodynamics. Quantum electrodynamics also deals with the quantization of the electron–positron field, the pion field, and other fields and describes their interaction with the quantized electromagnetic field (i.e. the quantized electromagnetic waves). To begin with, we briefly recapitulate Maxwell's classical equations.

1.1 Maxwell's Equations

The Maxwell equations of motion for an electromagnetic field read

$$\begin{aligned} \boldsymbol{\nabla} \times \boldsymbol{E} + \frac{1}{c}\frac{\partial \boldsymbol{B}}{\partial t} &= 0\,, \\ \boldsymbol{\nabla} \cdot \boldsymbol{D} &= 4\pi\varrho\,, \\ \boldsymbol{\nabla} \times \boldsymbol{H} - \frac{1}{c}\frac{\partial \boldsymbol{D}}{\partial t} &= \frac{4\pi}{c}\boldsymbol{j}\,, \\ \boldsymbol{\nabla} \cdot \boldsymbol{B} &= 0\,; \end{aligned} \tag{1.1}$$

here we have used the cms system. Taking the divergence of the second equation and combining it with the time derivative of the third equation, we deduce the continuity equation for the electric charge and current densities ρ and $\boldsymbol{j}$:

$$\boldsymbol{\nabla} \cdot \boldsymbol{j} + \frac{\partial \varrho}{\partial t} = 0\,. \tag{1.2}$$

The *electric* and *magnetic field strenghts*, $\boldsymbol{E}$ and $\boldsymbol{B}$, are expressable in terms of the *vector potential* $\boldsymbol{A}$ and the *scalar potential* φ,

$$\boldsymbol{E} = -\frac{1}{c}\frac{\partial \boldsymbol{A}}{\partial t} - \boldsymbol{\nabla}\varphi\,, \qquad \boldsymbol{B} = \boldsymbol{\nabla} \times \boldsymbol{A}\,. \tag{1.3}$$

As an immediate consequence the first and last of Maxwell's equations (1.1) are automatically fulfilled.

The potentials $\boldsymbol{A}$ and φ are not unique; the modified potentials

$$\boldsymbol{A}' = \boldsymbol{A} + \boldsymbol{\nabla}\chi\,, \qquad \varphi' = \varphi - \frac{1}{c}\frac{\partial \chi}{\partial t}\,, \tag{1.4}$$

where $\chi(\boldsymbol{r}, t)$ is an arbitrary function depending on position $\boldsymbol{r}$ and time t, yield the same fields $\boldsymbol{E}$ and $\boldsymbol{B}$. This modification of the potentials, which leaves the fields strenghts $\boldsymbol{E}$ and $\boldsymbol{B}$ unchanged, is called a *gauge transformation*. In our lectures about quantum mechanics it is proven that the wavefunction Ψ has to be replaced by

$$\Psi' = \Psi \exp\left(\frac{\mathrm{i}e}{\hbar c}\chi\right) \tag{1.5}$$

for a gauge transformation (1.4) once the electromagnetic potentials are introduced via the minimal coupling

$$\boldsymbol{p} \longrightarrow \boldsymbol{p} - \frac{e}{c}\boldsymbol{A} \tag{1.6}$$

into the Schrödinger equation. The form of the wave equation, which besides the Schrödinger equation could also be the Pauli equation or the Dirac equation or any other, remains unaltered. Sometimes the transformation (1.5) is called a *gauge transformation of first degree*, whereas the transformation (1.4) is denoted as a *gauge transformation of second degree.*

Inserting (1.3) into the second and third of Maxwell's equations (1.1), we arrive at

$$\begin{aligned} \boldsymbol{\nabla}\times(\boldsymbol{\nabla}\times\boldsymbol{A}) + \frac{1}{c^2}\frac{\partial^2\boldsymbol{A}}{\partial t^2} + \frac{1}{c}\boldsymbol{\nabla}\frac{\partial\varphi}{\partial t} &= \frac{4\pi}{c}\boldsymbol{j}\,, \\ \frac{1}{c}\frac{\partial}{\partial t}\boldsymbol{\nabla}\cdot\boldsymbol{A} + \boldsymbol{\nabla}^2\varphi &= -4\pi\varrho\,. \end{aligned} \tag{1.7}$$

Once the vector $\boldsymbol{A}$ is described in an Euclidean coordinate system, the first part of the left-hand side of the first equation can be rewritten as

$$\boldsymbol{\nabla}\times(\boldsymbol{\nabla}\times\boldsymbol{A}) = \boldsymbol{\nabla}(\boldsymbol{\nabla}\cdot\boldsymbol{A}) - \boldsymbol{\nabla}^2\boldsymbol{A}\,. \tag{1.8}$$

The last term $\boldsymbol{\nabla}^2\boldsymbol{A}$ represents a vector with components $\triangle A_i$, where the Laplace operator $\triangle$ acts on the components A_i of the vector potential separately. Making use of (1.8), we can further simplify the equations resulting from (1.7), once the gauge transformation (1.4) is applied, which leads to new potentials $\boldsymbol{A}'$ and φ'. We choose the *Lorentz gauge* fulfilling

$$\boldsymbol{\nabla}\cdot\boldsymbol{A}' + \frac{1}{c}\frac{\partial\varphi'}{\partial t} = 0\,. \tag{1.9}$$

Here the gauge function χ is determined from

$$\boldsymbol{\nabla}^2\chi - \frac{1}{c^2}\frac{\partial^2}{\partial t^2}\chi = -\left(\boldsymbol{\nabla}\cdot\boldsymbol{A} + \frac{1}{c}\frac{\partial\varphi}{\partial t}\right)\,. \tag{1.10}$$

Equations (1.7) finally read

$$\begin{aligned} \boldsymbol{\nabla}^2\boldsymbol{A}' - \frac{1}{c^2}\frac{\partial^2\boldsymbol{A}'}{\partial t^2} &= -\frac{4\pi}{c}\boldsymbol{j}\,, \\ \boldsymbol{\nabla}^2\varphi' - \frac{1}{c^2}\frac{\partial^2\varphi'}{\partial t^2} &= -4\pi\varrho\,. \end{aligned} \tag{1.11}$$

1.2 Electromagnetic Plane Waves

For a completely empty space $\boldsymbol{j} = 0$ and $\varrho = 0$. For this case we can always find a gauge function χ such that

$$\begin{aligned} \boldsymbol{\nabla} \cdot \boldsymbol{A}'(\boldsymbol{r},t) &= 0\,, \\ \varphi'(\boldsymbol{r},t) &= 0 \end{aligned} \tag{1.12}$$

for all $\boldsymbol{r}$ and t; see Exercise 1.1. This gauge is called the *Coulomb gauge.* As a consequence we find transverse plane waves as a solution for $\boldsymbol{A}'$ and consequently for $\boldsymbol{E}$ and $\boldsymbol{B}$. In the following we drop the primes attached to $\boldsymbol{A}'$ and φ' for convenience and deduce from (1.9) and (1.11) that

$$\begin{aligned} \boldsymbol{\nabla}^2 \boldsymbol{A} - \frac{1}{c^2}\frac{\partial^2 \boldsymbol{A}}{\partial t^2} &= 0\,, \\ \boldsymbol{\nabla} \cdot \boldsymbol{A} &= 0\,, \\ \varphi &= 0\,. \end{aligned} \tag{1.13}$$

A typical plane-wave solution of these equations is charaterized by a real vector potential $\boldsymbol{A}$ governed by the *wave-number vector* $\boldsymbol{k}$, called the wave vector for short, and the real *polarization vector* $\boldsymbol{\varepsilon}$:

$$\begin{aligned} \boldsymbol{A}(\boldsymbol{r},t) &= 2\boldsymbol{\varepsilon}|\boldsymbol{A}_0| \cos\left(\boldsymbol{k}\cdot\boldsymbol{r} - \omega\, t + \alpha\right) \\ &= \boldsymbol{A}_0\, \mathrm{e}^{\mathrm{i}(\boldsymbol{k}\cdot\boldsymbol{r}-\omega t)} + \boldsymbol{A}_0^*\, \mathrm{e}^{-\mathrm{i}(\boldsymbol{k}\cdot\boldsymbol{r}-\omega t)} \\ &= \boldsymbol{A}_0\, \mathrm{e}^{\mathrm{i}(\boldsymbol{k}\cdot\boldsymbol{r}-\omega t)} + \text{c.c.}\,. \end{aligned} \tag{1.14}$$

Here $\boldsymbol{A}_0 = |\boldsymbol{A}_0|\,\boldsymbol{\varepsilon}\,\mathrm{e}^{\mathrm{i}\alpha}$ is the amplitude and "c.c." stands for the "complex conjugate" of the first term. It is easy to see that the ansatz (1.14) is a solution of the first equation of (1.13) if

$$\omega = k\,c = |\boldsymbol{k}|\,c\,, \tag{1.15}$$

and of the second equation of (1.13) if the polarization vector $\boldsymbol{A}_0$ is perpendicular to the wave vector $\boldsymbol{k}$,

$$\boldsymbol{A}_0 \perp \boldsymbol{k}\,. \tag{1.16}$$

We say that $\boldsymbol{A}_0$ is transverse as it fulfills the last relation, i.e. $\boldsymbol{A}_0 \cdot \boldsymbol{k} = 0$. The corresponding electric and magnetic fields follow from (1.3) and (1.14):

$$\begin{aligned} \boldsymbol{E} &= -2k|\boldsymbol{A}_0|\boldsymbol{\varepsilon} \sin\left(\boldsymbol{k}\cdot\boldsymbol{r} - \omega t + \alpha\right)\,, \\ \boldsymbol{B} &= -2|\boldsymbol{A}_0|\boldsymbol{k}\times\boldsymbol{\varepsilon} \sin\left(\boldsymbol{k}\cdot\boldsymbol{r} - \omega t + \alpha\right)\,. \end{aligned} \tag{1.17}$$

The Poynting vector is defined as

$$\boldsymbol{S} = \frac{c}{4\pi}\boldsymbol{E}\times\boldsymbol{H} \tag{1.18}$$

and, obviously, is parallel to $\boldsymbol{k}$; remember that $\boldsymbol{H} = \boldsymbol{B}$ for a charge-free and current-free space. The time average of the Poynting vector, $\overline{S}$, over one period $T = 2\pi/\omega$ of oscillation is given by

$$\overline{S} = \frac{\omega^2}{2\pi c}\,|A_0|^2 \tag{1.19}$$

and represents the intensity of the electromagnetic plane wave.

EXERCISE

1.1 The Coulomb Gauge

Problem. Show that the general solution of the Maxwell equations can be expressed by potentials $\boldsymbol{A}'$ and φ' with $\boldsymbol{\nabla}\cdot\boldsymbol{A}' = 0$ and $\varphi' = 0$ if $\boldsymbol{\nabla}\cdot\boldsymbol{j} = 0$ and $\varrho = 0$.

Solution. According to the gauge transformation (1.4) the two requirements $\boldsymbol{\nabla}\cdot\boldsymbol{A}' = 0$ and $\varphi' = 0$ lead to

$$\frac{\partial\chi}{\partial t} = c\varphi(\boldsymbol{r},t)\,, \tag{1}$$

$$\boldsymbol{\nabla}^2\chi = -\boldsymbol{\nabla}\cdot\boldsymbol{A}(\boldsymbol{r},t)\,. \tag{2}$$

Differentiating the second equation with respect to time and then inserting the first equation yields

$$\boldsymbol{\nabla}^2\varphi(\boldsymbol{r},t) + \boldsymbol{\nabla}\cdot\frac{1}{c}\frac{\partial\boldsymbol{A}}{\partial t} = 0\,. \tag{3}$$

But this equation is only valid if $\varrho = 0$; in fact it follows from Maxwell's equation $\boldsymbol{\nabla}\cdot\boldsymbol{E} = 4\pi\,\varrho$:

$$\begin{aligned}\boldsymbol{\nabla}\cdot\boldsymbol{E} &= \boldsymbol{\nabla}\cdot\left(-\frac{1}{c}\frac{\partial\boldsymbol{A}}{\partial t} - \boldsymbol{\nabla}\varphi\right)\\ &= -\left(\boldsymbol{\nabla}\cdot\frac{\partial\boldsymbol{A}}{c\,\partial t} + \boldsymbol{\nabla}^2\varphi\right) = 4\pi\,\varrho\,.\end{aligned} \tag{4}$$

A consistency check with another one of Maxwell's equations then gives

$$\begin{aligned}\boldsymbol{\nabla}\cdot\left(\boldsymbol{\nabla}\times\boldsymbol{H} - \frac{1}{c}\frac{\partial\boldsymbol{D}}{\partial t}\right) &= \boldsymbol{\nabla}\cdot(\boldsymbol{\nabla}\times\boldsymbol{H}) - \frac{1}{c}\frac{\partial(\boldsymbol{\nabla}\cdot\boldsymbol{D})}{\partial t}\\ &= \frac{4\pi}{c}(\boldsymbol{\nabla}\cdot\boldsymbol{j})\,.\end{aligned} \tag{5}$$

Since the first term on the left-hand side is always equal to zero, i.e. $\mathrm{div}(\mathrm{curl}\,\boldsymbol{H}) = 0$, and the second term now vanishes because $\boldsymbol{\nabla}D = 4\pi\,\varrho = 0$, we conclude that

$$\boldsymbol{\nabla}\cdot\boldsymbol{j} = 0\,. \tag{6}$$

The explicit solution for the gauge function χ is obtained from a time integration of (1):

$$\chi(\boldsymbol{r},t) = c\int\varphi(\boldsymbol{r},t)\,\mathrm{d}t + \mathrm{const.} \tag{7}$$

Note that the two requirements $\boldsymbol{\nabla}\cdot\boldsymbol{A}' = 0$ and $\varphi' = 0$ automatically fulfill the Lorentz gauge (1.9), so that the wave equations (1.11) still hold, but only with $\varrho = \boldsymbol{\nabla}\cdot\boldsymbol{j} = 0$.

1.3 Quantization of Free Electromagnetic Fields

Once again we write down Maxwell's equations without source terms:

$$\begin{aligned} \nabla \cdot \boldsymbol{B} &= 0\,, \\ \nabla \cdot \boldsymbol{E} &= 0\,, \\ \nabla \times \boldsymbol{E} &= -\frac{1}{c}\frac{\partial \boldsymbol{B}}{\partial t}\,, \\ \nabla \times \boldsymbol{B} &= +\frac{1}{c}\frac{\partial \boldsymbol{E}}{\partial t}\,. \end{aligned} \tag{1.20}$$

The first three of these equations are automatically fulfilled if the Coulomb gauge (1.12)

$$\begin{aligned} \varphi &= 0\,, \\ \nabla \cdot \boldsymbol{A} &= 0 \end{aligned} \tag{1.21}$$

is considered and (1.3) is used. The last of (1.20) then leads to the wave equation

$$\nabla^2 \boldsymbol{A}(\boldsymbol{r},t) - \frac{1}{c^2}\frac{\partial^2}{\partial t^2}\boldsymbol{A}(\boldsymbol{r},t) = 0 \tag{1.22}$$

for the vector potential, which is already known from (1.13).

For the development of the quantum theory of an electromagnetic field it is of advantage to express the field by a set of discrete variables. We confine the electromagnetic field to a large box (cube) of volume $\Omega = L^3$; in this respect the normal modes of the electromagnetic field are discretized. The general $\boldsymbol{A}$ field is the superposition of these *normal modes* $\boldsymbol{A}_{\boldsymbol{k}\sigma}$. The Fourier coefficients $a_{\boldsymbol{k}\sigma}(t)$ of this expansion will be treated as the actual field variables, which describe the dynamics. We aspire to the quantization of these field variables.

The boundary conditions of the field at the box's sides now come into the problem. It is most convenient to require periodicity of $\boldsymbol{A}$ at the sides. This means that the normal modes are determined such that full wavelengths fit into the box. Mathematically, this translates into

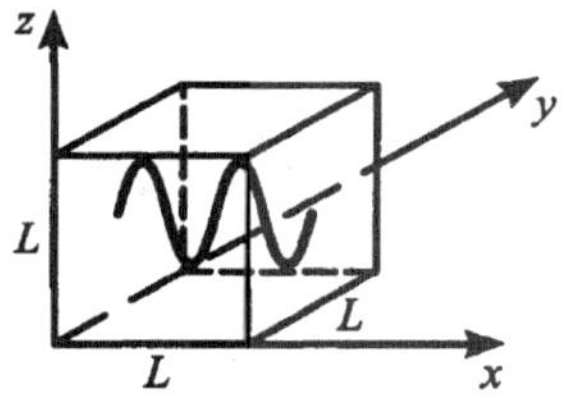

Fig. 1.1. The electromagnetic field is confined to a box; a normal wave mode with two full wavelengths along the x axis is indicated

$$\begin{aligned} \boldsymbol{A}(L,y,z,t) &= \boldsymbol{A}(0,y,z,t)\,, \\ \boldsymbol{A}(x,L,z,t) &= \boldsymbol{A}(x,0,z,t)\,, \\ \boldsymbol{A}(x,y,L,t) &= \boldsymbol{A}(x,y,0,t)\,. \end{aligned} \tag{1.23}$$

Remembering that for normal modes all degrees of freedom vibrate with the same frequency, we are lead to the ansatz

$$\boldsymbol{A}(x,y,z,t) = \boldsymbol{A}(x,y,z)\,\mathrm{e}^{\mathrm{i}\omega t}\quad. \tag{1.24}$$

Inserting it into the wave equation (1.22) we arrive at

$$\left(\Delta + \frac{\omega^2}{c^2}\right)\boldsymbol{A}(x,y,z) = 0\,. \tag{1.25}$$

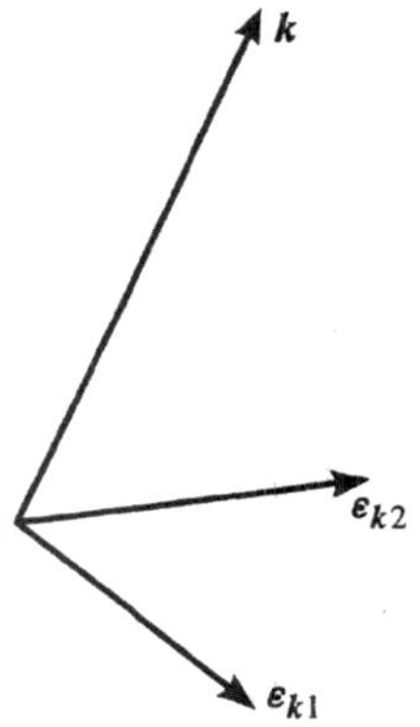

Fig. 1.2. The polarization vectors $\varepsilon_{k\sigma}$ are perpendicular to the wave vector k. This reflects the transverse character of electromagnetic waves

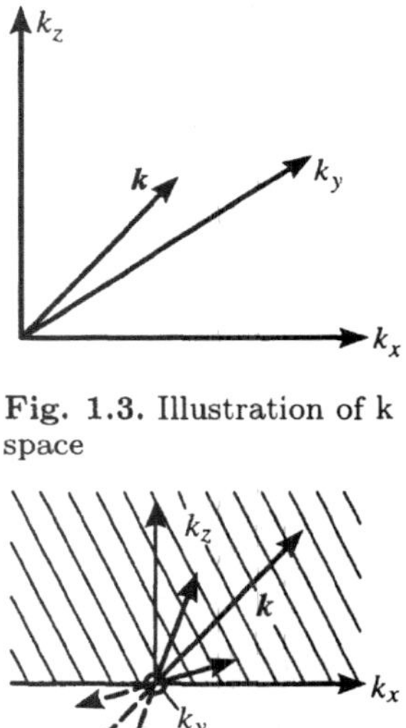

Fig. 1.3. Illustration of k space

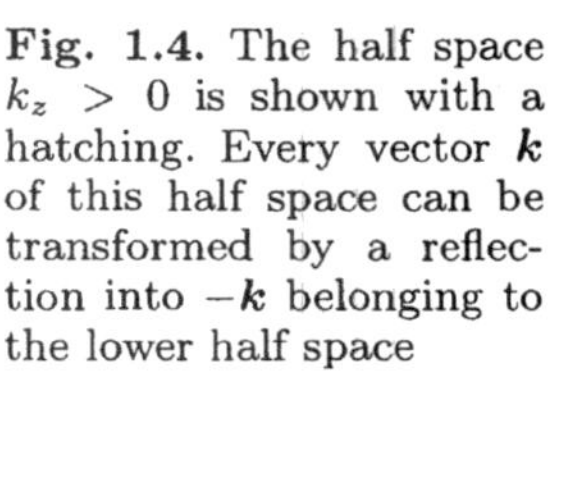

Fig. 1.4. The half space $k_z > 0$ is shown with a hatching. Every vector k of this half space can be transformed by a reflection into $-k$ belonging to the lower half space

A normal solution is given by

$$\boldsymbol{A}_{\boldsymbol{k}\sigma} = N_k\, \boldsymbol{\varepsilon}_{\boldsymbol{k}\sigma}\, \mathrm{e}^{\mathrm{i}\boldsymbol{k}\cdot\boldsymbol{x}} \qquad (\sigma = 1, 2)\,, \tag{1.26}$$

where

$$k^2 = k_x^2 + k_y^2 + k_z^2 = \frac{\omega_\perp^2}{c^2} \tag{1.27a}$$

or

$$\omega_k^2 = c^2 k^2\,. \tag{1.27b}$$

The two polarization vectors $\boldsymbol{\varepsilon}_{\boldsymbol{k}1}$ and $\boldsymbol{\varepsilon}_{\boldsymbol{k}2}$ are perpendicular to the wave vector $\boldsymbol{k}$, which determines the direction of propagation:

$$\boldsymbol{\varepsilon}_{\boldsymbol{k}\sigma}\cdot\boldsymbol{k} = 0\,. \tag{1.28}$$

This is a result of (1.21) and (1.26). We choose the two polarization vectors, which now both lie in the plane perpendicular to $\boldsymbol{k}$, as two linearly independent vectors orthogonal to each other (see Fig. 1.2):

$$\boldsymbol{\varepsilon}_{\boldsymbol{k}\sigma}\cdot\boldsymbol{\varepsilon}_{\boldsymbol{k}\sigma'} = \delta_{\sigma\sigma'}\,. \tag{1.29}$$

The normalization factor N_k appearing in (1.26) will be determined later. In order to fulfill the boundary conditions (1.23), the wave vector $\boldsymbol{k}$ has to take on the form

$$\boldsymbol{k} = (k_x, k_y, k_z) = \frac{2\pi}{L}\{n_1, n_2, n_3\}\,, \qquad n_1, n_2, n_3 \in \mathbf{Z}\,, \tag{1.30}$$

where n_i are integer numbers. $\boldsymbol{k}$ is a vector in k space (see Fig. 1.3).

The most general solution for the $\boldsymbol{A}$ field is now a superposition of all normal modes. Since the spatial part of the normal modes are pure imaginary exponentials, see (1.26), this superposition is equal to a Fourier series:

$$\begin{aligned}\boldsymbol{A}(\boldsymbol{x},t) &= \sum_{\substack{\boldsymbol{k}\\k_z>0}}\sum_{\sigma=1,2}\left[a_{\boldsymbol{k}\sigma}(t)\,\boldsymbol{A}_{\boldsymbol{k}\sigma}(\boldsymbol{x}) + a^*_{\boldsymbol{k}\sigma}(t)\,\boldsymbol{A}^*_{\boldsymbol{k}\sigma}(\boldsymbol{x})\right]\\ &= \sum_{\substack{\boldsymbol{k}\\k_z>0}}\sum_{\sigma=1,2} N_k\,\boldsymbol{\varepsilon}_{\boldsymbol{k}\sigma}\left(a_{\boldsymbol{k}\sigma}(t)\,\mathrm{e}^{\mathrm{i}\boldsymbol{k}\cdot\boldsymbol{x}} + a^*_{\boldsymbol{k}\sigma}(t)\,\mathrm{e}^{-\mathrm{i}\boldsymbol{k}\cdot\boldsymbol{x}}\right).\end{aligned} \tag{1.31}$$

We have made sure that the vector potential, which determines the real electromagnetic fields, stays real by adding the complex conjugate to each term of the sum; in order to avoid double counting the summation has to be restricted to those vectors $\boldsymbol{k} = \{k_x, k_y, k_z\}$ with $k_z > 0$. Each term in the sum (1.31) contains a term with $\mathrm{e}^{\mathrm{i}\boldsymbol{k}\cdot\boldsymbol{x}}$ and also one with $\mathrm{e}^{-\mathrm{i}\boldsymbol{k}\cdot\boldsymbol{x}}$, so that the summation has to be limited; see also Fig. 1.4. The factors $\mathrm{e}^{\mathrm{i}\omega_k t}$ have been absorbed into the time-dependent Fourier coefficients $a_{\boldsymbol{k}\sigma}$. Because of the normalization factor N_k, the Fourier coefficients do not represent the amplitudes of the $\boldsymbol{A}$ field alone, but nevertheless reflect a measure for the latter. Later on $a_{\boldsymbol{k}\sigma}$ and $a^*_{\boldsymbol{k}\sigma}$ will be interpreted as creation and annihilation operators for photons. Because $a_{\boldsymbol{k}\sigma}$ and $a^*_{\boldsymbol{k}\sigma}$ will not commute anymore as operators, we have to pay attention to their precise ordering from now on.

Inserting the expansion (1.31) into the wave equation (1.22) yields

$$\sum_{\substack{\boldsymbol{k}\\ k_z>0}} \sum_{\sigma=1,2} N_k\, \boldsymbol{\varepsilon}_{\boldsymbol{k}\sigma} \left[\left(-k^2 a_{\boldsymbol{k}\sigma}(t) - \frac{1}{c^2}\frac{\mathrm{d}^2 a_{\boldsymbol{k}\sigma}(t)}{\mathrm{d}t^2} \right) \mathrm{e}^{\mathrm{i}\boldsymbol{k}\cdot\boldsymbol{x}} + \left(-k^2 a^*_{\boldsymbol{k}\sigma}(t) - \frac{1}{c^2}\frac{\mathrm{d}^2 a^*_{\boldsymbol{k}\sigma}(t)}{\mathrm{d}t^2} \right) \mathrm{e}^{-\mathrm{i}\boldsymbol{k}\cdot\boldsymbol{x}} \right] = 0\,,$$

from which we deduce

$$\frac{\mathrm{d}^2 a_{\boldsymbol{k}\sigma}}{\mathrm{d}t^2} + \omega_k^2 a_{\boldsymbol{k}\sigma} = 0 \qquad \left(\omega_k^2 = k^2c^2\right)\,. \tag{1.32}$$

This differential equation has the solution

$$a_{\boldsymbol{k}\sigma}(t) = a^{(1)}_{\boldsymbol{k}\sigma}(0)\,\mathrm{e}^{-\mathrm{i}\omega_k t} + a^{(2)}_{\boldsymbol{k}\sigma}(0)\,\mathrm{e}^{\mathrm{i}\omega_k t}\,, \tag{1.33}$$

so that the most general $\boldsymbol{A}$ field (1.31) becomes

$$\begin{aligned}\boldsymbol{A}(\boldsymbol{x},t) = \sum_{\substack{\boldsymbol{k},\sigma\\ k_z>0}} N_k\, \boldsymbol{\varepsilon}_{\boldsymbol{k}\sigma} \Big[& a^{(1)}_{\boldsymbol{k}\sigma}(0)\,\mathrm{e}^{\mathrm{i}(\boldsymbol{k}\cdot\boldsymbol{x}-\omega_k t)} + a^{(1)*}_{\boldsymbol{k}\sigma}(0)\,\mathrm{e}^{-\mathrm{i}(\boldsymbol{k}\cdot\boldsymbol{x}-\omega_k t)} \\ & + a^{(2)}_{\boldsymbol{k}\sigma}(0)\,\mathrm{e}^{\mathrm{i}(\boldsymbol{k}\cdot\boldsymbol{x}+\omega_k t)} + a^{(2)*}_{\boldsymbol{k}\sigma}(0)\,\mathrm{e}^{-\mathrm{i}(\boldsymbol{k}\cdot\boldsymbol{x}+\omega_k t)} \Big]\,.\end{aligned} \tag{1.34}$$

Somehow, the restriction $k_z > 0$ of the sum over $\boldsymbol{k}$ is still disturbing. Therefore we redefine the free constants $a^{(1)}_{\boldsymbol{k}\sigma}(0)$ and $a^{(2)*}_{\boldsymbol{k}\sigma}(0)$ appearing in (1.33) as follows:

$$\left.\begin{aligned} a^{(1)}_{\boldsymbol{k}\sigma}(0) &= a_{\boldsymbol{k}\sigma}(0) \\ a^{(2)}_{\boldsymbol{k}\sigma}(0) &= -(-1)^\sigma\, a^*_{-\boldsymbol{k}\sigma}(0) \end{aligned}\right\} \quad \text{for} \quad k_z > 0\,. \tag{1.35}$$

Then (1.34) transforms into

$$\begin{aligned}\boldsymbol{A}(\boldsymbol{x},t) = \sum_{\substack{\boldsymbol{k},\sigma\\ k_z>0}} N_k \boldsymbol{\varepsilon}_{\boldsymbol{k}\sigma} \Big[& a_{\boldsymbol{k}\sigma}(0)\,\mathrm{e}^{\mathrm{i}(\boldsymbol{k}\cdot\boldsymbol{x}-\omega_k t)} + a^*_{\boldsymbol{k}\sigma}(0)\,\mathrm{e}^{-\mathrm{i}(\boldsymbol{k}\cdot\boldsymbol{x}-\omega_k t)} \\ & - (-1)^\sigma \left(a^*_{-\boldsymbol{k}\sigma}(0)\,\mathrm{e}^{\mathrm{i}(\boldsymbol{k}\cdot\boldsymbol{x}+\omega_k t)} + a_{-\boldsymbol{k}\sigma}(0)\,\mathrm{e}^{-\mathrm{i}(\boldsymbol{k}\cdot\boldsymbol{x}+\omega_k t)} \right) \Big]\,.\end{aligned} \tag{1.36}$$

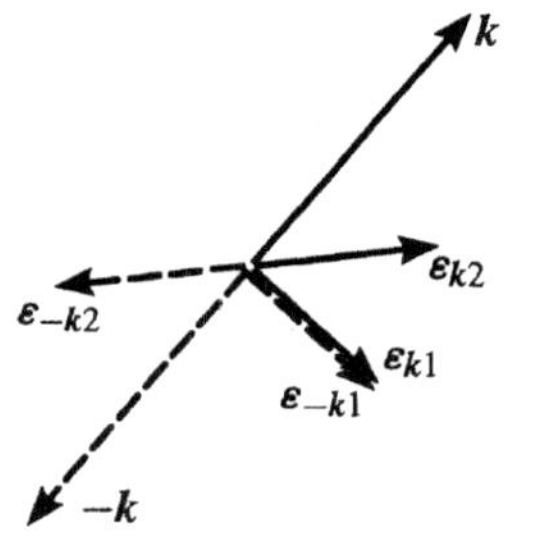

Fig. 1.5. Illustration of the relation $\boldsymbol{\varepsilon}_{\boldsymbol{k}\sigma} = -(-1)^\sigma \times \boldsymbol{\varepsilon}_{-\boldsymbol{k}\sigma}$, $\sigma = 1, 2$

Whereas the quantities N_k and ω_k do not depend on the direction of $\boldsymbol{k}$, the polarization vector $\boldsymbol{\varepsilon}_{\boldsymbol{k}\sigma}$ changes into $\boldsymbol{\varepsilon}_{-\boldsymbol{k}\sigma} = -(-1)^\sigma \boldsymbol{\varepsilon}_{\boldsymbol{k}\sigma}$ with $\sigma = 1, 2$ as $\boldsymbol{k}$ turns its direction around; in this respect the right handedness of the three vectors $\boldsymbol{k}$, $\boldsymbol{\varepsilon}_{\boldsymbol{k}1}$ and $\boldsymbol{\varepsilon}_{\boldsymbol{k}2}$ is guaranteed (see Fig. 1.5). We deduce further:

$$\begin{aligned}\boldsymbol{A}(\boldsymbol{x},t) = & \sum_{\substack{\boldsymbol{k},\sigma\\ k_z>0}} N_k\, \boldsymbol{\varepsilon}_{\boldsymbol{k}\sigma} \left[a_{\boldsymbol{k}\sigma}(0)\,\mathrm{e}^{\mathrm{i}(\boldsymbol{k}\cdot\boldsymbol{x}-\omega_k t)} + a^*_{\boldsymbol{k}\sigma}(0)\,\mathrm{e}^{-\mathrm{i}(\boldsymbol{k}\cdot\boldsymbol{x}-\omega_k t)} \right] \\ & + \sum_{\substack{\boldsymbol{k},\sigma\\ k_z>0}} N_k\, \boldsymbol{\varepsilon}_{-\boldsymbol{k}\sigma} \left[a^*_{-\boldsymbol{k}\sigma}(0)\,\mathrm{e}^{\mathrm{i}\omega_k t}\mathrm{e}^{\mathrm{i}\boldsymbol{k}\cdot\boldsymbol{x}} + a_{-\boldsymbol{k}\sigma}(0)\,\mathrm{e}^{\mathrm{i}(-\boldsymbol{k}\boldsymbol{x}-\omega_k t)} \right]\end{aligned}$$

$$
\begin{aligned}
&= \sum_{\substack{\boldsymbol{k},\sigma \\ k_z>0}} N_k\, \boldsymbol{\varepsilon}_{\boldsymbol{k}\sigma} \left[a_{\boldsymbol{k}\sigma}(0)\, \mathrm{e}^{\mathrm{i}(\boldsymbol{k}\cdot\boldsymbol{x}-\omega_k t)} + a^*_{\boldsymbol{k}\sigma}(0)\, \mathrm{e}^{-\mathrm{i}(\boldsymbol{k}\cdot\boldsymbol{x}-\omega_k t)}\right] \\
&\quad + \sum_{\substack{\boldsymbol{k},\sigma \\ k_z<0}} N_k\, \boldsymbol{\varepsilon}_{\boldsymbol{k}\sigma} \left[a_{\boldsymbol{k}\sigma}(0)\, \mathrm{e}^{-\mathrm{i}\omega_k t}\mathrm{e}^{\mathrm{i}\boldsymbol{k}\cdot\boldsymbol{x}} + a^*_{\boldsymbol{k}\sigma}(0)\, \mathrm{e}^{\mathrm{i}\omega_k t}\mathrm{e}^{-\mathrm{i}\boldsymbol{k}\cdot\boldsymbol{x}}\right] \\
&= \sum_{\boldsymbol{k},\sigma} N_k\, \boldsymbol{\varepsilon}_{\boldsymbol{k}\sigma} \left[a_{\boldsymbol{k}\sigma}(t)\, \mathrm{e}^{\mathrm{i}\boldsymbol{k}\cdot\boldsymbol{x}} + a^*_{\boldsymbol{k}\sigma}(t)\, \mathrm{e}^{-\mathrm{i}\boldsymbol{k}\cdot\boldsymbol{x}}\right] . \qquad (1.37)
\end{aligned}
$$

The second part of the summation in (1.36) is identical in form to the first part, except that the summation now runs over the half space $k_z < 0$. In the last step of (1.37) the relation

$$
a_{\boldsymbol{k}\sigma}(t) = a_{\boldsymbol{k}\sigma}(0)\, \mathrm{e}^{-\mathrm{i}\omega_k t} \qquad (1.38)
$$

has also been introduced, which is consistent with (1.33) and (1.35). From this relation it follows that

$$
\frac{\mathrm{d}a_{\boldsymbol{k}\sigma}(t)}{\mathrm{d}t} = -\mathrm{i}\omega_k\, a_{\boldsymbol{k}\sigma}(t) , \qquad (1.39)
$$

which for all $\boldsymbol{k}$ and σ can be interpreted as the equation of motion for the field. In fact, we will soon show that they can be derived from a Hamiltonian that is identical to the total energy of the field.

Let us first calculate the energy. We know from the lectures on classical electrodynamics,[1] that the energy of the electromagnetic field is given by

$$
\begin{aligned}
H &= \frac{1}{8\pi}\int_{L^3} \mathrm{d}^3x\, \left(\boldsymbol{E}^2 + \boldsymbol{B}^2\right) \\
&= \frac{1}{8\pi}\int_{L^3} \mathrm{d}^3x \left[\frac{1}{c^2}\frac{\partial \boldsymbol{A}}{\partial t}\cdot\frac{\partial \boldsymbol{A}^*}{\partial t} + (\boldsymbol{\nabla}\times\boldsymbol{A})\cdot(\boldsymbol{\nabla}\times\boldsymbol{A}^*)\right] , \qquad (1.40)
\end{aligned}
$$

where we have used expression (1.3) and the Coulomb gauge (1.12) in the last step. Now we insert the Fourier expansion (1.37). The first term in (1.40) with use of (1.39) then gives

$$
\begin{aligned}
\int_{L^3} \mathrm{d}^3x \frac{1}{8\pi c^2}\frac{\partial \boldsymbol{A}}{\partial t}\cdot\frac{\partial \boldsymbol{A}^*}{\partial t} &= -\sum_{\boldsymbol{k},\sigma}\sum_{\boldsymbol{k}',\sigma'} \frac{\omega_k\omega_k'}{8\pi c^2} N_k\, N_{k'}\, \boldsymbol{\varepsilon}_{\boldsymbol{k}\sigma}\cdot\boldsymbol{\varepsilon}_{\boldsymbol{k}'\sigma'} \\
&\quad \times \int_{L^3} \mathrm{d}^3x \left(a_{\boldsymbol{k}\sigma}\, \mathrm{e}^{\mathrm{i}\boldsymbol{k}\cdot\boldsymbol{x}} - a^*_{\boldsymbol{k}\sigma}\, \mathrm{e}^{-\mathrm{i}\boldsymbol{k}\cdot\boldsymbol{x}}\right) \\
&\quad \times \left(a_{\boldsymbol{k}'\sigma'}\, \mathrm{e}^{\mathrm{i}\boldsymbol{k}'\cdot\boldsymbol{x}} - a^*_{\boldsymbol{k}'\sigma'}\, \mathrm{e}^{-\mathrm{i}\boldsymbol{k}'\cdot\boldsymbol{x}}\right) . \qquad (1.41)
\end{aligned}
$$

We exploit the orthogonality relations

$$
\int_{L^3} \mathrm{d}^3x\, \mathrm{e}^{\mathrm{i}\boldsymbol{k}\cdot\boldsymbol{x}}\, \mathrm{e}^{-\mathrm{i}\boldsymbol{k}'\cdot\boldsymbol{x}} = L^3\, \delta_{\boldsymbol{k}\boldsymbol{k}'} \qquad (1.42)
$$

and

$$
\begin{aligned}
\boldsymbol{\varepsilon}_{\boldsymbol{k}\sigma}\cdot\boldsymbol{\varepsilon}_{\boldsymbol{k}\sigma'} &= \delta_{\sigma\sigma'} , \\
\boldsymbol{\varepsilon}_{\boldsymbol{k}\sigma}\cdot\boldsymbol{\varepsilon}_{-\boldsymbol{k}\sigma'} &= -(-1)^{\sigma}\, \delta_{\sigma\sigma'} ,
\end{aligned} \qquad (1.43)
$$

[1] See W. Greiner: *Classical Electrodynamics* (Springer, New York 1996).

so that (1.41) transforms into

$$\int_{L^3} \mathrm{d}^3x \, \frac{1}{8\pi c^2} \frac{\partial \boldsymbol{A}}{\partial t} \cdot \frac{\partial \boldsymbol{A}^*}{\partial t} = \sum_{\boldsymbol{k},\,\sigma} \frac{\omega_k^2}{8\pi c^2} N_k^2 \, L^3 \left\{ \left[a_{\boldsymbol{k}\sigma} \, a_{\boldsymbol{k}\sigma}^* + a_{\boldsymbol{k}\sigma}^* \, a_{\boldsymbol{k}\sigma} \right] \right.$$
$$\left. + (-1)^\sigma \left[a_{\boldsymbol{k}\sigma} \, a_{-\boldsymbol{k}\sigma} + a_{\boldsymbol{k}\sigma}^* \, a_{-\boldsymbol{k}\sigma}^* \right] \right\} . \tag{1.44}$$

Analogously the computation of the second line of (1.40) becomes

$$\int_{L^3} \mathrm{d}^3x \, \frac{1}{8\pi} (\boldsymbol{\nabla} \times \boldsymbol{A}) \cdot (\boldsymbol{\nabla} \times \boldsymbol{A}^*) = \sum_{\boldsymbol{k},\,\sigma} \frac{k^2}{8\pi} N_k^2 \, L^3 \left\{ \left[a_{\boldsymbol{k}\sigma} \, a_{\boldsymbol{k}\sigma}^* + a_{\boldsymbol{k}\sigma}^* \, a_{\boldsymbol{k}\sigma} \right] \right.$$
$$\left. - (-1)^\sigma \left[a_{\boldsymbol{k}\sigma} \, a_{-\boldsymbol{k}\sigma} + a_{\boldsymbol{k}\sigma}^* \, a_{-\boldsymbol{k}\sigma}^* \right] \right\} ; \tag{1.45}$$

consult Exercise 1.2. This is identical to the expression (1.44) except for the sign of the second term. Therefore the addition of both expressions, which is equal to the relation (1.40), yields

$$H = \sum_{\boldsymbol{k},\,\sigma} \frac{\omega_k^2}{4\pi c^2} N_k^2 \, L^3 \left(a_{\boldsymbol{k}\sigma} \, a_{\boldsymbol{k}\sigma}^* + a_{\boldsymbol{k}\sigma}^* \, a_{\boldsymbol{k}\sigma} \right) . \tag{1.46}$$

This resembles very much the energy of a system of uncoupled harmonic oscillators, which becomes even more apparent if we choose the normalization constant of the normal modes (1.26) to be

$$N_k = \sqrt{\frac{2\pi\hbar c^2}{L^3 \omega_k}} . \tag{1.47}$$

Then the energy of the electromagnetic field becomes equal to a sum of energies corresponding to "harmonic field oscillators":

$$H = \frac{1}{2} \sum_{\boldsymbol{k},\,\sigma} \hbar\omega_k \, \left(a_{\boldsymbol{k}\sigma} \, a_{\boldsymbol{k}\sigma}^* + a_{\boldsymbol{k}\sigma}^* \, a_{\boldsymbol{k}\sigma} \right) . \tag{1.48}$$

In other words, the normal modes of the electromagnetic field normalized with (1.47) behave like classical oscillators with energy $\hbar\omega_k$.

Up to now, $a_{\boldsymbol{k}\sigma}$ and $a_{\boldsymbol{k}\sigma}^*$ have been treated as classical amplitudes. On this level, we could rewrite (1.48) to give

$$H = \sum_{\boldsymbol{k},\,\sigma} \hbar\omega_k \, a_{\boldsymbol{k}\sigma}^* \, a_{\boldsymbol{k}\sigma} . \tag{1.49}$$

However, we prefer to stick to the form (1.48), where we have emphasized the exact ordering of the amplitudes. In a moment we will interpret $a_{\boldsymbol{k}\sigma}$ and $a_{\boldsymbol{k}\sigma}^*$ as noncommuting operators; as a consequence of which the two expressions (1.48) and (1.49) will not be identical anymore.

EXERCISE

1.2 Computation of the Magnetic Contributions to the Energy of an Electromagnetic Field

Problem. Compute the expression (1.45), i.e. the integral

$$I = \int_{L^3} \mathrm{d}^3x \, \frac{1}{8\pi} (\boldsymbol{\nabla} \times \boldsymbol{A}) (\boldsymbol{\nabla} \times \boldsymbol{A}^*) \,,$$

using the Fourier expansion (1.37).

Solution. With

$$\boldsymbol{A}(\boldsymbol{x},t) = \sum_{\boldsymbol{k},\sigma} N_k \, \boldsymbol{\varepsilon}_{\boldsymbol{k}\sigma} \left[a_{\boldsymbol{k}\sigma} \, \mathrm{e}^{\mathrm{i}\boldsymbol{k}\cdot\boldsymbol{x}} + a^*_{\boldsymbol{k}\sigma} \, \mathrm{e}^{-\mathrm{i}\boldsymbol{k}\cdot\boldsymbol{x}} \right]$$

and (1.37) we calculate

$$\boldsymbol{\nabla} \times \boldsymbol{A}(\boldsymbol{x},t) = \mathrm{i} \sum_{\boldsymbol{k},\sigma} N_k \, \boldsymbol{k} \times \boldsymbol{\varepsilon}_{\boldsymbol{k}\sigma} \left[a_{\boldsymbol{k}\sigma} \, \mathrm{e}^{\mathrm{i}\boldsymbol{k}\cdot\boldsymbol{x}} - a^*_{\boldsymbol{k}\sigma} \, \mathrm{e}^{-\mathrm{i}\boldsymbol{k}\cdot\boldsymbol{x}} \right]$$

and, analogously,

$$\boldsymbol{\nabla} \times \boldsymbol{A}^*(\boldsymbol{x},t) = \mathrm{i} \sum_{\boldsymbol{k}',\sigma'} N_{k'} \, \boldsymbol{k}' \times \boldsymbol{\varepsilon}_{\boldsymbol{k}'\sigma'} \left[a_{\boldsymbol{k}'\sigma'} \, \mathrm{e}^{\mathrm{i}\boldsymbol{k}'\cdot\boldsymbol{x}} - a^*_{\boldsymbol{k}'\sigma'} \, \mathrm{e}^{-\mathrm{i}\boldsymbol{k}'\cdot\boldsymbol{x}} \right] .$$

Altogether we get for the integral I

$$I = -\frac{1}{8\pi} \sum_{\boldsymbol{k},\sigma} \sum_{\boldsymbol{k}',\sigma'} N_k N_{k'} \, (\boldsymbol{k} \times \boldsymbol{\varepsilon}_{\boldsymbol{k}\sigma}) \cdot (\boldsymbol{k}' \times \boldsymbol{\varepsilon}_{\boldsymbol{k}'\sigma'})$$
$$\times \int_{L^3} \mathrm{d}^3x \left[a_{\boldsymbol{k}\sigma} \mathrm{e}^{\mathrm{i}\boldsymbol{k}\cdot\boldsymbol{x}} - a^*_{\boldsymbol{k}\sigma} \mathrm{e}^{-\mathrm{i}\boldsymbol{k}\cdot\boldsymbol{x}} \right] \left[a_{\boldsymbol{k}'\sigma'} \, \mathrm{e}^{\mathrm{i}\boldsymbol{k}'\cdot\boldsymbol{x}} - a^*_{\boldsymbol{k}'\sigma'} \mathrm{e}^{-\mathrm{i}\boldsymbol{k}'\cdot\boldsymbol{x}} \right] .$$

Using the identity

$$(\boldsymbol{k} \times \boldsymbol{\varepsilon}_{\boldsymbol{k}\sigma}) \cdot (\boldsymbol{k}' \times \boldsymbol{\varepsilon}_{\boldsymbol{k}'\sigma'}) = (\boldsymbol{k} \cdot \boldsymbol{k}') \, (\boldsymbol{\varepsilon}_{\boldsymbol{k}\sigma} \cdot \boldsymbol{\varepsilon}_{\boldsymbol{k}'\sigma'}) - (\boldsymbol{\varepsilon}_{\boldsymbol{k}\sigma} \cdot \boldsymbol{k}') \, (\boldsymbol{\varepsilon}_{\boldsymbol{k}'\sigma'} \cdot \boldsymbol{k})$$

and the orthogonality relation

$$\int_{L^3} \mathrm{d}^3x \, \mathrm{e}^{\mathrm{i}(\boldsymbol{k}-\boldsymbol{k}')\cdot\boldsymbol{x}} = L^3 \, \delta_{\boldsymbol{k}\boldsymbol{k}'}$$

we deduce further that

$$I = -\frac{1}{8\pi} \sum_{\boldsymbol{k},\sigma} \sum_{\boldsymbol{k}',\sigma'} L^3 N_k N_{k'} \left[(\boldsymbol{k} \cdot \boldsymbol{k}') \, (\boldsymbol{\varepsilon}_{\boldsymbol{k}\sigma} \cdot \boldsymbol{\varepsilon}_{\boldsymbol{k}'\sigma'}) - (\boldsymbol{\varepsilon}_{\boldsymbol{k}\sigma} \cdot \boldsymbol{k}') \, (\boldsymbol{\varepsilon}_{\boldsymbol{k}'\sigma'} \cdot \boldsymbol{k}) \right]$$
$$\times \left[(a_{\boldsymbol{k}\sigma} a_{\boldsymbol{k}'\sigma'} + a^*_{\boldsymbol{k}\sigma} a^*_{\boldsymbol{k}'\sigma'}) \, \delta_{\boldsymbol{k},-\boldsymbol{k}'} - (a_{\boldsymbol{k}\sigma} a^*_{\boldsymbol{k}'\sigma'} + a^*_{\boldsymbol{k}\sigma} a_{\boldsymbol{k}'\sigma'}) \, \delta_{\boldsymbol{k}\boldsymbol{k}'} \right] .$$

The relation $\boldsymbol{\varepsilon}_{\boldsymbol{k}\sigma} \cdot \boldsymbol{k}' = 0$ holds not only if $\boldsymbol{k} = \boldsymbol{k}'$, but also if $\boldsymbol{k} = -\boldsymbol{k}'$, because both $\boldsymbol{k}$ and $-\boldsymbol{k}$ are perpendicular to $\boldsymbol{\varepsilon}_{\boldsymbol{k}\sigma}$. In addition we have $\boldsymbol{\varepsilon}_{\boldsymbol{k}\sigma} \cdot \boldsymbol{\varepsilon}_{\boldsymbol{k}\sigma'} = \delta_{\sigma\sigma'}$ as well as $\boldsymbol{\varepsilon}_{\boldsymbol{k}\sigma} \cdot \boldsymbol{\varepsilon}_{-\boldsymbol{k}\sigma'} = -(-1)^\sigma \, \delta_{\sigma\sigma'}$; consult Fig. 1.5. Moreover $\omega_k = \omega_{-k}$ and $N_k = N_{-k}$. Therefore we get

$$I = \sum_{\boldsymbol{k},\sigma} \frac{1}{8\pi} k^2 N_k^2 L^3 \left[(a_{\boldsymbol{k}\sigma} a^*_{\boldsymbol{k}\sigma} + a^*_{\boldsymbol{k}\sigma} a_{\boldsymbol{k}\sigma}) - (-1)^\sigma \left(a_{\boldsymbol{k}\sigma} a_{-\boldsymbol{k}\sigma} + a^*_{\boldsymbol{k}\sigma} a^*_{-\boldsymbol{k}\sigma} \right) \right] .$$

Exercise 1.2.

With $k^2 = \omega_k^2/c^2$ we finally obtain

$$I = \sum_{k,\,\sigma} \frac{1}{8\pi}\frac{\omega_k^2}{c^2} N_k^2 L^3 \left[\left(a_{k\sigma}a_{k\sigma}^* + a_{k\sigma}^* a_{k\sigma}\right) - (-1)^\sigma \left(a_{k\sigma}a_{-k\sigma} + a_{k\sigma}^* a_{-k\sigma}^*\right)\right] ,$$

which is identical to the stated result (1.45).

Now we introduce the quantization. The expression (1.48) for the energy of an electromagnetic field motivates us to proceed in an analogous manner as with the quantization of phonons of the harmonic oscillator.[2] From now on we designate $a_{k\sigma}^*$ as $\hat{a}_{k\sigma}^\dagger$ and interpret $\hat{a}_{k\sigma}^\dagger$ as the Hermitian adjoint of $\hat{a}_{k\sigma}$. If the operators $\hat{a}_{k\sigma}^\dagger$ and $\hat{a}_{k'\sigma'}$ possess different quantum numbers, i.e. belong to different oscillators, it is reasonable to assume that they commute with each other. We require that

$$\left[\hat{a}_{k\sigma}, \hat{a}_{k'\sigma'}^\dagger\right]_- = \delta_{kk'}\delta_{\sigma\sigma'} . \tag{1.50}$$

In addition

$$\left[\hat{a}_{k\sigma}, \hat{a}_{k'\sigma'}\right]_- = \left[\hat{a}_{k\sigma}^\dagger, \hat{a}_{k'\sigma'}^\dagger\right]_- = 0 \tag{1.51}$$

should hold. Then the Hamiltonian (1.48) should be interpreted as an operator describing the electromagnetic field:

$$\begin{aligned} \hat{H} &= \sum_{k,\,\sigma} \frac{\hbar\omega_k}{2}\left(\hat{a}_{k\sigma}^\dagger \hat{a}_{k\sigma} + \hat{a}_{k\sigma}\hat{a}_{k\sigma}^\dagger\right) \\ &= \sum_{k,\,\sigma} \hbar\omega_k \left(\hat{a}_{k\sigma}^\dagger \hat{a}_{k\sigma} + \frac{1}{2}\right) . \end{aligned} \tag{1.52}$$

Obviously the *zero-point energy* of the electromagnetic field

$$\sum_{k,\,\sigma} \frac{\hbar\omega_k}{2}$$

is infinite because of the existence of infinitely many field oscillators. However, we will discover that this infinite contribution plays no role in most of the physical problems. Only in special cases is it necessary to consider the zero-point energy, for example, in the change in the zero-point energy resulting from a change in the volume or the boundary conditions of the physical system. As an explicit example we will discuss the Casimir effect.

Heisenberg's equations of motion for the operators $\hat{a}_{k\sigma}$ read

$$\mathrm{i}\hbar \frac{\partial \hat{a}_{k\sigma}}{\partial t} = \left[\hat{a}_{k\sigma}, \hat{H}\right] \qquad \text{or} \qquad \dot{\hat{a}}_{k\sigma} = -\mathrm{i}\omega_k \, \hat{a}_{k\sigma} . \tag{1.53}$$

Formally they are identical to the classical equations of motion (1.39).

[2] See W. Greiner: *Quantum Mechanics – An Introduction*, 3rd ed. (Springer, Berlin, Heidelberg 1995).

1.4 Eigenstates of Electromagnetic Fields

The Hamiltonian (1.52) splits into separate contributions resulting from the various normal modes of the electromagnetic field. As a consequence the eigenstate has to factorize with respect to different normal modes. The eigenstate $|\ldots, n_{\boldsymbol{k}\sigma}, \ldots, n_{\boldsymbol{k}'\sigma'}, \ldots\rangle$ of the electromagnetic field is a direct product of eigenstates $|n_{\boldsymbol{k}\sigma}\rangle$ belonging to separate field oscillators. We write

$$|\ldots, n_{\boldsymbol{k}\sigma}, \ldots, n_{\boldsymbol{k}'\sigma'}, \ldots\rangle = \ldots |n_{\boldsymbol{k}\sigma}\rangle \ldots |n_{\boldsymbol{k}'\sigma'}\rangle \ldots . \tag{1.54}$$

The relation

$$\begin{aligned} \hat{N}_{\boldsymbol{k}\sigma}|n_{\boldsymbol{k}\sigma}\rangle &\equiv \hat{a}^{\dagger}_{\boldsymbol{k}\sigma}\hat{a}_{\boldsymbol{k}\sigma}|n_{\boldsymbol{k}\sigma}\rangle \\ &= n_{\boldsymbol{k}\sigma}|n_{\boldsymbol{k}\sigma}\rangle \qquad (n_{\boldsymbol{k}\sigma} = 0, 1, 2, \ldots) \end{aligned} \tag{1.55}$$

represents the eigenvalue equation for the number operator $\hat{N}_{\boldsymbol{k}\sigma}$ of the normal mode $\boldsymbol{k}_\sigma$; in a moment we will realize which particles are to be counted here.

This and the two following relations have been adopted directly from our introductory course in quantum mechanics and need no further specification. As a direct consequence of the commutation relations (1.50) we note

$$\hat{a}_{\boldsymbol{k}\sigma}|\ldots, n_{\boldsymbol{k}\sigma}, \ldots\rangle = \sqrt{n_{\boldsymbol{k}\sigma}}\,|\ldots, n_{\boldsymbol{k}\sigma} - 1, \ldots\rangle \tag{1.56}$$

and

$$\hat{a}^{\dagger}_{\boldsymbol{k}\sigma}|\ldots, n_{\boldsymbol{k}\sigma}, \ldots\rangle = \sqrt{n_{\boldsymbol{k}\sigma} + 1}\,|\ldots, n_{\boldsymbol{k}\sigma} + 1, \ldots\rangle . \tag{1.57}$$

Using (1.54) and (1.55) together with (1.52), we observe that the states $|\ldots, n_{\boldsymbol{k}\sigma}, \ldots\rangle$ are indeed eigenstates of the Hamiltonian H:

$$\begin{aligned} \hat{H}|\ldots, n_{\boldsymbol{k}\sigma}, \ldots\rangle &= \sum_{\boldsymbol{k},\,\sigma} \hbar\omega_k \,(\hat{a}^{\dagger}_{\boldsymbol{k}\sigma}\hat{a}_{\boldsymbol{k}\sigma} + \frac{1}{2})|\ldots, n_{\boldsymbol{k}\sigma}, \ldots\rangle \\ &= \sum_{\boldsymbol{k},\,\sigma} \hbar\omega_k \left(\hat{n}_{\boldsymbol{k}\sigma} + \frac{1}{2}\right)|\ldots, n_{\boldsymbol{k}\sigma}, \ldots\rangle \\ &= \sum_{\boldsymbol{k},\,\sigma} \hbar\omega_k \left(n_{\boldsymbol{k}\sigma} + \frac{1}{2}\right)|\ldots, n_{\boldsymbol{k}\sigma}, \ldots\rangle \\ &= E\,|\ldots, n_{\boldsymbol{k}\sigma}, \ldots\rangle , \end{aligned} \tag{1.58}$$

where the energy E is given by

$$E = \sum_{\boldsymbol{k},\,\sigma} \hbar\omega_k \left(n_{\boldsymbol{k}\sigma} + \frac{1}{2}\right) . \tag{1.59}$$

$\sum_{\boldsymbol{k},\,\sigma} \hbar\omega_k/2$ represents the now-known infinitely large zero-point energy.

The momentum of the classical electromagnetic field is given by

$$\boldsymbol{p} = \int_{L^3} \mathrm{d}^3x \,\frac{1}{4\pi c}\boldsymbol{E} \times \boldsymbol{B} ; \tag{1.60}$$

compare with the lectures on classical electrodynamics. In Exercise 1.3 we will transform this expression into

$$p = \sum_{k,\,\sigma} \hbar k \; \hat{a}^*_{k\sigma}\hat{a}_{k\sigma} \,, \tag{1.61}$$

where (1.37), (1.3), and (1.12) will be used. This enables us to define the momentum operator of the electromagnetic field as

$$\hat{p} = \sum_{k,\,\sigma} \hbar k \, \hat{a}^{\dagger}_{k\sigma}\hat{a}_{k\sigma} = \sum_{k,\,\sigma} \hbar k \, \hat{n}_{k\sigma} \,. \tag{1.62}$$

Because of (1.55), the states (1.54) are also eigenstates of the momentum operator:

$$\begin{aligned}\hat{p}\,|\ldots,n_{k\sigma},\ldots\rangle &= \sum_{k,\,\sigma} \hbar k \, \hat{n}_{k\sigma}\,|\ldots,n_{k\sigma},\ldots\rangle \\ &= \sum_{k,\,\sigma} \hbar k \, n_{k\sigma}\,|\ldots,n_{k\sigma},\ldots\rangle\end{aligned} \tag{1.63}$$

with eigenvalues

$$\sum_{k,\,\sigma} \hbar k \, n_{k\sigma} \,. \tag{1.64}$$

Based on relations (1.55) to (1.64) the electromagnetic field may be interpreted as a state consisting of $n_{k\sigma}$ photons with energy $\hbar\omega_k$, momentum $\hbar k$, and polarization $\varepsilon_{k\sigma}$. The quanta of the field are the photons themselves. The electromagnetic field as a whole consists of many of these photons; depending on the occupation numbers $n_{k\sigma}$ the photon states describe various states of the field. According to the relation (1.57) the operator $\hat{a}^{\dagger}_{k\sigma}$ increases the number of photons with quantum number $k\sigma$ by one; therefore this operator is called a photon *creation operator*. On the other hand the operator $\hat{a}_{k\sigma}$ annihilates a photon with quantum number $k\sigma$, which follows from (1.56); it is called an *annihilation operator*. The operator $\hat{N}_{k\sigma}$ of (1.55) counts the number of photons belonging to the mode $k\sigma$ and is called a *number operator*.

EXERCISE

1.3 Momentum Operator of Electromagnetic Fields

Problem. Show that the momentum operator of the electromagnetic field can be brought into the form

$$\hat{p} = \sum_{k,\,\sigma} \hbar k \, \hat{a}^{\dagger}_{k\sigma}\hat{a}_{k\sigma} \,.$$

Start with the classical expression for p and use the Fourier expansion (1.37) in the Coulomb gauge.

Solution. The expressions for the classical fields read $E = -\frac{1}{c}\frac{\partial A}{\partial t} - \nabla\varphi$ and $B = \nabla \times A$. In the Coulomb gauge (without any sources) the potentials become $\varphi = 0$ and $\nabla \cdot A = 0$; the Fourier expansion of the vector potential yields

Exercise 1.3.

$$\hat{\boldsymbol{A}}(\boldsymbol{r},t) = \sum_{\boldsymbol{k},\sigma} N_k \boldsymbol{\varepsilon}_{\boldsymbol{k}\sigma}\left(\hat{a}_{\boldsymbol{k}\sigma}\mathrm{e}^{\mathrm{i}\boldsymbol{k}\cdot\boldsymbol{r}} + \hat{a}^\dagger_{\boldsymbol{k}\sigma}\mathrm{e}^{-\mathrm{i}\boldsymbol{k}\cdot\boldsymbol{r}}\right) ,$$

where N_k is given by (1.47). The momentum of the electromagnetic field is

$$\boldsymbol{p} = \int_{L^3} \mathrm{d}^3x \frac{\boldsymbol{E}\times\boldsymbol{B}}{4\pi c} .$$

From Heisenberg's equations of motion (1.53) we get $\dot{\hat{a}}_{\boldsymbol{k}\sigma} = -\mathrm{i}\omega_k \hat{a}_{\boldsymbol{k}\sigma}$, so that the electric field $\boldsymbol{E}$ becomes

$$\hat{\boldsymbol{E}} = -\frac{1}{c}\frac{\partial\hat{\boldsymbol{A}}}{\partial t} = \frac{\mathrm{i}}{c}\sum_{\boldsymbol{k},\sigma} N_k\omega_k\boldsymbol{\varepsilon}_{\boldsymbol{k}\sigma}\left(\hat{a}_{\boldsymbol{k}\sigma}\mathrm{e}^{\mathrm{i}\boldsymbol{k}\cdot\boldsymbol{r}} - \hat{a}^\dagger_{\boldsymbol{k}\sigma}\mathrm{e}^{-\mathrm{i}\boldsymbol{k}\cdot\boldsymbol{r}}\right) . \quad (1)$$

Similiarly we get for the magnetic field $\boldsymbol{B}$

$$\begin{aligned}\hat{\boldsymbol{B}} &= \sum_{\boldsymbol{k}',\sigma'} N_{k'}\boldsymbol{\nabla}\times\boldsymbol{\varepsilon}_{\boldsymbol{k}'\sigma'}\left(\hat{a}_{\boldsymbol{k}'\sigma'}\mathrm{e}^{\mathrm{i}\boldsymbol{k}'\cdot\boldsymbol{r}} + \hat{a}^\dagger_{\boldsymbol{k}'\sigma'}\mathrm{e}^{-\mathrm{i}\boldsymbol{k}'\cdot\boldsymbol{r}}\right) \\ &= \sum_{\boldsymbol{k}',\sigma'} N_{k'}\mathrm{i}\boldsymbol{k}'\times\boldsymbol{\varepsilon}_{\boldsymbol{k}'\sigma'}\left(\hat{a}_{\boldsymbol{k}'\sigma'}\mathrm{e}^{\mathrm{i}\boldsymbol{k}'\cdot\boldsymbol{r}} - \hat{a}^\dagger_{\boldsymbol{k}'\sigma'}\mathrm{e}^{-\mathrm{i}\boldsymbol{k}'\cdot\boldsymbol{r}}\right) . \quad (2)\end{aligned}$$

Observe that the electromagnetic fields have become operators. Then the momentum operator $\hat{\boldsymbol{p}}$ results:

$$\begin{aligned}\hat{\boldsymbol{p}} &= \frac{1}{4\pi c}\int_{L^3}\mathrm{d}^3x\boldsymbol{E}\times\boldsymbol{B} \\ &= -\frac{1}{4\pi c^2}\sum_{\boldsymbol{k},\sigma}\sum_{\boldsymbol{k}',\sigma'} N_kN_{k'}\omega_k\boldsymbol{\varepsilon}_{\boldsymbol{k}\sigma}\times(\boldsymbol{k}'\times\boldsymbol{\varepsilon}_{\boldsymbol{k}'\sigma'}) \\ &\quad\times\int_{L^3}\mathrm{d}^3x\left(\hat{a}_{\boldsymbol{k}\sigma}\mathrm{e}^{\mathrm{i}\boldsymbol{k}\cdot\boldsymbol{r}} - \hat{a}^\dagger_{\boldsymbol{k}\sigma}\mathrm{e}^{-\mathrm{i}\boldsymbol{k}\cdot\boldsymbol{r}}\right)\left(\hat{a}_{\boldsymbol{k}'\sigma'}\mathrm{e}^{\mathrm{i}\boldsymbol{k}'\cdot\boldsymbol{r}} - \hat{a}^\dagger_{\boldsymbol{k}'\sigma'}\mathrm{e}^{-\mathrm{i}\boldsymbol{k}'\cdot\boldsymbol{r}}\right) .\end{aligned}$$

We multiply the last two square brackets explicitly, use

$$\int_{L^3}\mathrm{d}^3x\,\mathrm{e}^{\mathrm{i}(\boldsymbol{k}-\boldsymbol{k}')\cdot\boldsymbol{r}} = L^3\delta_{\boldsymbol{k}\boldsymbol{k}'} ,$$

transform

$$\boldsymbol{\varepsilon}_{\boldsymbol{k}\sigma}\times(\boldsymbol{k}'\times\boldsymbol{\varepsilon}_{\boldsymbol{k}'\sigma'}) = \boldsymbol{k}'\left(\boldsymbol{\varepsilon}_{\boldsymbol{k}\sigma}\cdot\boldsymbol{\varepsilon}_{\boldsymbol{k}'\sigma'}\right) - \boldsymbol{\varepsilon}_{\boldsymbol{k}'\sigma'}\left(\boldsymbol{\varepsilon}_{\boldsymbol{k}\sigma}\cdot\boldsymbol{k}'\right) ,$$

and arrive at

$$\begin{aligned}\hat{\boldsymbol{p}} &= -\frac{L^3}{4\pi c^2}\sum_{\boldsymbol{k},\sigma}\sum_{\boldsymbol{k}',\sigma'} N_kN_{k'}\omega_k\left[\boldsymbol{k}'\left(\boldsymbol{\varepsilon}_{\boldsymbol{k}\sigma}\cdot\boldsymbol{\varepsilon}_{\boldsymbol{k}'\sigma'}\right) - \boldsymbol{\varepsilon}_{\boldsymbol{k}'\sigma'}\left(\boldsymbol{\varepsilon}_{\boldsymbol{k}\sigma}\cdot\boldsymbol{k}'\right)\right] \\ &\quad\times\left[\left(\hat{a}_{\boldsymbol{k}\sigma}\hat{a}_{\boldsymbol{k}'\sigma'} + \hat{a}^\dagger_{\boldsymbol{k}\sigma}\hat{a}^\dagger_{\boldsymbol{k}'\sigma'}\right)\delta_{-\boldsymbol{k},\boldsymbol{k}'} - \left(\hat{a}^\dagger_{\boldsymbol{k}\sigma}\hat{a}_{\boldsymbol{k}'\sigma'} + \hat{a}_{\boldsymbol{k}\sigma}\hat{a}^\dagger_{\boldsymbol{k}'\sigma'}\right)\delta_{\boldsymbol{k}\boldsymbol{k}'}\right] . \quad (3)\end{aligned}$$

First we concentrate on the term $\sim\delta_{\boldsymbol{k}\boldsymbol{k}'}$ symmetric in $\boldsymbol{k}$ and $\boldsymbol{k}'$:

$$\begin{aligned}\hat{\boldsymbol{p}}_{\mathrm{s}} &= \frac{L^3}{4\pi c^2}\sum_{\boldsymbol{k},\sigma}\sum_{\boldsymbol{k}',\sigma'} N_kN_{k'}\omega_k\left[\boldsymbol{k}'\left(\boldsymbol{\varepsilon}_{\boldsymbol{k}\sigma}\cdot\boldsymbol{\varepsilon}_{\boldsymbol{k}'\sigma'}\right) - \boldsymbol{\varepsilon}_{\boldsymbol{k}'\sigma'}\left(\boldsymbol{\varepsilon}_{\boldsymbol{k}\sigma}\cdot\boldsymbol{k}'\right)\right] \\ &\quad\times\left(\hat{a}^\dagger_{\boldsymbol{k}\sigma}\hat{a}_{\boldsymbol{k}'\sigma'} + \hat{a}_{\boldsymbol{k}\sigma}\hat{a}^\dagger_{\boldsymbol{k}'\sigma'}\right)\delta_{\boldsymbol{k}\boldsymbol{k}'} .\end{aligned}$$

Exercise 1.3.

The term $\sim \boldsymbol{\varepsilon}_{k\sigma} \cdot \boldsymbol{k}'$ vanishes for $\boldsymbol{k} = \boldsymbol{k}'$ since $\boldsymbol{\varepsilon}_{k\sigma} \perp \boldsymbol{k}$. Furthermore $\boldsymbol{\varepsilon}_{k\sigma} \cdot \boldsymbol{\varepsilon}_{k\sigma'} = \delta_{\sigma\sigma'}$, so that the double sum in $\hat{\boldsymbol{p}}_s$ breaks down:

$$\hat{\boldsymbol{p}}_s = \frac{L^3}{4\pi c^2} \sum_{\boldsymbol{k},\,\sigma} N_k^2 \omega_k \boldsymbol{k} \left(\hat{a}_{k\sigma} \hat{a}^\dagger_{k\sigma} + \hat{a}^\dagger_{k\sigma} \hat{a}_{k\sigma} \right) .$$

With

$$N_k^2 = \frac{2\pi\hbar c^2}{L^3 \omega_k} ,$$

i.e.

$$\frac{L^3}{4\pi c^2} N_k^2 \omega_k = \frac{L^3}{4\pi c^2} \frac{2\pi\hbar c^2}{L^3 \omega_k} \omega_k = \frac{\hbar}{2} ,$$

and $\hat{a}_{k\sigma} \hat{a}^\dagger_{k\sigma} - \hat{a}^\dagger_{k\sigma} \hat{a}_{k\sigma} = 1$, we deduce

$$\hat{\boldsymbol{p}}_s = \sum_{\boldsymbol{k},\,\sigma} \frac{\hbar}{2} \boldsymbol{k} \left(1 + 2 \hat{a}^\dagger_{k\sigma} \hat{a}_{k\sigma} \right) .$$

The sum $\sum_{\boldsymbol{k},\,\sigma} \frac{\hbar}{2} \boldsymbol{k}$ over all momenta is equal to zero, because positive and negative contributions cancel pairwise. Thus

$$\hat{\boldsymbol{p}}_s = \sum_{\boldsymbol{k},\,\sigma} \hbar \boldsymbol{k} \hat{a}^\dagger_{k\sigma} \hat{a}_{k\sigma} . \tag{4}$$

We still have to work out the antisymmetric term $\sim \delta_{\boldsymbol{k},-\boldsymbol{k}'}$ in (3):

$$\begin{aligned} \hat{\boldsymbol{p}}_{as} &= -\frac{L^3}{4\pi c^2} \sum_{\boldsymbol{k},\sigma,\sigma'} N_k N_{-k} \omega_k \left[-\boldsymbol{k} \left(\boldsymbol{\varepsilon}_{k\sigma} \cdot \boldsymbol{\varepsilon}_{-k\sigma'} \right) + \boldsymbol{\varepsilon}_{-k\sigma} \left(\boldsymbol{\varepsilon}_{k\sigma} \cdot \boldsymbol{k} \right) \right] \\ &\quad \times \left(\hat{a}_{k\sigma} \hat{a}_{-k\sigma'} + \hat{a}^\dagger_{k\sigma} \hat{a}^\dagger_{-k\sigma'} \right) . \end{aligned}$$

Again $\boldsymbol{\varepsilon}_{k\sigma} \cdot \boldsymbol{k} = 0$ vanishes and $\boldsymbol{\varepsilon}_{k\sigma} \cdot \boldsymbol{\varepsilon}_{-k\sigma'} = -(-1)^\sigma \delta_{\sigma\sigma'}$; see Fig. 1.5. We deduce

$$\begin{aligned} \hat{\boldsymbol{p}}_{as} &= -\frac{L^3}{4\pi c^2} \sum_{\boldsymbol{k}} \Big[N_k N_{-k} \omega_k (-\boldsymbol{k}) \left(\hat{a}_{k1} \hat{a}_{-k1} + \hat{a}^\dagger_{k1} \hat{a}^\dagger_{-k1} \right) \\ &\quad + N_k N_{-k} \omega_k (+\boldsymbol{k}) \left(\hat{a}_{k2} \hat{a}_{-k2} + \hat{a}^\dagger_{k2} \hat{a}^\dagger_{-k2} \right) \Big] . \end{aligned}$$

The two terms both include a factor $\boldsymbol{k}$ multiplied with other quantities, which are invariant under the interchange $\boldsymbol{k} \to -\boldsymbol{k}$. Because the sum goes over all positive and negative components of $\boldsymbol{k}$, terms cancel pairwise again. Therefore

$$\hat{\boldsymbol{p}}_{as} = 0 .$$

Then the total momentum of the electromagnetic field finally reads

$$\hat{\boldsymbol{p}} = \hat{\boldsymbol{p}}_s + \hat{\boldsymbol{p}}_{as} = \sum_{\boldsymbol{k},\,\sigma} \hbar \boldsymbol{k} \hat{a}^\dagger_{k\sigma} \hat{a}_{k\sigma} .$$

1.5 Coherent States (Glauber States) of Electromagnetic Fields

The electric field is determined by

$$\boldsymbol{E} = -\frac{1}{c}\frac{\partial \boldsymbol{A}}{\partial t} = -\frac{1}{c}\frac{\partial}{\partial t}\sum_{\boldsymbol{k},\,\sigma}\sqrt{\frac{2\pi\hbar c^2}{L^3\omega_k}}\boldsymbol{\varepsilon}_{\boldsymbol{k}\sigma}\left[a_{\boldsymbol{k}\sigma}(t)\,\mathrm{e}^{\mathrm{i}\boldsymbol{k}\cdot\boldsymbol{x}} + a^*_{\boldsymbol{k}\sigma}(t)\,\mathrm{e}^{-\mathrm{i}\boldsymbol{k}\cdot\boldsymbol{x}}\right]$$

$$= +\mathrm{i}\sum_{\boldsymbol{k},\,\sigma}\sqrt{\frac{2\pi\hbar c^2}{L^3\omega_k}}\frac{\omega_k}{c}\,\boldsymbol{\varepsilon}_{\boldsymbol{k}\sigma}\left[a_{\boldsymbol{k}\sigma}(t)\,\mathrm{e}^{\mathrm{i}\boldsymbol{k}\cdot\boldsymbol{x}} - a^*_{\boldsymbol{k}\sigma}(t)\,\mathrm{e}^{-\mathrm{i}\boldsymbol{k}\cdot\boldsymbol{x}}\right], \tag{1.65}$$

where (1.3), (1.12), (1.37), and (1.39) have been used. It is tiresome to carry along the sum over $\boldsymbol{k}$ and σ all the time; therefore we will suppress this sum in the following and only track one representative term out of (1.65):

$$\boldsymbol{E} = +\mathrm{i}\sqrt{\frac{2\pi\hbar\omega}{L^3}}\boldsymbol{\varepsilon}\left[a(t)\,\mathrm{e}^{\mathrm{i}\boldsymbol{k}\cdot\boldsymbol{x}} - a^*(t)\,\mathrm{e}^{-\mathrm{i}\boldsymbol{k}\cdot\boldsymbol{x}}\right]. \tag{1.66}$$

This term describes one normal mode of the electromagnetic field. We are tempted to change the classical electric field of $\boldsymbol{E}$ in (1.66) straight forwardly into an operator $\hat{\boldsymbol{E}}$ by writing

$$\hat{\boldsymbol{E}}(\boldsymbol{x},t) = +\mathrm{i}\sqrt{\frac{2\pi\hbar\omega}{L^3}}\boldsymbol{\varepsilon}\left[\hat{a}(t)\,\mathrm{e}^{\mathrm{i}\boldsymbol{k}\cdot\boldsymbol{x}} - \hat{a}^\dagger(t)\,\mathrm{e}^{-\mathrm{i}\boldsymbol{k}\cdot\boldsymbol{x}}\right]. \tag{1.67}$$

However, some care has to be taken. We start again with the original classical equation $\boldsymbol{E} = -\frac{1}{c}\frac{\partial \boldsymbol{A}}{\partial t}$ and define the operator for the electric field by the operator equation

$$\hat{\boldsymbol{E}} = -\frac{1}{c}\frac{\partial \hat{\boldsymbol{A}}}{\partial t} = -\frac{1}{c}\left(-\frac{\mathrm{i}}{\hbar}\right)\left[\hat{\boldsymbol{A}},\hat{H}\right]_- . \tag{1.68}$$

For the computation of the commutator we need to consider only the terms in the expression for $\hat{H}$ and $\hat{A}$ with the same $\hat{a}$ and $\hat{a}^\dagger$; they characterize one normal mode. The operators belonging to different normal modes commute because of (1.50). We get

$$\hat{\boldsymbol{E}} = -\frac{1}{c}\sqrt{\frac{2\pi\hbar c^2}{L^3\omega_k}}\hbar\omega_k\left(-\frac{\mathrm{i}}{\hbar}\right)\boldsymbol{\varepsilon}\left[\hat{a}\mathrm{e}^{\mathrm{i}\boldsymbol{k}\cdot\boldsymbol{x}} + \hat{a}^\dagger\,\mathrm{e}^{-\mathrm{i}\boldsymbol{k}\cdot\boldsymbol{x}}, \hat{a}^\dagger\hat{a}\right]_-$$

$$= \mathrm{i}\sqrt{\frac{2\pi\hbar\omega_k}{L^3}}\boldsymbol{\varepsilon}\left[\hat{a}\mathrm{e}^{\mathrm{i}\boldsymbol{k}\cdot\boldsymbol{x}} - \hat{a}^\dagger\,\mathrm{e}^{-\mathrm{i}\boldsymbol{k}\cdot\boldsymbol{x}}\right], \tag{1.69}$$

which is identical with the straightforward proposition (1.67). The last approach makes use of the definition of the vector potential as a quantum-mechanical operator and of Heisenberg's equations of motion. It represents the correct quantum-mechanical approach.

Generally the electromagnetic field contains many photons. It is characterized by an N-photon state like the one of (1.54), which is a direct product of states $|n_{\boldsymbol{k}\sigma}\rangle$ describing $n_{\boldsymbol{k}\sigma}$ photons of mode $\boldsymbol{k}\sigma$. With $|n\rangle$ we simply denote the state corresponding to the mode characterized by the operators $\hat{a}, \hat{a}^\dagger$. In order to determine the *classical value of the electromagnetic field* $\hat{\boldsymbol{E}}$, we determine the expectation value of the $\boldsymbol{E}$ operator within the n-photon state $|n\rangle$. With (1.67) we find

$$\langle n|\hat{\boldsymbol{E}}|n\rangle = \mathrm{i}\sqrt{\frac{2\pi\hbar\omega}{L^3}}\boldsymbol{\varepsilon}\langle n|\hat{a}\,\mathrm{e}^{\mathrm{i}\boldsymbol{k}\cdot\boldsymbol{x}} - \hat{a}^\dagger\,\mathrm{e}^{-\mathrm{i}\boldsymbol{k}\cdot\boldsymbol{x}}|n\rangle = 0\,, \tag{1.70}$$

where

$$\langle n|\hat{a}|n\rangle = \langle n|\hat{a}^\dagger|n\rangle = 0 \tag{1.71}$$

enters because of relations (1.56) and (1.57). On the other hand the expectation value of the square of the electric field, which apart from a factor of 8π describes part of the energy density, is equal to

$$\begin{aligned}\langle n|\frac{\hat{\boldsymbol{E}}\hat{\boldsymbol{E}}^\dagger}{8\pi}|n\rangle &= \frac{2\pi\hbar\omega}{8\pi L^3}\boldsymbol{\varepsilon}\cdot\boldsymbol{\varepsilon}\,\langle n|\left(\hat{a}\mathrm{e}^{\mathrm{i}\boldsymbol{k}\cdot\boldsymbol{x}} - \hat{a}^\dagger\,\mathrm{e}^{-\mathrm{i}\boldsymbol{k}\cdot\boldsymbol{x}}\right)\\ &\qquad\times\left(\hat{a}^\dagger\mathrm{e}^{-\mathrm{i}\boldsymbol{k}\cdot\boldsymbol{x}} - \hat{a}\,\mathrm{e}^{\mathrm{i}\boldsymbol{k}\cdot\boldsymbol{x}}\right)|n\rangle\\ &= \frac{2\pi\hbar\omega}{8\pi L^3}\,\langle n|\hat{a}\hat{a}^\dagger + \hat{a}^\dagger\hat{a}|n\rangle\\ &= \frac{\hbar\omega}{4L^3}\,\langle n|2\,\hat{a}^\dagger\hat{a}+1|n\rangle = \frac{\hbar\omega}{2L^3}\left(n+\frac{1}{2}\right)\,.\end{aligned} \tag{1.72}$$

This result is surprising. The expectation value of the $\boldsymbol{E}$ field in the n-photon state vanishes, whereas, except for the zero-point energy, the energy density is equal to the half of n photons times $\hbar\omega$ per volume L^3. The other half of the expected energy density results from the term $\hat{\boldsymbol{B}}\cdot\hat{\boldsymbol{B}}^\dagger/8\pi$. This puzzle is resolved once we realize that statistical phases are assigned to the n photons in the field. After we average over the phases, the expectation value of $\boldsymbol{E}$ vanishes. On the other hand the phases drop out for the energy density; the energy density is a real positive number for every space–time point.

It was Glauber,[3] who introduced a field state, for which the expectation value of the $\hat{\boldsymbol{E}}$ operator resembles the classical $\boldsymbol{E}$ field. It allows for an uncertainty in the photon number in order to define the phase of the field more precisely. The *Glauber state* $|c\rangle$, often also called the *coherent state*, reads

$$|c\rangle = \sum_{n=0}^{\infty} b_n\,|n\rangle\,, \tag{1.73}$$

where

$$b_n = \frac{c^n\,\mathrm{e}^{-\frac{1}{2}|c|^2}}{\sqrt{n!}} \tag{1.74}$$

and c represents a complex number. The probability of finding n photons in the Glauber state is given by the square of the amplitude:

$$|b_n|^2 = \frac{|c|^{2n}\,\mathrm{e}^{-|c|^2}}{n!}\,. \tag{1.75}$$

The sum over all these probabilities has to be unity; indeed, we find

[3] R.J. Glauber: Phys. Rev. Lett. **10** (1963) 84.

$$\sum_{n=0}^{\infty} |b_n|^2 = \mathrm{e}^{-|c|^2} \sum_{n=0}^{\infty} \frac{|c|^{2n}}{n!} = \mathrm{e}^{-|c|^2} \mathrm{e}^{+|c|^2} = 1 . \tag{1.76}$$

Since the Glauber state has no definite photon number, the expectation value of the annihilation operator is not zero; this is in contrast to the incoherent state (1.54) with definite photon number. We calculate

$$\begin{aligned} \langle c|\hat{a}|c\rangle &= \sum_{m=0}^{\infty}\sum_{n=0}^{\infty} b_m^* b_n \,\langle m|\hat{a}|n\rangle \\ &= \sum_{m,n=0}^{\infty} b_m^* b_n \,\sqrt{n}\,\delta_{m,n-1} \\ &= \sum_{m=0}^{\infty} b_m^* b_{m+1}\,\sqrt{m+1} \\ &= c\,\mathrm{e}^{-|c|^2} \sum_{m=0}^{\infty} \frac{c^{*m}c^m}{m!} = c\,\mathrm{e}^{-|c|^2}\,\mathrm{e}^{+|c|^2} = c . \end{aligned} \tag{1.77}$$

In the same manner we get

$$\langle c|\hat{a}^\dagger|c\rangle = c^* . \tag{1.78}$$

With these two results we deduce the expectation value of the $\boldsymbol{E}$ field (1.67) in the Glauber state:

$$\begin{aligned} \langle c|\hat{\boldsymbol{E}}|c\rangle &= +\mathrm{i}\sqrt{\frac{2\pi\hbar\omega}{L^3}}\,\boldsymbol{\varepsilon}\langle c|\hat{a}\,\mathrm{e}^{\mathrm{i}\boldsymbol{k}\cdot\boldsymbol{x}} - \hat{a}^\dagger\,\mathrm{e}^{-\mathrm{i}\boldsymbol{k}\cdot\boldsymbol{x}}|c\rangle \\ &= +\mathrm{i}\sqrt{\frac{2\pi\hbar\omega}{L^3}}\,\boldsymbol{\varepsilon}\left[c\,\mathrm{e}^{\mathrm{i}\boldsymbol{k}\cdot\boldsymbol{x}} - c^*\,\mathrm{e}^{-\mathrm{i}\boldsymbol{k}\cdot\boldsymbol{x}}\right] \\ &= +\mathrm{i}\sqrt{\frac{2\pi\hbar\omega}{L^3}}\,\boldsymbol{\varepsilon}\,2\,\mathrm{i}\,|c|\,\sin(\boldsymbol{k}\cdot\boldsymbol{x}+\delta_c) \\ &= -2\,|c|\sqrt{\frac{2\pi\hbar\omega}{L^3}}\,\boldsymbol{\varepsilon}\,\sin(\boldsymbol{k}\cdot\boldsymbol{x}+\delta_c) . \end{aligned} \tag{1.79}$$

This expression has the form of a classical electromagnetic wave; see (1.17). Its amplitude is given by $|c|$ and its phase is determined from the phase of $c = |c|\,\mathrm{e}^{\mathrm{i}\delta_c}$. It is worth mentioning that the second line of (1.79) is formally identical to the expression (1.67); the creation and annihilation operators $\hat{a}$, $\hat{a}^\dagger$ have been substituted by the complex amplitudes c, c^*.

EXERCISE

1.4 Matrix Elements with Coherent States

Problem. Show the validity of the following relations:

(a) $\langle c|\hat{a}^\dagger\hat{a}|c\rangle = |c|^2 = \langle\hat{n}\rangle$

(b) $\langle c|\hat{a}\hat{a}^\dagger|c\rangle = |c|^2 + 1 = \langle\hat{n}\rangle + 1$

Exercise 1.4.

(c) $\langle c|\hat{a}^\dagger\hat{a}\hat{a}^\dagger\hat{a}|c\rangle = |c|^4 + |c|^2 = \langle\hat{n}^2\rangle$

(d) $\langle c|\hat{a}^2|c\rangle = c^2$

(e) $\langle c|\hat{a}^{\dagger 2}|c\rangle = c^{*2}$

Solution. We already know the action of $\hat{a}$ on $|c\rangle$:

$$\hat{a}\,|c\rangle = c\,|c\rangle\,; \tag{1}$$

that means that $|c\rangle$ is an eigenstate of $\hat{a}$ with eigenvalue c. The conjugate equation reads

$$(\hat{a}\,|c\rangle)^\dagger = (c\,|c\rangle)^\dagger = \langle c|\,\hat{a}^\dagger = \langle c|\,c^*\,,$$

or in short

$$\langle c|\,\hat{a}^\dagger = \langle c|\,c^*\,. \tag{2}$$

We also remember the commutation relation

$$\hat{a}\hat{a}^\dagger - \hat{a}^\dagger\hat{a} = 1\,. \tag{3}$$

With the help of (1)–(3) we prove relations (a)–(e):

(a) $\langle c|\hat{a}^\dagger\hat{a}|c\rangle = c^*\langle c|\hat{a}|c\rangle = c^*c\langle c|c\rangle = |c|^2\,,$ because $\langle c|c\rangle = 1\,.$

(b) $\langle c|\hat{a}\hat{a}^\dagger|c\rangle = \langle c|1+\hat{a}^\dagger\hat{a}|c\rangle = \langle c|c\rangle + \langle c|\hat{a}^\dagger\hat{a}|c\rangle = |c|^2+1\,,$ because of (a) and (3)

(c) $\langle c|\hat{a}^\dagger\hat{a}\hat{a}^\dagger\hat{a}|c\rangle = \langle c|\hat{a}^\dagger(1+\hat{a}^\dagger\hat{a})\hat{a}|c\rangle = \langle c|\hat{a}^\dagger\hat{a}|c\rangle + \langle c|\hat{a}^\dagger\hat{a}^\dagger\hat{a}\hat{a}|c\rangle = |c|^2 + c^*c^*cc = |c|^4+|c|^2$

(d) $\langle c|\hat{a}^2|c\rangle = \langle c|\hat{a}\hat{a}|c\rangle = c\langle c|\hat{a}|c\rangle = c^2$

(e) $\langle c|\hat{a}^{\dagger 2}|c\rangle = \langle c|\hat{a}^\dagger\hat{a}^\dagger|c\rangle = c^*\langle c|\hat{a}^\dagger|c\rangle = c^{*2}\,.$

The photon number operator $\hat{n} = \hat{a}^\dagger\hat{a}$ has the expectation value $\langle c|\hat{n}|c\rangle = |c|^2$ and $\langle c|\hat{n}^2|c\rangle = |c|^4 + |c|^2$ for its square. This shows again that the coherent state has no definite photon number, since c is an arbitrary complex number.

The Glauber state has no definite photon number n. We therefore ask for the uncertainty Δn of the photon number within the coherent state $|c\rangle$, which is defined as

$$\Delta n = \sqrt{\langle c|\,(\hat{n}-\langle\hat{n}\rangle)^2\,|c\rangle}\,. \tag{1.80}$$

With the results of Exercise 1.4 we determine

$$\begin{aligned}\Delta n &= \sqrt{\langle c|\hat{n}^2|c\rangle - \langle c|\hat{n}|c\rangle^2}\\ &= \sqrt{|c|^4+|c|^2-|c|^4}\\ &= |c| = \sqrt{\langle c|\hat{n}|c\rangle}\,.\end{aligned} \tag{1.81}$$

The so-called *relative uncertainty*

$$\begin{aligned}\frac{\Delta n}{n} &= \frac{\Delta n}{\langle c|\hat{n}|c\rangle} \\ &= \frac{1}{\sqrt{\langle c|\hat{n}|c\rangle}}\end{aligned} \tag{1.82}$$

becomes smaller the larger the average photon number $\langle c|\hat{n}|c\rangle$ becomes in the Glauber state. For coherent states with very many photons the $\boldsymbol{E}$ field still behaves like a classical field [see (1.79)] and, in general the relative uncertainty of the photon number becomes very small. Such states play an important role in laser physics. Coherent states have minimum uncertainty, so that the equality sign holds in Heisenberg's uncertainty relation.

EXERCISE

1.5 The Mean Quadratic Deviation of the Electric Field Within the Coherent State

Problem. Determine the mean quadratic deviation of the electric field within a coherent state and show that it vanishes in the classical limit.

Solution. The mean quadratic deviation of the electric field is defined by $\langle c|\hat{\boldsymbol{E}}\cdot\hat{\boldsymbol{E}}^\dagger|c\rangle - \langle c|\hat{\boldsymbol{E}}|c\rangle^2$. Using expression (1.67) for the $\hat{\boldsymbol{E}}$ operator we find

$$\begin{aligned}\langle c|\hat{\boldsymbol{E}}\cdot\hat{\boldsymbol{E}}^\dagger|c\rangle &= \frac{2\pi\hbar\omega}{L^3}\boldsymbol{\varepsilon}\cdot\boldsymbol{\varepsilon}\langle c|\left(\hat{a}\mathrm{e}^{\mathrm{i}\boldsymbol{k}\cdot\boldsymbol{x}} - \hat{a}^\dagger\mathrm{e}^{-\mathrm{i}\boldsymbol{k}\cdot\boldsymbol{x}}\right) \\ &\qquad\times\left(\hat{a}^\dagger\mathrm{e}^{-\mathrm{i}\boldsymbol{k}\cdot\boldsymbol{x}} - \hat{a}\mathrm{e}^{\mathrm{i}\boldsymbol{k}\cdot\boldsymbol{x}}\right)|c\rangle \\ &= \frac{2\pi\hbar\omega}{L^3}\langle c|\hat{a}\hat{a}^\dagger + \hat{a}^\dagger\hat{a} - \hat{a}\hat{a}\mathrm{e}^{\mathrm{i}2\boldsymbol{k}\cdot\boldsymbol{x}} - \hat{a}^\dagger\hat{a}^\dagger\mathrm{e}^{-\mathrm{i}2\boldsymbol{k}\cdot\boldsymbol{x}}|c\rangle\,.\end{aligned}$$

With the results derived in Exercise 1.4 it follows further that

$$\langle c|\hat{\boldsymbol{E}}\cdot\hat{\boldsymbol{E}}^\dagger|c\rangle = \frac{2\pi\hbar\omega}{L^3}\left(2\,|c|^2 + 1 - c^2\,\mathrm{e}^{\mathrm{i}\,2\,\boldsymbol{k}\cdot\boldsymbol{x}} - c^{*2}\,\mathrm{e}^{-\mathrm{i}\,2\,\boldsymbol{k}\cdot\boldsymbol{x}}\right). \tag{1}$$

The expectation value of $\hat{\boldsymbol{E}}$ within the Glauber state has already been calculated in (1.79). We recapitulate:

$$\begin{aligned}|\langle c|\hat{\boldsymbol{E}}|c\rangle|^2 &= \langle c|\hat{\boldsymbol{E}}|c\rangle\,\langle c|\hat{\boldsymbol{E}}|c\rangle^* \\ &= \frac{2\pi\hbar\omega}{L^3}\,\boldsymbol{\varepsilon}\cdot\boldsymbol{\varepsilon}\,\left(|c|^2 + |c^*|^2\right. \\ &\qquad\left. - c^2\,\mathrm{e}^{\mathrm{i}\,2\,\boldsymbol{k}\cdot\boldsymbol{x}} - (c^*)^2\,\mathrm{e}^{-\mathrm{i}\,2\,\boldsymbol{k}\cdot\boldsymbol{x}}\right).\end{aligned} \tag{2}$$

Substraction of (2) from (1) yields the mean quadratic deviation

$$\langle c|\hat{\boldsymbol{E}}\cdot\hat{\boldsymbol{E}}^\dagger|c\rangle - |\langle c|\hat{\boldsymbol{E}}|c\rangle|^2 = \frac{2\pi\hbar\omega}{L^3}\,. \tag{3}$$

In the classical limit, i.e. $\hbar\to 0$, the mean quadratic deviation goes to zero.

EXAMPLE

1.6 The Aharonov–Bohm Effect

Here we will discuss an effect that gives interesting insight into the nature of the electromagnetic field and its role for quantum mechanics. It is a typical interference phenomenon that can be understood at an elementary level; strangely enough, its importance was recognized only 30 years after the development of quantum mechanics.[4]

Let us recapitulate how the interaction between charged particles is described in the classical picture. Following the conception by Faraday and Maxwell this interaction is not given via a remote action, but as a local interaction with an electric field $\boldsymbol{E}(\boldsymbol{x},t)$ and a magnetic field $\boldsymbol{B}(\boldsymbol{x},t)$. It is sufficient to know the field strengths $\boldsymbol{E}$ and $\boldsymbol{B}$ at each space–time point of the particle's trajectory to describe the motion of a particle with charge e. Newton's equation of motion holds with the Lorentz force

$$\boldsymbol{F} = e\left(\boldsymbol{E} + \frac{1}{c}\dot{\boldsymbol{x}} \times \boldsymbol{B}\right) . \tag{1}$$

In classical electrodynamics the Coulomb potential $A_0(\boldsymbol{x},t)$ and the vector potential $\boldsymbol{A}(\boldsymbol{x},t)$ are merely introduced as technical quantities to facilitate formulations:

$$\begin{aligned} \boldsymbol{E} &= -\boldsymbol{\nabla} A_0 - \frac{1}{c}\frac{\partial \boldsymbol{A}}{\partial t} , \\ \boldsymbol{B} &= \boldsymbol{\nabla} \times \boldsymbol{A} . \end{aligned} \tag{2}$$

These potentials were not expected to have a physical interpretation. Beyond that, the potentials are not unique; gauge transformations

$$\begin{aligned} A_0' &= A_0 - \frac{1}{c}\frac{\partial \chi}{\partial t} , \\ \boldsymbol{A}' &= \boldsymbol{A} + \boldsymbol{\nabla}\chi \end{aligned} \tag{3}$$

with an arbitrary function $\chi(\boldsymbol{x},t)$ leave the electromagnetic fields and, consequently, also the Lorentz force invariant.

In quantum mechanics the notion of a force is not of much use. One starts instead with a Hamiltonian. The Hamiltonian of a nonrelativistic charged particle moving in an electromagnetic field contains the potentials A_0 and $\boldsymbol{A}$,

$$H = \frac{1}{2m}\left(\boldsymbol{p} - \frac{e}{c}\boldsymbol{A}\right)^2 + eA_0 , \tag{4}$$

and leads to the Schrödinger equation

$$\mathrm{i}\hbar\frac{\partial}{\partial t}\Psi(\boldsymbol{x},t) = \left[\frac{1}{2m}\left(-\mathrm{i}\hbar\boldsymbol{\nabla} - \frac{e}{c}\boldsymbol{A}\right)^2 + eA_0\right]\Psi(\boldsymbol{x},t) . \tag{5}$$

The wavefunction Ψ now directly depends on the potentials A_0 and $\boldsymbol{A}$ and only indirectly on the fields $\boldsymbol{E}$ and $\boldsymbol{B}$. Even when fields vanish, the wavefunction Ψ is still influenced by the potentials.

[4] Y. Aharonov, D. Bohm: Phys. Rev. **115** (1959) 485.

Example 1.6.

To begin with, we study the special case in which $\boldsymbol{A}$ leads to a stationary and curlfree field $\boldsymbol{B} = \boldsymbol{\nabla} \times \boldsymbol{A} = 0$. If the wavefunction Ψ_0 represents a solution of the Schrödinger equation (5) with $\boldsymbol{A} = 0$ but otherwise arbitrary $A_0(\boldsymbol{x}, t)$, then we can easily show that the wavefunction

$$\Psi(\boldsymbol{x}, t) = \Psi_0(\boldsymbol{x}, t) \exp\left(\mathrm{i}\frac{e}{\hbar c}\int_\Gamma \mathrm{d}\boldsymbol{x}' \cdot \boldsymbol{A}(\boldsymbol{x}')\right) \tag{6}$$

solves the Schrödinger equation including the vector potential $\boldsymbol{A}$. Here $\Gamma \equiv \Gamma(\boldsymbol{x})$ characterizes an arbitrary curve in three-dimensional space which ends at point $\boldsymbol{x}$. Let us investigate the momentum operator acting on a wavefunction of type (6). Now the gradient also acts on the phase factor: to be more precise, it acts on the upper boundary of the integration. We get

$$\boldsymbol{\nabla} \exp\left(\mathrm{i}\frac{e}{\hbar c}\int_\Gamma \mathrm{d}\boldsymbol{x}' \cdot \boldsymbol{A}(\boldsymbol{x}')\right) = \mathrm{i}\frac{e}{\hbar c}\boldsymbol{A}(\boldsymbol{x}) \exp\left(\mathrm{i}\frac{e}{\hbar c}\int_\Gamma \mathrm{d}\boldsymbol{x}' \cdot \boldsymbol{A}(\boldsymbol{x}')\right) . \tag{7}$$

Some care has to be taken when performing this step; this transformation only holds when the curl of $\boldsymbol{A}$ is vanishing. (We convince ourselves with a simple counter example. A homogeneous magnetic field in the z direction $B = \boldsymbol{B} \cdot \boldsymbol{e}_z = \boldsymbol{\nabla} \times \boldsymbol{A}$ is achieved $\boldsymbol{A} = \frac{1}{2}Br\boldsymbol{e}_\varphi$; we choose the integration path in the radial direction, so that $\boldsymbol{e}_r \cdot \boldsymbol{e}_\varphi = 0$ and the path integral vanishes. Then (7) cannot be fulfilled.) Only if the curl of the vector potential is equal to zero is the path integral independend of the chosen path. With the help of (7) we find

$$\begin{aligned}\left(-\mathrm{i}\hbar\boldsymbol{\nabla} - \frac{e}{c}\boldsymbol{A}\right)\Psi &= \exp\left(\mathrm{i}\frac{e}{\hbar c}\int_\Gamma \mathrm{d}\boldsymbol{x}' \cdot \boldsymbol{A}(\boldsymbol{x}')\right) \\ &\quad \times \left[\left(-\mathrm{i}\hbar\boldsymbol{\nabla} - \frac{e}{c}\boldsymbol{A}\right)\Psi_0 - \mathrm{i}\hbar\frac{ie}{\hbar c}\boldsymbol{A}\Psi_0\right]\end{aligned} \tag{8a}$$

and

$$\begin{aligned}\left(-\mathrm{i}\hbar\boldsymbol{\nabla} - \frac{e}{c}\boldsymbol{A}\right)^2\Psi &= \exp\left(\mathrm{i}\frac{e}{\hbar c}\int_\Gamma \mathrm{d}\boldsymbol{x}' \cdot \boldsymbol{A}(\boldsymbol{x}')\right) \\ &\quad \times \left[\left(-\mathrm{i}\hbar\boldsymbol{\nabla} - \frac{e}{c}\boldsymbol{A}\right)^2\Psi_0 + \left(-\mathrm{i}\hbar\boldsymbol{\nabla} - \frac{e}{c}\boldsymbol{A}\right)\frac{e}{c}\boldsymbol{A}\Psi_0\right. \\ &\quad \left. - \mathrm{i}\hbar\frac{\mathrm{i}e}{\hbar c}\boldsymbol{A}\left(\left(-\mathrm{i}\hbar\boldsymbol{\nabla} - \frac{e}{c}\boldsymbol{A}\right)\Psi_0 + \frac{e}{c}\boldsymbol{A}\Psi_0\right)\right] \\ &= \exp\left(\mathrm{i}\frac{e}{\hbar c}\int_\Gamma \mathrm{d}\boldsymbol{x}' \cdot \boldsymbol{A}(\boldsymbol{x}')\right)(-\mathrm{i}\hbar\boldsymbol{\nabla})^2\Psi_0 ,\end{aligned} \tag{8b}$$

so that (5) reduces to

$$\mathrm{i}\hbar\frac{\partial}{\partial t}\Psi_0 = \left(\frac{1}{2m}(-\mathrm{i}\hbar\boldsymbol{\nabla})^2 + eA_0\right)\Psi_0 . \tag{9}$$

According to (6) the presence of a vector potential leads to a change in the phase of the wavefunction.

All physical observables depend only on the absolute square of the wavefunction Ψ. At first, the additional factor from (6) seems to have no significance. But if two amplitudes superimpose coherently to a total wavefunction,

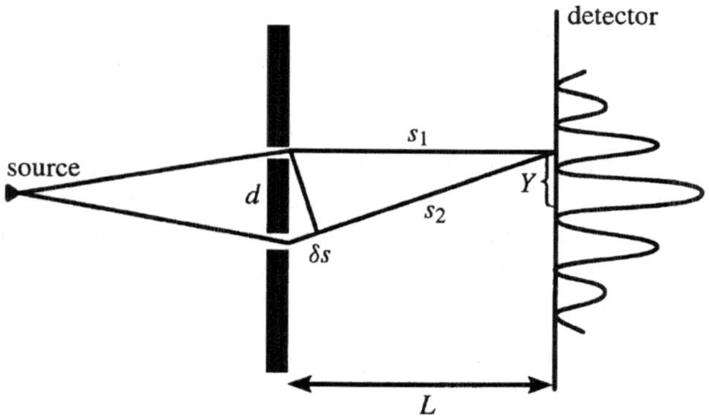

Fig. 1.6. Interference pattern behind two slits

phase factors are measurable. A diffraction experiment represents an ideal example. If particles, such as electrons, are projected onto an inpenetrable wall with two slits (see Fig. 1.6) an interference pattern arises for the probability distribution of the transmitted particles. The in-going beam can be described by a plane wave, which has the same phase at both slits. Then, strictly speaking, the Schrödinger equation should be solved with $\Psi = c\,\mathrm{e}^{\mathrm{i}\,\boldsymbol{p}\cdot\boldsymbol{x}}$ for $x < 0$ with the boundary condition $\Psi = 0$ at the wall. A simpler way to explain the interference pattern is to apply Huyghen's principle. After passing through the slits the two partial waves superimpose linearly and coherently at the detection screen:

$$\Psi = c\,\mathrm{e}^{\mathrm{i}\,\varphi_1} + c\,\mathrm{e}^{\mathrm{i}\,\varphi_2} . \tag{10}$$

The phases depend on the travelled distances,

$$\varphi_i \equiv \varphi_i^0 = \frac{p}{\hbar} s_i = \frac{s_i}{\lambda} , \tag{11}$$

where $\lambda = \hbar/p$ represents the de Broglie wavelength divided by 2π. The maxima of the interference pattern appear for phase differences of $(\varphi_2 - \varphi_1)/2$ in

$$\begin{aligned} |\Psi|^2 &= |c|^2 |\,\mathrm{e}^{\mathrm{i}\varphi_1} + \mathrm{e}^{\mathrm{i}\varphi_2}|^2 \\ &= |c|^2 \left| \exp\left[\mathrm{i}\left(\frac{\varphi_2 - \varphi_1}{2}\right)\right] + \exp\left[-\mathrm{i}\left(\frac{\varphi_2 - \varphi_1}{2}\right)\right] \right|^2 \\ &= 4|c|^2 \cos^2 \frac{\varphi_2 - \varphi_1}{2} , \end{aligned} \tag{12}$$

which are integer multiples of π. Assuming a large distance between the screen and slits, $L \gg d$, we approximate the angle of diffraction by $\tan\beta = Y/L \approx \delta s/d$ (see Fig. 1.6), so that

$$\delta\varphi^0 \equiv \varphi_2 - \varphi_1 = \frac{p}{\hbar}\delta s = \frac{p}{\hbar}\frac{Yd}{L} = \frac{\delta s}{\lambda} = \frac{Yd}{\lambda L} ; \tag{13}$$

compare with Fig. 1.6. For the interference maxima we then find

$$Y_n = n 2\pi \lambda \frac{L}{d} . \tag{14}$$

Example 1.6.

Because of the finite width of the slits a small destructive interference leads to a decrease in intensity for the interference maxima; the interference pattern is shown in Fig. 1.6.

What is going to change if a vector potential $\boldsymbol{A}(\boldsymbol{x})$ is introduced into the diffraction experiment? According to (6) new phase factors come into play. It seems plausible to identify the path from the source through the slits to the point $\boldsymbol{x}$ on the screen as the integration path $\Gamma(\boldsymbol{x})$. Different paths $\Gamma_1(\boldsymbol{x})$ and $\Gamma_2(\boldsymbol{x})$ belong to the two slits and correspond to different partial wavefunctions Ψ_1 and Ψ_2; see Fig. 1.7. Then the new phases result in

$$\varphi_i = \varphi_i^0 + \frac{e}{\hbar c}\int_{\Gamma_i} \mathrm{d}\boldsymbol{x}' \cdot \boldsymbol{A}(\boldsymbol{x}') \;, \tag{15}$$

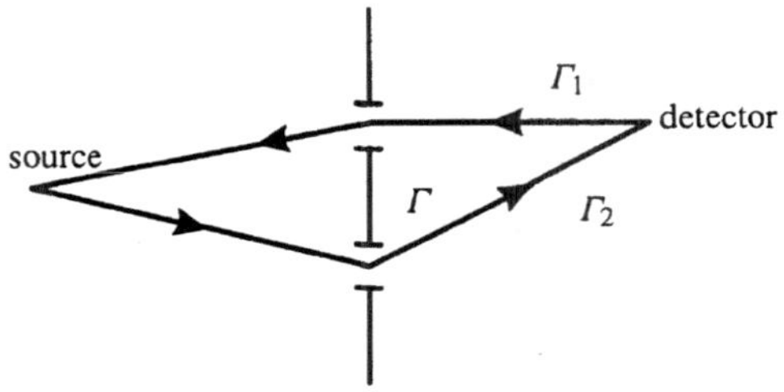

Fig. 1.7. Two paths are possible for the wave propagation from the source through the two slits to the detector. An additional phase difference occurs because of the presence of an $\boldsymbol{A}$ field; it corresponds to the path integral $\oint \boldsymbol{A} \cdot \mathrm{d}\boldsymbol{x}$ along the closed path $\Gamma = \Gamma_1 - \Gamma_2$

so that the phase difference becomes

$$\delta\varphi = \delta\varphi^0 + \frac{e}{\hbar c}\left(\int_{\Gamma_2} - \int_{\Gamma_1}\right) \mathrm{d}\boldsymbol{x}' \cdot \boldsymbol{A}(\boldsymbol{x}') \;. \tag{16}$$

Now we realize that the difference between two path integrals can be written as an integral over the closed path Γ. Furthermore, applying Stoke's theorem, we get

$$\begin{aligned}\left(\int_{\Gamma_2} - \int_{\Gamma_2}\right) \mathrm{d}\boldsymbol{x}' \cdot \boldsymbol{A}(\boldsymbol{x}') &= \oint_{\Gamma} \mathrm{d}\boldsymbol{x}' \cdot \boldsymbol{A}(\boldsymbol{x}') = \int \mathrm{d}\boldsymbol{F} \cdot (\boldsymbol{\nabla} \times \boldsymbol{A}) \\ &= \int \mathrm{d}\boldsymbol{F} \cdot \boldsymbol{B}(\boldsymbol{x}') = \Phi \,. \end{aligned} \tag{17}$$

This is just the magnetic flux going through the surface surrounded by the curve $\Gamma = \Gamma_2$ - Γ_1. The phase difference

$$\delta\varphi = \delta\varphi^0 + \frac{e}{\hbar c}\Phi \tag{18}$$

changes by an amount that is proportional to the magnetic flux inside the two possible trajectories (slit 1 or slit 2) of the electron. Only if the curl of $\boldsymbol{A}$ vanishes will there be no phase difference, $\delta\varphi = \delta\varphi^0$, in every part of this region and the vector potential will not influence the observed interference pattern. Such an $\boldsymbol{A}$ field really has no physical significance. Nevertheless, we could think of an experiment for which $\boldsymbol{B} = \boldsymbol{\nabla} \times \boldsymbol{A} \neq 0$ does not vanish in a tiny region behind the two slits. Everywhere else the magnetic field is zero; see Fig. 1.8.

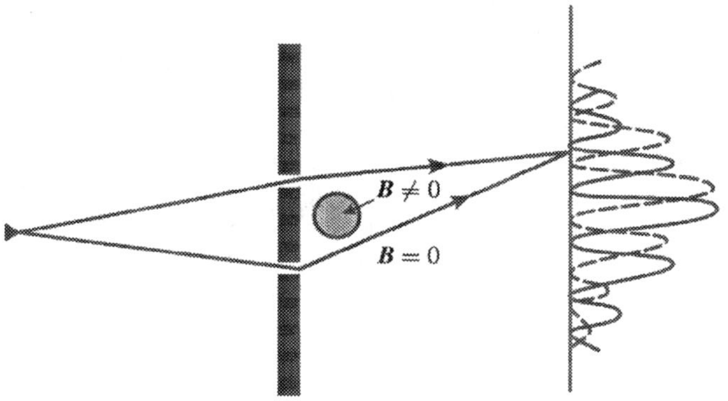

Fig. 1.8. A magnetic field $B = \nabla \times A \neq 0$ is introduced within a tiny region (*hatched*) in between the two beam paths. In this way an additional phase difference between the two beams occurs, which results in a displacement of the interference pattern (*dashed*)

This is easy to achieve. Think of a very long solenoid perpendicular to the scattering plane. A homogenous magnetic field

$$B = \frac{4\pi}{c}\nu I e_z \tag{19}$$

prevails in the interior of the solenoid carrying a current I and wound with ν turns per unit length; this follows directly from $\nabla \times B = (4\pi/c)j$. In the exterior region of the solenoid the magnetic field is negligibly small as the field lines close at infinity. The vector potential corresponding to an ideal solenoid with radius R reads

$$A(r) = \begin{cases} \frac{\phi}{2\pi R^2} r e_\varphi & \text{for} \quad r \leq R \\ \frac{\phi}{2\pi}\frac{1}{r} e_\varphi & \text{for} \quad r > R\,; \end{cases} \tag{20}$$

e_φ is the azimuthal unit vector in polar coordinates. $A_\varphi = A \cdot e_\varphi$ is illustrated in Fig. 1.9. It is straightforward to confirm, that $\nabla \times A = 0$ in the exterior region and that the magnetic field takes on the value $B = (\phi/\pi R^2)$ in the interior region, where $\Phi = \oint B \cdot dS = B\pi R^2$ holds for a homogenous magnetic field of the solenoid.

Because of (3) the vector potential is only determined up to a gauge transformation; however, due to the presence of a magnetic flux, it is not possible to find a gauge transformation such that $A(x)$ becomes zero everywhere in the exterior region. In the set-up illustrated in Fig. 1.8 the presence of the magnetic flux tube leads to a displacement of the interference pattern. The new maxima now lie at

$$Y_n' = Y_n + \lambda \frac{L}{d}\frac{e\phi}{\hbar c}\,. \tag{21}$$

A detailed analysis shows that the envelope of the interference pattern does not change. This shows that on average the electron beam is not deflected by the localized B field; only quantum-mechanical interference effects are influenced. Nevertheless the result seems to be stunning. Although the electron beam encounters no magnetic field at any place, its presence results in a

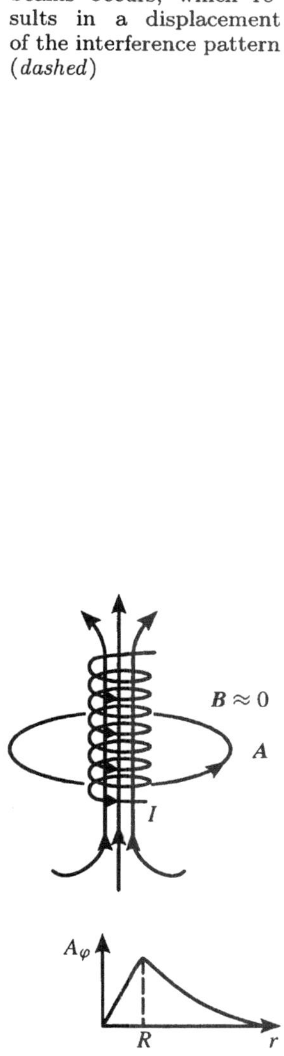

Fig. 1.9. Illustration of the vector potential $A = \{0, 0, A_\varphi\}$ of an ideal solenoid

Example 1.6.

displacement of the interference pattern. Referring to the concept of a local interaction, which we have good reasons to believe in, we come to the conclusion that the vector potential $\boldsymbol{A}(\boldsymbol{x},t)$ plays a more fundamental role than the magnetic field $\boldsymbol{B}(\boldsymbol{x},t)$. On the other side because, of the gauge freedom, $\boldsymbol{A}(\boldsymbol{x},t)$ contains too much information. The only true and decisive quantity remains the phase integral $\exp\left[\mathrm{i}\frac{e}{\hbar c}\int_\Gamma \mathrm{d}\boldsymbol{x}'\cdot\boldsymbol{A}(\boldsymbol{x}')\right]$.

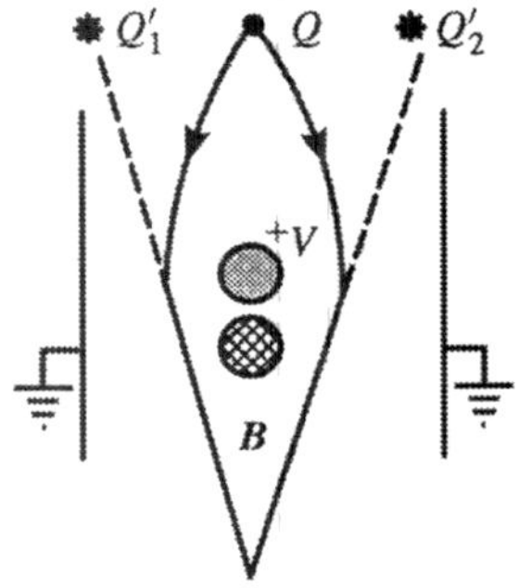

Fig. 1.10. Electrostatic biprism

The experiment sketched in Fig. 1.8 was performed in a slightly modified form soon after the proposal by Aharonov and Bohm.[5] Instead of two slits a biprism was used to achieve larger intensities. Such a biprism consists of a thin electrically charged wire between two grounded plates. It deflects the beam as if it came from two spatially separated sources Q_1' and Q_2'; see Fig. 1.10. Chambers[6] added a thin ferromagnetic metal fragment, a so-called whisker, behind the biprism and was then able to detect the change in the interference pattern. Somewhat more elegant was the experiment of Möllenstedt and Bayh.[7] With a clever electron optical device they succeeded in separating the two partial beams by a distance of about 60 μm. A freely suspended air solenoid of 5 mm length and 7 μm radius (!) was introduced into the gap. Varying the current through the solenoid, they were able to displace the interference pattern continuously; the envelope remained unaltered. The displacement was in full accordance with the expected result (21). Although the Aharonov–Bohm effect follows naturally from quantum mechanics, its existence had been questioned repeatedly. New experiments[8] use small toroidal magnets instead of solenoids; one of the two beams goes through their openings. In this manner the magnetic scattering field can be eliminated. In addition the torus was covered with a superconductor to exclude any penetration of the electrons into the magnetic region. None of these precautions changes the experimental results.

The Aharonov–Bohm effect (the influence of the potential on the phase of a wavefunction) does not only hold for magnetic fields. In fact, a relation analogous to (6) can be derived for the case of electric fields. The presence of a spatially constant but time-dependent electrostatic potential $A_0(t)$ alters the phase of a wavefunction according to

$$\Psi(\boldsymbol{x},t) = \Psi_0(\boldsymbol{x},t)\exp\left[-\mathrm{i}(e/\hbar)\int^t \mathrm{d}t' A_0(t')\right] . \qquad (22)$$

Because $\boldsymbol{E} = -\boldsymbol{\nabla} A_0 = 0$, no electric field is present. Again, the appearance of such a phase factor can be demonstrated with an interference experiment.

As sketched in Fig. 1.11 an electron beam is split, guided through two metalic boxes, which serve as Faraday cages, and finally brought to interfere. With a suitable device the beam is cut into separate wave pieces, which are

---

[5] In fact the experiment was proposed 10 years earlier by W. Ehrenberg and R.E. Siday, Proc. Roy. Soc. **62B** (1949) 8, but at that time it did not raise much interest. Independently, Aharonov and Bohm developed the idea and analyzed it on more profound theoretical grounds.

[6] R.G. Chambers: Phys. Rev. Lett. **5** (1960) 3.

[7] G. Möllenstedt, W. Bayh: Naturwiss. **49** (1962) 81.

[8] Osakabe et al.: Phys. Rev. A **43** (1986) 815.

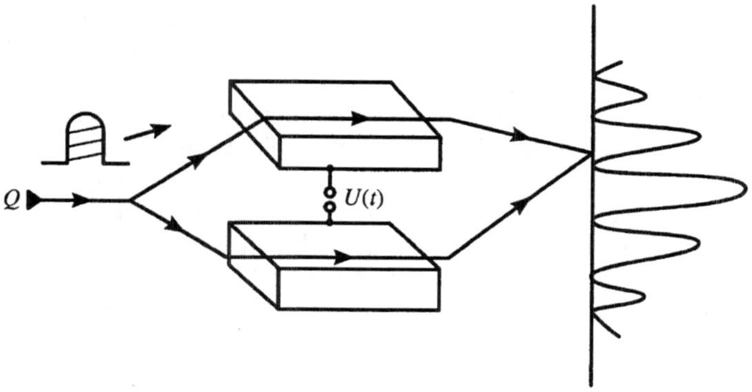

Fig. 1.11. Sketch of an experiment that demonstrates the dependence of the phase of a wavefunction on a constant but temporally changing electric potential $A_0(t)$

shorter than the length of the boxes. As soon as the wave is inside the box, a voltage U is applied for a short time. The phases then become

$$\varphi_i = \varphi_i^0 - \frac{e}{\hbar}\int_{t_1}^{t_2} \mathrm{d}t A_0{}^i(t) \; , \tag{23}$$

so that

$$\begin{aligned} \delta\varphi &= \delta\varphi^0 - \frac{e}{\hbar}\int_{t_1}^{t_2} \mathrm{d}t\left[A_0^2(t) - A_0^1(t)\right] \\ &= \delta\varphi^0 + \frac{e}{\hbar}\oint \mathrm{d}t\,\mathrm{d}y\,E_y \end{aligned} \tag{24}$$

is obtained for the relative phase. In order to have as much similarity as possible to (18) for the magnetic field case, the relation $\boldsymbol{E} = -\boldsymbol{\nabla} A_0$ has been employed and an integral over a spatiotemporal region has been introduced. The electric flux defined by the double integral of (24) is gauge invariant as in the case for the magnetic flux of (17). The electric flux also leads to a translation (shift) of the interference pattern. This holds although the electrons are inside the Faraday cages and thus do not encounter any $\boldsymbol{E}$ field as the potential is switched on. On the other side the electron wavefunction is spread all over space with, of course, different strengths.

From these considerations about the Aharanov–Bohm effect we conclude that in quantum mechanics the interaction of a charged particle with an electromagnetic field results in a phase factor

$$R_{12} = \exp\left[\mathrm{i}\frac{e}{\hbar}\int_1^2\left(\frac{1}{c}\boldsymbol{A}\cdot\mathrm{d}\boldsymbol{x} - A_0\,\mathrm{d}t\right)\right] = \exp\left(\mathrm{i}\frac{e}{\hbar c}\int_1^2 A_\mu\,\mathrm{d}x^\mu\right) \; . \tag{25}$$

This contains the potentials A_0 and $\boldsymbol{A}$ as fundamental quantities and not the field strengths $\boldsymbol{E}$ and $\boldsymbol{B}$. The factor R_{12} specifies the probability amplitude for a particle to propagate from space–time point $(\boldsymbol{x}_1, t_1)$ to point $(\boldsymbol{x}_2, t_2)$. The phase factor depends on the trajectory $\boldsymbol{x}(t)$. In principle, all possible trajectories have to be considered, each of them with its own phase factor.

Example 1.6.

In fact, this conception provides the basis for an alternative, yet equivalent, formulation of quantum mechanics; i.e. the "path integral" approach[9] to quantum mechanics.

Finally we now show that the expression (25) also incorporates "macroscopic effects" of the electromagnetic field as the deflection of a charged particle due to the Lorentz force. Again we concentrate on the case of the simple two-slits experiment and on small deflection angles. The considerations leading to (19) and (21) are repeated with the additional assumption that the electrons run through a sector of width D, which is penetrated by a homogenous magnetic field $\boldsymbol{B}$; see Fig. 1.12. The magnetic flux determines the deflection; it is given by

$$\Phi = \oint \mathrm{d}\boldsymbol{F} \cdot \boldsymbol{B} \approx \mathrm{d}DB \,. \tag{26}$$

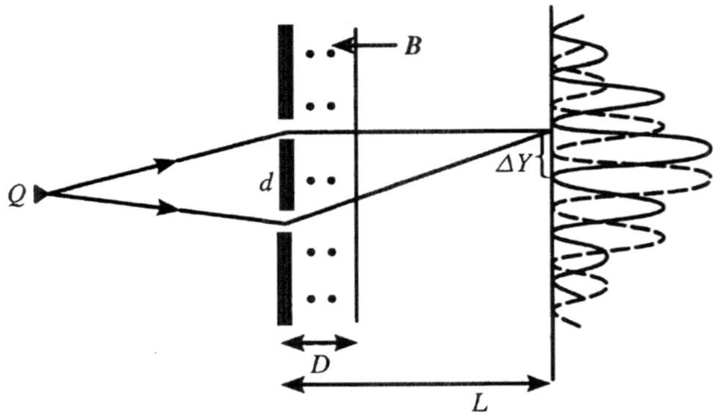

Fig. 1.12. In the region of width D behind the two slits a magnetic field $\boldsymbol{B}$ acts on the electrons, which not only leads to a shift of the interference pattern but also to a shift of its envelope. This is caused by the Lorentz force, which acts on the electrons as they move through the $\boldsymbol{B}$ region

From (21) the shift of the interference pattern is obtained as

$$\Delta Y = \frac{L}{d}\lambda\!\!\!^{-}\frac{e\Phi}{\hbar c} = DL\frac{eB}{pc} \,. \tag{27}$$

In a classical argumentation, the Lorentz force $F = (e/c)vB$ acts during the short period $\tau \approx D/v$; according to Newton's equation of motion this force results for small deflection angles in a momentum $p_y = \tau(e/c)vB$ in the y direction. The deflection angle then is $\tan\beta = p_y/p \approx \Delta Y/L$. The deflection due to the Lorentz force becomes

$$\Delta Y = L\frac{p_y}{p} = DL\frac{eB}{pc} \,. \tag{28}$$

This proves (in first order in β) that the phase factor (25) includes the action of the Lorentz force. A more detailed analysis would show that the total probability distribution is shifted as sketched in the Fig. 1.12; this is in contrast to the Aharanov–Bohm effect, for which the envelope of the interference

[9] See for example R.P. Feynman, A.R. Hibbs: *Quantum Mechanics and Path Integrals* (MacGraw-Hill, NY 1965) and Chap. 13 of this book.

patter remains unchanged, indicating that on average no action of a force has occured.[10]

Example 1.6.

1.6 Biographical Notes

PAULI, Wolfgang, *Vienna 25.4.1900, †Zürich 15.12.1958, professor at the "Eidgenössische Technische Hochschule" (Federal Technical University) in Zürich since 1928. Pauli was a student of Arnold Sommerfeld and Max Born. In 1945 he was awarded the Nobel prize for the discovery of the exclusion principle that carries his name. He also developed the first theory of the electron spin, which led to the Pauli equation.

[10] As a supplement to our presentation we recommend the short treatise by M. Danos: Amer. Journ. of Physics Teachers **50** (1982) 64 on the Aharanov–Bohm effect.

2. Interaction of Electromagnetic Fields with Matter

In most cases we can think of matter as being build up of particles of mass m_i and charge e_i. The interaction of these particles with each other can often be described in terms of a potential $V(\dots, \boldsymbol{x}_i, \dots, \boldsymbol{x}_j, \dots)$. For example, for the case of Coulomb interaction this potential would be

$$V_{\text{Coulomb}}(\dots, \boldsymbol{x}_i, \dots, \boldsymbol{x}_j, \dots) = \frac{1}{2} \sum_{\substack{i,j \\ i \neq j}} \frac{e_i e_j}{|\boldsymbol{x}_i - \boldsymbol{x}_j|} . \tag{2.1}$$

Such a modeling of many-particle systems is widely used in atomic, molecular, and solid-state physics (atoms, molecules, and their interaction), in nuclear physics (protons and neutrons and their interaction), and also in elementary particle physics (quarks and their interaction). The Hamiltonian for such a many-particle system for nonrelativistic particles reads

$$\hat{H}_{\text{mp}} = \sum_i \frac{\hat{\boldsymbol{p}}_i{}^2}{2m_i} + V . \tag{2.2}$$

The index 'mp' stands for 'many particles'.

How does this many-particle system interact with electromagnetic fields? From the lectures on quantum mechanics[1] we remember that it is necessary for a quantum theory to remain gauge invariant. The interaction between radiation and matter has to be in such a way that gauge transformations of the form (1.4) do not change the observable quantities predicted from theory (eigenvalues, expectation values, transition probabilities, and so on). This important requirement for gauge invariance is fulfilled with the minimal coupling

$$\hat{\boldsymbol{p}}_i \longrightarrow \hat{\boldsymbol{p}}_i - \frac{e_i}{c} \hat{\boldsymbol{A}}(\boldsymbol{x}_i) . \tag{2.3}$$

Together with the Hamiltonian for the radiation field, (1.40) and (1.48), the Hamiltonian for the many-particle system (2.2) then becomes

$$\begin{aligned} \hat{H} &= \sum_i \frac{\left(\hat{\boldsymbol{p}}_i - \frac{e_i}{c}\hat{\boldsymbol{A}}(\boldsymbol{x}_i)\right)\left(\hat{\boldsymbol{p}}_i - \frac{e_i}{c}\hat{\boldsymbol{A}}(\boldsymbol{x}_i)\right)}{2m_i} + V \\ &\quad + \int \frac{\hat{\boldsymbol{E}} \cdot \hat{\boldsymbol{E}}^* + \hat{\boldsymbol{B}} \cdot \hat{\boldsymbol{B}}^*}{8\pi} \, \mathrm{d}^3x \\ &= \hat{H}_{\text{mp}} + \hat{H}_{\text{rad}} + \hat{H}_{\text{int}} . \end{aligned} \tag{2.4}$$

[1] W. Greiner: *Quantum Mechanics – An Introduction*, 3rd ed. (Springer, Berlin, Heidelberg 1994).

This is the Hamiltonian for the total system "matter + radiation". The interaction operator $\hat{H}_{\text{int}}$ for interaction between the many-particle system and the electromagnetic field follows from calculating the square of the expression inside the sum of (2.4),

$$\hat{H}_{\text{int}} = \sum_i \left[-\frac{e_i}{m_i c} \hat{\boldsymbol{p}}_i \cdot \hat{\boldsymbol{A}}(\boldsymbol{x}_i) + \frac{e_i^2}{2m_i c^2} \hat{\boldsymbol{A}}^2(\boldsymbol{x}_i) \right] . \tag{2.5}$$

Here, the so-called *Coulomb gauge* $\boldsymbol{\nabla} \cdot \boldsymbol{A} = 0$ has been employed, which allows us to write $\hat{\boldsymbol{p}} \cdot \boldsymbol{A} + \boldsymbol{A} \cdot \hat{\boldsymbol{p}}$ as $2\hat{\boldsymbol{p}} \cdot \boldsymbol{A}$. In the following, we suppress the sum and the corresponding index i for simplicity, so that

$$\hat{H}_{\text{int}} = -\frac{e}{mc} \hat{\boldsymbol{p}} \cdot \hat{\boldsymbol{A}}(\boldsymbol{x}) + \frac{e^2}{2mc^2} \hat{\boldsymbol{A}}^2(\boldsymbol{x}) \tag{2.6}$$

represents the interaction of one particle with the radiation field. We will come back to the explicit sum (2.5) when necessary. Now we separate the interaction (2.6) into a term proportional to $\boldsymbol{A}$ and another term proportional to $\boldsymbol{A}^2$,

$$\hat{H}_{\text{int}} = \hat{H}'_{\text{int}} + \hat{H}''_{\text{int}} , \tag{2.7}$$

and deduce further with (1.31) that

$$\hat{H}'_{\text{int}} = -\frac{e}{mc} \sum_{\boldsymbol{k},\sigma} \sqrt{\frac{2\pi\hbar c^2}{L^3 \omega_k}} \hat{\boldsymbol{p}} \cdot \boldsymbol{\varepsilon}_{\boldsymbol{k}\sigma} \left(\hat{a}_{\boldsymbol{k}\sigma} \, \mathrm{e}^{\mathrm{i}\boldsymbol{k}\cdot\boldsymbol{x}} + \hat{a}^\dagger_{\boldsymbol{k}\sigma} \, \mathrm{e}^{-\mathrm{i}\boldsymbol{k}\cdot\boldsymbol{x}} \right) \tag{2.8}$$

and

$$\begin{aligned} \hat{H}''_{\text{int}} = {} & \frac{e^2}{2mc^2} \sum_{\boldsymbol{k}\sigma} \sum_{\boldsymbol{k}'\sigma'} \left(\frac{2\pi\hbar c^2}{L^3} \right) \frac{\boldsymbol{\varepsilon}_{\boldsymbol{k}\sigma} \cdot \boldsymbol{\varepsilon}_{\boldsymbol{k}'\sigma'}}{(\omega_k \omega_{k'})^{\frac{1}{2}}} \\ & \times \left(\hat{a}_{\boldsymbol{k}\sigma} \hat{a}_{\boldsymbol{k}'\sigma'} \, \mathrm{e}^{\mathrm{i}(\boldsymbol{k}+\boldsymbol{k}')\cdot\boldsymbol{x}} + \hat{a}_{\boldsymbol{k}\sigma} \hat{a}^\dagger_{\boldsymbol{k}'\sigma'} \, \mathrm{e}^{\mathrm{i}(\boldsymbol{k}-\boldsymbol{k}')\cdot\boldsymbol{x}} \right. \\ & \left. + \, \hat{a}^\dagger_{\boldsymbol{k}\sigma} \hat{a}_{\boldsymbol{k}'\sigma'} \, \mathrm{e}^{\mathrm{i}(-\boldsymbol{k}+\boldsymbol{k}')\cdot\boldsymbol{x}} + \hat{a}^\dagger_{\boldsymbol{k}\sigma} \hat{a}^\dagger_{\boldsymbol{k}'\sigma'} \, \mathrm{e}^{\mathrm{i}(-\boldsymbol{k}-\boldsymbol{k}')\cdot\boldsymbol{x}} \right) . \end{aligned} \tag{2.9}$$

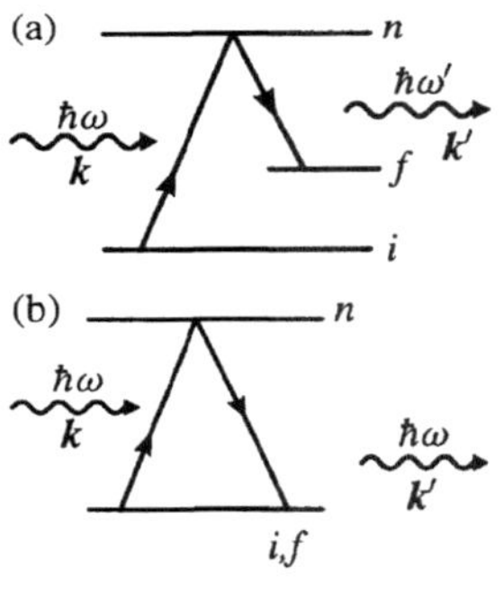

Fig. 2.1a,b. Absorption and reemission of a photon (light scattering): (**a**) inelastic photon scattering (Raman scattering), (**b**) elastic photon scattering. For the process in (**a**) both energy ($\hbar\omega \to \hbar\omega'$) and momentum ($\hbar\boldsymbol{k} \to \hbar\boldsymbol{k}'$) change, whereas for the process in (**b**) only momentum (its direction) changes

The interaction $\hat{H}_{\text{int}}$ between the electromagnetic field and the many-particle system will remain small. Therefore we will treat it as a perturbation. Hence, the unperturbed system consists of the many-particle system and the radiation field, i.e.

$$\hat{H}_0 = \hat{H}_{\text{mp}} + \hat{H}_{\text{rad}} \tag{2.10}$$

with state vectors

$$|\text{mp} + \text{rad}\rangle = |\text{mp}\rangle| \ldots n_{\boldsymbol{k}\sigma} \ldots \rangle_{\text{rad}} . \tag{2.11}$$

Here, mp represents the quantum numbers of the many-particle system. Later we will replace $|\text{mp}\rangle$ by $|a\rangle$, indicating that we consider atoms specifically. The interaction (2.7) will generate transitions between the states (2.11). We will calculate the transition amplitudes according to formulas known from quantum mechanics.

A few more words about the interactions (2.8) and (2.9) are in order. Since $\hat{H}'_{\text{int}}$ contains only creation operators $\hat{a}^\dagger$ and annihilation operators $\hat{a}$

for photons linearly, $\hat{H}'_{\text{int}}$ induces transitions for which one photon is either created or annhilated. On the other side $\hat{H}''_{\text{int}}$ induces transitions where *two photons* are involved. For example the processes $(\hat{a}^\dagger\hat{a}, \hat{a}\hat{a}^\dagger)$ describe absorption and reemission of one photon or vice versa (see Fig. 2.1). We will demonstrate this point further with various examples.

2.1 Emission of Radiation from an Excited Atom

An atom is in the initial state $|a_{\mathrm{i}}\rangle$ and decays into the final state $|a_{\mathrm{f}}\rangle$ by emitting a photon with wavevector $\boldsymbol{k}$ and polarization σ. Subscripts i and f stand for "initial" and "final", respectively. The initial and final states of the total system "atom + radiation field" are denoted by $|i\rangle$ and $|f\rangle$, respectively, and, more explicitly, are given by

$$\begin{aligned} |i\rangle &= |a_{\mathrm{i}}\rangle|\ldots n_{\boldsymbol{k}\sigma}\ldots\rangle\,, \\ |f\rangle &= |a_{\mathrm{f}}\rangle|\ldots n_{\boldsymbol{k}\sigma}+1\ldots\rangle\,. \end{aligned} \tag{2.12}$$

They are graphically depicted in Fig. 2.2.

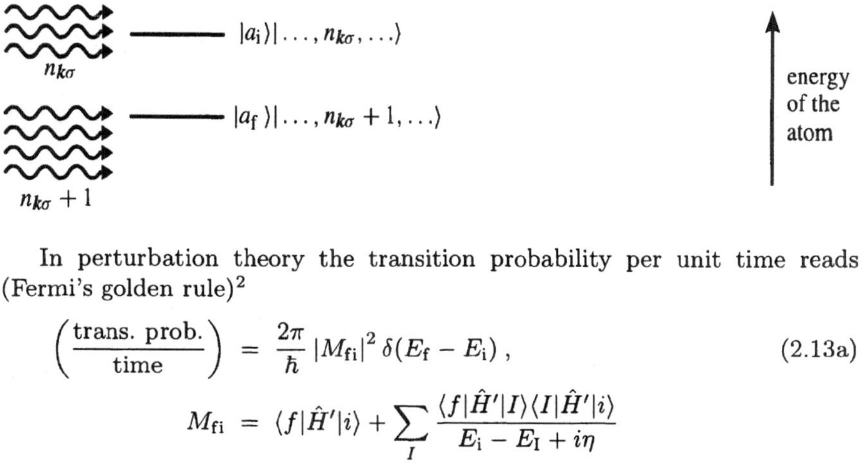

Fig. 2.2. Illustration of the initial and final state of the total system "atom + radiation field". For the initial state the many-particle system occupies the state $|a_{\mathrm{i}}\rangle$ and the radiation field contains $n_{\boldsymbol{k}\sigma}$ photons characterized by $(\boldsymbol{k}, \sigma)$. For the final state the many-particle system occupies the energetically lower state $|a_{\mathrm{f}}\rangle$ and the radiation field contains one additional photon with quantum numbers $(\boldsymbol{k}, \sigma)$

In perturbation theory the transition probability per unit time reads (Fermi's golden rule)[2]

$$\left(\frac{\text{trans. prob.}}{\text{time}}\right) = \frac{2\pi}{\hbar}|M_{\mathrm{fi}}|^2\,\delta(E_{\mathrm{f}} - E_{\mathrm{i}})\,, \tag{2.13a}$$

$$\begin{aligned} M_{\mathrm{fi}} &= \langle f|\hat{H}'|i\rangle + \sum_I \frac{\langle f|\hat{H}'|I\rangle\langle I|\hat{H}'|i\rangle}{E_{\mathrm{i}} - E_{\mathrm{I}} + i\eta} \\ &\quad + \sum_{I,II} \frac{\langle f|\hat{H}'|I\rangle\langle I|\hat{H}'|II\rangle\langle II|\hat{H}'|i\rangle}{(E_{\mathrm{i}} - E_{\mathrm{I}} + i\eta)(E_{\mathrm{i}} - E_{\mathrm{II}} + i\eta)} + \ldots\,, \end{aligned} \tag{2.13b}$$

where I and II describe a complete set of states. Here, M_{fi} represents the transition matrix element. In our case the *contribution of first order* to M_{fi} results merely from $\hat{H}'_{\text{int}}$ and we get with (2.8)

[2] W. Greiner: *Quantum Mechanics – An Introduction*, 3rd ed. (Springer, Berlin, Heidelberg 1994).

$$
\begin{aligned}
&\langle f|\hat{H}'_{\text{int}}|i\rangle \\
&= \langle a_f|\,\langle \ldots, n_{k\sigma}+1, \ldots| \\
&\quad \left[-\frac{e}{mc}\sum_{k',\sigma'}\sqrt{\frac{2\pi\hbar c^2}{L^3\omega_{k'}}}\hat{p}\cdot\varepsilon_{k\sigma}\left(\hat{a}_{k'\sigma'}\,\mathrm{e}^{\mathrm{i}k'\cdot x}+\hat{a}^\dagger_{k'\sigma'}\,\mathrm{e}^{-\mathrm{i}k'\cdot x}\right)\right] \\
&\quad |a_i\rangle\,|\ldots n_{k\sigma}\ldots\rangle \\
&= -\frac{e}{mc}\sqrt{\frac{2\pi\hbar c^2}{L^3\omega_k}}\langle a_\mathrm{f}|\hat{p}\cdot\varepsilon_{k\sigma}\mathrm{e}^{-\mathrm{i}k\cdot x}|a_\mathrm{i}\rangle \\
&\quad \times\langle \ldots, n_{k\sigma}+1, \ldots|\hat{a}^\dagger_{k\sigma}|\ldots, n_{k\sigma}, \ldots\rangle \\
&= -\frac{e}{mc}\sqrt{\frac{2\pi\hbar c^2}{L^3\omega_k}}\langle a_\mathrm{f}|\hat{p}\cdot\varepsilon_{k\sigma}\,\mathrm{e}^{-\mathrm{i}k\cdot x}|a_\mathrm{i}\rangle\sqrt{n_{k\sigma}+1}\,. \qquad (2.14)
\end{aligned}
$$

In the second step we have taken into account that only the term with the creation operator $\hat{a}^\dagger_{k\sigma}$ contributes within the infinite sum. Equation (1.57) has been employed for the last step. The δ function appearing in (2.13a) guarantees energy conservation:

$$
E_\mathrm{f}-E_\mathrm{i} = \underbrace{E_{a_\mathrm{f}}+\hbar\omega_k}_{=E_\mathrm{f}} - \underbrace{E_{a_\mathrm{i}}}_{=E_\mathrm{i}} = E_{a_\mathrm{f}}-E_{a_\mathrm{i}}+\hbar\omega_k = 0\,. \qquad (2.15)
$$

The photon with energy $\hbar\omega_k$ takes away the energy $E_{a_\mathrm{i}}-E_{a_\mathrm{f}}$ from the atom. With (2.13a) and (2.14) we obtain the transition probability per unit time for the transition of the atom from $|a_\mathrm{i}\rangle$ to $|a_\mathrm{f}\rangle$ *by emission of a photon in first-order perturbation theory*:

$$
\begin{aligned}
\left(\frac{\text{trans. prob.}}{\text{time}}\right)_{\text{emission}} &= \frac{2\pi}{\hbar}\left|\langle f|\hat{H}'_{\text{int}}|i\rangle\right|^2\delta(E_\mathrm{f}-E_\mathrm{i}) \\
&= \frac{2\pi}{\hbar}\left(\frac{e}{mc}\right)^2\left(\frac{2\pi\hbar c^2}{L^3\omega_k}\right)(n_{k\sigma}+1) \\
&\quad\times\left|\langle a_\mathrm{f}|\hat{p}\cdot\varepsilon_{k\sigma}\,\mathrm{e}^{-\mathrm{i}k\cdot x}|a_\mathrm{i}\rangle\right|^2 \\
&\quad\times\delta(E_{a_\mathrm{f}}-E_{a_\mathrm{i}}+\hbar\omega_k)\,. \qquad (2.16)
\end{aligned}
$$

The factor $n_{k\sigma}+1$ in the radiation matrix element has deep and far-reaching consequences. The probability for emission is not solely proportional to the number of photons $n_{k\sigma}$ already contained in the initial state; this would be *stimulated emission* alone. Emission also exists if initially no photon is present at all ($n_{k\sigma}=0$); this is known as *spontaneous emission.* Once again, the induced emission is proportional to the number of photons $n_{k\sigma}$ already present in the radiation field. In some way they shake the excited state and stimulate the transition. It is the zero-point oscillations of the radiation field that effect the transition because of their "shaking". These zero-point oscillations are of pure quantum-mechanical origin and are always present. The 1 appearing in $n_{k\sigma}+1$ results from the commutation relations (1.50); this is for the same reasons as the term $\frac{1}{2}$ appears in (1.52) for the zero-point energy of the radiation field. It is really remarkable and very satisfying indeed that the theory we have developed so far automatically includes spontaneous emission. We will first concentrate on the spontaneous emission process. Later on we will come back to the induced emission in connection with the absorption process.

2.2 Lifetime of an Excited State

An excited state may decay via spontaneous emission of light. If no photon is present at the beginning, we have to set $n_{k\sigma} = 0$ in (2.16) and then we can calculate the transition probability per unit time for this process. We also sum over all directions $\boldsymbol{k}$ and polarizations σ of the photon. The *lifetime* τ of state $|a_{\mathrm{i}}\rangle$ is defined as

$$\begin{aligned} \left(\frac{1}{\tau}\right)_{\mathrm{i}\to\mathrm{f}} &= \frac{2\pi}{\hbar}\sum_{\boldsymbol{k},\sigma}\left|\langle f|\hat{H}'_{\mathrm{int}}|i\rangle\right|^2 \delta(E_{a_\mathrm{f}} - E_{a_\mathrm{i}} + \hbar\omega_k) \\ &= \frac{4\pi^2 e^2}{m^2 L^3}\sum_{\boldsymbol{k},\sigma}\frac{1}{\omega_k}\left|\langle a_\mathrm{f}|\hat{\boldsymbol{p}}\cdot\boldsymbol{\varepsilon}_{\boldsymbol{k}\sigma}\,\mathrm{e}^{-\mathrm{i}\boldsymbol{k}\cdot\boldsymbol{x}}|a_\mathrm{i}\rangle\right|^2 \\ &\quad\times\delta(E_{a_\mathrm{f}} - E_{a_\mathrm{i}} + \hbar\omega_k)\,. \end{aligned} \tag{2.17}$$

We investigate the limiting case $L^3 \to \infty$ for this equation. The volume of the box to which the electromagnetic field is confined goes to ∞; the electromagnetic field then occupies the whole space. For this limiting case we will show that

$$\sum_{\boldsymbol{k}} \longrightarrow \frac{L^3}{(2\pi)^3}\int \mathrm{d}^3k \tag{2.18}$$

becomes valid. For this we consider the three-dimensional lattice space with axes n_x, n_y, n_z; $n_i \in Z$. Every lattice point $\{n_x, n_y, n_z\}$ represents a normal mode of the electromagnetic field with a certain polarization.

Every vector $\boldsymbol{n} = \{n_x, n_y, n_z\}$ represents such a normal mode; each vector begins at the origin. Into a lattice volume $\Delta n_x\,\Delta n_y\,\Delta n_z$, which reaches from n_x to $n_x+\Delta n_x$, n_y to $n_y+\Delta n_y$ and n_z to $n_z+\Delta n_z$, fall exactly $\Delta n_x\,\Delta n_y\,\Delta n_z$ arrow heads (lattice points). Therefore, exactly $\Delta n_x\,\Delta n_y\,\Delta n_z$ electromagnetic normal modes exist in this volume. According to (1.30)

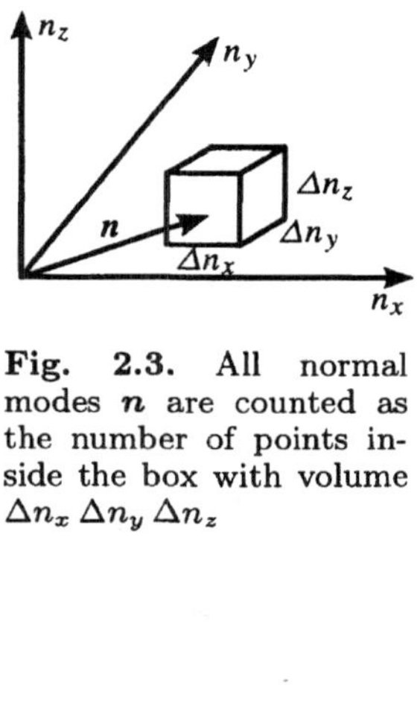

Fig. 2.3. All normal modes $\boldsymbol{n}$ are counted as the number of points inside the box with volume $\Delta n_x\,\Delta n_y\,\Delta n_z$

$$Lk_i = 2\pi n_i \quad (i = 1,2,3 \text{ for } x,y,z) \tag{2.19}$$

and thus

$$L\,\Delta k_i = 2\pi\,\Delta n_i\,, \tag{2.20}$$

from which

$$\begin{aligned} \Delta n_1\,\Delta n_2\,\Delta n_3 &= \Delta n_x\,\Delta n_y\,\Delta n_z \\ &= \frac{L^3}{(2\pi)^3}\Delta k_x\,\Delta k_y\,\Delta k_z \end{aligned} \tag{2.21}$$

follows for the number of normal modes falling into the interval $[k_i, k_i + \Delta k_i]$. In the limiting case $L \to \infty$ and $\Delta k_i \to 0$ (2.21) goes over to

$$\frac{L^3}{(2\pi)^3}\,\mathrm{d}^3k\,, \tag{2.22}$$

from which (2.18) follows.

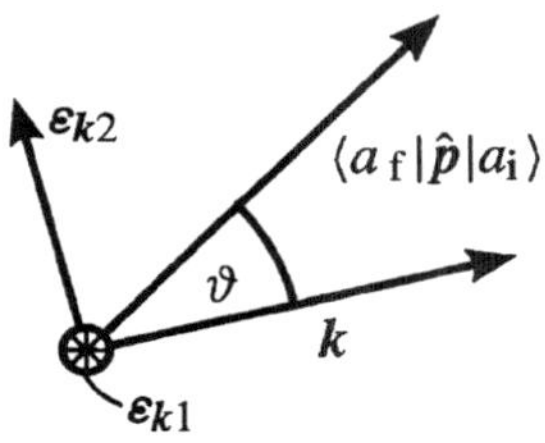

Fig. 2.4. The polarization vector ε_{k1} points into the plane of the page; ε_{k2} falls into the plane given by p and k and is of course perpendicular to k. The k_z coordinate axis is chosen to be parallel to the vector $\langle a_\mathrm{f}|\hat{p}|a_\mathrm{i}\rangle$; hence, ϑ represents the polar angle in k space

We return now to the determination of the lifetime (2.17) and choose the two polarization vectors $\varepsilon_{k\sigma}$ as depicted in Fig. 2.4. It follows that

$$\sum_{\sigma=1,2} \left|\langle a_\mathrm{f}|\hat{p}\cdot\varepsilon_{k\sigma}\,\mathrm{e}^{-\mathrm{i}k\cdot x}|a_\mathrm{i}\rangle\right|^2 = \left|\langle a_\mathrm{f}|\hat{p}\cdot\varepsilon_{k_2}\,\mathrm{e}^{-\mathrm{i}k\cdot x}|a_\mathrm{i}\rangle\right|^2 = \left|\varepsilon_{k2}\cdot\langle a_\mathrm{f}|\hat{p}\,\mathrm{e}^{-\mathrm{i}k\cdot x}|a_\mathrm{i}\rangle\right|^2 . \tag{2.23}$$

Light emitted from atoms typically has energies of $\hbar\omega \approx 10\,\mathrm{eV}$. We conclude that

$$\begin{aligned} k\cdot x &\approx \frac{2\pi}{\lambda}a_\mathrm{Bohr} = \frac{\hbar\omega}{\hbar c}a_\mathrm{Bohr} \\ &= \frac{10\,\mathrm{eV}\,\frac{1}{2}\times 10^{-8}\,\mathrm{cm}}{1.97\times 10^{-5}\,\mathrm{eV\,cm}} \\ &= 2.7\times 10^{-3} \ll 1 \end{aligned} \tag{2.24}$$

and expand the exponential appearing in (2.23) into a Taylor series:

$$\mathrm{e}^{-\mathrm{i}k\cdot x} = 1 - \mathrm{i}k\cdot x - \tfrac{1}{2}(k\cdot x)^2 + \cdots . \tag{2.25}$$

Because of (2.24) we restrict the expansion to the first term only; thus (2.23) becomes

$$\sum_{\sigma=1,2} \left|\langle a_\mathrm{f}|\hat{p}\cdot\varepsilon_{k\sigma}\,\mathrm{e}^{-\mathrm{i}k\cdot x}|a_\mathrm{i}\rangle\right|^2 \approx \left|\varepsilon_{k2}\cdot\langle a_\mathrm{f}|\hat{p}|a_\mathrm{i}\rangle\right|^2 . \tag{2.26}$$

For reasons that will become clear in a moment this approximation is called the *dipole approximation.* The next terms in expansion (2.25) would lead to *magnetic dipole radiation, electric quadrupole radiation,* and so on. The quantum-mechanical theory of *multipole radiation* represents an important concept in nuclear physics; see the specialist literature.[3] Looking again at Fig. 2.4 we can further simplify expression (2.26) into

$$\Big||\varepsilon_{k2}|\,|\langle a_\mathrm{f}|\hat{p}\,|a_\mathrm{i}\rangle|\cos(90^\circ-\vartheta)\Big|^2 = |\langle a_\mathrm{f}|\hat{p}\,|a_\mathrm{i}\rangle|^2\sin^2\vartheta . \tag{2.27}$$

Using also (2.23) and (2.22), we find that the lifetime (2.17) now becomes

$$\left(\frac{1}{\tau}\right)_{\mathrm{i}\to\mathrm{f}} = \frac{e^2}{2\pi m^2}\int \mathrm{d}^3k\,\frac{1}{\omega_k}\,|\langle a_\mathrm{f}|\hat{p}\,|a_\mathrm{i}\rangle|^2\sin^2\vartheta\,\delta(E_{a_\mathrm{f}} - E_{a_\mathrm{i}} + \hbar\omega_k) . \tag{2.28}$$

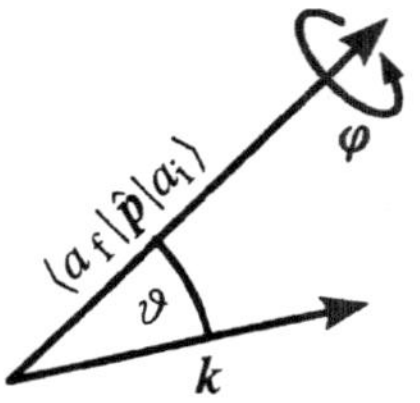

Fig. 2.5. Spherical coordinates in k space

We introduce spherical coordinates in k space and set the k_z axis along the direction of the vector $\langle a_\mathrm{f}|\hat{p}|a_\mathrm{i}\rangle$ (see Fig. 2.5). Then it follows that

$$\begin{aligned} \mathrm{d}^3k &= k^2\,\mathrm{d}k\,\sin\vartheta\,\mathrm{d}\vartheta\,\mathrm{d}\varphi \\ &\to 2\pi\frac{\omega_k^2\,\mathrm{d}\omega_k}{c^3}\sin\vartheta\,\mathrm{d}\vartheta . \end{aligned} \tag{2.29}$$

In the last step, the integration $\int \mathrm{d}\varphi = 2\pi$ has been performed because expression (2.28) does not dependent on φ. Thus we deduce

[3] J.M. Eisenberg, W. Greiner: *Nuclear Theory,* Vol. 2. *Excitation Mechanism of the Nucleus,* 3rd ed. (North-Holland, Amsterdam 1975).

$$\left(\frac{1}{\tau}\right)_{i\to f} = \frac{e^2 2\pi}{2\pi m^2 c^3}\int d\omega_k\,\omega_k\,|\langle a_f|\hat{\boldsymbol{p}}\,|a_i\rangle|^2 \times \delta(E_{a_f}-E_{a_i}+\hbar\omega_k)\int_0^\pi \sin^3\vartheta\,d\vartheta = \frac{4}{3}\frac{e^2}{m^2c^3\hbar}\omega_{fi}\,|\langle a_f|\hat{\boldsymbol{p}}\,|a_i\rangle|^2 \quad (2.30)$$

with $\omega_{fi} = (E_{a_i} - E_{a_f})/\hbar$ as the transition frequency. The matrix element $\langle a_f|\hat{\boldsymbol{p}}\,|a_i\rangle$ can be cast in another form, i.e.

$$\begin{aligned}\langle a_f|\hat{\boldsymbol{p}}\,|a_i\rangle &= \langle a_f|m\frac{d\boldsymbol{x}}{dt}|a_i\rangle \\ &= -\frac{i}{\hbar}m\langle a_f|\hat{\boldsymbol{x}}\hat{H}_{mp} - \hat{H}_{mp}\hat{\boldsymbol{x}}|a_i\rangle \\ &= -\frac{im}{\hbar}(E_{a_i}-E_{a_f})\langle a_f|\hat{\boldsymbol{x}}|a_i\rangle \\ &= -im\,\omega_{fi}\langle a_f|\hat{\boldsymbol{x}}|a_i\rangle\,, \end{aligned} \quad (2.31)$$

so that (2.30) becomes

$$\left(\frac{1}{\tau}\right)_{i\to f} = \frac{4e^2\omega_{fi}^3}{3\hbar c^3}\,|\langle a_f|\hat{\boldsymbol{x}}|a_i\rangle|^2\,. \quad (2.32)$$

The Hamiltonian $\hat{H}_{mp}$ from (2.2) represents the unperturbed many-particle system and has been used in (2.31) for the operator differentiation

$$\frac{d\hat{\boldsymbol{x}}}{dt} = -\frac{i}{\hbar}[\hat{\boldsymbol{x}}, \hat{H}_{mp}]_-\,.$$

Applying the trick (2.31), we recognize that the momentum matrix element $e\langle a_f|\,\hat{\boldsymbol{p}}\,|a_i\rangle$ can be expressed as a matrix element of the *dipole operator* $e\hat{\boldsymbol{x}}$. By now the designations *dipole approximation* and *dipole radiation* introduced earlier should have become clear.

Yet another form can be derived for the expression (2.32) of the lifetime once Heisenberg's equations of motion are applied twice,

$$\begin{aligned}\langle a_f|m\frac{d^2\hat{\boldsymbol{x}}}{dt^2}|a_i\rangle &= \langle a_f|\frac{d\hat{\boldsymbol{p}}}{dt}|a_i\rangle \\ &= -\frac{i}{\hbar}\langle a_f|\hat{\boldsymbol{p}}\hat{H}_{mp} - \hat{H}_{mp}\hat{\boldsymbol{p}}\,|a_i\rangle \\ &= -\frac{i}{\hbar}(E_{a_i}-E_{a_f})\langle a_f|\hat{\boldsymbol{p}}\,|a_i\rangle \\ &= -\frac{i}{\hbar}(E_{a_i}-E_{a_f})m\langle a_f|\frac{d\hat{\boldsymbol{x}}}{dt}|a_i\rangle \\ &= \left(-\frac{i}{\hbar}\right)\left(-\frac{i}{\hbar}\right)(E_{a_i}-E_{a_f})m\langle a_f|\hat{\boldsymbol{x}}\hat{H}_{mp} - \hat{H}_{mp}\hat{\boldsymbol{x}}|a_i\rangle \\ &= -\frac{m}{\hbar^2}(E_{a_i}-E_{a_f})^2\langle a_f|\hat{\boldsymbol{x}}|a_i\rangle\,, \end{aligned}$$

so that

$$\left|\langle a_f|\frac{d^2\hat{\boldsymbol{x}}}{dt^2}|a_i\rangle\right|^2 = \omega_{fi}^4\,|\langle a_f|\hat{\boldsymbol{x}}|a_i\rangle|^2\,. \quad (2.33)$$

Finally this yields for (2.32)

$$\left(\frac{\hbar\omega_{\rm fi}}{\tau}\right)_{\rm i\to f} = \frac{4e^2}{3c^3}\left|\langle a_{\rm f}|\frac{{\rm d}^2\hat{x}}{{\rm d}t^2}|a_{\rm i}\rangle\right|^2 . \tag{2.34}$$

This relation can be regarded as the quantum-mechanical analogue to the Larmor equation; it states that the energy radiated by an electron accelerated nonrelativistically is given by

$$\frac{\text{energy}}{\text{time}} = \frac{2e^2}{3c^3}\left|\frac{{\rm d}^2\boldsymbol{x}}{{\rm d}t^2}\right|^2 . \tag{2.35}$$

We illustrate the theory developed so far with a couple of examples.

EXERCISE

2.1 Selection Rules for Electric Dipole Transitions

Problem. Show that the selection rules for dipole transitions are given by $\Delta l = \pm 1$ and $\Delta m = \pm 1, 0$, where l and m represent the angular momentum quantum numbers of the electron.

Solution. With spherical coordinates the wavefunction of the electron takes on the form

$$\psi_{nlm}(\boldsymbol{r}) = R_{nl}(r)\,Y_{lm}(\vartheta,\varphi) .$$

The matrix element describing electric dipole transitions is given by

$$M \sim \int {\rm d}^3r\,\psi^*_{n'l'm'}(\boldsymbol{r})\,\hat{\boldsymbol{r}}\,\psi_{nlm}(\boldsymbol{r}) .$$

To begin with we express the components of the position vector in terms of spherical harmonics. We need

$$\begin{aligned}
Y_{11}(\vartheta,\varphi) &= -\sqrt{\frac{3}{8\pi}}\sin\vartheta\,{\rm e}^{{\rm i}\varphi} ,\\
Y_{1-1}(\vartheta,\varphi) &= \sqrt{\frac{3}{8\pi}}\sin\vartheta\,{\rm e}^{-{\rm i}\varphi} ,\\
Y_{10}(\vartheta,\varphi) &= \sqrt{\frac{3}{4\pi}}\cos\vartheta .
\end{aligned}$$

Addition and subtraction yield

$$\begin{aligned}
\sin\vartheta\,\cos\varphi &= \frac{1}{2}\sqrt{\frac{8\pi}{3}}\left(-Y_{11}(\vartheta,\varphi)+Y_{1-1}(\vartheta,\varphi)\right)\\
&= a_xY_{11}(\vartheta,\varphi)+b_xY_{1-1}(\vartheta,\varphi) ,\\
\sin\vartheta\,\sin\varphi &= -\frac{1}{2i}\sqrt{\frac{8\pi}{3}}\left(Y_{11}(\vartheta,\varphi)+Y_{1-1}(\vartheta,\varphi)\right)\\
&= a_yY_{11}(\vartheta,\varphi)+b_yY_{1-1}(\vartheta,\varphi) ,
\end{aligned}$$

Exercise 2.1.

$$\begin{aligned}\cos\vartheta &= \sqrt{\frac{4\pi}{3}}Y_{10}(\vartheta,\varphi)\\ &= a_z Y_{10}(\vartheta,\varphi)\,.\end{aligned}$$

Here a_i, b_i $(i = x, y)$, and a_z are constants, which can be read off immediately, i.e.

$$\begin{aligned}a_x &= -b_x &&= -\frac{1}{2}\sqrt{\frac{8\pi}{3}}\,,\\ a_y &= b_y &&= -\frac{1}{2i}\sqrt{\frac{8\pi}{3}}\,,\\ a_z &= \sqrt{\frac{4\pi}{3}}\,.\end{aligned}$$

We deduce further that

$$\begin{aligned}x &= r\sin\vartheta\cos\varphi &&= r(a_x Y_{11} + b_x Y_{1-1})\\ y &= r\sin\vartheta\sin\varphi &&= r(a_y Y_{11} + b_y Y_{1-1})\\ z &= r\cos\vartheta &&= r a_z Y_{10}\,.\end{aligned}$$

Referring to the matrix element M above, we multiply these spherical harmonics further with $Y_{lm}(\vartheta,\varphi)$. The following integrals appear:

$$\int \mathrm{d}\Omega\, Y^*_{l'm'}(\vartheta,\varphi)Y_{1,\pm 1}(\vartheta,\varphi)Y_{lm}(\vartheta,\varphi)$$

and

$$\int \mathrm{d}\Omega\, Y^*_{l'm'}(\vartheta,\varphi)Y_{10}(\vartheta,\varphi)Y_{lm}(\vartheta,\varphi)$$

with

$$\mathrm{d}\Omega = \sin\vartheta\,\mathrm{d}\vartheta\,\mathrm{d}\varphi\,.$$

Generally

$$\begin{aligned}&\int \mathrm{d}\Omega\, Y^*_{l_3 m_3}(\vartheta,\varphi)Y_{l_2 m_2}(\vartheta,\varphi)Y_{l_1 m_1}(\vartheta,\varphi)\\ &\quad = \left(\frac{(2l_1+1)(2l_2+1)}{4\pi(2l_3+1)}\right)^{\frac{1}{2}} (l_1 l_2 l_3|m_1 m_2 m_3)(l_1 l_2 l_3|000)\end{aligned}$$

holds. The *Clebsch–Gordan coefficients* $(l_1 l_2 l_3|m_1 m_2 m_3)$ vanish except for[4]

1) $m_1 + m_2 = m_3$,

$$\begin{aligned}&\text{i.e.} \quad m \pm 1 = m' \quad \text{or } \Delta m = m' - m = \pm 1,\\ &\qquad\qquad\qquad\qquad \text{for the } x \text{ and } y \text{ components}\\ &\text{and} \quad m + 0 = m' \quad \text{or } \Delta m = m' - m = 0,\\ &\qquad\qquad\qquad\qquad \text{for the } z \text{ component.}\end{aligned}$$

[4] See e.g. W. Greiner: *Quantum Mechanics – Symmetries*, 2nd ed. (Springer, Berlin, Heidelberg 1994).

Exercise 2.1.

2) The Clebsch–Gordan coefficient $(l1l'|000)$ is only nonzero if $l+l'+1$ is an even number and if l and l' differ at most by 1. Therefore $l-l'=\pm 1$.

Hence, we get the following selection rules:

$$\left.\begin{aligned}\Delta l &= \pm 1\\ \Delta m &= \pm 1\end{aligned}\right\}\ \text{for the } x \text{ and } y \text{ components,}$$

$$\left.\begin{aligned}\Delta l &= \pm 1\\ \Delta m &= 0\end{aligned}\right\}\ \text{for the } z \text{ component.}$$

For all other cases the angular integrals vanish, so that the matrix element is zero.

EXERCISE

2.2 Lifetime of the 2p State with $m=0$ in the Hydrogen Atom with Respect to Decay Into the 1s State

Problem. Calculate the lifetime τ of the 2p state with $m=0$ of the hydrogen atom with respect to decay into the 1s state. The wavefunctions are given by[5]

$$\begin{aligned}\psi(1\mathrm{s}) &= \frac{1}{\sqrt{\pi a^3}}\,\mathrm{e}^{-r/a}\,,\\ \psi(2\mathrm{p}, m=0) &= \frac{1}{8\sqrt{\pi a^3}}\frac{r}{a}\,\mathrm{e}^{-r/2a}\sqrt{2}\cos\vartheta\,,\end{aligned}$$

where $a=\hbar^2/me^2$ denotes the Bohr radius.

Solution. We remember that

$$\frac{1}{\tau} = \frac{4e^2\omega^3_{1\mathrm{s}\,2\mathrm{p}}}{3\hbar c^3}\,|\langle 2\mathrm{p}, m = 0|\boldsymbol{r}\,|1\mathrm{s}\rangle|^2\,,$$

where

$$\hbar\omega_{1\mathrm{s}\,2\mathrm{p}} = E_{1\mathrm{s}} - E_{2\mathrm{p}} = \frac{me^4}{2\hbar^2}\left(1-\frac{1}{4}\right) = \frac{3}{8}\frac{e^2}{a}$$

has been used. In spherical coordinates the unit vectors are given by $\boldsymbol{e}_{\pm 1} = \mp\frac{1}{\sqrt{2}}(\boldsymbol{e}_x \pm \mathrm{i}\boldsymbol{e}_y)$ and $\boldsymbol{e}_0 = \boldsymbol{e}_z$. As a consequence, for $m=0$ only the z component of the matrix element of $\boldsymbol{r}$ does not vanish. Thus, it follows that

$$\frac{1}{\tau} = \frac{4}{3}\left(\frac{3}{8}\right)^3\frac{c}{a^3}\left(\frac{e^2}{\hbar c}\right)^4 |\langle 2\mathrm{p}, m = 0|z|1\mathrm{s}\rangle|^2\,.$$

This matrix element becomes

[5] W. Greiner: *Quantum Mechanics – An Introduction*, 3rd ed. (Springer, Berlin, Heidelberg 1994).

Excercise 2.2.

$$\begin{aligned} M &= \langle 2\mathrm{p}|z|1\mathrm{s}\rangle \\ &= \int \mathrm{d}^3r\, \psi^*(2\mathrm{p})\, z\, \psi(1\mathrm{s}) \\ &= \frac{1}{8\pi a^3}\int \frac{r}{a}\,\mathrm{e}^{-r/2a}\sqrt{2}\cos\vartheta(r\cos\vartheta)\mathrm{e}^{-r/a}r^2\,\mathrm{d}r\,\sin\vartheta\,\mathrm{d}\vartheta\,\mathrm{d}\varphi\,. \end{aligned}$$

The integration over φ is simple:

$$\begin{aligned} M &= \frac{\sqrt{2}}{4a^4}\int_0^\infty r^4\,\mathrm{d}r\,\mathrm{e}^{-3r/2a}\int_0^\pi \cos^2\vartheta\,\sin\vartheta\,\mathrm{d}\vartheta \\ &= \frac{2}{3}\frac{\sqrt{2}}{4a^4}\int_0^\infty \mathrm{e}^{-3r/2a}r^4\,\mathrm{d}r \\ &= 4\sqrt{2}\left(\frac{2}{3}\right)^5 a\,, \end{aligned}$$

so that

$$\frac{1}{\tau} = \frac{4}{3}\left(\frac{3}{8}\right)^3\frac{c}{a}\left(\frac{e^2}{\hbar c}\right)^4 2^5\left(\frac{2}{3}\right)^{10}$$

or

$$\tau = \left(\frac{3}{2}\right)^8\frac{a}{c}\left(\frac{\hbar c}{e^2}\right)^4 = \left(\frac{3}{2}\right)^8\frac{a}{c\alpha^4} \qquad \left(\alpha = \frac{e^2}{\hbar c} = \frac{1}{137}\right).$$

Inserting numbers yields

$$\tau \approx 1.6\times 10^{-9}\,\mathrm{s}\,.$$

EXERCISE

2.3 Impossibility of the Decay of the 2s State of the Hydrogen Atom via the $\boldsymbol{p}\cdot\boldsymbol{A}$ Interaction

Problem. Show that the 2s state of the hydrogen atom cannot decay via the $\boldsymbol{p}\cdot\boldsymbol{A}$ interaction into the 1s state by emitting a photon, i.e.

$$\langle 2\mathrm{s}|\boldsymbol{\varepsilon}_{\boldsymbol{k}\sigma}\cdot\boldsymbol{p}\,\mathrm{e}^{-\mathrm{i}\boldsymbol{k}\cdot\boldsymbol{r}}|1\mathrm{s}\rangle = 0\,.$$

Solution. We choose $\boldsymbol{k}$ to be parallel to the z axis, so that

$$\mathrm{e}^{-\mathrm{i}\boldsymbol{k}\cdot\boldsymbol{r}} = \mathrm{e}^{-\mathrm{i}kz}\,.$$

Then the matrix element becomes ($\hat{\boldsymbol{p}} = -\mathrm{i}\hbar\boldsymbol{\nabla}$):

$$\langle 2\mathrm{s}|\boldsymbol{\varepsilon}\cdot\boldsymbol{p}\,\mathrm{e}^{-\mathrm{i}\boldsymbol{k}\cdot\boldsymbol{r}}|1\mathrm{s}\rangle = \frac{\hbar}{\mathrm{i}}\int \mathrm{d}^3r\,\psi_{2\mathrm{s}}\,\mathrm{e}^{-\mathrm{i}kz}\boldsymbol{\varepsilon}\cdot\boldsymbol{\nabla}\psi_{1\mathrm{s}}\,.$$

Since

$$\psi_{1\mathrm{s}} = \frac{1}{\sqrt{\pi a^3}}\mathrm{e}^{-r/a}\,,$$

we deduce that

Excercise 2.3.

$$\nabla\psi_{1s} = \frac{\partial\psi_{1s}}{\partial r}\frac{\boldsymbol{r}}{r} = -\frac{1}{a}\frac{\boldsymbol{r}}{r}\psi_{1s}$$

and derive

$$\langle 2s|\boldsymbol{\varepsilon}\cdot\boldsymbol{p}\,\mathrm{e}^{-\mathrm{i}\boldsymbol{k}\cdot\boldsymbol{r}}|1s\rangle = -\frac{\hbar}{\mathrm{i}a}\int \mathrm{d}^3r\,\psi_{2s}\psi_{1s}\frac{\boldsymbol{\varepsilon}\cdot\boldsymbol{r}}{r}\,\mathrm{e}^{-\mathrm{i}k\cdot z}\,.$$

Since $\boldsymbol{k}$ points in the z direction and $\boldsymbol{\varepsilon}$ is perpendicular to $\boldsymbol{k}$, the vector $\boldsymbol{\varepsilon}$ has to lie in the x-y plane. Then, the scalar product $\boldsymbol{\varepsilon}\cdot\boldsymbol{r}$ represents the projection of $\boldsymbol{r}$ into the x-y plane and the integrand factorizes into a part that depends only on x and y, and into another part depending only on z. Because ψ_{2s} and ψ_{1s} are spherically symmetric, the integrations over x and y vanish. As a consequence the matrix element vanishes.

EXERCISE

2.4 The Hamiltonian for Interaction Between the Electron Spin and the Electromagnetic Field

Problem. Up to now we have neglected the electron spin. The magnetic moment $\hat{\boldsymbol{\mu}} = (e\hbar/2mc)\hat{\boldsymbol{\sigma}}$ of the electron means an additional spin-dependent term has to be taken into account for the interaction of an electron with the electromagnetic field, in addition to the usual terms $\hat{H}' \sim \hat{\boldsymbol{p}}\cdot\hat{\boldsymbol{A}}$ and $\hat{H}'' \sim \hat{\boldsymbol{A}}^2$ discussed up to now. Derive an expression for this additional term.

Solution. The interaction energy of a magnetic dipole $\boldsymbol{\mu}$ with a magnetic field $\boldsymbol{B}$ is equal to

$$H''' = -\boldsymbol{\mu}\cdot\boldsymbol{B}\,.$$

Now $\boldsymbol{B} = \mathrm{rot}\,\boldsymbol{A} = \nabla\times\boldsymbol{A}$ and $\hat{p} = -\mathrm{i}\hbar\nabla = \hbar\hat{\boldsymbol{k}}$ hold, so that

$$\hat{\boldsymbol{k}} = -\mathrm{i}\nabla \qquad \text{or} \qquad \nabla = \mathrm{i}\hat{\boldsymbol{k}}\,,$$

$$\hat{\boldsymbol{B}} = \mathrm{i}\hat{\boldsymbol{k}}\times\hat{\boldsymbol{A}}\,,$$

$$\hat{H}''' = -\frac{\mathrm{i}e\hbar}{2mc}\left(\hat{\boldsymbol{k}}\times\hat{\boldsymbol{A}}\right)\cdot\boldsymbol{\sigma}\,.$$

Here plane-wave photons have been assumed, as usual. With

$$\hat{\boldsymbol{A}} = \sum_{\boldsymbol{k},\sigma} N_{\boldsymbol{k}}\boldsymbol{\varepsilon}_{\boldsymbol{k}\sigma}\left[\hat{a}_{\boldsymbol{k}\sigma}(t)\,\mathrm{e}^{\mathrm{i}\boldsymbol{k}\cdot\boldsymbol{r}} + \hat{a}^{\dagger}_{\boldsymbol{k}\sigma}(t)\,\mathrm{e}^{-\mathrm{i}\boldsymbol{k}\cdot\boldsymbol{r}}\right]$$

$\hat{H}'''$ becomes

$$\hat{H}''' = -\frac{\mathrm{i}e\hbar}{2mc}\sum_{\boldsymbol{k},\sigma} N_{\boldsymbol{k}}\,\hat{\boldsymbol{\sigma}}\cdot\left(\hat{\boldsymbol{k}}\times\boldsymbol{\varepsilon}_{\boldsymbol{k}\sigma}\right)\left[\hat{a}_{\boldsymbol{k}\sigma}\,\mathrm{e}^{\mathrm{i}\boldsymbol{k}\cdot\boldsymbol{r}} + \hat{a}^{\dagger}_{\boldsymbol{k}\sigma}\,\mathrm{e}^{-\mathrm{i}\boldsymbol{k}\cdot\boldsymbol{r}}\right]\,.$$

The operator $\hat{\boldsymbol{k}} = -\mathrm{i}\nabla$ acts on the exponential function:

$$\hat{H}''' = -\frac{\mathrm{i}e\hbar}{2mc}\sum_{\boldsymbol{k},\sigma} N_{\boldsymbol{k}}\,\boldsymbol{\sigma}\cdot\left(\boldsymbol{k}\times\boldsymbol{\varepsilon}_{\boldsymbol{k}\sigma}\right)\left[\hat{a}_{\boldsymbol{k}\sigma}\,\mathrm{e}^{\mathrm{i}\boldsymbol{k}\cdot\boldsymbol{r}} - \hat{a}^{\dagger}_{\boldsymbol{k}\sigma}\,\mathrm{e}^{-\mathrm{i}\boldsymbol{k}\cdot\boldsymbol{r}}\right]\,.$$

Here $\boldsymbol{k}$ is no longer an operator; the minus sign in the second term results from differentiation of the exponential.

EXERCISE

2.5 Lifetime of the Ground State of the Hydrogen Atom with Hyperfine Splitting

Problem. The magnetic interaction between the spin of the electron and the nuclear spin splits the ground state of the hydrogen atom into two levels: one with total spin 1 and one with total spin 0. The photon, which is emitted from the transition between these two states, comes with a wavelength of $\lambda \approx 21\text{cm}$. Use the results of Exercise 2.4 to calculate the lifetime of this transition. For the calculation of the lifetime you may choose one particular state of the degenerate spin-1 triplet. Why?

Solution. The spin of the electron $|\mu_1\rangle_\mathrm{e}$ and that of the proton $|\mu_2\rangle_\mathrm{p}$ couple to the total spin wavefunction

$$|S\mu\rangle = \sum_{\mu_1,\mu_2} \left(\frac{1}{2}\frac{1}{2}S|\mu_1\mu_2\mu\right) |\mu_1\rangle_\mathrm{e}\,|\mu_2\rangle_\mathrm{p} \,.$$

S takes the values $S = 1$ (triplet) or $S = 0$ (singlet). $(j_1 j_2 j|m_1 m_2 m)$ denote the Clebsch–Gordan coefficients.[6] As the initial state we choose

$$|i\rangle = |1\mathrm{s}\rangle_\mathrm{e}\,|\uparrow\rangle_\mathrm{e}\,|\uparrow\rangle_\mathrm{nucleus}|\text{no photon}\rangle_\mathrm{rad} \,;$$

a 1s electron $|1\mathrm{s}\rangle_\mathrm{e}$ comes with spin up $|\uparrow\rangle_\mathrm{e}$, the nucleus has spin up $|\uparrow\rangle_\mathrm{nucleus}$, and the radiation field is initially represented by the 0-photon state.

This initial state has total spin $S = 1$ and its z projection is also $\mu = 1$. It is a member of the spin triplet. Since each of these degenerate spin-triplet states should have the same width (decay time) – because of the rotational symmetry of the problem – we may restrict ourselves to calculating the decay of this particular member of the spin triplet. In the final state $|f\rangle$ the electron is still in the 1s state, but the spin orientations of the electron and the nucleus have changed. In addition, one photon exists in the state $|\ldots, 1_{k\sigma}, \ldots\rangle_\mathrm{rad}$.

$$|f\rangle = |1\mathrm{s}\rangle_\mathrm{e}\frac{1}{\sqrt{2}}\left(|\uparrow\rangle_\mathrm{e}\,|\downarrow\rangle_\mathrm{nucleus} - |\downarrow\rangle_\mathrm{e}\,|\uparrow\rangle_\mathrm{nucleus}\right)|\ldots, 1_{k\sigma}, \ldots\rangle_\mathrm{rad} \,.$$

This final state carries angular momentum 0 for the electron–nucleus system. The one photon carries the total angular momentum of the state, i.e. 1. The lifetime is given by the expression

$$\frac{1}{\tau} = \sum_{\text{final states}} \frac{2\pi}{\hbar}\left|\langle f|\hat{H}'''|i\rangle\right|^2 \delta(\Delta E - \hbar ck) \,.$$

The δ function guarantees energy conservation; here $\Delta E = \hbar\omega_k = \hbar ck_0$, with $k_0 = 2\pi/21\,\mathrm{cm}^{-1}$, represents the energy difference between the two states, which is equal to the energy $\hbar\omega_k = \hbar ck$ of the emitted photon. Furthermore (see Exercise 2.4),

[6] W. Greiner, B. Müller: *Quantum Mechanics – Symmetries*, 2nd ed. (Springer, Berlin, Heidelberg 1994).

Exercise 2.5.

$$\hat{H}''' = -\frac{\mathrm{i}e\hbar}{2mc}\sum_{\boldsymbol{k},\sigma} N_k \boldsymbol{\sigma}\cdot(\boldsymbol{k}\times\boldsymbol{\varepsilon}_{\boldsymbol{k}\sigma})\left(\hat{a}_{\boldsymbol{k}\sigma}\,\mathrm{e}^{\mathrm{i}\boldsymbol{k}\cdot\boldsymbol{r}} - \hat{a}^{\dagger}_{\boldsymbol{k}\sigma}\,\mathrm{e}^{-\mathrm{i}\boldsymbol{k}\cdot\boldsymbol{r}}\right),$$

with $N_k^2 = 2\pi\hbar c^2/L^3\omega_k$. The term including $\hat{a}_{\boldsymbol{k}\sigma}$ annihilates a photon; since we only calculate the process of creating a photon, we discard this term. With $\omega_k = ck$ it follows that

$$\hat{H}''' = \frac{\mathrm{i}e\hbar}{2mc}\left(\frac{2\pi\hbar c^2}{L^3}\right)^{\frac{1}{2}}\sum_{\boldsymbol{k},\sigma}\frac{1}{\sqrt{ck}}\boldsymbol{\sigma}\cdot(\boldsymbol{k}\times\boldsymbol{\varepsilon}_{\boldsymbol{k}\sigma})\,\hat{a}^{\dagger}_{\boldsymbol{k}\sigma}\,\mathrm{e}^{-\mathrm{i}\boldsymbol{k}\cdot\boldsymbol{r}}.$$

With $\delta(ax) = \frac{1}{|a|}\delta(x)$ this yields

$$\frac{1}{\tau} = \left(\frac{e\hbar}{2mc}\right)^2\left(\frac{2\pi\hbar c^2}{L^3}\right)\frac{2\pi}{\hbar c}\sum_{\boldsymbol{k},\sigma}\frac{1}{k}\left|\langle f|\hat{\boldsymbol{\sigma}}\cdot(\boldsymbol{k}\times\boldsymbol{\varepsilon}_{\boldsymbol{k}\sigma})\hat{a}^{\dagger}_{\boldsymbol{k}\sigma}\,\mathrm{e}^{-\mathrm{i}\boldsymbol{k}\cdot\boldsymbol{r}}|i\rangle\right|^2 \times \frac{1}{\hbar c}\delta(k_0 - k).$$

Again, we employ the dipole approximation

$$\mathrm{e}^{\mathrm{i}\boldsymbol{k}\cdot\boldsymbol{r}} \approx 1$$

and replace the summation over k for the final photon states with an integration

$$\sum_{\boldsymbol{k}} \longrightarrow \frac{L^3}{(2\pi)^3}\int \mathrm{d}^3k.$$

Now k has become a continuous variable; compare again with (2.18). It follows that

$$\frac{1}{\tau} = \frac{e^2\hbar}{4m^2c^2}\frac{1}{2\pi}\sum_{\sigma}\left(\int\frac{\mathrm{d}^3k}{k}\left|\langle f|\hat{\boldsymbol{\sigma}}\cdot(\boldsymbol{k}\times\boldsymbol{\varepsilon}_{\boldsymbol{k}\sigma})\hat{a}^{\dagger}_{\boldsymbol{k}\sigma}|i\rangle\right|^2\delta(k_0 - k)\right).$$

Furthermore

$$\hat{a}^{\dagger}_{\boldsymbol{k}\sigma}|i\rangle = |1\mathrm{s}\rangle_{\mathrm{e}}|\uparrow\rangle_{\mathrm{e}}|\uparrow\rangle_{\mathrm{nucleus}}|\dots,1_{\boldsymbol{k}\sigma},\dots\rangle_{\mathrm{rad}},$$

so that we derive the matrix element as

$$M = \left\langle\frac{1}{\sqrt{2}}(\langle\uparrow|_{\mathrm{e}}\langle\downarrow|_{\mathrm{nucleus}} - \langle\downarrow|_{\mathrm{e}}\langle\uparrow|_{\mathrm{nucleus}})\langle 1\mathrm{s}|_{\mathrm{e}}\langle 1_{\boldsymbol{k}\sigma}|\right| \times \hat{\boldsymbol{\sigma}}\cdot(\boldsymbol{k}\times\boldsymbol{\varepsilon}_{\boldsymbol{k}\sigma})\big||1\mathrm{s}\rangle_{\mathrm{e}}|\uparrow\rangle_{\mathrm{e}}|\uparrow\rangle_{\mathrm{nucleus}}|1_{\boldsymbol{k}\sigma}\rangle\big\rangle.$$

The scalar product is

$$\hat{\boldsymbol{\sigma}}\cdot(\boldsymbol{k}\times\boldsymbol{\varepsilon}_{\boldsymbol{k}\sigma}) = \hat{\sigma}_x(\boldsymbol{k}\times\boldsymbol{\varepsilon}_{\boldsymbol{k}\sigma})_x + \hat{\sigma}_y(\boldsymbol{k}\times\boldsymbol{\varepsilon}_{\boldsymbol{k}\sigma})_y + \hat{\sigma}_z(\boldsymbol{k}\times\boldsymbol{\varepsilon}_{\boldsymbol{k}\sigma})_z.$$

We use the Pauli matrices

$$\hat{\sigma}_x = \begin{pmatrix}0 & 1\\ 1 & 0\end{pmatrix}, \quad \hat{\sigma}_y = \begin{pmatrix}0 & -i\\ i & 0\end{pmatrix}, \quad \hat{\sigma}_z = \begin{pmatrix}1 & 0\\ 0 & -1\end{pmatrix}$$

and deduce

$$\hat{\sigma}_x|\uparrow\rangle_{\mathrm{e}} = |\downarrow\rangle_{\mathrm{e}}, \quad \hat{\sigma}_y|\uparrow\rangle_{\mathrm{e}} = i|\downarrow\rangle_{\mathrm{e}}, \quad \hat{\sigma}_z|\uparrow\rangle_{\mathrm{e}} = |\uparrow\rangle_{\mathrm{e}}.$$

Exercise 2.5.

Because of the orthogonality of $|f\rangle$ and $|i\rangle$ only those configurations contribute for which the electron switches from the initial state $|\uparrow\rangle_e$ to the state $|\downarrow\rangle_e$ due to $\hat{\boldsymbol{\sigma}}$. Hence, only σ_x and σ_y contribute:

$$M = -\frac{1}{\sqrt{2}}\left[(\boldsymbol{k}\times\boldsymbol{\varepsilon}_{\boldsymbol{k}\sigma})_x + \mathrm{i}(\boldsymbol{k}\times\boldsymbol{\varepsilon}_{\boldsymbol{k}\sigma})_y\right] .$$

Initially we have chosen the spin to be along the z axis. In such a system it is not convenient to evaluate the cross products appearing in M. Thus, we rotate the coordinate system in such a way, that the z' axis coincides with the direction of $\boldsymbol{k}$ (see Fig. 2.6):

$$\begin{aligned} x &= x'\cos\vartheta - z'\sin\vartheta , \\ y &= y' , \\ z &= x'\sin\vartheta + z'\cos\vartheta . \end{aligned}$$

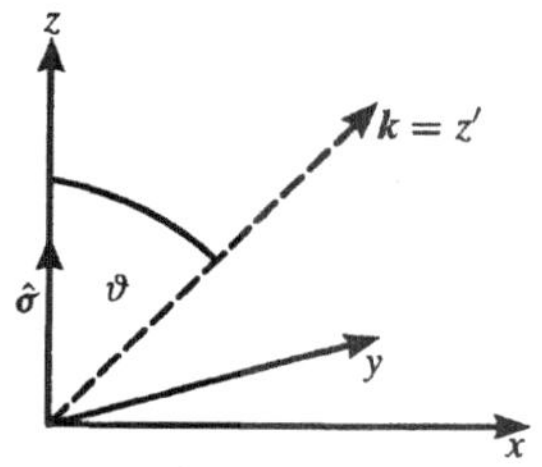

Fig. 2.6. Illustration of the rotation of the coordinate system

Via these transformations,

$$\sum_\sigma |M|^2 = \frac{1}{2}\left[(\boldsymbol{k}\times\boldsymbol{\varepsilon}_{\boldsymbol{k}\sigma})_x^2 + (\boldsymbol{k}\times\boldsymbol{\varepsilon}_{\boldsymbol{k}\sigma})_y^2\right]$$

becomes

$$\begin{aligned} \sum_\sigma |M|^2 = \sum_\sigma \frac{1}{2}\Big\{&\left[(\boldsymbol{k}\times\boldsymbol{\varepsilon}_{\boldsymbol{k}\sigma})_{x'}\cos\vartheta - (\boldsymbol{k}\times\boldsymbol{\varepsilon}_{\boldsymbol{k}\sigma})_{z'}\sin\vartheta\right]^2 \\ &+ \left[(\boldsymbol{k}\times\boldsymbol{\varepsilon}_{\boldsymbol{k}\sigma})_{y'}\right]^2\Big\} \end{aligned}$$

in the rotated coordinate system. The latter has been chosen such that

$$\boldsymbol{\varepsilon}_{\boldsymbol{k}1} = \boldsymbol{e}_{x'} , \qquad \boldsymbol{\varepsilon}_{\boldsymbol{k}2} = \boldsymbol{e}_{y'} = \boldsymbol{e}_y , \qquad \boldsymbol{k} = k\boldsymbol{e}_{z'} ,$$

so that

$$\begin{aligned} (\boldsymbol{k}\times\boldsymbol{\varepsilon}_{\boldsymbol{k}\sigma})_{x'}^2 &= k^2 , \\ (\boldsymbol{k}\times\boldsymbol{\varepsilon}_{\boldsymbol{k}\sigma})_{z'} &= 0 , \\ (\boldsymbol{k}\times\boldsymbol{\varepsilon}_{\boldsymbol{k}\sigma})_{y'}^2 &= k^2 . \end{aligned}$$

We get

$$\sum_\sigma |M|^2 = \frac{1}{2}k^2(1+\cos^2\vartheta) ;$$

ϑ represents the angle between the spin vector and $\boldsymbol{k}$. Finally we obtain

$$\frac{1}{\tau} = \frac{e^2\hbar}{4m^2c^2}\frac{1}{2\pi}\left(\int\frac{\mathrm{d}^3k}{k}\frac{1}{2}k^2(1+\cos^2\vartheta)\delta(k_0-k)\right) ,$$

and inserting $\mathrm{d}^3k = k^2\,\mathrm{d}k\,\sin\vartheta\,\mathrm{d}\vartheta\,\mathrm{d}\varphi$ we have the final result:

$$\begin{aligned} \frac{1}{\tau} &= \frac{e^2\hbar}{4m^2c^2}\frac{1}{2}\int(1+\cos^2\vartheta)k^3\,\mathrm{d}k\,\sin\vartheta\,\mathrm{d}\vartheta\,\delta(k_0-k) \\ &= \frac{e^2\hbar}{4m^2c^2}\frac{1}{2}\frac{8}{3}\int k^3\delta(k_0-k)\,\mathrm{d}k \\ &= \frac{1}{3}\frac{e^2\hbar}{m^2c^2}k_0^3 . \end{aligned}$$

Exercise 2.5.

Inserting the value $k_0 = 2\pi/21\,\mathrm{cm}^{-1}$, we calculate a lifetime of

$$\tau \approx 3.4 \times 10^{14}\,\mathrm{s} \approx 10^7\,\text{years}\,.$$

Now it has become obvious that these kinds of transitions are not observable in a laboratory. However, such photons are well known in radioastronomy.

EXERCISE

2.6 One-Photon Decay of the 2s State in the Hydrogen Atom

Problem. Use the results of Exercise 2.5 to calculate the lifetime of the 2s state of the hydrogen atom. The spins of the electron and proton couple as in Exercise 2.5, i.e. to total spin 1. Assume that, because of the spin-dependent interaction, the 2s state decays into the ground state by emitting *one* photon. Later on we will demonstrate, in Exercise 2.10, that the two-photon decay process is much faster.

Solution. In a manner analogous to Exercise 2.5 we arrive at

$$\begin{aligned}\frac{1}{\tau} &= \frac{L^3}{(2\pi)^3}\left(\frac{e\hbar}{2mc}\right)^2\left(\frac{2\pi\hbar c}{L^3}\right)\frac{2\pi}{\hbar^2c^2}\\ &\quad\times\sum_\sigma\int\frac{\mathrm{d}^3k}{k}\delta(k_0-k)\left|\langle 1\mathrm{s}\downarrow|\boldsymbol{\sigma}\cdot(\boldsymbol{k}\times\boldsymbol{\varepsilon}_{\boldsymbol{k}\sigma})\,\mathrm{e}^{-\mathrm{i}\boldsymbol{k}\cdot\boldsymbol{r}}|2\mathrm{s}\uparrow\rangle\right|^2\\ &= \frac{e^2\hbar}{8\pi m^2c^2}\int\frac{\mathrm{d}^3k}{k}k^2(1+\cos^2\vartheta)\delta(k_0-k)\left|\langle 1\mathrm{s}|\,\mathrm{e}^{-\mathrm{i}\boldsymbol{k}\cdot\boldsymbol{r}}|2\mathrm{s}\rangle\right|^2\,.\end{aligned}$$

Here we have analyzed the spin-dependent part of the matrix element in a way similar to that in Exercise 2.5, except that the dipole approximation has not been employed. The spatial part of the matrix element remains; in fact, this part may be drawn out of the integral:

$$\begin{aligned}\frac{1}{\tau} &= \frac{e^2\hbar}{8\pi m^2c^2}\left(\int k^3\,\mathrm{d}k\,\sin\vartheta\,\mathrm{d}\vartheta\,\mathrm{d}\varphi\,(1+\cos^2\vartheta)\right)\\ &\quad\times\delta(k_0-k)\left|\langle 1\mathrm{s}|\,\mathrm{e}^{-\mathrm{i}\boldsymbol{k}\cdot\boldsymbol{r}}|2\mathrm{s}\rangle\right|^2\\ &= \frac{2}{3}\frac{e^2\hbar}{m^2c^2}k_0^3\left|\langle 1\mathrm{s}|\,\mathrm{e}^{-\mathrm{i}\boldsymbol{k}\cdot\boldsymbol{r}}|2\mathrm{s}\rangle\right|^2\,.\end{aligned}$$

We expand:

$$\mathrm{e}^{-\mathrm{i}\boldsymbol{k}\cdot\boldsymbol{r}} = 1-\mathrm{i}\boldsymbol{k}\cdot\boldsymbol{r}+\tfrac{1}{2}\mathrm{i}^2(\boldsymbol{k}\cdot\boldsymbol{r})^2+\cdots\,.$$

Since $\langle 1\mathrm{s}|2\mathrm{s}\rangle = 0$ the first term of the expansion does not contribute. The second term leads to the matrix element

$$M = \langle 1\mathrm{s}|\,\boldsymbol{k}\cdot\boldsymbol{r}\,|2\mathrm{s}\rangle\,.$$

The 1s and 2s wavefunctions depend only on r. The scalar product yields $\boldsymbol{k}\cdot\boldsymbol{r} = kr\cos\vartheta$ as the z axis is chosen parallel to $\boldsymbol{k}$. Thus the matrix element is proportional to

Exercise 2.6.

$$M \sim \int_0^\pi \sin\vartheta \, \cos\vartheta \, d\vartheta = 0 \, ;$$

as a consequence, the second term of the expansion does not contribute. Therefore we start all over with

$$\langle 1s| \, e^{-ik\cdot r} |2s\rangle \approx -\tfrac{1}{2}\langle 1s|(\boldsymbol{k}\cdot\boldsymbol{r})^2|2s\rangle \, .$$

The corresponding wavefunctions are given by

$$\psi_{1s} = \frac{1}{\sqrt{a^3\pi}} \, e^{-r/a}$$

and

$$\psi_{2s} = \frac{1}{4\sqrt{2\pi a^3}} \left(2 - \frac{r}{a}\right) e^{-r/a} \, ,$$

so that the matrix element now becomes

$$\begin{aligned} M &= \langle 1s| \, e^{-ik\cdot r} |2s\rangle \\ &\approx -\frac{1}{2}\int r^2 \, dr \, \frac{1}{\sqrt{a^3\pi}} \, e^{-r/a}(\boldsymbol{k}\cdot\boldsymbol{r})^2 \\ &\quad \times \frac{1}{4\sqrt{2\pi a^3}} \left(2 - \frac{r}{a}\right) e^{-r/a} \sin\vartheta \, d\vartheta \, d\varphi \, . \end{aligned}$$

Performing the φ integration and using $\boldsymbol{k}\cdot\boldsymbol{r} = kr\cos\vartheta$ gives

$$\begin{aligned} M &= -\frac{k^2}{4\sqrt{2}a^3}\int_0^\infty r^4 \, e^{-2r/a}\left(2 - \frac{r}{a}\right) dr \int_0^\pi \sin\vartheta \, \cos^2\vartheta \, d\vartheta \, , \\ &= -\frac{k^2}{6\sqrt{2}a^3}\int_0^\infty r^4 \, e^{-2r/a}\left(2 - \frac{r}{a}\right) dr \, , \\ &= \frac{1}{16\sqrt{2}}(ka)^2 \, . \end{aligned}$$

Because of energy conservation the energy of the emitted photon is $|E_{2s} - E_{1s}| = \hbar c k_0$. We thus arrive at

$$\frac{1}{\tau} \approx \frac{2}{3}\frac{e^2\hbar}{m^2c^2}k_0^3\left(\frac{1}{16\sqrt{2}}(k_0 a)^2\right)^2 \, ,$$

or

$$\frac{1}{\tau} \approx \frac{1}{768}\frac{e^2\hbar}{m^2c^2}k_0^7 a^4 \, .$$

From

$$E_{1s} - E_{2s} = \hbar c k_0 = \frac{3}{8}\frac{e^2}{a} \, , \qquad a = \frac{\hbar^2}{me^2} \, , \qquad \alpha = \frac{e^2}{\hbar c}$$

it follows that $k_0 = \frac{3}{8}\frac{\alpha}{a}$ and

$$\frac{1}{\tau} \approx \frac{1}{768}\left(\frac{3}{8}\right)^7 \alpha^{10}\frac{c}{a} \, .$$

Finally, using $\alpha \approx 1/137$ and $a = 5.292 \times 10^{-9}$ cm we get

Exercise 2.6.

$$\tau \approx 3\times 10^8\,\mathrm{s} \approx 9.5\,\mathrm{years}\,.$$

The large lifetime underlines that the one-photon decay is strongly suppressed. This is in contrast to the two-photon decay, which we will encounter in Exercise 2.10.

2.3 Absorption of Photons

For the process of absorption of photons, the initial and final states are given by

$$|i\rangle = |a_\mathrm{i}\rangle\,|\dots,n_{\boldsymbol{k}\sigma},\dots\rangle \tag{2.36}$$

and

$$|f\rangle = |a_\mathrm{f}\rangle\,|\dots,n_{\boldsymbol{k}\sigma}-1,\dots\rangle\,, \tag{2.37}$$

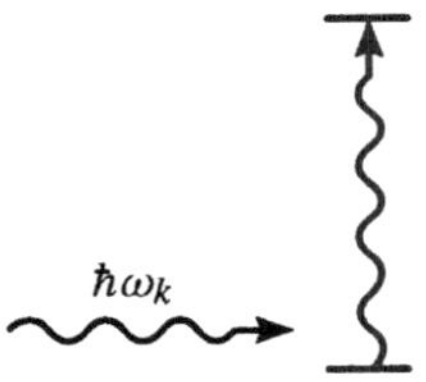

Fig. 2.7. As it absorbs a photon the many-particle system changes from $|a_\mathrm{i}\rangle$ to $|a_\mathrm{f}\rangle$

respectively. For most cases of interest $|a_\mathrm{i}\rangle$ represents the ground state of the atom or the nucleus. The final state differs from the initial state in that the many-particle system is in an excited state and the radiation field has lost one photon characterized by $\boldsymbol{k}\sigma$.

The relevant matrix element now reads

$$\begin{aligned}\langle f|\hat{H}'_{\mathrm{int}}|i\rangle &= \langle a_\mathrm{f}|\,\langle\dots,n_{\boldsymbol{k}\sigma}-1,\dots|\\ &\quad\times\left(-\frac{e}{mc}\right)\sum_{\boldsymbol{k},\sigma}\sqrt{\frac{2\pi\hbar c^2}{L^3\omega_k}}\hat{\boldsymbol{p}}\cdot\boldsymbol{\varepsilon}_{\boldsymbol{k}\sigma}\left(\hat{a}_{\boldsymbol{k}\sigma}\,\mathrm{e}^{\mathrm{i}\boldsymbol{k}'\cdot\boldsymbol{x}}+\hat{a}^\dagger_{\boldsymbol{k}\sigma}\,\mathrm{e}^{-\mathrm{i}\boldsymbol{k}'\cdot\boldsymbol{x}}\right)\\ &\quad\times|a_\mathrm{i}\rangle\,|\dots,n_{\boldsymbol{k}\sigma},\dots\rangle\\ &= -\frac{e}{mc}\sqrt{\frac{2\pi\hbar c^2}{L^3\omega_k}}\langle a_\mathrm{f}|\hat{\boldsymbol{p}}\cdot\boldsymbol{\varepsilon}_{\boldsymbol{k}\sigma}\,\mathrm{e}^{\mathrm{i}\boldsymbol{k}\cdot\boldsymbol{x}}|a_\mathrm{i}\rangle\\ &\quad\times\langle\dots,n_{\boldsymbol{k}\sigma}-1,\dots|\hat{a}_{\boldsymbol{k}\sigma}|\dots,n_{\boldsymbol{k}\sigma},\dots\rangle\\ &= -\frac{e}{mc}\sqrt{\frac{2\pi\hbar c^2}{L^3\omega_k}}\langle a_\mathrm{f}|\hat{\boldsymbol{p}}\cdot\boldsymbol{\varepsilon}_{\boldsymbol{k}\sigma}\,\mathrm{e}^{\mathrm{i}\boldsymbol{k}\cdot\boldsymbol{x}}|a_\mathrm{i}\rangle\sqrt{n_{\boldsymbol{k}\sigma}}\,;\end{aligned} \tag{2.38}$$

compare this with (2.14). Here we have assumed again that only $\hat{H}'_{\mathrm{int}}$ contributes to this process in first order and that (1.56) determines the photon-matrix element. According to (2.13a) the transition probability per unit time for the absorption process is given by

$$\begin{aligned}\left(\frac{\text{trans. prob.}}{\text{time}}\right)_{\text{absorption}} &= \frac{2\pi}{\hbar}\left(\frac{e}{mc}\right)^2\left(\frac{2\pi\hbar c^2}{L^3\omega_k}\right)n_{\boldsymbol{k}\sigma}\\ &\quad\times\left|\langle a_\mathrm{f}|\hat{\boldsymbol{p}}\cdot\boldsymbol{\varepsilon}_{\boldsymbol{k}\sigma}\,\mathrm{e}^{\mathrm{i}\boldsymbol{k}\cdot\boldsymbol{x}}|a_\mathrm{i}\rangle\right|^2\delta(E_\mathrm{f}-E_\mathrm{i})\,.\end{aligned} \tag{2.39}$$

Here

$$\begin{aligned}E_\mathrm{f} &= E_{a_\mathrm{f}}+(n_{\boldsymbol{k}\sigma}-1)\hbar\omega_{\boldsymbol{k}\sigma}\,,\\ E_\mathrm{i} &= E_{a_\mathrm{i}}+n_{\boldsymbol{k}\sigma}\hbar\omega_{\boldsymbol{k}\sigma}\,,\end{aligned}$$

so that the δ function can be rewritten as

$$\begin{aligned}\delta(E_{\mathrm{f}} - E_{\mathrm{i}}) &= \delta(E_{a_{\mathrm{f}}} - E_{a_{\mathrm{i}}} - \hbar\omega_{\boldsymbol{k}\sigma}) \\ &= \delta(E_{a_{\mathrm{i}}} + \hbar\omega_{\boldsymbol{k}\sigma} - E_{a_{\mathrm{f}}}) \,. \end{aligned} \tag{2.40}$$

Inserting this into (2.39) and making use of the relation

$$\langle a_{\mathrm{f}}|\hat{\boldsymbol{p}}\cdot\boldsymbol{\varepsilon}_{\boldsymbol{k}\sigma}\,\mathrm{e}^{\mathrm{i}\boldsymbol{k}\cdot\boldsymbol{x}}|a_{\mathrm{i}}\rangle = \langle a_{\mathrm{i}}|\hat{\boldsymbol{p}}\cdot\boldsymbol{\varepsilon}_{\boldsymbol{k}\sigma}\,\mathrm{e}^{-\mathrm{i}\boldsymbol{k}\cdot\boldsymbol{x}}|a_{\mathrm{f}}\rangle^*$$

yields

$$\begin{aligned}&\left(\frac{\text{trans. prob.}}{\text{time}}\right)_{\text{absorption}} \\ &= \frac{2\pi}{\hbar}\left(\frac{e}{mc}\right)^2\left(\frac{2\pi\hbar c^2}{L^3\omega_k}\right) n_{\boldsymbol{k}\sigma}\left|\langle a_{\mathrm{i}}|\hat{\boldsymbol{p}}\cdot\boldsymbol{\varepsilon}_{\boldsymbol{k}\sigma}\,\mathrm{e}^{-\mathrm{i}\boldsymbol{k}\cdot\boldsymbol{x}}|a_{\mathrm{f}}\rangle\right|^2 \\ &\times\ \delta(E_{a_{\mathrm{i}}} + \hbar\omega_{\boldsymbol{k}\sigma} - E_{a_{\mathrm{f}}}) \,. \end{aligned} \tag{2.41}$$

A comparison with (2.16) for the emission process shows that the *transition probability for stimulated emission, which is the term in* (2.16) *proportional to* $n_{\boldsymbol{k}\sigma}$, *is identical to the absorption probability per unit time* (2.41). Observe that $|a_{\mathrm{i}}\rangle, |a_{\mathrm{f}}\rangle$ in (2.16) are the same states as $|a_{\mathrm{f}}\rangle$ and $|a_{\mathrm{i}}\rangle$ in (2.41), respectively. The final state (ground state of the atom) for emission is identical to the initial state (again the ground state of the atom) for absorption.

We will calculate the *cross section* $\sigma_{\mathrm{i}\to\mathrm{f}}(\boldsymbol{k}\sigma)$ for the absorption of a photon with momentum $\hbar\boldsymbol{k}$ and polarization σ. It is defined as the *transition probability per unit time for the absorption of a photon divided by the incoming photon flux* j_{photon}:

$$j_{\text{photon}} = \frac{n_{\boldsymbol{k}\sigma}}{L^3}c \,. \tag{2.42}$$

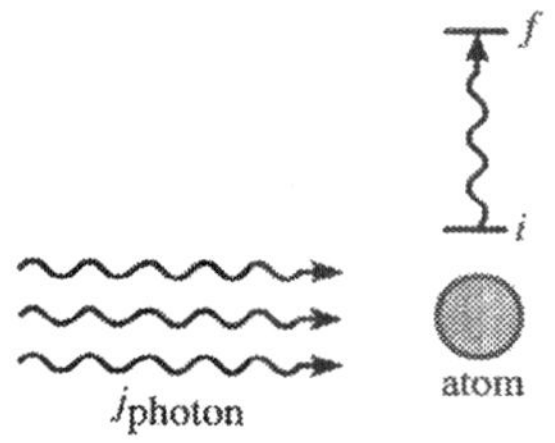

Fig. 2.8. Illustration of the absorption process

This yields

$$\begin{aligned}\sigma_{\mathrm{i}\to\mathrm{f}}(\boldsymbol{k}\sigma) &= \frac{4\pi^2 e^2}{m^2\omega_k c}\left|\langle a_{\mathrm{f}}|\hat{\boldsymbol{p}}\cdot\boldsymbol{\varepsilon}_{\boldsymbol{k}\sigma}\,\mathrm{e}^{\mathrm{i}\boldsymbol{k}\cdot\boldsymbol{x}}|a_{\mathrm{i}}\rangle\right|^2 \\ &\times \delta(E_{a_{\mathrm{i}}} + \hbar\omega_k - E_{a_{\mathrm{f}}}) \,. \end{aligned} \tag{2.43}$$

The incoming radiation will possess a certain frequency spectrum. If we integrate over this spectrum, the δ function appearing in (2.43) will disappear and, instead, the intensity of the absorbed photons from the radiation will appear in (2.43). Soon we will recognize that the atomic levels (states of the particle system) are not sharp either; so far we have assumed the opposite, which expresses itself in terms of the δ function appearing in (2.43). The atomic levels have a *natural linewidth*; see Sect. 2.7.

EXERCISE

2.7 Differential Cross Section $\mathrm{d}\sigma/\mathrm{d}\Omega$ for Photoelectric Emission of an Electron in the Hydrogen Atom (Dipole Approximation)

Problem. Consider the photoelectric emission of an electron out of the ground state of a hydrogen atom. Assume that the incoming photon has a very large energy, so that the wavefunction of the electron to be emitted can be approximated as a plane wave. Furthermore, assume that the photon momentum is parallel to the x axis and that the polarization vector points in the direction of the z axis. Make use of the dipole approximation and calculate the differential cross section for the emission of an electron into the solid angle element $\mathrm{d}\Omega$. Neglect the electron spin, because it is irrelevant in this process.

Solution. According to (2.41) the transition probability per unit time is given by

$$\frac{2\pi}{\hbar}\left(\frac{e}{mc}\right)^2\left(\frac{2\pi\hbar c}{L^3\omega_k}\right) n_{\boldsymbol{k}\sigma}\left|\langle a|\hat{\boldsymbol{p}}\cdot\boldsymbol{\varepsilon}_{\boldsymbol{k}\sigma}\,\mathrm{e}^{\mathrm{i}\boldsymbol{k}\cdot\boldsymbol{r}}|b\rangle\right|^2\delta(E_b+\hbar\omega_k-E_a)\,.$$

Here $|b\rangle=|1\mathrm{s}\rangle$ and $|a\rangle$ is described by a plane wave

$$|a\rangle \;=\; \psi_q \;=\; \langle\boldsymbol{r}\,|\,\boldsymbol{q}\rangle \;=\; \frac{\mathrm{e}^{\mathrm{i}\boldsymbol{q}\cdot\boldsymbol{r}}}{\sqrt{L^3}}\,.$$

The energy of the latter is purely kinetic and is given by

$$E_a \;=\; \frac{\hbar^2q^2}{2m}\,,$$

so that we arrive at

$$\frac{2\pi}{\hbar}\left(\frac{e}{mc}\right)^2\left(\frac{2\pi\hbar c}{L^3\omega_k}\right) n_{\boldsymbol{k}\sigma}\left|\langle q|\hat{\boldsymbol{p}}\cdot\boldsymbol{\varepsilon}_{\boldsymbol{k}\sigma}\,\mathrm{e}^{\mathrm{i}\boldsymbol{k}\cdot\boldsymbol{r}}|1\mathrm{s}\rangle\right|^2\delta\left(\frac{\hbar^2q^2}{2m}-E_{1\mathrm{s}}-\hbar\omega_k\right)\,,$$

with the 1s wavefunction

$$\psi_{1\mathrm{s}}(\boldsymbol{r}) \;=\; \langle\boldsymbol{r}\,|\,1\mathrm{s}\rangle \;=\; \frac{1}{\sqrt{\pi a^3}}\,\mathrm{e}^{-r/a}\,.$$

The cross section reads [compare with (2.43)]

$$\sigma(\boldsymbol{k}\sigma) \;=\; \sum_{\text{final states}}\frac{4\pi^2e^2}{m^2\omega_k c}\left|\langle q|\hat{\boldsymbol{p}}\cdot\boldsymbol{\varepsilon}_{\boldsymbol{k}\sigma}\,\mathrm{e}^{\mathrm{i}\boldsymbol{k}\cdot\boldsymbol{r}}|1\mathrm{s}\rangle\right|^2 \times\delta\left(\frac{\hbar^2q^2}{2m}-E_{1\mathrm{s}}-\hbar\omega_k\right)\,.$$

We replace the sum over the final states by

$$\sum_k \;\longrightarrow\; \frac{L^3}{(2\pi)^3}\int q^2\,\mathrm{d}q\,\mathrm{d}\Omega_\mathrm{e}\,,$$

where $\mathrm{d}^3q=q^2\,\mathrm{d}q\,\mathrm{d}\Omega_\mathrm{e}$ represents the volume element in q space and $\mathrm{d}\Omega_\mathrm{e}=\sin\vartheta\,\mathrm{d}\vartheta\,\mathrm{d}\varphi$ the solid angle element for the electron. Thus,

Exercise 2.7.

$$\frac{\mathrm{d}\sigma}{\mathrm{d}\Omega}(\boldsymbol{k},\sigma) = \sum_{\sigma} \frac{L^3}{2\pi} \frac{e^2}{m^2 \omega_k c} \int q^2 \,\mathrm{d}q \, |\langle q|\hat{\boldsymbol{p}} \cdot \boldsymbol{\varepsilon}_{\boldsymbol{k}\sigma}|1\mathrm{s}\rangle|^2 \times \delta\left(\frac{\hbar^2 q^2}{2m} - E_{1\mathrm{s}} - \hbar\omega_k\right) .$$

We have employed the dipole approximation, i.e. $\boldsymbol{k} \cdot \boldsymbol{r} \ll 1$, and therefore

$$\mathrm{e}^{\mathrm{i}\boldsymbol{k}\cdot\boldsymbol{r}} \approx 1 .$$

Choosing $\boldsymbol{\varepsilon}_{\boldsymbol{k}\sigma} = \boldsymbol{e}_z$, so that the photon is linearly polarized in the z direction, one obviously gets

$$\hat{\boldsymbol{p}} \cdot \boldsymbol{\varepsilon}_{\boldsymbol{k}\sigma} = \frac{\hbar}{i} \boldsymbol{\nabla} \cdot \boldsymbol{e}_z = \frac{\hbar}{i} \frac{\partial}{\partial z}$$

and

$$\frac{\mathrm{d}\sigma}{\mathrm{d}\Omega}(\boldsymbol{k},\sigma) = \frac{L^3}{2\pi} \frac{e^2}{m^2 \omega_k c} \int q^2 \,\mathrm{d}q \left|\left\langle q \left| \frac{\hbar}{\mathrm{i}} \frac{\partial}{\partial z} \right| 1\mathrm{s} \right\rangle\right|^2 \times \delta\left(\frac{\hbar^2 q^2}{2m} - E_{1\mathrm{s}} - \hbar\omega\right) .$$

Now

$$\frac{\partial}{\partial z} = \cos\vartheta \frac{\partial}{\partial r} ,$$

$$r_\mathrm{e} = \frac{e^2}{mc^2} ,$$

$$\omega_k = ck ,$$

and

$$\frac{\mathrm{d}\sigma}{\mathrm{d}\Omega}(\boldsymbol{k},\sigma) = \frac{L^3}{2\pi} r_\mathrm{e} \frac{\hbar^2}{m} \frac{1}{ka^2} \int q^2 \,\mathrm{d}q \, |\langle q| \cos\vartheta |1\mathrm{s}\rangle|^2 \times \delta\left(\frac{\hbar^2 q^2}{2m} - E_{1\mathrm{s}} - \hbar\omega\right) .$$

Since $\delta(\alpha x) = (1/|\alpha|)\delta(x)$ we obtain

$$\delta\left(\frac{\hbar^2 q^2}{2m} - E_{1\mathrm{s}} - \hbar\omega\right) = \frac{2m}{\hbar^2} \delta\left(q^2 - \frac{2m}{\hbar^2}(E_{1\mathrm{s}} + \hbar\omega)\right) = \frac{2m}{\hbar^2} \delta(q^2 - q_0^2)$$

for the δ function, where q_0^2 stands for $q_0^2 = (2m/\hbar^2)(E_{1\mathrm{s}} + \hbar\omega)$. Furthermore,

$$\delta(q^2 - q_0^2) = \frac{1}{2q_0} \left[\delta(q - q_0) + \delta(q + q_0)\right] .$$

Because q, $q_0 > 0$, it is impossible for their sum, $q + q_0$, to become zero; as a consequence, $\delta(q + q_0) = 0$ and

$$\delta\left(\frac{\hbar^2 q^2}{2m} - E_{1\mathrm{s}} - \hbar\omega\right) = \frac{2m}{\hbar^2} \frac{1}{2q_0} \delta(q - q_0) .$$

It follows that

Exercise 2.7.

$$\frac{\mathrm{d}\sigma}{\mathrm{d}\Omega}=\frac{L^3}{2\pi}r_\mathrm{e}\frac{2}{ka^2}\int q^2\,\mathrm{d}q\,|\langle q|\cos\vartheta|1\mathrm{s}\rangle|^2\frac{1}{2q_0}\delta(q-q_0)$$

or

$$\frac{\mathrm{d}\sigma}{\mathrm{d}\Omega_\mathrm{e}}=\frac{L^3}{2\pi}r_\mathrm{e}\frac{q_0}{ka^2}|\langle q_0|\cos\vartheta|1\mathrm{s}\rangle|^2\,.$$

We expand the plane waves into spherical harmonics[7]

$$\mathrm{e}^{\mathrm{i}\boldsymbol{q}_0\cdot\boldsymbol{x}}=4\pi\sum_{l=0}^{\infty}\sum_{m=-l}^{+l}i^l j_l(q_0r)Y_{lm}(\vartheta,\varphi)Y_{lm}(\vartheta',\varphi')\,,$$

where ϑ,φ and ϑ',φ' represent the corresponding polar angles of $\boldsymbol{q}_0$ and $\boldsymbol{x}$ with respect to the z axis. We notice also that

$$\cos\vartheta'=\left(\frac{3}{4\pi}\right)^{\frac{1}{2}}Y_{10}(\vartheta',\varphi')\,.$$

Then, we deduce for the matrix element

$$\begin{aligned}M&=\langle q|\cos\vartheta|1\mathrm{s}\rangle\\&=\frac{4\pi}{\sqrt{\pi a^3L^3}}\int\mathrm{d}^3x\sum_{l=0}^{\infty}\sum_{m=-l}^{+l}(-i)^l j_l(q_0r)\\&\quad\times Y^*_{lm}(\vartheta,\varphi)Y^*_{lm}(\vartheta',\varphi')\left(\frac{3}{4\pi}\right)^{\frac{1}{2}}Y_{10}(\vartheta',\varphi')\,\mathrm{e}^{-r/a}\end{aligned}$$

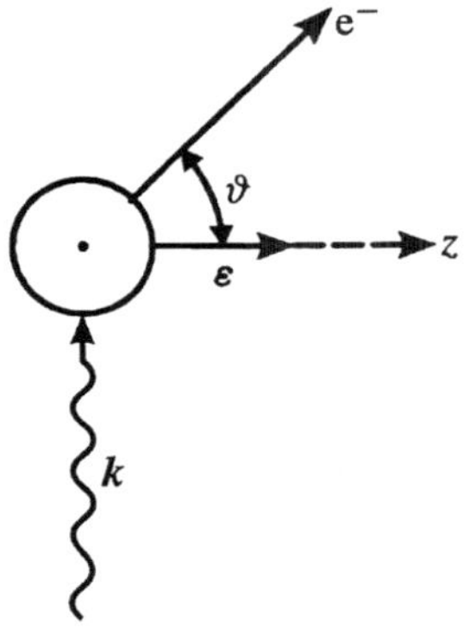

Fig. 2.9. Illustration of the geometry for the photoelectric emission of a photon

or

$$\begin{aligned}M&=\frac{4\pi}{\sqrt{\pi a^3L^3}}\int r^2\,\mathrm{e}^{-r/a}\,\mathrm{d}r\sum_{l=0}^{\infty}\sum_{m=-l}^{+l}(-i)^l j_l(q_0r)Y^*_{lm}(\vartheta,\varphi)\\&\quad\times\int\mathrm{d}\Omega'\,Y^*_{lm}(\vartheta',\varphi')\left(\frac{3}{4\pi}\right)^{\frac{1}{2}}Y_{10}(\vartheta',\varphi')\,,\end{aligned}$$

respectively. Because of the orthonormality relation

$$\int\mathrm{d}\Omega'\,Y^*_{lm}(\vartheta',\varphi')Y_{l'm'}(\vartheta',\varphi')=\delta_{ll'}\delta_{mm'}\,,$$

the summations break down:

$$M=\frac{4\pi}{\sqrt{\pi a^3L^3}}\int r^2\,\mathrm{e}^{-r/a}\,\mathrm{d}r\left[(-i)^l j_l(q_0r)Y^*_{lm}(\vartheta,\varphi)\left(\frac{3}{4\pi}\right)^{\frac{1}{2}}\right]\delta_{l1}\delta_{m0}\,.$$

Since

$$\begin{aligned}\cos\vartheta&=\left[Y_{lm}\left(\frac{3}{4\pi}\right)^{\frac{1}{2}}\right]\delta_{l1}\delta_{m0}\\&=Y_{10}\left(\frac{3}{4\pi}\right)^{\frac{1}{2}}\\&=Y^*_{10}(\vartheta,\varphi)\left(\frac{3}{4\pi}\right)^{\frac{1}{2}}\,,\end{aligned}$$

[7] W. Greiner: *Quantum Mechanics – An Introduction*, 3rd ed. (Springer, Berlin, Heidelberg 1994).

Exercise 2.7.

we get

$$M = -\frac{4\pi \mathrm{i}\cos\vartheta}{\sqrt{\pi a^3 L^3}}\int_0^\infty r^2\,\mathrm{d}r\, j_1(q_0 r)\,\mathrm{e}^{-r/a}\,.$$

Substituting $x = r/a$ yields

$$M = -4\pi \mathrm{i}\cos\vartheta\sqrt{\frac{a^3}{\pi L^3}}f(q_0 a)$$

with

$$\begin{aligned} f(q_0 a) &= \int_0^\infty x^2\,\mathrm{d}x\, j_1(q_0 a x)\,\mathrm{e}^{-x} \\ &= \frac{2q_0 a}{(1+q_0^2 a^2)^2}\,. \end{aligned}$$

With the abbreviations $r_\mathrm{e} = e^2/mc^2$ and $a = \hbar^2/me^2$ the scattering cross section finally takes on the form

$$\begin{aligned} \frac{\mathrm{d}\sigma}{\mathrm{d}\Omega_e} &= \frac{L^3}{2\pi} r_\mathrm{e}\frac{q_0}{ka^2}\frac{16\pi^2\cos^2\vartheta\, a^3}{\pi L^3} f^2(q_0 a) \\ &= 8a^2\left(\frac{e^2}{\hbar c}\right)^2 \cos^2\vartheta\,\frac{q_0}{k} f^2(q_0 a)\,. \end{aligned}$$

The scattering cross section is largest for $\vartheta = 0$, i.e. in the direction of the incoming photon. Most of the electrons are obviously kicked out directly by the photon. Observe, however, that the solid angle $\mathrm{d}\Omega_\mathrm{e} = \sin\vartheta\,\mathrm{d}\vartheta\,\mathrm{d}\varphi$ is zero in this direction; as a consequence the maximum of the electron intensity is shifted towards angles $\vartheta \neq 0$.

EXAMPLE

2.8 Spectrum of Black-Body Radiation

Here we come back to the derivation of Planck's radiation formula, which we considered at the very beginning of our lectures on quantum mechanics.[8] At that time we pursued Einstein's derivation; right now we develop an advanced point of view and explicitly work out some of the assumptions that were employed during the former approach.

We consider a sample of atoms in thermal equilibrium, which means that owing to thermal collisions the number of atoms that are excited is equal to the number of atoms that are deexcited. The number of atoms in state $|f\rangle$ is denoted by N_f and for those in state $|i\rangle$ the number is N_i. Transitions occur between the two states; photons may be absorbed or emitted from the radiation field. Already with this background we understand the following two differential equations:

[8] W. Greiner: *Quantum Mechanics – An Introduction*, 3rd ed. (Springer, Berlin, Heidelberg 1994).

Example 2.8.

$$
\begin{aligned}
\frac{\mathrm{d}N_\mathrm{f}}{\mathrm{d}t} &= -N_\mathrm{f}\left(\frac{\text{trans. prob.}}{\text{time}}\right)_{\text{abs.}} + N_\mathrm{i}\left(\frac{\text{trans. prob.}}{\text{time}}\right)_{\text{em.}} ,\\
\frac{\mathrm{d}N_\mathrm{i}}{\mathrm{d}t} &= -N_\mathrm{i}\left(\frac{\text{trans. prob.}}{\text{time}}\right)_{\text{em.}} + N_\mathrm{f}\left(\frac{\text{trans. prob.}}{\text{time}}\right)_{\text{abs.}} .
\end{aligned}
$$

They express the coupling between the accumulations of both states according to the radiation. We assume that a coupling of atoms takes place only via the radiation field and that collisions between atoms do not contribute to the transitions. Equilibrium requires that

$$\dot{N}_\mathrm{f} = \dot{N}_\mathrm{i} = 0$$

and

$$\frac{N_\mathrm{f}}{N_\mathrm{i}} = \frac{\exp\left(-\frac{E_\mathrm{f}}{k_\mathrm{B}T}\right)}{\exp\left(-\frac{E_\mathrm{i}}{k_\mathrm{B}T}\right)} = \exp\left(-\frac{(E_\mathrm{f}-E_\mathrm{i})}{k_\mathrm{B}T}\right) .$$

Here T represents temperature and k_B is Boltzmann's constant. With the first two equations it then follows that

$$\frac{N_\mathrm{f}}{N_\mathrm{i}} = \frac{(\text{trans. prob./time})_{\text{em.}}}{(\text{trans. prob./time})_{\text{abs.}}} = \frac{n_{\boldsymbol{k}\sigma}+1}{n_{\boldsymbol{k}\sigma}} = \mathrm{e}^{\hbar\omega_k/k_\mathrm{B}T} ,$$

where we have used (2.41), (2.16), and $E_\mathrm{i} - E_\mathrm{f} = \hbar\omega_k$. Solving for $n_{\boldsymbol{k}\sigma}$ results in

$$n_{\boldsymbol{k}\sigma} = \frac{1}{\exp\left(\frac{\hbar\omega_k}{k_\mathrm{B}T}\right) - 1} .$$

This is Planck's distribution, from which the radiation energy E_rad in the volume L^3 immediately follows:

$$\mathrm{d}E_\mathrm{rad} = 2\frac{\hbar\omega_k}{L^3} n_{\boldsymbol{k}\sigma}\frac{L^3}{(2\pi)^3}\,\mathrm{d}^3k ,$$

where $\boldsymbol{k}$ is the wavevector in d^3k. Furthermore, a factor 2 for the polarization has been inserted and (2.22) has been used. Now we can define a *spectral function* $u(\omega)$, which describes the radiation energy per volume in the frequency range between ω and $\omega + \mathrm{d}\omega$. Making use of (2.29) we obtain

$$
\begin{aligned}
u(\omega)\,\mathrm{d}\omega &= \int_{\vartheta,\varphi} 2\hbar\omega_k n_{\boldsymbol{k}\sigma}\frac{1}{(2\pi)^3}\frac{\omega_k^2\,\mathrm{d}\omega_k}{c^3}\sin\vartheta\,\mathrm{d}\vartheta\,\mathrm{d}\varphi\\
&= \frac{8\pi\hbar\omega_k^3\,\mathrm{d}\omega_k}{(2\pi)^3c^3\left(\exp\left(\frac{\hbar\omega}{k_\mathrm{B}T}\right)-1\right)}\\
&= \frac{\hbar\omega_k^3}{\pi^2c^3}\frac{1}{\left(\exp\left(\frac{\hbar\omega}{k_BT}\right)-1\right)}\,\mathrm{d}\omega_k . \qquad (1)
\end{aligned}
$$

This agrees with an earlier result derived in a more elementary way.[9]

[9] W. Greiner: *Quantum Mechanics – An Introduction*, 3rd ed. (Springer, Berlin, Heidelberg 1994).

2.4 Photon Scattering from Free Electrons

Free, noninteracting particles are described by the Hamiltonian

$$\hat{H} = \frac{\hat{\boldsymbol{p}}^2}{2m} . \tag{2.44}$$

Its normalized eigenstates are denoted by $|\boldsymbol{q}\rangle$ and are confined to a volume L^3:

$$\langle \boldsymbol{x}|\boldsymbol{q}\rangle = \psi_{\boldsymbol{q}}(\boldsymbol{x}) = \frac{1}{\sqrt{L^3}} \mathrm{e}^{\mathrm{i}\boldsymbol{q}\cdot\boldsymbol{x}} . \tag{2.45}$$

The momentum vectors $\hbar\boldsymbol{q}$ obey periodic boundary conditions at the surface of L^3, which is in close analogy to the case of electromagnetic waves [compare with (1.30)].

The normalization is given by

$$\int_{L^3} \mathrm{d}^3x\, \psi_{\boldsymbol{q}}^*(\boldsymbol{x})\psi_{\boldsymbol{q}'}(\boldsymbol{x}) = \delta_{\boldsymbol{q},\boldsymbol{q}'} . \tag{2.46}$$

The energy eigenvalues belonging to (2.44) are

$$E_q = \frac{(\hbar q)^2}{2m} . \tag{2.47}$$

At first, we consider the process of *a free electron absorbing a photon.* The initial and final states are given by

$$\begin{aligned} |i\rangle &= |\boldsymbol{q}_\mathrm{i}\rangle|\ldots, n_{\boldsymbol{k}\sigma}, \ldots\rangle , \\ |f\rangle &= |\boldsymbol{q}_\mathrm{f}\rangle|\ldots, n_{\boldsymbol{k}\sigma} - 1, \ldots\rangle . \end{aligned} \tag{2.48}$$

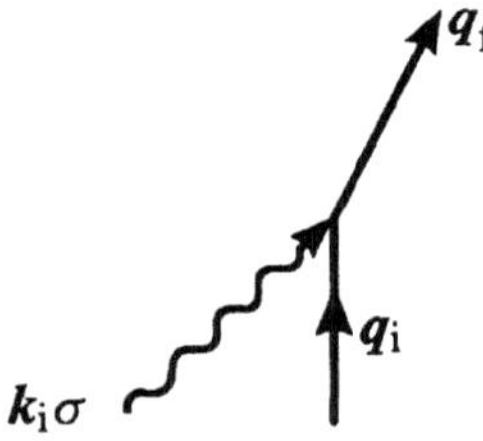

Fig. 2.10. Absorption and emission of a photon by a free electron is prohibited. The photon is indicated by a wavy line, the electron by a straight line. The interaction point is called vertex

Transitions between these two states can only be caused by $\hat{H}'_\mathrm{int}$ from (2.8). We deduce that

$$\begin{aligned} \langle f|\hat{H}'_\mathrm{int}|i\rangle &= \langle \boldsymbol{q}_\mathrm{f}|\langle\ldots, n_{\boldsymbol{k}\sigma} - 1, \ldots| \left(-\frac{e}{mc}\right) \hat{\boldsymbol{p}} \cdot \boldsymbol{\varepsilon}_{\boldsymbol{k}\sigma}\sqrt{\frac{2\pi\hbar c}{L^3\omega_k}} \\ &\quad \times \left(\hat{a}_{\boldsymbol{k}\sigma}\,\mathrm{e}^{\mathrm{i}\boldsymbol{k}\cdot\boldsymbol{x}} + \hat{a}^\dagger_{\boldsymbol{k}\sigma}\,\mathrm{e}^{-\mathrm{i}\boldsymbol{k}\cdot\boldsymbol{x}}\right)|\boldsymbol{q}_\mathrm{i}\rangle|\ldots, n_{\boldsymbol{k}\sigma}, \ldots\rangle \\ &= -\frac{e}{mc}\sqrt{\frac{2\pi\hbar c}{L^3\omega_k}}\langle \boldsymbol{q}_\mathrm{f}|\hat{\boldsymbol{p}} \cdot \boldsymbol{\varepsilon}_{\boldsymbol{k}\sigma}\,\mathrm{e}^{\mathrm{i}\boldsymbol{k}\cdot\boldsymbol{x}}|\boldsymbol{q}_i\rangle\sqrt{n_{\boldsymbol{k}\sigma}} \\ &= -\frac{e}{mc}\sqrt{\frac{2\pi\hbar c}{L^3\omega_k}}(-\boldsymbol{q}_\mathrm{f} \cdot \boldsymbol{\varepsilon}_{\boldsymbol{k}\sigma})\sqrt{n_{\boldsymbol{k}\sigma}} \\ &\quad \times \int_{L^3} \frac{\mathrm{e}^{-\mathrm{i}\boldsymbol{q}_\mathrm{f}\cdot\boldsymbol{x}}}{L^{3/2}}\,\mathrm{e}^{\mathrm{i}\boldsymbol{k}\cdot\boldsymbol{x}}\,\frac{\mathrm{e}^{\mathrm{i}\boldsymbol{q}_\mathrm{i}\cdot\boldsymbol{x}}}{L^{3/2}}\,\mathrm{d}^3x . \end{aligned} \tag{2.49}$$

The integral yields

$$\int_{L^3} \frac{\mathrm{e}^{-\mathrm{i}\boldsymbol{q}_\mathrm{f}\cdot\boldsymbol{x}}}{\sqrt{L^3}}\,\frac{\mathrm{e}^{\mathrm{i}(\boldsymbol{q}_\mathrm{i}+\boldsymbol{k})\cdot\boldsymbol{x}}}{\sqrt{L^3}}\,\mathrm{d}^3x = \delta_{\boldsymbol{q}_\mathrm{f},\boldsymbol{q}_\mathrm{i}+\boldsymbol{k}} . \tag{2.50}$$

This reflects momentum conservation, i.e.

$$\hbar\boldsymbol{q}_\mathrm{f} = \hbar\boldsymbol{q}_\mathrm{i} + \hbar\boldsymbol{k} . \tag{2.51}$$

Furthermore, another restriction has to be considered for the transition probability: According to (2.13a) for the transition probability per unit time conservation of energy

$$E_f = E_i \quad \text{or} \quad \frac{(\hbar q_f)^2}{2m} = \hbar\omega_k + \frac{(\hbar q_i)^2}{2m} \tag{2.52}$$

is a result of the δ function.

It is straightforward to show that the two equations (2.51) and (2.52) cannot be fulfilled simultaneously. The best way to prove this is to argue relativistically. We remember that the norm of four-vectors is constant. Also we simplify the situation with $c = 1$. For four-vectors it holds that

$$(p_e + p_\gamma) = p_{e'}, \qquad (p_e + p_\gamma)^2 = p_{e'}^2 = p_e^2 .$$

Since $p_e^2 = m_e^2$ and $p_\gamma^2 = 0$ it follows that $0 + 2p_\gamma p_e + m_e^2 = m_e^2$, so that $p_\gamma p_e = 0$. In the rest frame of the electron the four-vectors become

$$p_e = (m_e, \mathbf{0}), \qquad p_\gamma = (E_\gamma, \mathbf{p}_\gamma), \qquad p_\gamma p_e = m_e E_\gamma = 0 .$$

Hence, it follows that

$$E_\gamma = 0 ;$$

which means that no photon is present. Therefore it is impossible for a free electron to absorb a photon. We have to draw the conclusion that processes of first order caused by $\hat{H}'_{\text{int}}$ do not exist. As a consequence we investigate now the *processes of first order* with $\hat{H}''_{\text{int}}$; see (2.9). This part of the interaction contains terms of the form $\hat{a}^\dagger_{\boldsymbol{k}\sigma}\hat{a}_{\boldsymbol{k}'\sigma'}$. Obviously, they describe processes in which a photon of the kind $\boldsymbol{k}'\sigma'$ is absorbed and another photon $\boldsymbol{k}\sigma$ is emitted. Such photon scattering is best illustrated with Feynman diagrams as in Fig. 2.11.

The full lines describe the electron, whereas the wave-like lines represent the photon. The graph is to be read from bottom to top in accordance with the direction of time. The incoming photon $\boldsymbol{k}_i\sigma_i$ is absorbed (annihilated) by the incoming electron. Another photon $\boldsymbol{k}_f\sigma_f$ is produced. In this annihilation–creation process the electron scatters into the state $\boldsymbol{q}_f$. The initial and final states for this process are

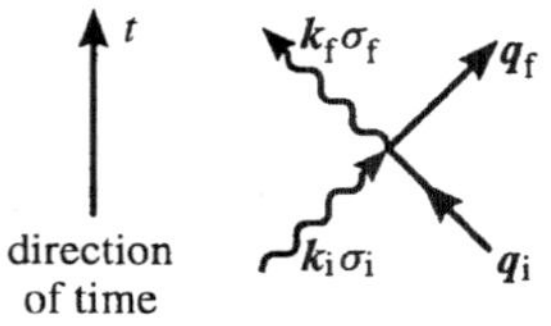

Fig. 2.11. Sketch of a process, where an electron in state $|\boldsymbol{q}_i\rangle$ absorbs a photon $\boldsymbol{k}_i\sigma_i$ and emits another photon $\boldsymbol{k}_f\sigma_f$, so that the electron scatters into state $|\boldsymbol{q}_f\rangle$

$$\begin{aligned} |i\rangle &= |\boldsymbol{q}_i\rangle\, |\ldots, n_{\boldsymbol{k}_i\sigma_i}, \ldots, n_{\boldsymbol{k}_f\sigma_f}, \ldots\rangle , \\ |f\rangle &= |\boldsymbol{q}_f\rangle\, |\ldots, n_{\boldsymbol{k}_i\sigma_i} - 1, \ldots, n_{\boldsymbol{k}_f\sigma_f} + 1, \ldots\rangle . \end{aligned} \tag{2.53}$$

The transition from $|i\rangle$ to $|f\rangle$ is caused by the two terms with $\hat{a}^\dagger_{\boldsymbol{k}_f\sigma_f}\hat{a}_{\boldsymbol{k}_i\sigma_i}$ in (2.9). According to (2.9) and (2.13a) the transition probability per unit time becomes

$$\begin{aligned} &\left(\frac{\text{trans. prob.}}{\text{time}}\right)_{\text{photon scattering}} \\ &= \frac{2\pi}{\hbar}\left(\frac{e^2}{2mc^2}\right)^2 \left(\frac{2\pi\hbar c^2}{L^3}\right)^2 \frac{|\boldsymbol{\varepsilon}_{\boldsymbol{k}_f\sigma_f}\cdot\boldsymbol{\varepsilon}_{\boldsymbol{k}_i\sigma_i}|^2}{\omega_{k_i}\omega_{k_f}} \\ &\times \left|\langle\boldsymbol{q}_f|\,\langle\ldots, n_{\boldsymbol{k}_i\sigma_i} - 1, \ldots, n_{\boldsymbol{k}_f\sigma_f} + 1, \ldots\right| \end{aligned}$$

$$\times \left(\hat{a}_{\boldsymbol{k}_\mathrm{i}\sigma_\mathrm{i}}\hat{a}^\dagger_{\boldsymbol{k}_\mathrm{f}\sigma_\mathrm{f}}\mathrm{e}^{\mathrm{i}(\boldsymbol{k}_\mathrm{i}-\boldsymbol{k}_\mathrm{f})\cdot\boldsymbol{x}} + \hat{a}^\dagger_{\boldsymbol{k}_\mathrm{f}\sigma_\mathrm{f}}\hat{a}_{\boldsymbol{k}_\mathrm{i}\sigma_\mathrm{i}}\mathrm{e}^{\mathrm{i}(\boldsymbol{k}_\mathrm{i}-\boldsymbol{k}_\mathrm{f})\cdot\boldsymbol{x}}\right)$$
$$\times |\boldsymbol{q}_\mathrm{i}\rangle\, |\ldots, n_{\boldsymbol{k}_\mathrm{i}\sigma_\mathrm{i}}, \ldots, n_{\boldsymbol{k}_\mathrm{f}\sigma_\mathrm{f}}, \ldots\rangle|^2$$
$$\times\, \delta\left(\hbar\omega_\mathrm{i} + \frac{(\hbar\boldsymbol{q}_\mathrm{i})^2}{2m} - \hbar\omega_\mathrm{f} - \frac{(\hbar\boldsymbol{q}_\mathrm{f})^2}{2m}\right) . \tag{2.54}$$

Owing to (1.56) and (1.57) the matrix element becomes

$$\begin{aligned} |M_\mathrm{fi}|^2 &= \left|\langle\boldsymbol{q}_\mathrm{f}|\,\mathrm{e}^{\mathrm{i}(\boldsymbol{k}_\mathrm{i}-\boldsymbol{k}_\mathrm{f})\cdot\boldsymbol{x}}\,|\boldsymbol{q}_\mathrm{i}\rangle\right|^2 \\ &\quad \times \left(\sqrt{n_{\boldsymbol{k}_\mathrm{i}\sigma_\mathrm{i}}}\sqrt{n_{\boldsymbol{k}_\mathrm{f}\sigma_\mathrm{f}}+1} + \sqrt{n_{\boldsymbol{k}_\mathrm{i}\sigma_\mathrm{i}}}\sqrt{n_{\boldsymbol{k}_\mathrm{f}\sigma_\mathrm{f}}+1}\right)^2 \\ &= 4n_{\boldsymbol{k}_\mathrm{i}\sigma_\mathrm{i}}(n_{\boldsymbol{k}_\mathrm{f}\sigma_\mathrm{f}}+1)\left|\langle\boldsymbol{q}_\mathrm{f}|\mathrm{e}^{\mathrm{i}(\boldsymbol{k}_\mathrm{i}-\boldsymbol{k}_\mathrm{f})\cdot\boldsymbol{x}}|\boldsymbol{q}_\mathrm{i}\rangle\right|^2 , \end{aligned} \tag{2.55}$$

so that (2.54) results in

$$\begin{aligned} &\left(\frac{\text{trans. prob.}}{\text{time}}\right)_{\text{photon scattering}} \\ &= \frac{8\pi}{\hbar}\left(\frac{e^2}{2mc^2}\right)^2\left(\frac{2\pi\hbar c^2}{L^3}\right)^2\frac{|\boldsymbol{\varepsilon}_{\boldsymbol{k}_\mathrm{f}\sigma_\mathrm{f}}\cdot\boldsymbol{\varepsilon}_{\boldsymbol{k}_\mathrm{i}\sigma_\mathrm{i}}|^2}{\omega_{\boldsymbol{k}_\mathrm{i}}\omega_{\boldsymbol{k}_\mathrm{f}}} \\ &\quad\times \left|\langle\boldsymbol{q}_\mathrm{f}|\mathrm{e}^{\mathrm{i}(\boldsymbol{k}_\mathrm{i}-\boldsymbol{k}_\mathrm{f})\cdot\boldsymbol{x}}|\boldsymbol{q}_\mathrm{i}\rangle\right|^2 n_{\boldsymbol{k}_\mathrm{i}\sigma_\mathrm{i}}(n_{\boldsymbol{k}_\mathrm{f}\sigma_\mathrm{f}}+1) \\ &\quad\times\,\delta\left(\hbar\omega_{\boldsymbol{k}_\mathrm{i}} + \frac{(\hbar\boldsymbol{q}_\mathrm{i})^2}{2m} - \hbar\omega_{\boldsymbol{k}_\mathrm{f}} - \frac{(\hbar\boldsymbol{q}_\mathrm{f})^2}{2m}\right) . \end{aligned} \tag{2.56}$$

The matrix element between the electron states is calculated to be

$$\begin{aligned} \langle\boldsymbol{q}_f|\,\mathrm{e}^{\mathrm{i}(\boldsymbol{k}_\mathrm{i}-\boldsymbol{k}_\mathrm{f})\cdot\boldsymbol{x}}\,|\boldsymbol{q}_\mathrm{i}\rangle &= \int \mathrm{d}^3x\,\frac{\mathrm{e}^{-\mathrm{i}(\boldsymbol{q}_\mathrm{f}+\boldsymbol{k}_\mathrm{f})\cdot\boldsymbol{x}}}{\sqrt{L^3}}\,\frac{\mathrm{e}^{\mathrm{i}(\boldsymbol{q}_\mathrm{i}+\boldsymbol{k}_\mathrm{i})\cdot\boldsymbol{x}}}{\sqrt{L^3}} \\ &= \delta_{\boldsymbol{q}_\mathrm{f}+\boldsymbol{k}_\mathrm{f},\boldsymbol{k}_\mathrm{i}+\boldsymbol{q}_\mathrm{i}}\,. \end{aligned} \tag{2.57}$$

It expresses momentum conservation:

$$\hbar(\boldsymbol{q}_\mathrm{f}+\boldsymbol{k}_\mathrm{f}) = \hbar(\boldsymbol{q}_\mathrm{i}+\boldsymbol{k}_\mathrm{i})\,. \tag{2.58}$$

Energy conservation is contained in the δ function of (2.56):

$$\hbar\omega_{\boldsymbol{k}_\mathrm{i}} + \frac{(\hbar\boldsymbol{q}_\mathrm{i})^2}{2m} = \hbar\omega_{\boldsymbol{k}_\mathrm{f}} + \frac{(\hbar\boldsymbol{q}_\mathrm{f})^2}{2m}\,. \tag{2.59}$$

The occurence of the factor $n_{\boldsymbol{k}_\mathrm{f}\sigma_\mathrm{f}}+1$ in (2.56) is interpreted as stimulated emission of photons $\boldsymbol{k}_\mathrm{f}\sigma_\mathrm{f}$ in the final state; apparently the scattering is enhanced if photons $\boldsymbol{k}_\mathrm{f}\sigma_\mathrm{f}$ of the final state are already present. Often this is not the case, so we set $n_{\boldsymbol{k}_\mathrm{f}\sigma_\mathrm{f}} = 0$.

2.5 Calculation of the Total Photon Scattering Cross Section

To calculate the total photon scattering cross section σ_{total}, we sum (2.56) over all the final states of the photon as well as the final states of the electrons; this result is then divided by the incoming photon flux $n_{\boldsymbol{k}_\mathrm{i}\sigma}c/L^3$. With $n_{\boldsymbol{k}_\mathrm{f}\sigma_\mathrm{f}} = 0$, this yields

$$\begin{aligned}\sigma_{\text{total}} &= \frac{1}{n_{k_i\sigma}c/L^3}\frac{2\pi}{\hbar}\left(\frac{e^2}{2mc^2}\right)^2\left(\frac{2\pi\hbar c^2}{L^3}\right)^2 \\ &\times 4\sum_{k_f\sigma_f}\sum_{q_f}\frac{|\varepsilon_{k_f\sigma_f}\cdot\varepsilon_{k_i\sigma_i}|^2}{\omega_{k_i}\omega_{k_f}}n_{k_i\sigma_i}\delta_{q_f+k_f,q_i+k_i} \\ &\times \delta\left(\hbar\omega_{k_i}+\frac{(\hbar q_i)^2}{2m}-\hbar\omega_{k_f}-\frac{(\hbar q_f)^2}{2m}\right). \qquad (2.60)\end{aligned}$$

If we employ the Kronecker delta function $q_f = q_i + k_i - k_f$, the sum over q_f breaks down. Furthermore, we make use of the relation

$$\sum_{k_f} \underset{L^3\to\infty}{\longrightarrow} \frac{L^3}{(2\pi)^3}\int d^3k_f\,,$$

which is proven after (2.18), and get

$$\begin{aligned}\sigma_{\text{total}} &= \frac{\hbar e^4}{m^2c}\sum_{\sigma_f}\int d^3k_f\,\frac{|\varepsilon_{k_f\sigma_f}\cdot\varepsilon_{k_i\sigma_i}|^2}{\omega_{k_i}\omega_{k_f}} \\ &\times \delta\left(\hbar c(k_i-k_f)+\frac{(\hbar q_i)^2}{2m}-\frac{\hbar^2(q_i+k_i-k_f)^2}{2m}\right). \qquad (2.61)\end{aligned}$$

From the two conservation laws, (2.58) and (2.59), it follows that the wavelength of the scattered photon differs from that of the incoming photon by (see Exercise 2.9):

$$\Delta\lambda = \lambda_f - \lambda_i = \left(\frac{h}{mc}\right)(1-\cos\vartheta)\,; \qquad (2.62)$$

here $\cos\vartheta = k_i\cdot k_f/k_ik_f$ represents the cosine of the angle between the incoming and the scattered photon. Evidently this small shift is of pure quantum-mechanical origin; it vanishes for the limiting case $\hbar\to 0$. Hence, we pursue the classical approximation and set $\lambda_i = \lambda_f$. As a consequence the δ function in (2.61) becomes

$$\delta\big(\hbar c(k_i-k_f)\big) = \frac{1}{\hbar c}\delta(k_i-k_f)\,. \qquad (2.63)$$

Here $k = k_i - k_f \approx 0$ has been employed for the energy of the electron:

$$\frac{\hbar^2}{2m}(q_i+k)^2 = \frac{\hbar^2}{2m}q_i^2+\frac{\hbar^2}{2m}k^2+\frac{\hbar^2}{m}q_i\cdot k \approx \frac{\hbar^2}{2m}q_i^2\,. \qquad (2.64)$$

With (2.63) the total photon scattering cross section (2.61) becomes

$$\begin{aligned}\sigma_{\text{total}} &= \left(\frac{\hbar e^4}{m^2c}\right)\sum_{\sigma_f}\int k_f^2\,dk_f\,d\Omega_{k_f}\,\frac{|\varepsilon_{k_i\sigma_i}\cdot\varepsilon_{k_f\sigma_f}|^2}{c^2k_ik_f}\frac{1}{\hbar c}\delta(k_f-k_i) \\ &= \left(\frac{e^2}{mc^2}\right)^2\sum_{\sigma_f}\int d\Omega_{k_f}\,|\varepsilon_{k_i\sigma_i}\cdot\varepsilon_{k_f\sigma_f}|^2\,. \qquad (2.65)\end{aligned}$$

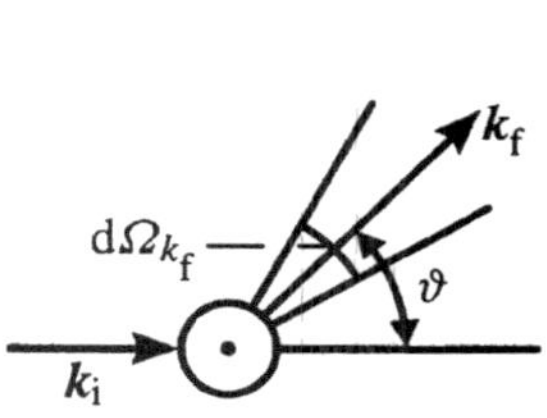

Fig. 2.12. Geometry of the photon scattering

The photon is scattered into the solid angle $d\Omega_f$ (see Fig. 2.12).

The differential cross section for an incoming photon with, polarization $\varepsilon_{k_i\sigma_i}$, to be scattered into the solid angle $d\Omega_{k_f}$ with polarization $\varepsilon_{k_f\sigma_f}$ is given by

$$\frac{d\sigma_{\text{total}}}{d\Omega_{k_f}} = \left(\frac{e^2}{mc^2}\right)^2 |\varepsilon_{k_i\sigma_i} \cdot \varepsilon_{k_f\sigma_f}|^2 . \tag{2.66}$$

If the photons are unpolarized, expression (2.65) has to be averaged over the initial polarizations and has to be summed over the final polarizations to obtain the total cross section. We get (see Fig. 2.13)

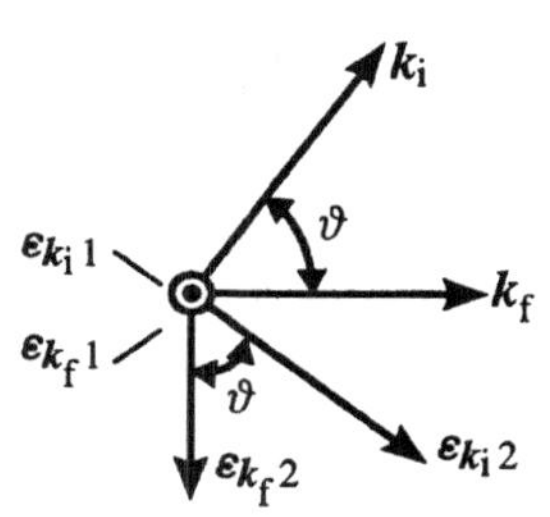

Fig. 2.13. Wavenumber and polarization vectors of the absorbed (i) and emitted (f) photons

$$\begin{aligned}\frac{1}{2}\sum_{\sigma_i}\sum_{\sigma_f} |\varepsilon_{k_i\sigma_i} \cdot \varepsilon_{k_f\sigma_f}|^2 &= \frac{1}{2}\left[|\varepsilon_{k_i1} \cdot \varepsilon_{k_f1}|^2 + |\varepsilon_{k_i1} \cdot \varepsilon_{k_f2}|^2 \right. \\ &\quad \left. + |\varepsilon_{k_i2} \cdot \varepsilon_{k_f1}|^2 + |\varepsilon_{k_i2} \cdot \varepsilon_{k_f2}|^2\right] \\ &= \frac{1}{2}\left[1 + 0 + 0 + \cos^2\vartheta\right] \\ &= \frac{1}{2}\left(1 + \cos^2\vartheta\right)\end{aligned} \tag{2.67}$$

and perform the integration over $d\Omega_f = \sin\vartheta \, d\vartheta \, d\varphi$ in (2.65). The outcome is

$$\sigma_{\text{total}} = \frac{8\pi}{3}\left(\frac{e^2}{mc^2}\right) = \frac{8\pi}{3} r_e^2 \approx 0.66\, b . \tag{2.68}$$

Here, $r_e = e^2/mc^2 \approx 2.8$ fm represents the classical electron radius. Expression (2.68) can also be derived classically and is known as *Thomson's scattering cross section.*

EXERCISE

2.9 The Compton Effect

Problem. Show that for photon scattering a (small) shift in wavelength $\Delta\lambda = \lambda_f - \lambda_i = h(1 - \cos\vartheta)/mc$ occurs as a result of the nonrelativisitic conservation laws for momentum and energy, (2.58) and (2.59). Assume that the electron is initially at rest.

Solution. Since the electron is at rest initially we have $\boldsymbol{q}_i = 0$. With $\hbar ck = \hbar\omega$ we derive from (2.58) and (2.59) that

$$\hbar ck_i = \hbar ck_f + \frac{\hbar^2 q_f^2}{2m} \tag{1}$$

and

$$\boldsymbol{k}_i = \boldsymbol{k}_f + \boldsymbol{q}_f . \tag{2}$$

From the second equation we directly get

$$q_f^2 = (\boldsymbol{k}_i - \boldsymbol{k}_f)^2 = \boldsymbol{k}_i^2 - 2\boldsymbol{k}_i \cdot \boldsymbol{k}_f + \boldsymbol{k}_f^2 .$$

Inserting this into the first equation gives

$$\hbar ck_i = \hbar ck_f + \frac{\hbar^2}{2m}(k_i^2 - 2k_ik_f\cos\vartheta + k_f^2) ,$$

where ϑ represents the angle between $\boldsymbol{k}_i$ and $\boldsymbol{k}_f$. With $\hbar k = h/\lambda$ it follows that

Exercise 2.9.

$$c\frac{h}{\lambda} = c\frac{h}{\lambda'} + \frac{1}{2m}\left[\left(\frac{h}{\lambda}\right)^2 - 2\frac{h}{\lambda}\frac{h}{\lambda'}\cos\vartheta + \left(\frac{h}{\lambda'}\right)^2\right] .$$

Multiplication with $\lambda\lambda'$ results in

$$ch\underbrace{(\lambda' - \lambda)}_{=\Delta\lambda} = \frac{h^2}{2m}\left(\frac{\lambda'}{\lambda} + \frac{\lambda}{\lambda'}\right) - \frac{h^2}{m}\cos\vartheta ,$$

$$\Delta\lambda = \frac{h}{mc}\left[\frac{1}{2}\left(\frac{\lambda'}{\lambda} + \frac{\lambda}{\lambda'}\right) - \cos\vartheta\right] .$$

If we assume that the shift in wavelength stays small, i.e. $\lambda' \approx \lambda$, then

$$\frac{\lambda'}{\lambda} + \frac{\lambda}{\lambda'} = \frac{\lambda'^2 + \lambda^2}{\lambda\lambda'} \approx \frac{2\lambda^2}{\lambda^2} = 2$$

and

$$\Delta\lambda = \frac{h}{mc}(1 - \cos\vartheta) .$$

EXERCISE

2.10 Two-Photon Decay of the 2s State of the Hydrogen Atom

Problem. Determine the lifetime of the 2s state of the hydrogen atom assuming a decay with the emission of *two* photons. It is necessary to combine the matrix element of second order, $\hat{H}' \sim \hat{\boldsymbol{p}} \cdot \hat{\boldsymbol{A}}$, with the matrix element of first order, $\hat{H}'' \sim \hat{\boldsymbol{A}}^2$. Do not solve the problem exactly, only estimate the order of magnitude.

Solution. The lifetime is given by the expression

$$\frac{1}{\tau} = \sum_{\text{final states}} \frac{2\pi}{\hbar}|M|^2\delta(E_{2\text{s}} - E_{1\text{s}} - \hbar ck_1 - \hbar ck_2) ,$$

where $\hbar ck_1$ and $\hbar ck_2$ represent the energies of the emitted photons. The sum over the final states goes with a summation over the polarizations and an integration over the momenta of the *two* photons,

$$\sum_{\text{final states}} \longrightarrow \frac{L^3}{(2\pi)^3}\sum_{\sigma_1}\int \mathrm{d}^3k_1 \frac{L^3}{(2\pi)^3}\sum_{\sigma_2}\int \mathrm{d}^3k_2$$

(L^3 represents the normalization volume) and, consequently,

$$\frac{1}{\tau} = \frac{L^6}{(2\pi)^6}\frac{2\pi}{\hbar^2 c}\sum_{\sigma_1}\sum_{\sigma_2}\int \mathrm{d}^3k_1 \int \mathrm{d}^3k_2 |M|^2 \delta(k_0 - k_1 - k_2) ,$$

where $(E_{2\text{s}} - E_{1\text{s}}) = \hbar ck_0$ and $\delta(\alpha x) = 1/(|\alpha|)\delta(x)$ have been used.

The Hamiltonian entering the matrix element $M = \langle 1\text{s}|\hat{H}_\text{I}|2\text{s}\rangle$ splits into $\hat{H}_\text{I} = \hat{H}' + \hat{H}''$, where according to (2.8) and (2.9)

Exercise 2.10.

$$\hat{H}' = -\frac{e}{mc}\sum_{\boldsymbol{k},\sigma}\left(\frac{2\pi\hbar c^2}{L^3\omega_k}\right)^{\frac{1}{2}}\hat{\boldsymbol{p}}\cdot\boldsymbol{\varepsilon}_{\boldsymbol{k}\sigma}\left(\hat{a}_{\boldsymbol{k}\sigma}\,\mathrm{e}^{\mathrm{i}\boldsymbol{k}\cdot\boldsymbol{x}}+\hat{a}^{\dagger}_{\boldsymbol{k}\sigma}\,\mathrm{e}^{-\mathrm{i}\boldsymbol{k}\cdot\boldsymbol{x}}\right)$$

and

$$\begin{aligned}\hat{H}'' &= \frac{e^2}{2mc^2}\sum_{\boldsymbol{k},\sigma}\sum_{\boldsymbol{k}',\sigma'}\left(\frac{2\pi\hbar c^2}{L^3}\right)\frac{\boldsymbol{\varepsilon}_{\boldsymbol{k}\sigma}\cdot\boldsymbol{\varepsilon}_{\boldsymbol{k}'\sigma'}}{(\omega_k\omega_{k'})^{1/2}}\\ &\quad\times\left(\hat{a}_{\boldsymbol{k}\sigma}\hat{a}_{\boldsymbol{k}'\sigma'}\,\mathrm{e}^{\mathrm{i}(\boldsymbol{k}+\boldsymbol{k}')\cdot\boldsymbol{x}}+\hat{a}_{\boldsymbol{k}\sigma}\hat{a}^{\dagger}_{\boldsymbol{k}'\sigma'}\mathrm{e}^{\mathrm{i}(\boldsymbol{k}-\boldsymbol{k}')\cdot\boldsymbol{x}}\right.\\ &\quad\left.+\,\hat{a}^{\dagger}_{\boldsymbol{k}\sigma}\hat{a}_{\boldsymbol{k}'\sigma'}\,\mathrm{e}^{\mathrm{i}(-\boldsymbol{k}+\boldsymbol{k}')\cdot\boldsymbol{x}}+\hat{a}^{\dagger}_{\boldsymbol{k}\sigma}\hat{a}^{\dagger}_{\boldsymbol{k}'\sigma'}\,\mathrm{e}^{\mathrm{i}(-\boldsymbol{k}-\boldsymbol{k}')\cdot\boldsymbol{x}}\right)\end{aligned}$$

represent the interaction terms $\sim\hat{\boldsymbol{p}}\cdot\hat{\boldsymbol{A}}$ and $\sim\hat{\boldsymbol{A}}^2$, respectively.

We are interested only in photon creation, so we restrict our consideration to

$$\hat{H}' = -\frac{e}{mc}\left(\frac{2\pi\hbar c^2}{L^3}\right)^{1/2}\sum_{\boldsymbol{k},\sigma}\frac{1}{\omega_k^{1/2}}\hat{\boldsymbol{p}}\cdot\boldsymbol{\varepsilon}_{\boldsymbol{k}\sigma}\hat{a}^{\dagger}_{\boldsymbol{k}\sigma}\,\mathrm{e}^{-\mathrm{i}\boldsymbol{k}\cdot\boldsymbol{x}}$$

and

$$\hat{H}'' = \frac{e^2}{2mc^2}\sum_{\boldsymbol{k},\sigma,\boldsymbol{k}',\sigma'}\left(\frac{2\pi\hbar c^2}{L^3}\right)\frac{\boldsymbol{\varepsilon}_{\boldsymbol{k}\sigma}\cdot\boldsymbol{\varepsilon}_{\boldsymbol{k}'\sigma'}}{(\omega_k\omega_{k'})^{1/2}}\hat{a}^{\dagger}_{\boldsymbol{k}\sigma}\hat{a}^{\dagger}_{\boldsymbol{k}'\sigma'}\,\mathrm{e}^{-\mathrm{i}(\boldsymbol{k}_1+\boldsymbol{k}_2)\cdot\boldsymbol{x}}\,.$$

$\hat{H}''$ creates two photons, so it is sufficient to consider the matrix element of first order perturbation theory. On the other side $\hat{H}'$ creates only one photon in first order. Hence, the two photons have to be created successively, i.e. via a process of second order in perturbation theory. We arrive at two different matrix elements:

$$\begin{aligned}M_{\mathrm{fi}}^{(1)} &= \langle f|\hat{H}''|i\rangle \quad \text{(first order)}\,,\\ M_{\mathrm{fi}}^{(2)} &= \sum_n\frac{\langle f|\hat{H}'|n\rangle\langle n|\hat{H}'|i\rangle}{E_{\mathrm{i}}-E_n} \quad \text{(second order)}\,.\end{aligned}$$

In the latter term the (time) order of the creation of the two photons is important:

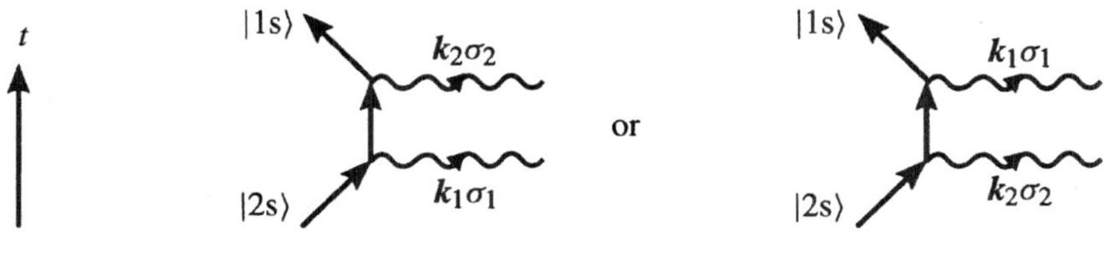

Fig. 2.14. Feynman diagrams of second order

Therefore, we obtain two contributions of second order. Both amplitudes have to be added coherently. They are illustrated in Fig. 2.14. With $\omega = ck$, terms carrying dimensions can be explicitly factorized out:

$$M = \frac{e^2}{2mc^2}\left(\frac{2\pi\hbar c^2}{L^3}\right)\frac{1}{c\sqrt{k_1k_2}}M'\,.$$

Exercise 2.10.

Thus the matrix element M' is dimensionless:

$$M' = \boldsymbol{\varepsilon}_1 \cdot \boldsymbol{\varepsilon}_2 \langle 1\mathrm{s}| \,\mathrm{e}^{-\mathrm{i}(\boldsymbol{k}_1+\boldsymbol{k}_2)\cdot\boldsymbol{x}}\,|2\mathrm{s}\rangle + \frac{2}{m}\sum_n \left(\frac{\langle 1\mathrm{s}|\hat{\boldsymbol{p}}\cdot\boldsymbol{\varepsilon}_2\,\mathrm{e}^{-\mathrm{i}\boldsymbol{k}_2\cdot\boldsymbol{x}}|n\rangle\langle n|\hat{\boldsymbol{p}}\cdot\boldsymbol{\varepsilon}_1\,\mathrm{e}^{-\mathrm{i}\boldsymbol{k}_1\cdot\boldsymbol{x}}|2\mathrm{s}\rangle}{E_{2\mathrm{s}} - E_n - \hbar c k_1} + \frac{\langle 1\mathrm{s}|\hat{\boldsymbol{p}}\cdot\boldsymbol{\varepsilon}_1\,\mathrm{e}^{-\mathrm{i}\boldsymbol{k}_1\cdot\boldsymbol{x}}|n\rangle\langle n|\hat{\boldsymbol{p}}\cdot\boldsymbol{\varepsilon}_2\,\mathrm{e}^{-\mathrm{i}\boldsymbol{k}_2\cdot\boldsymbol{x}}|2\mathrm{s}\rangle}{E_{2\mathrm{s}} - E_n - \hbar c k_2} \right) .$$

Here $\boldsymbol{\varepsilon}_1$ and $\boldsymbol{\varepsilon}_2$ are the polarization vectors of the first and second photons, respectively. We then get

$$\frac{1}{\tau} = \frac{r_\mathrm{e}^2 c}{4(2\pi)^3} \sum_{\sigma_1}\sum_{\sigma_2} \int \frac{\mathrm{d}^3k_1}{k_1} \int \frac{\mathrm{d}^3k_2}{k_2} |M'|^2 \delta(k_0 - k_1 - k_2) ,$$

where $r_\mathrm{e} = e^2/mc^2$ represents the classical electron radius. Since we are interested only in an order of magnitude estimation, we approximate the dimensionless matrix element M' as

$$\sum_{\sigma_1}\sum_{\sigma_2} |M'|^2 \approx 1 .$$

As a consequence we obtain

$$\frac{1}{\tau} = \frac{r_\mathrm{e}^2 c}{4(2\pi)^3} \int \frac{\mathrm{d}^3k_1}{k_1} \int \frac{\mathrm{d}^3k_2}{k_2} \delta(k_0 - k_1 - k_2) .$$

Because of spherical symmetry $[f(\boldsymbol{k}) = f(|\boldsymbol{k}|)]$ the relation

$$\int \mathrm{d}^3k\, f(\boldsymbol{k}) = \int k^2\,\mathrm{d}k \int \mathrm{d}\Omega\, f(|\boldsymbol{k}|) = 4\pi \int k^2\,\mathrm{d}k\, f(|\boldsymbol{k}|)$$

holds, so that

$$\frac{1}{\tau} = \frac{r_\mathrm{e}^2 c}{4(2\pi)^3} 16\pi^2 \int k_1\,\mathrm{d}k_1 \int k_2\,\mathrm{d}k_2\, \delta\left[(k_0 - k_1) - k_2\right] .$$

Owing to the δ function, the outcome of the integration over k_2 is

$$\begin{aligned} \frac{1}{\tau} &= \frac{r_\mathrm{e}^2 c}{2\pi} \int_0^{k_0} k_1(k_0 - k_1)\,\mathrm{d}k_1 \\ &= \frac{r_\mathrm{e}^2 c}{2\pi} \left(\frac{1}{2}k_0^3 - \frac{1}{3}k_0^3 \right) \\ &= \frac{r_\mathrm{e}^2 c}{12\pi} k_0^3 , \end{aligned}$$

where k_0 is the maximum energy that can be emitted. Since

$$\hbar c k_0 = E_{2\mathrm{s}} - E_{1\mathrm{s}} = \frac{3}{8}\frac{Z^2 e^2}{a} ,$$

where Z is the central charge of the nucleus (for hydrogen $Z = 1$), we deduce

$$k_0 = \frac{3}{8} Z^2 \frac{\alpha}{a}$$

with $\alpha = e^2/\hbar c \approx \frac{1}{137}$. It then follows that

$$\frac{1}{\tau} = \frac{Z^6}{12\pi}\left(\frac{3}{8}\right)^3 \alpha^7 \frac{c}{a} \approx 8.75\, Z^6\, \mathrm{s}^{-1}\,,$$

Exercise 2.10.

where the Bohr radius $a = \hbar^2/me^2$ has been used. We obtain $\tau = 0.114\, Z^{-6}$ s. Hence, we deduce for the hydrogen atom $\tau_{\mathrm{H}} = 0.114$ s. Indeed, the two-photon decay is much faster than the one-photon decay with $\tau = 9.5$ years (compare with Exercise 2.6).

With an exact treatment of the matrix element M' we would derive

$$\frac{1}{\tau} = 8.226\, Z^6\, \mathrm{s}^{-1}\,.$$

We realize that the approximation we have used is well justified!

Historical Remark. Over many years two-photon decay has attracted much interest. Qualitatively it is different from the one-photon transition and represents another confirmation of the quantum field theory of radiation. We refer the interested reader to the publications by M. Göppert-Mayer: Ann. Phys. **9** (1931) 273 and J. Shapiro, G. Breit: Phys. Rev. **113** (1959) 179.

2.6 Cherenkov Radiation of a Schrödinger Electron

In the previous subsections we have seen that an electron moving freely in the vacuum cannot radiate, i.e. cannot emit a photon. Also, such an electron cannot absorb a photon, because – as for the photon emission – the conservation laws of momentum and energy cannot be fulfilled simultaneously; see again the discussion following (2.51) and (2.52). At least one additional particle has to be involved to absorb the recoil.

Let us now turn to the motion of a charged particle (electron) in a *dielectric medium.* From classical electrodynamics we know that the speed of light c' in such a medium is given by

$$c' = \frac{c}{n} = \frac{c}{\sqrt{\varepsilon\mu}} \approx \frac{c}{\sqrt{\varepsilon}}\,. \tag{2.69}$$

The variable $\varepsilon(\omega)$ is known as the *dielectric constant* and depends on the frequency ω of the electromagnetic wave. $\mu \approx 1$ is called the *magnetic permeability* of the medium. $n(\omega) \approx \sqrt{\varepsilon(\omega)}$ represents the *refractive index.* The frequency (ω) dependence of these quantities indicates the *dispersion* of electromagnetic waves in the medium. In general the speed of light c' in the medium is smaller than the speed of light c in vacuum, i.e. $c' < c$. For this reason a particle can move with a velocity $v > c'$ in the medium, which is beyond the speed of light (in a medium). We will convince ourselves that under these circumstances a charged particle is able to emit and absorb photons. The medium acts as an additional particle that absorbs the recoil and in this way permits the conservation laws to be fulfilled.

The energy of the electromagnetic field is given by expression (1.40):

$$H_{\mathrm{rad}} = \frac{1}{8\pi}\int_{L^3} \mathrm{d}^3x\,(\boldsymbol{E}^2 + \boldsymbol{B}^2)\,. \tag{2.70}$$

However, this is no longer the total energy of the system. A contribution to the energy from the medium is missing that expresses the reaction of the particles in the medium to the electromagnetic wave. First of all we illuminate this problem with the following example.

EXAMPLE

2.11 The Field Energy in Media with Dispersion

The expression for the energy flux density (Poynting vector),

$$S = \frac{c}{4\pi} E \times H , \tag{1}$$

is valid for any arbitrary electromagnetic field. This also holds when dispersion is present. We can see this in the following way.[10] Directly at an interface (for example medium–vacuum) the tangential components of E and H are continuous (see the lectures on classical electrodynamics). Therefore we decompose them into tangential (t) and normal (n) components:

$$E^{(1)} = E_{\mathrm{t}}^{(1)} + E_{\mathrm{n}}^{(1)} , \tag{2}$$

$$E^{(2)} = E_{\mathrm{t}}^{(2)} + E_{\mathrm{n}}^{(2)} , \tag{3}$$

$$H^{(1)} = H_{\mathrm{t}}^{(1)} + H_{\mathrm{n}}^{(1)} , \tag{4}$$

$$H^{(2)} = H_{\mathrm{t}}^{(2)} + H_{\mathrm{n}}^{(2)} . \tag{5}$$

Consequently, the Poynting vector at side 1 of the interface can also be decomposed into a normal and a tangential component:

$$\begin{aligned} S^{(1)} &= \frac{c}{4\pi}\left[(E_{\mathrm{t}}^{(1)} + E_{\mathrm{n}}^{(1)}) \times (H_{\mathrm{t}}^{(1)} + H_{\mathrm{n}}^{(1)})\right] \\ &= \frac{c}{4\pi}\left[E_{\mathrm{t}}^{(1)} \times H_{\mathrm{t}}^{(1)} + E_{\mathrm{t}}^{(1)} \times H_{\mathrm{n}}^{(1)} + E_{\mathrm{n}}^{(1)} \times H_{\mathrm{t}}^{(1)}\right] \\ &= \frac{c}{4\pi}\left[S_{\mathrm{n}} + S_{\mathrm{t}}^{(1)}\right] , \end{aligned} \tag{6}$$

where we have distinguished between tangential and normal components, i.e.

$$S_{\mathrm{n}}^{(1)} = E_{\mathrm{t}}^{(1)} \times H_{\mathrm{t}}^{(1)} , \tag{7}$$

$$S_{\mathrm{t}}^{(1)} = E_{\mathrm{t}}^{(1)} \times H_{\mathrm{n}}^{(1)} + E_{\mathrm{n}}^{(1)} \times H_{\mathrm{t}}^{(1)} . \tag{8}$$

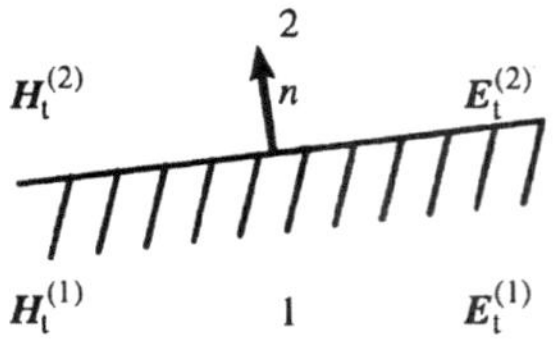

Fig. 2.15. At the interface $H_t^{(1)} = H_t^{(2)}$ and $E_t^{(1)} = E_t^{(2)}$ hold, i.e. the tangential components of E and H are continuous

The term

$$E_{\mathrm{n}}^{(1)} \times H_{\mathrm{n}}^{(1)} = 0 \tag{9}$$

vanishes. Similarly we arrive at the result

$$S^{(2)} = \frac{c}{4\pi}\left[S_{\mathrm{n}} + S_{\mathrm{t}}^{(2)}\right] \tag{10}$$

for side 2 of the interface. Obviously, because of the continuity of the tangential components, we have

[10] We follow here arguments given by L.D. Landau, E.M. Lifshitz: *Electrodynamics of Continuous Media* (Pergamon, Oxford 1960).

$$\left(\boldsymbol{S}^{(1)}\right)_{\mathrm{n}} = \left(\boldsymbol{S}^{(2)}\right)_{\mathrm{n}} ; \tag{11}$$

Example 2.11.

the normal component of the energy flux is continuous at the interface. This is what one expects. We would not have obtained this result if we had used $\boldsymbol{E}\times\boldsymbol{B}$ or $\boldsymbol{D}\times\boldsymbol{H}$ or $\boldsymbol{D}\times\boldsymbol{B}$ in (1). We realize that (1) represents the correct general expression for the energy flux density; moreover, in vacuum it coincides with the already known result (1.18).

The change of energy per second in a normalized volume of the medium is described by $\operatorname{div}\boldsymbol{S}$. Applying Maxwell's equations in a medium,

$$\begin{aligned} \operatorname{rot}\boldsymbol{H} - \frac{1}{c}\frac{\partial \boldsymbol{D}}{\partial t} &= \frac{4\pi}{c}\boldsymbol{j} , \\ \operatorname{rot}\boldsymbol{E} + \frac{1}{c}\frac{\partial \boldsymbol{B}}{\partial t} &= 0 , \\ \operatorname{div}\boldsymbol{B} &= 0 , \\ \operatorname{div}\boldsymbol{D} &= 4\pi\rho , \end{aligned} \tag{12}$$

we obtain after a short calculation

$$\operatorname{div}\boldsymbol{S} = -\frac{1}{4\pi}\left(\boldsymbol{E}\cdot\frac{\partial \boldsymbol{D}}{\partial t} + \boldsymbol{H}\cdot\frac{\partial \boldsymbol{B}}{\partial t}\right) . \tag{13}$$

For the case of a dielectric medium without dispersion, for which ε and μ are pure constants, (13) can be interpreted as the change in the electromagnetic energy density,

$$u = \frac{1}{8\pi}(\varepsilon\boldsymbol{E}^2 + \mu\boldsymbol{H}^2) , \tag{14}$$

with time, i.e.

$$\frac{\partial u}{\partial t} = -\operatorname{div}\boldsymbol{S} . \tag{15}$$

Such a simple interpretation is not possible for media with dispersion. Moreover, dispersion present in a medium in general causes *dissipation of energy*; a medium showing dispersion is absorbing. In order to calculate this dissipation we consider a monochromatic electromagnetic wave and determine the average of expression (13) over time. Because (13) is quadratic in the real field quantities $\boldsymbol{E}$ and $\boldsymbol{H}$ we have to be careful with a transition to complex-valued fields and have to rewrite all expressions with the real forms. To avoid a misunderstanding we now call the monochromatic fields $\underline{\boldsymbol{E}}$ and $\underline{\boldsymbol{H}}$ and denote them in complex-valued form:

$$\begin{aligned} \underline{\boldsymbol{E}} &\longrightarrow \frac{1}{2}(\boldsymbol{E} + \boldsymbol{E}^*) , \\ \frac{\partial \underline{\boldsymbol{D}}}{\partial t} &\longrightarrow \frac{1}{2}(-\mathrm{i}\omega\varepsilon\boldsymbol{E} + \mathrm{i}\omega\varepsilon^*\boldsymbol{E}^*) , \\ \underline{\boldsymbol{H}} &\longrightarrow \frac{1}{2}(\boldsymbol{H} + \boldsymbol{H}^*) , \\ \frac{\partial \underline{\boldsymbol{B}}}{\partial t} &\longrightarrow \frac{1}{2}(-\mathrm{i}\omega\mu\boldsymbol{H} + \mathrm{i}\omega\mu^*\boldsymbol{H}^*) . \end{aligned} \tag{16}$$

Example 2.11.

Products like $\boldsymbol{E} \cdot \boldsymbol{E}$ and $\boldsymbol{E}^* \cdot \boldsymbol{E}^*$ vanish with time averaging, because they contain factors $\mathrm{e}^{\mp 2\mathrm{i}\omega t}$. We deduce that

$$\begin{aligned} \overline{\frac{1}{4\pi}\left(\underline{\boldsymbol{E}} \cdot \frac{\partial \underline{\boldsymbol{D}}}{\partial t} + \underline{\boldsymbol{H}} \cdot \frac{\partial \underline{\boldsymbol{B}}}{\partial t}\right)} &= \frac{\mathrm{i}\omega}{16\pi}\left[(\varepsilon^* - \varepsilon)\boldsymbol{E} \cdot \boldsymbol{E}^* + (\mu^* - \mu)\boldsymbol{H} \cdot \boldsymbol{H}^*\right] \\ &= \frac{\omega}{8\pi}\left(\varepsilon''|\boldsymbol{E}|^2 + \mu''|\boldsymbol{H}|^2\right) \\ &\equiv Q\,. \end{aligned} \tag{17}$$

The quantity Q can be understood as a mean amount of heat produced per normalized time and normalized volume; according to (13) it describes the disappearing energy flux averaged over time and per unit time and volume element. Furthermore, we can write Q also in the form

$$Q = \frac{\omega}{4\pi}\left(\varepsilon''\overline{\underline{\boldsymbol{E}}^2} + \mu''\overline{\underline{\boldsymbol{H}}^2}\right) \tag{18}$$

with

$$\varepsilon'' = \frac{\mathrm{i}}{2}(\varepsilon^* - \varepsilon)\,, \qquad \mu'' = \frac{\mathrm{i}}{2}(\mu^* - \mu)\,.$$

Again, $\underline{\boldsymbol{E}}$ and $\underline{\boldsymbol{H}}$ stand for the real field quantities and the bar indicates a time average over a period $T = 2\pi/\omega$.

Equation (18) is very important as it demonstrates that *energy dissipation* (absorption) is governed by the imaginary parts of ε and μ. The two terms in (18) are interpreted as *electric and magnetic losses.* The sign of Q must always be positive because the second law of thermodynamics means the total entropy of a closed system, which is a measure for the equipartitioning of energy, must increase. The dissipation of energy results in production of energy. Therefore, $Q > 0$ has to hold for all times. From this and from (18) we conclude that the imaginary parts of ε and μ must always be positive:

$$\varepsilon'' > 0\,, \qquad \mu'' > 0\,. \tag{19}$$

This is a very general statement. It is a result of the second law of thermodynamics and is valid for all materials and frequencies. The real parts ε' and μ' of ε and μ are given by

$$\varepsilon' = \frac{\varepsilon + \varepsilon^*}{2}\,, \qquad \mu' = \frac{\mu + \mu^*}{2}\,. \tag{20}$$

No restrictions exist for them. Depending on the particular physical circumstances, they could be either positive or negative.

For a real physical material, nonstationary processes are always thermodynamically irreversible. Thus a variable electromagnetic field always suffers from electric and magnetic losses in a real material. Often they remain small, but they are always present. As a conclusion, the imaginary parts, $\varepsilon''(\omega)$ and $\mu''(\omega)$, do not become rigorously zero in any frequency domain. This finding is of fundamental importance, as we will recognize in the following. Obviously this does not exclude the existance of frequency domains for which $\varepsilon''(\omega)$ and $\mu''(\omega)$ take on very small numbers, so that only minor losses occur. Such frequency domains are called *domains of transparency of the material.* For them it makes sense to introduce the notion of the *intrinsic energy U* of a material

Example 2.11.

in an electromagnetic field, presuming, of course, absorption to be neglected. It then has the same significance as for a constant field.

To determine U we have to consider a field formed by a *bunch of monochromatic waves* whose frequencies fall into a narrow interval around the mean frequency ω_0. Owing to its strict periodicity, a purely monochromatic field would not allow the electromagnetic energy to be localized in time. This can be achieved only with a *wave packet.* Hence, we introduce

$$\begin{aligned} \boldsymbol{E} &= \boldsymbol{E}_0(t)\,\mathrm{e}^{-\mathrm{i}\omega_0 t}\,, \\ \boldsymbol{H} &= \boldsymbol{H}_0(t)\,\mathrm{e}^{-\mathrm{i}\omega_0 t}\,, \end{aligned} \tag{21}$$

where $\boldsymbol{E}_0(t)$ and $\boldsymbol{H}_0(t)$ are slowly varying functions in time compared to $\mathrm{e}^{-\mathrm{i}\omega_0 t}$.

To determine the intrinsic energy we have to calculate the right-hand side of (13) for transparent materials. We insert the real parts of (21) into (13) and obtain

$$\operatorname{div}\boldsymbol{S} = -\frac{1}{4\pi}\left[\frac{(\boldsymbol{E}+\boldsymbol{E}^*)}{2}\cdot\frac{(\dot{\boldsymbol{D}}+\dot{\boldsymbol{D}}^*)}{2}+\frac{(\boldsymbol{H}+\boldsymbol{H}^*)}{2}\cdot\frac{(\dot{\boldsymbol{B}}+\dot{\boldsymbol{B}}^*)}{2}\right]. \tag{22}$$

Averaging this expression over time, i.e. with respect to the period $T = 2\pi/\omega_0$, leads to the products $\boldsymbol{E}\cdot\dot{\boldsymbol{D}}$, $\boldsymbol{E}^*\cdot\dot{\boldsymbol{D}}^*$, $\boldsymbol{H}\cdot\dot{\boldsymbol{B}}$, and $\boldsymbol{H}^*\cdot\dot{\boldsymbol{B}}^*$ vanishing and results in

$$\overline{\operatorname{div}\boldsymbol{S}} = -\frac{1}{16\pi}\left[(\boldsymbol{E}\cdot\dot{\boldsymbol{D}}^*+\boldsymbol{E}^*\cdot\dot{\boldsymbol{D}})+(\boldsymbol{H}\cdot\dot{\boldsymbol{B}}^*+\boldsymbol{H}^*\cdot\dot{\boldsymbol{B}})\right]. \tag{23}$$

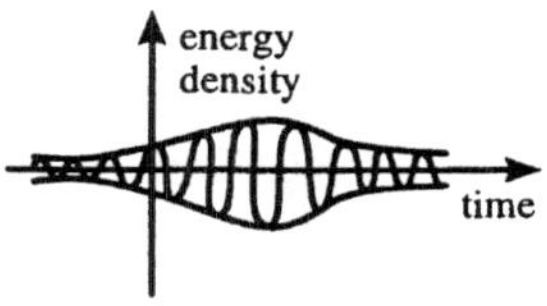

Fig. 2.16. A wave packet consisting of electromagnetic waves allows us to study the intrinsic energy of a material because of dispersion

The derivative $\partial\boldsymbol{D}(t)/\partial t$ will be a function of $\boldsymbol{E}(t)$. We write

$$\frac{\partial\boldsymbol{D}}{\partial t} = \hat{f}\boldsymbol{E} \quad\text{and analogously}\quad \frac{\partial\boldsymbol{B}}{\partial t} = \hat{g}\boldsymbol{H}\,, \tag{24}$$

where $\hat{f}$ and $\hat{g}$ have to be thought of as operators. If $\boldsymbol{E}_0$ in (21) were constant in time such that $\boldsymbol{D}(t) = \varepsilon(\omega)\boldsymbol{E} = \varepsilon(\omega)\boldsymbol{E}_0\,\mathrm{e}^{-\mathrm{i}\omega_0 t}$, we would deduce that

$$\frac{\partial\boldsymbol{D}}{\partial t} = -\mathrm{i}\omega_0\varepsilon(\omega_0)\boldsymbol{E} \equiv \hat{f}\boldsymbol{E} = f(\omega_0)\boldsymbol{E}\,, \tag{25}$$

and thus

$$f(\omega) = -\mathrm{i}\omega\varepsilon(\omega)\,. \tag{26}$$

We expand the function $\boldsymbol{E}_0(t)$ and, analogously, $\boldsymbol{H}_0(t)$ from (21) into a Fourier integral:

$$\boldsymbol{E}_0(t) = \int \boldsymbol{E}_0(\alpha)\,\mathrm{e}^{-\mathrm{i}\alpha t}\,\mathrm{d}\alpha\,. \tag{27}$$

Compared to $\mathrm{e}^{-\mathrm{i}\omega_0 t}$, the amplitude $\boldsymbol{E}_0(t)$ changes only slowly with time; therefore we expect only those components with

$$\alpha \ll \omega_0 \tag{28}$$

to appear in the Fourier integral. Then we can write for each single component

Exampel 2.11.

$$\begin{aligned}\hat{f}\boldsymbol{E}_0(\alpha)\,\mathrm{e}^{-\mathrm{i}(\omega_0+\alpha)t} &= f(\alpha+\omega_0)\boldsymbol{E}_0(\alpha)\,\mathrm{e}^{-\mathrm{i}(\omega_0+\alpha)t}\\ &\approx f(\omega_0)\boldsymbol{E}_0(\alpha)\,\mathrm{e}^{-\mathrm{i}(\omega_0+\alpha)t}\\ &\quad+\frac{\mathrm{d}f(\omega_0)}{\mathrm{d}\omega_0}\,\alpha\boldsymbol{E}_0(\alpha)\,\mathrm{e}^{-\mathrm{i}(\omega_0+\alpha)t}\\ &= \left(f(\omega_0)+\alpha\frac{\mathrm{d}f(\omega_0)}{\mathrm{d}\omega_0}\right)\boldsymbol{E}_0(\alpha)\,\mathrm{e}^{-\mathrm{i}(\omega_0+\alpha)t}\,. \end{aligned}\tag{29}$$

Inserting (27) and (29) into (21) we deduce that

$$\begin{aligned}\hat{f}\boldsymbol{E}(t) &= \hat{f}\boldsymbol{E}_0(t)\,\mathrm{e}^{-\mathrm{i}\omega_0 t}\\ &= \hat{f}\int \boldsymbol{E}_0(\alpha)\,\mathrm{e}^{-\mathrm{i}(\omega_0+\alpha)t}\,\mathrm{d}\alpha\\ &= \int \hat{f}\boldsymbol{E}_0(\alpha)\,\mathrm{e}^{-\mathrm{i}(\omega_0+\alpha)t}\,\mathrm{d}\alpha\\ &= \int \mathrm{d}\alpha\left(f(\omega_0)+\alpha\frac{\mathrm{d}f(\omega_0)}{\mathrm{d}\omega_0}\right)\boldsymbol{E}_0(\alpha)\,\mathrm{e}^{-\mathrm{i}(\omega_0+\alpha)t}\\ &= f(\omega_0)\boldsymbol{E}(t)+\frac{\mathrm{d}f(\omega_0)}{\mathrm{d}\omega_0}\left(\int \alpha\boldsymbol{E}_0(\alpha)\,\mathrm{e}^{-\mathrm{i}\alpha t}\,\mathrm{d}\alpha\right)\mathrm{e}^{-\mathrm{i}\omega_0 t}\\ &= f(\omega_0)\boldsymbol{E}(t)+\frac{\mathrm{d}f(\omega_0)}{\mathrm{d}\omega_0}\,\mathrm{i}\,\frac{\partial\boldsymbol{E}_0(t)}{\partial t}\,\mathrm{e}^{-\mathrm{i}\omega_0 t}\,. \end{aligned}\tag{30}$$

Now we omit the index zero for ω_0 and with the help of (24) and (21) we are led to the result

$$\frac{\partial\boldsymbol{D}(t)}{\partial t} = -\,\mathrm{i}\omega\varepsilon(\omega)\boldsymbol{E}(t)+\frac{\mathrm{d}(\omega\varepsilon)}{\mathrm{d}\omega}\frac{\partial\boldsymbol{E}_0(t)}{\partial t}\,\mathrm{e}^{-\mathrm{i}\omega t}\,.\tag{31}$$

From here we calculate the first term in (23). We assume transparency, neglect the imaginary part of ε (transparency of the material is assumed!), and finally obtain

$$\begin{aligned}-\frac{1}{16\pi}\left(\overline{\boldsymbol{E}\cdot\dot{\boldsymbol{D}}^*}+\overline{\boldsymbol{E}^*\cdot\dot{\boldsymbol{D}}}\right) &= -\frac{1}{16\pi}\left(\mathrm{i}\omega\varepsilon(\omega)\overline{\boldsymbol{E}\cdot\boldsymbol{E}^*(t)}+\frac{\mathrm{d}(\omega\varepsilon)}{\mathrm{d}\omega}\overline{\boldsymbol{E}_0\cdot\frac{\partial\boldsymbol{E}^*}{\partial t}}\right.\\ &\quad\left.-\,\mathrm{i}\omega\varepsilon(\omega)\overline{\boldsymbol{E}^*\cdot\boldsymbol{E}(t)}+\frac{\mathrm{d}(\omega\varepsilon)}{\mathrm{d}\omega}\overline{\boldsymbol{E}_0^*\cdot\frac{\partial\boldsymbol{E}_0}{\partial t}}\right)\\ &= -\frac{1}{16\pi}\frac{\mathrm{d}(\omega\varepsilon)}{\mathrm{d}\omega}\overline{\left(\boldsymbol{E}_0^*\cdot\frac{\partial\boldsymbol{E}_0(t)}{\partial t}+\boldsymbol{E}_0\cdot\frac{\partial\boldsymbol{E}_0^*(t)}{\partial t}\right)}\\ &= -\frac{1}{16\pi}\frac{\mathrm{d}(\omega\varepsilon)}{\mathrm{d}\omega}\overline{\frac{\mathrm{d}}{\mathrm{d}t}(\boldsymbol{E}\cdot\boldsymbol{E}^*)}\,. \end{aligned}\tag{32}$$

For the last step we have used $\boldsymbol{E}_0\cdot\boldsymbol{E}_0^* = \boldsymbol{E}\cdot\boldsymbol{E}^*$, which follows from (21). Analogously we deduce an expression for the second term in (23), the only difference is that ε has to be replaced by μ and $\boldsymbol{E}$ by $\boldsymbol{H}$. Hence, we write (23) in the form

$$\begin{aligned}\overline{\operatorname{div}\boldsymbol{S}} &= -\frac{\mathrm{d}}{\mathrm{d}t}\left\{\frac{1}{16\pi}\left[\left(\frac{\mathrm{d}(\omega\varepsilon)}{\mathrm{d}\omega}\right)\overline{\boldsymbol{E}\cdot\boldsymbol{E}^*}+\left(\frac{\mathrm{d}(\omega\mu)}{\mathrm{d}\omega}\right)\overline{\boldsymbol{H}\cdot\boldsymbol{H}^*}\right]\right\}\\ &= -\frac{\mathrm{d}}{\mathrm{d}t}\overline{u}\,, \end{aligned}\tag{33}$$

Example 2.11.

where

$$\overline{u} = \frac{1}{16\pi}\left(\frac{\mathrm{d}(\omega\varepsilon)}{\mathrm{d}\omega}\overline{\boldsymbol{E}\cdot\boldsymbol{E}^*} + \frac{\mathrm{d}(\omega\mu)}{\mathrm{d}\omega}\overline{\boldsymbol{H}\cdot\boldsymbol{H}^*}\right) \tag{34}$$

represents the time average of the electromagnetic part of the intrinsic energy per normalized volume in a transparent medium. Introducing again the *real fields* $\underline{\boldsymbol{E}}$ and $\underline{\boldsymbol{H}}$ via

$$\underline{\boldsymbol{E}} = \frac{\boldsymbol{E}+\boldsymbol{E}^*}{2}, \qquad \underline{\boldsymbol{H}} = \frac{\boldsymbol{H}+\boldsymbol{H}^*}{2}, \tag{35}$$

we have

$$\overline{\underline{\boldsymbol{E}}\cdot\underline{\boldsymbol{E}}} = \tfrac{1}{4}\overline{(\boldsymbol{E}\cdot\boldsymbol{E}+2\boldsymbol{E}\cdot\boldsymbol{E}^*+\boldsymbol{E}^*\cdot\boldsymbol{E}^*)} = \tfrac{1}{2}\overline{\boldsymbol{E}\cdot\boldsymbol{E}^*} \tag{36}$$

and an analogous result for the $\boldsymbol{H}$ term. This yields for (34)

$$\overline{u} = \frac{1}{8\pi}\overline{\left(\frac{\mathrm{d}(\omega\varepsilon)}{\mathrm{d}\omega}\underline{\boldsymbol{E}}\cdot\underline{\boldsymbol{E}} + \frac{\mathrm{d}(\omega\mu)}{\mathrm{d}\omega}\underline{\boldsymbol{H}}\cdot\underline{\boldsymbol{H}}\right)}\,. \tag{37}$$

For media without dispersion ε und μ do not depend on frequency and, as expected, (37) becomes identical to the expression (14). However, although small, absorbtion is always present. As a consequence the total energy density $\overline{u}$ transforms into heat as the electromagnetic energy flux is turned off from outside. According to the second law of thermodynamics this heat is produced and is not transformed back into electromagnetic energy. Therefore,

$$\overline{u} > 0\,,$$

from which

$$\frac{\mathrm{d}(\omega\varepsilon)}{\mathrm{d}\omega} > 0 \quad\text{and}\quad \frac{\mathrm{d}(\omega\mu)}{\mathrm{d}\omega} > 0 \tag{38}$$

follow in general.

We return to (2.70) for the energy of the free electromagnetic field. This expression has to be modified for media with dispersion according to (37) of the last exercise:

$$\begin{aligned} H_{\mathrm{rad}} &= \int \mathrm{d}^3x\, u \\ &= \int \mathrm{d}^3x\,\frac{1}{8\pi}\left(\frac{\mathrm{d}\,(\omega\varepsilon(\omega))}{\mathrm{d}\omega}\boldsymbol{E}\cdot\boldsymbol{E} + \frac{\mathrm{d}\,(\omega\mu(\omega))}{\mathrm{d}\omega}\boldsymbol{H}\cdot\boldsymbol{H}\right)\,. \end{aligned} \tag{2.71}$$

In comparison with (2.70) the additional terms of the form $\omega(\mathrm{d}\varepsilon(\omega)/\mathrm{d}\omega)\boldsymbol{E}^2$ and $\omega(\mathrm{d}\mu(\omega)/\mathrm{d}\omega)\boldsymbol{H}^2$ describe the reaction of the medium to the energy contained in the electromagnetic waves; to be more precise, they describe the effect on the energy stored by the molecular currents induced by the electromagnetic waves. For most materials the permeability is $\mu \approx 1$, which indeed is a very good approximation, and, as a consequence, $\boldsymbol{H} = \boldsymbol{B}$. Then, the field energy (2.71) in the medium becomes

$$H_{\text{rad}} = \int d^3x \, \frac{1}{8\pi} \left(\frac{d\left(\omega\varepsilon(\omega)\right)}{d\omega} \boldsymbol{E} \cdot \boldsymbol{E} + \boldsymbol{B} \cdot \boldsymbol{B} \right) . \tag{2.72}$$

For the case of media with no frequency dependent dielectric constant $\varepsilon(\omega) = \varepsilon = \text{const}$ we deduce the expression

$$H_{\text{rad}} = \int d^3x \, \frac{1}{8\pi} \left(\varepsilon \boldsymbol{E} \cdot \boldsymbol{E} + \boldsymbol{B} \cdot \boldsymbol{B} \right) , \tag{2.73}$$

which is well known from electrodynamics. We have set $\mu = 1$.

We use the results (1.44) and (1.45) obtained earlier to derive the quantum-mechanical operator for the field energy corresponding to (2.73). The expression (1.41) for $\int_{L^3} \boldsymbol{E}^2 \, d^3x$ was given by

$$\begin{aligned} \int_{L^3} \frac{1}{8\pi} \boldsymbol{E} \cdot \boldsymbol{E} \, d^3x &= \sum_{\boldsymbol{k},\sigma} \frac{\omega_k^2}{8\pi c^2} N_k^2 L^3 \Big[\left(\hat{a}_{\boldsymbol{k}\sigma} \hat{a}^\dagger_{\boldsymbol{k}\sigma} + \hat{a}^\dagger_{\boldsymbol{k}\sigma} \hat{a}_{\boldsymbol{k}\sigma} \right) \\ &\quad + (-1)^\sigma \left(\hat{a}_{\boldsymbol{k}\sigma} \hat{a}_{-\boldsymbol{k}\sigma} + \hat{a}^\dagger_{\boldsymbol{k}\sigma} \hat{a}^\dagger_{-\boldsymbol{k}\sigma} \right) \Big] . \end{aligned}$$

Then the classical expression of (2.73),

$$\int d^3x \, \frac{1}{8\pi} \varepsilon \boldsymbol{E} \cdot \boldsymbol{E} ,$$

becomes

$$\sum_{\boldsymbol{k},\sigma} \frac{\omega_k^2}{8\pi c^2} \varepsilon N_k^2 L^3 \Big[\left(\hat{a}_{\boldsymbol{k}\sigma} \hat{a}^\dagger_{\boldsymbol{k}\sigma} + \hat{a}^\dagger_{\boldsymbol{k}\sigma} \hat{a}_{\boldsymbol{k}\sigma} \right) + (-1)^\sigma \left(\hat{a}_{\boldsymbol{k}\sigma} \hat{a}_{-\boldsymbol{k}\sigma} + \hat{a}^\dagger_{\boldsymbol{k}\sigma} \hat{a}^\dagger_{-\boldsymbol{k}\sigma} \right) \Big] . \tag{2.74}$$

Similarily we deduce from (1.45) that

$$\begin{aligned} \int d^3x \, \frac{1}{8\pi} \boldsymbol{B} \cdot \boldsymbol{B} &\longrightarrow \sum_{\boldsymbol{k},\sigma} \frac{k^2}{8\pi} N_k^2 L^3 \Big[\left(\hat{a}_{\boldsymbol{k}\sigma} \hat{a}^\dagger_{\boldsymbol{k}\sigma} + \hat{a}^\dagger_{\boldsymbol{k}\sigma} \hat{a}_{\boldsymbol{k}\sigma} \right) \\ &\quad - (-1)^\sigma \left(\hat{a}_{\boldsymbol{k}\sigma} \hat{a}_{-\boldsymbol{k}\sigma} + \hat{a}^\dagger_{\boldsymbol{k}\sigma} \hat{a}^\dagger_{-\boldsymbol{k}\sigma} \right) \Big] . \end{aligned} \tag{2.75}$$

Hence, we derive, with $k^2 = \varepsilon \omega_k^2 / c^2$ for the energy of the radiation field (2.72),

$$\begin{aligned} H_{\text{rad}} &= \sum_{\boldsymbol{k},\sigma} \frac{\omega_k^2}{8\pi c^2} 2\varepsilon(\omega_k) N_k^2 L^3 \left(\hat{a}_{\boldsymbol{k}\sigma} \hat{a}^\dagger_{\boldsymbol{k}\sigma} + \hat{a}^\dagger_{\boldsymbol{k}\sigma} \hat{a}_{\boldsymbol{k}\sigma} \right) \\ &\quad + \sum_{\boldsymbol{k},\sigma} \frac{\omega_k^2}{8\pi c^2} \left(\frac{d\left(\omega_k \varepsilon(\omega_k)\right)}{d\omega_k} - \varepsilon \right) N_k^2 L^3 (-1)^\sigma \\ &\quad \times \left(\hat{a}_{\boldsymbol{k}\sigma} \hat{a}_{-\boldsymbol{k}\sigma} + \hat{a}^\dagger_{\boldsymbol{k}\sigma} \hat{a}^\dagger_{-\boldsymbol{k}\sigma} \right) . \end{aligned} \tag{2.76}$$

We realize that terms not conserving the photon number [second sum in (2.76)] vanish for a constant dielectric constant. With the refractive index, $n^2 = \varepsilon$, the remaining term becomes

$$\hat{H}_{\text{rad}} = \sum_{\boldsymbol{k},\sigma} \frac{\omega_k^2}{4\pi c^2} n^2 N_k^2 L^3 \left[\hat{a}_{\boldsymbol{k}\sigma} \hat{a}^\dagger_{\boldsymbol{k}\sigma} + \hat{a}^\dagger_{\boldsymbol{k}\sigma} \hat{a}_{\boldsymbol{k}\sigma} \right] . \tag{2.77}$$

This result has to be compared with (1.46). To obtain a far analogous to (1.48), we have to choose the normalization constant as

$$N_k = \sqrt{\frac{2\pi\hbar c^2}{L^3\omega_k n^2}} , \tag{2.78}$$

so that

$$\hat{H}_{\rm rad} = \frac{1}{2}\sum_{k,\sigma} \hbar\omega_k \left(\hat{a}_{k\sigma}\hat{a}^\dagger_{k\sigma} + \hat{a}^\dagger_{k\sigma}\hat{a}_{k\sigma}\right) . \tag{2.79}$$

This change in the normalization becomes noticeable for the expression for the vector potential. From (1.37) we deduce with (2.78) that

$$\hat{\boldsymbol{A}}(\boldsymbol{x},t) = \sum_{k,\sigma} \sqrt{\frac{2\pi\hbar c^2}{L^3\omega_k n^2}} \boldsymbol{\varepsilon}_{k\sigma} \left(\hat{a}_{k\sigma}\,\mathrm{e}^{\mathrm{i}\boldsymbol{k}\cdot\boldsymbol{x}} + \hat{a}^\dagger_{k\sigma}\,\mathrm{e}^{-\mathrm{i}\boldsymbol{k}\cdot\boldsymbol{x}}\right) . \tag{2.80}$$

Then, only a normalization constant is changing for the interaction Hamiltonian $\hat{H}'_{\rm int}$ [see (2.8)]:

$$\hat{H}'_{\rm int} = -\frac{e}{mc}\sum_{k,\sigma} \sqrt{\frac{2\pi\hbar c^2}{L^3\omega_k n^2}} \hat{\boldsymbol{p}}\cdot\boldsymbol{\varepsilon}_{k\sigma} \left(\hat{a}_{k\sigma}\,\mathrm{e}^{\mathrm{i}\boldsymbol{k}\cdot\boldsymbol{x}} + \hat{a}^\dagger_{k\sigma}\,\mathrm{e}^{-\mathrm{i}\boldsymbol{k}\cdot\boldsymbol{x}}\right) . \tag{2.81}$$

With this result we now determine the transition amplitude per unit time for a free electron with momentum $\hbar\boldsymbol{q}$ emitting a photon with momentum $\hbar\boldsymbol{k}$. After the emission of the photon the electron has momentum $\hbar\boldsymbol{q}' = \hbar(\boldsymbol{q}-\boldsymbol{k})$. We find

$$\left(\frac{\text{trans. prob.}}{\text{time}}\right)_{q\to q'} = \frac{2\pi}{\hbar}|M_{\rm fi}|^2\delta\left(\frac{\hbar^2 q^2}{2m} - \frac{\hbar^2 q'^2}{2m} - \hbar\omega_k\right) , \tag{2.82}$$

where the relevant matrix element is given with the help of (2.45) and (2.81) as

$$\begin{aligned} M_{\rm fi} &= \langle\psi_{q'}(\boldsymbol{x})|\langle\ldots,1_{k\sigma},\ldots|(-\frac{e}{mc})\sum_{k',\sigma'}\sqrt{\frac{2\pi\hbar c^2}{L^3\omega_{k'}n^2}}\hat{\boldsymbol{p}}\cdot\boldsymbol{\varepsilon}_{k'\sigma'} \\ &\quad\times\left[\hat{a}_{k'\sigma'}\,\mathrm{e}^{\mathrm{i}\boldsymbol{k}'\cdot\boldsymbol{x}} + \hat{a}^\dagger_{k'\sigma'}\,\mathrm{e}^{-\mathrm{i}\boldsymbol{k}'\cdot\boldsymbol{x}}\right]|\psi_q(\boldsymbol{x})\rangle\,|\ldots,0_{k\sigma},\ldots\rangle \\ &= -\frac{e}{mc}\sqrt{\frac{2\pi\hbar c^2}{L^3\omega_k n^2}}\langle\psi_{q'}(\boldsymbol{x})|\,\hat{\boldsymbol{p}}\cdot\boldsymbol{\varepsilon}_{k\sigma}\,\mathrm{e}^{-\mathrm{i}\boldsymbol{k}\cdot\boldsymbol{x}}\,|\psi_q(\boldsymbol{x})\rangle . \end{aligned} \tag{2.83}$$

Obviously,

$$\langle\psi_{q'}(\boldsymbol{x})|\,\hat{\boldsymbol{p}}\cdot\boldsymbol{\varepsilon}_{k\sigma}\,\mathrm{e}^{-\mathrm{i}\boldsymbol{k}\cdot\boldsymbol{x}}\,|\psi_q(\boldsymbol{x})\rangle = \hbar\boldsymbol{q}\cdot\boldsymbol{\varepsilon}_{k\sigma}\delta_{q',q-k} , \tag{2.84}$$

and we obtain the following preliminary result for (2.82):

$$\begin{aligned} \left(\frac{\text{trans. prob.}}{\text{time}}\right)_{q\to q-k} &= \frac{2\pi}{\hbar}\left(\frac{e}{mc}\right)^2\left(\frac{2\pi\hbar c^2}{L^3\omega_k n^2}\right)|\hbar\boldsymbol{q}'\cdot\boldsymbol{\varepsilon}_{k\sigma}|^2 \\ &\quad\times\delta\left(\frac{\hbar^2 q^2}{2m} - \frac{\hbar^2(\boldsymbol{q}-\boldsymbol{k})^2}{2m} - \hbar\omega_k\right) . \end{aligned} \tag{2.85}$$

Since $(\boldsymbol{q}-\boldsymbol{k})^2 = q^2 + k^2 - 2kq\cos\vartheta$, it follows for the δ function that

$$\begin{aligned}
\delta\left(\frac{\hbar^2 q^2}{2m} - \frac{\hbar^2(\boldsymbol{q}-\boldsymbol{k})^2}{2m} - \hbar\omega_k\right) &= \delta\left(\frac{2\hbar^2 qk\cos\vartheta}{2m} - \frac{\hbar^2 k^2}{2m} - \hbar\omega_k\right) \\
&= \frac{m}{\hbar^2 qk}\delta\left(\cos\vartheta - \frac{\hbar k}{2\hbar q} - \frac{\hbar\omega_k m}{\hbar q \hbar k}\right) \\
&= \frac{1}{v\hbar k}\delta\left(\cos\vartheta - \frac{n\hbar\omega_k}{2mvc} - \frac{\hbar\omega_k m}{mv\hbar\frac{\omega_k}{c}n}\right) \\
&= \frac{1}{v\hbar k}\delta\left(\cos\vartheta - \frac{c}{nv} - \frac{n\hbar\omega_k}{2mc}\right) .
\end{aligned}$$

Here we have used $\delta(ax) = (1/|a|)\delta(x)$ and $\hbar\boldsymbol{q} = m\boldsymbol{v}$. The final result then reads

$$\begin{aligned}
\left(\frac{\text{trans. prob.}}{\text{time}}\right)_{q\to q'=q-k} &= \frac{4\pi^2 e^2\hbar^2|\boldsymbol{q}'\cdot\boldsymbol{\varepsilon}_{k\sigma}|^2}{m^2 L^3 v\hbar k\omega_k n^2} \\
&\quad \times\delta\left(\cos\vartheta - \frac{c}{nv} - \frac{n\hbar\omega_k}{2mcv}\right) . \qquad (2.86)
\end{aligned}$$

The photon is emitted at an *angle* ϑ with respect to the electron trajectory. This *Mach angle* is given by

$$\cos\vartheta = \frac{c}{nv}\left[1 + \frac{\hbar\omega n^2}{2mc^2}\right] . \qquad (2.87)$$

If the photon energy is much smaller than the rest energy of the electron, $\hbar\omega \ll mc^2$ ($mc^2 \approx 0.5\,\text{MeV}$), we obtain approximately

$$\cos\vartheta = \frac{c}{nv} = \frac{c'}{v} . \qquad (2.88)$$

This is the classical Mach angle (see Fig. 2.17), which in this context is also named after its Russian discoverer as the *Cherenkov angle*. Equation (2.88) can be fulfilled only if the particle velocity v is larger than the wave velocity $c' = c/n$, i.e. $v > c/n$. For a vacuum, $n = 1$ and v can never become larger than c. Therefore a photon emission from a free particle is not allowed in vacuum; see again the discussion about photon absorption after (2.52).

In the following we are interested in the loss of energy of the electron per unit length along its trajectory (see Fig. 2.18). It is given by

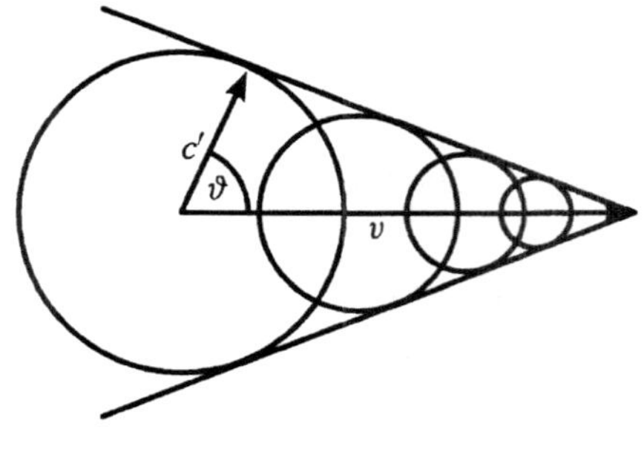

Fig. 2.17. $\cos\vartheta = c'/v$ determines the classical Mach angle. Cherenkov radiation can be interpreted as a Mach phenomenon: supersonic flight of an electron through a medium with $c' < c$

$$\frac{dW}{dx} = \frac{1}{v}\frac{dW}{dt} = \frac{1}{v}\sum_{k\sigma} \hbar\omega_k \left(\frac{\text{trans. prob.}}{\text{time}}\right)_{q\to q-k} . \tag{2.89}$$

We calculate

$$\begin{aligned}\sum_\sigma |(q-k)\cdot \varepsilon_{k\sigma}|^2 &= \sum_\sigma |q\cdot\varepsilon_{k\sigma}|^2 \\ &= q^2\sin^2\vartheta \\ &= q^2(1-\cos^2\vartheta) \\ &= \frac{m^2v^2}{\hbar^2}(1-\cos^2\vartheta)\end{aligned}$$

and use (2.18), which can be expressed in spherical coordinates in k space:

$$\sum_k \longrightarrow \frac{L^3}{(2\pi)^3}\int_0^\infty k^2\,dk\int_{-1}^1 d(\cos\vartheta)\int_0^{2\pi} d\varphi\,. \tag{2.90}$$

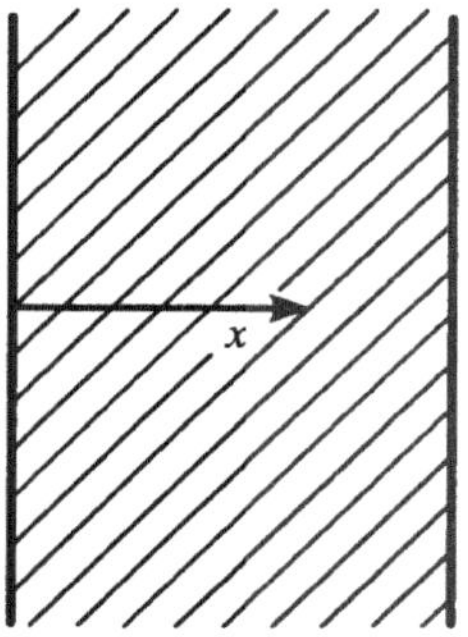

Fig. 2.18. Illustration for the energy loss of a particle in a medium along the trajectory x

We then derive

$$\begin{aligned}\frac{dW}{dx} &= e^2\int_0^\infty k\,dk\int_{-1}^1 d(\cos\vartheta)\,(\hbar\omega_k) \\ &\quad\times\frac{(1-\cos^2\vartheta)\delta\left[\cos\vartheta-(c/nv)-(n\hbar\omega_k/2mcv)\right]}{\hbar\omega_k n^2} \\ &= \frac{e^2}{c^2}\int_0^\infty d\omega_k\,\omega_k\left(1-\frac{c^2}{n^2v^2}\left(1+\frac{\hbar\omega n^2}{2mc^2}\right)^2\right) .\end{aligned} \tag{2.91}$$

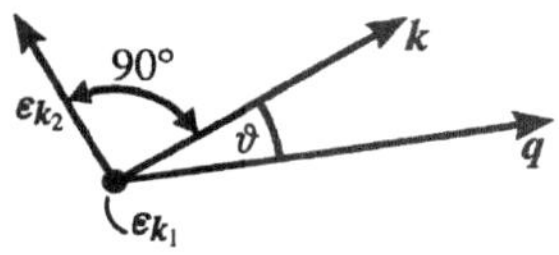

Fig. 2.19. Illustration of the polarization vectors relative to the particle momentum $hq = mv$

In the last step we have not been careful enough; as it is, the integral would diverge. But this is not the case because, as a result of (2.87), $\cos\vartheta$ can only range between -1 and 1 and thus limits the photon energy to

$$\hbar\omega \leq \frac{(nv/c-1)\,2mc^2}{n^2} . \tag{2.92}$$

The integration domain in (2.91) does not reach out to infinity, only to the limiting energy (2.92).

EXERCISE

2.12 The Cherenkov Angle

Problem. Show that the Cherenkov angle is given by the expression

$$\cos\vartheta = \frac{c}{nv}\left[1+\frac{\hbar\omega}{2mc^2}(n^2-1)\sqrt{1-\frac{v^2}{c^2}}\right]$$

if the relativistic instead of the nonrelativistic expression for the particle energy is applied.

Exercise 2.12.

Solution. As $E^2 = p^2c^2 + m^2c^4$, it follows from energy conservation that

$$\sqrt{\hbar^2c^2q^2 + m^2c^4} = \sqrt{\hbar^2c^2|\boldsymbol{q}-\boldsymbol{k}|^2 + m^2c^4} + \frac{\hbar kc}{n}\,.$$

Taking the square yields

$$\hbar^2c^2|\boldsymbol{q}-\boldsymbol{k}|^2 = \hbar^2c^2q^2 - 2E\frac{\hbar kc}{n} + \frac{\hbar^2c^2k^2}{n^2}\,,$$

where the electron energy is given by $E = \sqrt{\hbar^2c^2q^2 + m^2c^4}$. From

$$|\boldsymbol{q}-\boldsymbol{k}|^2 = q^2 - 2qk\cos\vartheta + k^2$$

we derive

$$\cos\vartheta = \frac{E}{n\hbar cq} + \frac{k}{2qn^2}(n^2-1)\,.$$

Now

$$E = \frac{mc^2}{\sqrt{1-v^2/c^2}} \quad \text{and} \quad \hbar q = \frac{mv}{\sqrt{1-v^2/c^2}}$$

as well as $ck = n\omega$ and it follows that

$$\cos\vartheta = \frac{c}{vn}\left[1 + \frac{\hbar\omega}{2mc^2}\sqrt{1-\frac{v^2}{c^2}}(n^2-1)\right]\,.$$

2.7 Natural Linewidth and Self-energy

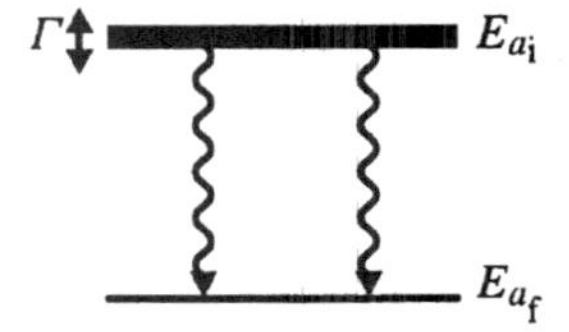

Fig. 2.20. A finite lifetime means a finite level width Γ and a finite spectral line width result from the uncertainty relation $\Delta t \Delta E \approx \hbar$

The photon emitted from a quantum system corresponds to a wave of finite duration and of finite length. The spectral line emitted from the system has a (natural) linewidth because the initial state has only a finite lifetime as a result of its decay. According to Heisenberg's uncertainty relation the energy uncertainty ΔE is related to the lifetime τ by $\Delta E \approx \hbar/\tau$. Up to now we have considered only sharp spectral lines. This is reflected in the δ function $\delta(E_{a_i} - E_{a_f} - \hbar\omega_k)$ appearing in the expression for the transition probability (2.16). Evidently something must be wrong in our considerations employed so far. In fact the mistake comes from perturbation theory as we have taken it over from the Introduction to Quantum Mechanics.[11] Now we will modify this formalism and include a finite linewidth.[12]

Hence, we again consider the emission of a photon from an atom. Initially, the atom is in state $|a_i\rangle$ and, to keep things simple, is only allowed to decay into one final state $|a_f\rangle$. As previously no photon should be present in the initial state and only one in the final state. We write

[11] W. Greiner: *Quantum Mechanics – An Introduction*, 3rd ed. (Springer, Berlin, Heidelberg 1994).

[12] This was first examined by V. Weisskopf, E. Wigner: Z. Phys. **63** (1930) 54 and **65** (1930) 18.

$$\begin{aligned} |i\rangle &= |a_i\rangle|\ldots,0_{k\sigma},\ldots\rangle\,, \\ |f\rangle &= |a_f\rangle|\ldots,1_{k\sigma},\ldots\rangle\,. \end{aligned} \tag{2.93}$$

With states $|\psi\rangle$ of the total system expanded in terms of unperturbed states $|\phi_n\rangle$,

$$|\psi\rangle = \sum_n c_n(t)\exp\left(-\frac{\mathrm{i}E_n}{\hbar}t\right)|\phi_n\rangle\,, \tag{2.94}$$

the time-dependent Schrödinger equation reads

$$-\frac{\hbar}{\mathrm{i}}\frac{\partial|\psi\rangle}{\partial t} = (\hat{H}_0 + \hat{H}_{\mathrm{int}})|\psi\rangle\,. \tag{2.95}$$

Here $|\phi_n\rangle$ represent the solutions of the unperturbed system, which consists of the atomic many particle system and the radiation field $\hat{H}_0 = \hat{H}_{\mathrm{mp}} + \hat{H}_{\mathrm{rad}}$ [compare with (2.4)]:

$$\hat{H}_0|\phi_n\rangle = E_n|\phi_n\rangle\,. \tag{2.96}$$

Inserting (2.94) into (2.95) and considering (2.96) leads to a system of coupled differential equations:

$$\frac{\mathrm{d}c_m(t)}{\mathrm{d}t} = -\frac{\mathrm{i}}{\hbar}\sum_n c_n(t)\langle\phi_m|\hat{H}_{\mathrm{int}}|\phi_n\rangle\exp\left(\frac{\mathrm{i}(E_m-E_n)t}{\hbar}\right)\,. \tag{2.97}$$

We denote the amplitude of state $|i\rangle$ with c_{i0} and those of states $|f\rangle$ with $c_{fk\sigma}$. The indices indicate that $|i\rangle$ has no photon and $|f\rangle$ has one photon of the kind $k\sigma$. Then (2.97) read

$$\begin{aligned} \frac{\mathrm{d}c_{i0}}{\mathrm{d}t} &= -\frac{\mathrm{i}}{\hbar}\sum_{k\sigma}\langle i0|\hat{H}_{int}|fk\sigma\rangle\exp\left(\frac{\mathrm{i}}{\hbar}(E_{\mathrm{i}}-E_{\mathrm{f}}-\hbar\omega_k)t\right)c_{fk\sigma}\,, \\ \frac{\mathrm{d}c_{fk\sigma}}{\mathrm{d}t} &= -\frac{\mathrm{i}}{\hbar}\langle fk\sigma|\hat{H}_{\mathrm{int}}|i0\rangle\exp\left(-\frac{\mathrm{i}}{\hbar}(E_{\mathrm{i}}-E_{\mathrm{f}}-\hbar\omega_k)t\right)c_{i0}\,. \end{aligned} \tag{2.98}$$

Be aware that we consider only the two atomic states $|i\rangle$ and $|f\rangle$. On the right-hand side of the first equation all states $|a_f\rangle|k\sigma\rangle$ are considered, into which the initial state $|a_i\rangle|0\rangle$ can decay. On the right-hand side of the second equation only the term proportional to c_{i0} is taken into account; in principle also matrix elements with two-photon states $|a_f\rangle|\ldots,2_{k\sigma},\ldots\rangle$ appear, but the amplitudes of these states stay negligibly small. For the amplitudes c_{i0} of the decaying state $|a_i\rangle$ we make the ansatz

$$c_{i0}(t) = \exp\left(-\frac{\mathrm{i}}{\hbar}\Delta E_{\mathrm{i}}t\right)\,. \tag{2.99}$$

The energy shift ΔE_{i} remains unknown for the moment and still has to be determined. ΔE_{i} can be a real, imaginary, or complex-valued quantity. We insert (2.99) into the second equation of (2.98) and deduce

$$\frac{\mathrm{d}c_{fk\sigma}}{\mathrm{d}t} = -\frac{\mathrm{i}}{\hbar}\langle fk\sigma|\hat{H}_{\mathrm{int}}|i0\rangle\exp\left(-\frac{\mathrm{i}}{\hbar}(E_{\mathrm{i}}+\Delta E_{\mathrm{i}}-E_{\mathrm{f}}-\hbar\omega_k)t\right)\,.$$

After an integration over time from 0 to t we get

$$c_{fk\sigma}(t) = \langle f\boldsymbol{k}\sigma|\hat{H}_{\text{int}}|i0\rangle \frac{\exp\left[-\frac{\mathrm{i}}{\hbar}(E_\mathrm{i} + \Delta E_\mathrm{i} - E_\mathrm{f} - \hbar\omega_k)t\right] - 1}{E_\mathrm{i} + \Delta E_\mathrm{i} - E_\mathrm{f} - \hbar\omega_k} . \tag{2.100}$$

So far we have not used the first equation of (2.98) at all. Also, we did not determine ΔE_i. We choose ΔE_i such that the first equation of (2.98) is fulfilled. We insert (2.99) into the left side of the first equation of (2.98), substitute the amplitudes (2.100) into the right side, and get

$$\begin{aligned}
&-\frac{\mathrm{i}}{\hbar}\Delta E_\mathrm{i} \exp\left(-\frac{\mathrm{i}}{\hbar}\Delta E_\mathrm{i} t\right) \\
&= -\frac{\mathrm{i}}{\hbar}\sum_{k\sigma}\langle i0|\hat{H}_{\text{int}}|f\boldsymbol{k}\sigma\rangle\langle f\boldsymbol{k}\sigma|\hat{H}_{\text{int}}|i0\rangle \\
&\quad \times \frac{\exp\left[-\frac{\mathrm{i}}{\hbar}(E_\mathrm{i} + \Delta E_\mathrm{i} - E_\mathrm{f} - \hbar\omega_k)t\right] - 1}{E_\mathrm{i} + \Delta E_\mathrm{i} - E_\mathrm{f} - \hbar\omega_k} \exp\left(\frac{\mathrm{i}}{\hbar}(E_\mathrm{i} - E_\mathrm{f} - \hbar\omega_k)t\right) \\
&= -\frac{\mathrm{i}}{\hbar}\sum_{k\sigma}\left|\langle f\boldsymbol{k}\sigma|\hat{H}_{\text{int}}|i0\rangle\right|^2 \frac{1 - \exp\left[\frac{\mathrm{i}}{\hbar}(E_\mathrm{i} + \Delta E_\mathrm{i} - E_\mathrm{f} - \hbar\omega_k)t\right]}{E_\mathrm{i} + \Delta E_\mathrm{i} - E_\mathrm{f} - \hbar\omega_k} \\
&\quad \times \exp\left(-\frac{\mathrm{i}}{\hbar}(E_\mathrm{i} + \Delta E_\mathrm{i} - E_\mathrm{f} - \hbar\omega_k)t + \frac{\mathrm{i}}{\hbar}(E_\mathrm{i} + \Delta E_\mathrm{i} - E_\mathrm{f} - \hbar\omega_k)t\right) .
\end{aligned}$$

Division by $(-\mathrm{i}/\hbar)\exp(-\frac{\mathrm{i}}{\hbar}\Delta E_\mathrm{i} t)$ on both sides results in

$$\begin{aligned}
\Delta E_\mathrm{i} &= \sum_{k\sigma}\left|\langle f\boldsymbol{k}\sigma|\hat{H}_{\text{int}}|i0\rangle\right|^2 \\
&\quad \times \left(\frac{1 - \exp\left(\frac{\mathrm{i}}{\hbar}(E_\mathrm{i} + \Delta E_\mathrm{i} - E_\mathrm{f} - \hbar\omega_k)t\right)}{E_\mathrm{i} + \Delta E_\mathrm{i} - E_\mathrm{f} - \hbar\omega_k}\right) .
\end{aligned} \tag{2.101}$$

For large times $t \to \infty$ it holds that

$$\lim_{t\to\infty}\left(\frac{1 - \mathrm{e}^{\mathrm{i}xt}}{x}\right) = \frac{1}{x} - \mathrm{i}\pi\delta(x) . \tag{2.102}$$

We prove this important relation in Exercise 2.13. Applied to (2.101) this leads to

$$\begin{aligned}
\Delta E_\mathrm{i} &= \sum_{k\sigma}\frac{\left|\langle f\boldsymbol{k}\sigma|\hat{H}'_{\text{int}}|i0\rangle\right|^2}{E_\mathrm{i} + \Delta E_\mathrm{i} - E_\mathrm{f} - \hbar\omega_k} \\
&\quad - \mathrm{i}\pi\sum_{k\sigma}\left|\langle f\boldsymbol{k}\sigma|\hat{H}'_{\text{int}}|i0\rangle\right|^2\delta(E_\mathrm{i} + \Delta E_\mathrm{i} - E_\mathrm{f} - \hbar\omega_k) .
\end{aligned}$$

Since we want to determine ΔE_i only up to second order in e, i.e. in $\hat{H}'_{\text{int}}$, it is consistent to leave out ΔE_i on the right-hand side of the last equation. Consequently, we obtain

$$\begin{aligned}
\Delta E_\mathrm{i} &= \sum_{k,\sigma}\frac{\left|\langle f\boldsymbol{k}\sigma|\hat{H}'_{\text{int}}|i0\rangle\right|^2}{E_\mathrm{i} - E_\mathrm{f} - \hbar\omega_k} \\
&\quad - \mathrm{i}\pi\sum_{k,\sigma}\left|\langle f\boldsymbol{k}\sigma|\hat{H}'_{\text{int}}|i0\rangle\right|^2\delta(E_\mathrm{i} - E_\mathrm{f} - \hbar\omega_k) .
\end{aligned} \tag{2.103}$$

Hence, the energy shift ΔE_i is complex valued. The real and imaginary parts can be read off right away from (2.103):

$$\Re(\Delta E_i) = \sum_{k,\sigma} \frac{|\langle f k\sigma|\hat{H}'_{int}|i0\rangle|^2}{E_i - E_f - \hbar\omega_k}, \tag{2.104a}$$

$$\Im(\Delta E_i) = -\pi \sum_{k,\sigma} |\langle f k\sigma|\hat{H}'_{int}|i0\rangle|^2 \delta(E_i - E_f - \hbar\omega_k). \tag{2.104b}$$

The sum $\sum_{k\sigma}$ is over all photons with all possible momenta $\hbar k$ and polarizations σ. It is straightforward to see that a consideration of several atomic states $|a_f\rangle$ leads to an additional summation $\sum_f$ for the right-hand side of the first equation of (2.98); hence, the result (2.104) modifies to

$$\Re(\Delta E_i) = \sum_{k,\sigma} \sum_f \frac{|\langle f k\sigma|\hat{H}'_{int}|i0\rangle|^2}{E_i - E_f - \hbar\omega_k}, \tag{2.105a}$$

$$\Im(\Delta E_i) = -\pi \sum_{k,\sigma} \sum_f |\langle f k\sigma|\hat{H}'_{int}|i0\rangle||^2 \delta(E_i - E_f - \hbar\omega_k). \tag{2.105b}$$

The sum over final states ($\sum_f$) is not restricted in (2.105a); on the other hand, because of the δ function, the sum in the second equation is restricted by $E_f < E_i$.

We draw the conclusion that the processes of emission and absorption of photons as described by (2.105) generally do not need to fulfill energy conservation. Such processes are called *virtual processes.* We can think in terms of the physical atom having dissociated for a short time into "atom + virtual photon". According to (2.104) and (2.105) it is these virtual photons of all possible momenta and polarizations that cause the real energy shift. The diagrams in Fig. 2.21 illustrate this point again.

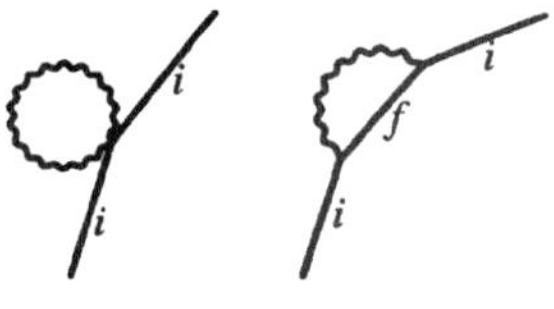

Fig. 2.21. Self-energy of a bound electron. Photons are emitted and reabsorbed. Such photons are called virtual photons. Also the electron in the intermediate state f (between emission and reabsorption of the photon) is virtual

Contrary to this the sums in (2.104b) and (2.105b) only cover photons that have energy conservation:

$$E_i = E_f + \hbar\omega.$$

In contrast to the virtual photons discussed before these photons are *real.* Obviously the energy shift ΔE_i possesses an imaginary part only if the state $|a_i\rangle$ is allowed to decay into $|f\rangle$ via spontaneous emission without *violation* of energy conservation; see Fig. 2.22. This is expressed by the δ-function $\delta(E_i - E_f - \hbar\omega)$.

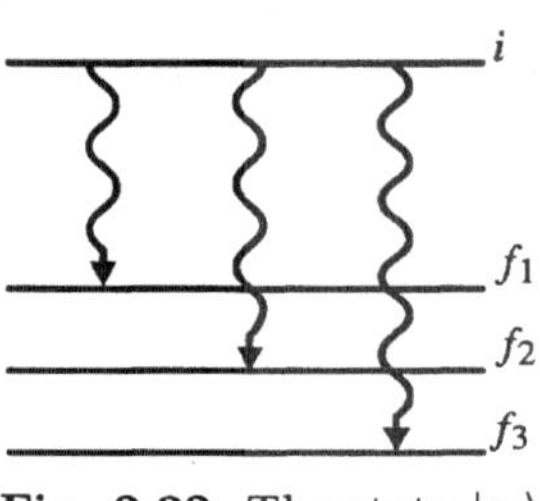

Fig. 2.22. The state $|a_i\rangle$ can only decay into energetically lower states $|a_f\rangle$

Equation (2.105b) can also be cast into the form

$$-\frac{2}{\hbar}\Im(\Delta E_i) = \sum_{k\sigma} \sum_f \frac{2\pi}{\hbar} |\langle f k\sigma|\hat{H}'_{int}|i0\rangle|^2 \delta(E_i - E_f - \hbar\omega),$$

where we recognize the expression (2.17) for the reciprocal lifetime $1/\tau_i$ of the state $|a_i\rangle$ on the right-hand side. Therefore, we identify

$$\frac{1}{\tau_i} = \gamma_i = -\frac{2}{\hbar}\Im(\Delta E_i). \tag{2.106}$$

The physical interpretation of $\Im(\Delta E_i)$ has now become clear. We go back to (2.99) and combine it with the "usual" time dependence $\exp(-\frac{i}{\hbar}E_i t)$ of the state $|a_i\rangle$; we deduce

$$\begin{aligned}\psi_{a_i} &= u_{a_i}(\boldsymbol{x})\exp\left(-\frac{i}{\hbar}(E_i+\Delta E_i)t\right)\\ &= u_{a_i}(\boldsymbol{x})\exp\left(-\frac{i}{\hbar}(E_i+\Re(\Delta E_i))t\right)\exp\left(-\frac{\gamma_i}{2}t\right)\end{aligned}\tag{2.107}$$

or

$$|\psi_{a_i}|^2 \sim \exp(-\gamma_i t) = \exp\left(-\frac{t}{\tau_i}\right).\tag{2.108}$$

The state $|a_i\rangle$ *decays* with lifetime τ. On the same footing the state obtains a *level shift* $\Re(\Delta E)$, which results from the emission and reabsorption of (virtual) photons (see Fig. 2.21).

Another consequence from the decay of state $|a_i\rangle$ becomes clear by looking at (2.100). Inserting

$$\Delta E_i = \Re(\Delta E_i) - \frac{i}{2}\gamma_i\hbar$$

yields

$$c_{f\boldsymbol{k}\sigma}(t) = \langle f\boldsymbol{k}\sigma|\hat{H}'_{\text{int}}|i0\rangle\frac{\exp\left(-\frac{i}{\hbar}(E_i+\Re(\Delta E_i)-\hbar\omega_k)t\right)\exp(-\gamma_i t/2)-1}{E_i+\Re(\Delta E_i)-E_f-\hbar\omega_k+(i/2)\hbar\gamma_i}.$$

The probability of finding a photon with frequency ω_k, momentum $\hbar\boldsymbol{k}$, and polarization σ in the radiation field after a long time $t \gg 1/\gamma_i$ is given by

$$\begin{aligned}\lim_{t\to\infty}|c_{f\boldsymbol{k}\sigma}|^2 &= \left|\langle f\boldsymbol{k}\sigma|\hat{H}'_{\text{int}}|i0\rangle\right|^2\\ &\quad\times\frac{1}{(E_i+\Re(\Delta E_i)-E_f-\hbar\omega_k)^2+\hbar^2\gamma_i^2/4}.\end{aligned}\tag{2.109}$$

Of course this reflects the intensity distribution of the emitted line: The spectral line has a *Breit–Wigner distribution* with center $\hbar\omega_k = E_i + \Re(\Delta E_i) - E_f$ and width $\hbar\gamma_i$; see Fig. 2.23. Because of the self-energy of the electron, emission and reabsorption of photons results in the spectral line being shifted by $\Re(\Delta E_i)$. In Sect. 5 we will come back to the determination and the necessary renormalization of the self-energy in more detail.

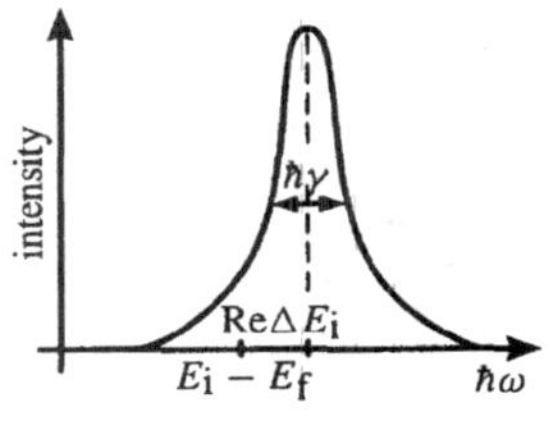

Fig. 2.23. Breit–Wigner form of the spectral line

EXERCISE

2.13 Plemlj's Formula

Problem. Prove the relation

$$\lim_{t\to\infty}\left(\frac{1-e^{ixt}}{x}\right) = P\left(\frac{1}{x}\right) - i\pi\delta(x).\tag{1}$$

Exercise 2.13.

Solution. The relation

$$\lim_{t\to\infty}\left(\frac{1-\mathrm{e}^{\mathrm{i}xt}}{x}\right) = \lim_{t\to\infty}\left(-\mathrm{i}\int_0^t \mathrm{e}^{\mathrm{i}xt'}\,\mathrm{d}t'\right) \tag{2}$$

holds because

$$-\mathrm{i}\int_0^t \mathrm{e}^{\mathrm{i}xt'}\,\mathrm{d}t' = -\mathrm{i}\,\frac{1}{\mathrm{i}x}\,\mathrm{e}^{\mathrm{i}xt'}\Big|_{t'=0}^{t'=t} = \frac{1}{x}-\frac{\mathrm{e}^{\mathrm{i}xt}}{x}\,. \tag{3}$$

Now, let us introduce a small imaginary part and let ε go to 0^+:

$$-\lim_{t\to\infty}\left(\mathrm{i}\int_0^t \mathrm{e}^{\mathrm{i}xt'}\mathrm{d}t'\right) = -\lim_{t\to\infty}\lim_{\varepsilon\to 0^+}\left(\mathrm{i}\int_0^t \mathrm{e}^{\mathrm{i}(x+\mathrm{i}\varepsilon)t'}\,\mathrm{d}t'\right)\,; \tag{4}$$

uniform convergence in ε is assumed. We interchange the two limiting procedures

$$\begin{aligned}
-\lim_{t\to\infty}\lim_{\varepsilon\to 0^+}\left(\mathrm{i}\int_0^t \mathrm{e}^{\mathrm{i}(x+\mathrm{i}\varepsilon)t'}\,\mathrm{d}t'\right) &= -\lim_{\varepsilon\to 0^+}\lim_{t\to\infty}\left(\mathrm{i}\int_0^t \mathrm{e}^{\mathrm{i}(x+\mathrm{i}\varepsilon)t'}\,\mathrm{d}t'\right)\\
&= -\lim_{\varepsilon\to 0^+}\left(\mathrm{i}\int_0^\infty \mathrm{e}^{\mathrm{i}(x+\mathrm{i}\varepsilon)t'}\,\mathrm{d}t'\right)\,.
\end{aligned} \tag{5}$$

The last step is possible because the integral is defined as long as $\varepsilon > 0$ and also remains finite for $t \to \infty$.

This integration can be performed:

$$\int_0^\infty \mathrm{e}^{\mathrm{i}(x+\mathrm{i}\varepsilon)t'}\,\mathrm{d}t' = \frac{1}{\mathrm{i}(x+\mathrm{i}\varepsilon)}\,\mathrm{e}^{\mathrm{i}(x+\mathrm{i}\varepsilon)t'}\Big|_{t'=0}^{t'=\infty}\,. \tag{6}$$

Since this integral also converges at the upper bound because of the convergence factor $\exp(-\varepsilon t')$, i.e. $\exp(\mathrm{i}(x+\mathrm{i}\varepsilon)t')|_{t'=\infty} = 0$, we deduce that

$$\begin{aligned}
-\lim_{\varepsilon\to 0^+}\left(\mathrm{i}\int_0^\infty \mathrm{e}^{\mathrm{i}(x+\mathrm{i}\varepsilon)t'}\,\mathrm{d}t'\right) &= \lim_{\varepsilon\to 0^+}\left(\frac{1}{x+\mathrm{i}\varepsilon}\right)\\
&= \lim_{\varepsilon\to 0^+}\left(\frac{x-\mathrm{i}\varepsilon}{(x+\mathrm{i}\varepsilon)(x-\mathrm{i}\varepsilon)}\right)\\
&= \lim_{\varepsilon\to 0^+}\left(\frac{x}{x^2+\varepsilon^2}-\mathrm{i}\frac{\varepsilon}{x^2+\varepsilon^2}\right)\,.
\end{aligned} \tag{7}$$

The imaginary part represents a well-known representation of the δ function:

$$\lim_{\varepsilon\to 0^+}\left(\frac{\varepsilon}{x^2+\varepsilon^2}\right) = \pi\delta(x)\,. \tag{8}$$

To prove this we check the defining properties of the δ function:

$$\begin{aligned}
\int_{-\infty}^\infty \delta(x)\,\mathrm{d}x &= \lim_{\varepsilon\to 0^+}\left(\frac{1}{\pi}\int_{-\infty}^\infty \frac{\varepsilon}{x^2+\varepsilon^2}\,\mathrm{d}x\right)\\
&= \lim_{\varepsilon\to 0^+}\left(\frac{1}{\pi}\int_{-\infty}^\infty \frac{\mathrm{d}(x/\varepsilon)}{1+(x/\varepsilon)^2}\right)\\
&= \lim_{\varepsilon\to 0^+}\left(\frac{1}{\pi}\int_{-\infty}^\infty \frac{\mathrm{d}y}{1+y^2}\right)\\
&= 1\,,
\end{aligned} \tag{9}$$

Exercise 2.13.

$$\begin{aligned}\int_{-\infty}^{\infty} \delta(x) f(x)\,\mathrm{d}x &= \lim_{\varepsilon\to 0+}\left[\frac{1}{\pi}\int_{-\infty}^{\infty}\frac{\varepsilon}{x^2+\varepsilon^2}\right.\\ &\quad \left.\times\left(f(0)+xf'(0)+\frac{x^2}{2}f''(0)+\ldots\right)\mathrm{d}x\right]\\ &= f(0)\cdot 1+\lim_{\varepsilon\to 0+}\left(\frac{f'(0)}{\pi}\int_{-\infty}^{\infty}\frac{x\varepsilon\,\mathrm{d}x}{x^2+\varepsilon^2}+\ldots\right)\\ &= f(0)+\lim_{\varepsilon\to 0+}\left(\frac{f'(0)}{\pi}\varepsilon\int_{-\infty}^{\infty}\frac{y\,\mathrm{d}y}{1+y^2}\right)\\ &= f(0)\,. \end{aligned} \tag{10}$$

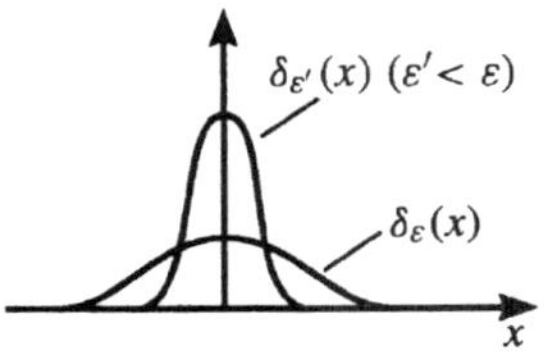

Fig. 2.24. Approximation of the δ function

Thus, $\delta_\varepsilon(x) = \varepsilon/\pi(x^2+\varepsilon^2)$ approximates the δ function for small and decreasing ε; see Fig. 2.24. The real part of the last part of (7),

$$\lim_{\varepsilon\to 0+}\left(\frac{x}{x^2+\varepsilon^2}\right) = P\left(\frac{1}{x}\right) \tag{11}$$

represents the so-called *principal value* of $1/x$, which is defined by

$$\int_{-\infty}^{\infty} P\left(\frac{1}{x}\right) f(x)\,\mathrm{d}x = \lim_{\varepsilon\to 0+}\left(\int_{-\infty}^{-\varepsilon}\frac{f(x)}{x}\,\mathrm{d}x+\int_{\varepsilon}^{\infty}\frac{f(x)}{x}\,\mathrm{d}x\right). \tag{12}$$

Hence, we obtain

$$\lim_{t\to\infty}\left(-\mathrm{i}\int_0^t \mathrm{e}^{\mathrm{i}xt'}\,\mathrm{d}t'\right) = P\left(\frac{1}{x}\right) - \mathrm{i}\pi\delta(x)\,. \tag{13}$$

3. Noninteracting Fields

So far we have learnt that the classical radiation field takes on particle properties (quanta with energy $\hbar\omega$ and momentum $\hbar k$) as soon as it has been quantized. Thus, we might suspect that all wave fields show particle character as soon as they have been quantized and that, on the other side, all particles appearing in nature could be understood as quanta of a field. With certain particles, such as electrons, the question arises, what is the wave field that contains these particles as quanta? We presume that it might be the wavefunction $\psi(x,t)$ and begin with the nonrelativistic Schrödinger equation

$$\mathrm{i}\hbar\frac{\partial\psi(\boldsymbol{x},t)}{\partial t} = -\frac{\hbar^2}{2m}\boldsymbol{\nabla}^2\psi(\boldsymbol{x},t) + V(\boldsymbol{x})\psi(\boldsymbol{x},t) = \hat{\overline{H}}\psi(\boldsymbol{x},t) \tag{3.1}$$

for a particle with mass m in a potential $V(x)$. Equation (3.1) is the fundamental equation of elementary quantum mechanics. As with the Maxwell equations representing the field equations for the electromagnetic fields (combined in the vector potential $\boldsymbol{A}$ and Coulomb potential ϕ), we consider the Schrödinger equation to be a field equation for the wave field $\psi(x,t)$. Now we quantize Schrödinger's wave field $\psi(x,t)$ in complete analogy to the quantization (1.50) of the vector potential (and with that of the electric and magnetic fields) in the electromagnetic case. We assume $\{\psi_n(x)\}$ to be a complete set of solutions of $\hat{\overline{H}}$,

$$\hat{\overline{H}}\psi_n(\boldsymbol{x}) = E_n\psi_n(\boldsymbol{x})\,, \tag{3.2}$$

so that we can write the most general solution of (3.1) as

$$\psi(\boldsymbol{x},t) = \sum_n b_n(t)\psi_n(\boldsymbol{x})\,. \tag{3.3}$$

The b_n can be understood as normal coordinates of the system; then $\psi_n(\boldsymbol{x})$ are called normal states. Insertion of (3.3) into (3.1) yields

$$\frac{\mathrm{d}b_n(t)}{\mathrm{d}t} = -\frac{\mathrm{i}}{\hbar}E_n b_n\,. \tag{3.4}$$

This represents the equations of motion for the normal coordinates. It is called *first* (i.e. the usual) *quantization.* In contrast to that formulation we will now learn how to introduce creation and annihilation operators for field quanta (which may be particles, photons, and so on). The formulation of quantum mechanics in terms of such creation and annihilation operators is called *second quantization.* To achieve this, our goal is now to find a Hamiltonian depending only on the normal coordinates b_n, introduced above, and yielding the equations of motion (3.4). It could be that H is given by

$$\begin{aligned} H &= \int d^3x\psi^*(\boldsymbol{x},t)\left[-\frac{\hbar^2}{2m}\boldsymbol{\nabla}^2 + V(\boldsymbol{x})\right]\psi(\boldsymbol{x},t) \\ &= \int d^3x\, \psi^*(\boldsymbol{x},t)\hat{\overline{H}}\psi(\boldsymbol{x},t)\,, \end{aligned} \tag{3.5}$$

because it represents the expectation value of the energy. The quantity

$$\psi^*\left[-\frac{\hbar^2}{2m}\boldsymbol{\nabla}^2 + V(x)\right]\psi$$

represents the energy density. With the help of (3.2) and (3.3) and employing the orthonormality $\langle\psi_n|\psi_m\rangle = \delta_{nm}$ we straightforwardly calculate for (3.5)

$$\begin{aligned} H &= \int d^3x\psi^*(\boldsymbol{x},t)\hat{\overline{H}}\psi(\boldsymbol{x},t) \\ &= \sum_n E_n b_n^* b_n\,. \end{aligned} \tag{3.6}$$

This resembles a Hamiltonian consisting of an infinite sum of harmonic oscillators with energies E_n and frequencies $\omega_n = E_n/\hbar$. Hence, we interpret

$$\begin{aligned} b_n^* &\to \hat{b}_n^\dagger\,, \\ b_n &\to \hat{b}_n \end{aligned} \tag{3.7}$$

as *creation and annhilation operators* of these oscillator quanta and require the commutation relations

$$\begin{aligned} \left[\hat{b}_n,\hat{b}_{n'}\right]_- &= \left[\hat{b}_n^\dagger,\hat{b}_{n'}^\dagger\right]_- = 0\,, \\ \left[\hat{b}_n,\hat{b}_{n'}^\dagger\right]_- &= \delta_{nn'}\,. \end{aligned} \tag{3.8}$$

Then the Hamiltonian of (3.6) becomes

$$\hat{H} = \sum_n E_n\hat{b}_n^\dagger\hat{b}_n \tag{3.9}$$

and the equations of motion for $\hat{b}_n$ read

$$i\hbar\frac{d\hat{b}_n}{dt} = \left[\hat{b}_n,\hat{H}\right]_- \tag{3.10a}$$

or

$$\begin{aligned} i\hbar\frac{d\hat{b}_n}{dt} &= \sum_{n'}\left[\hat{b}_n, E_{n'}\hat{b}_{n'}^\dagger\hat{b}_{n'}\right] \\ &= \sum_{n'} E_{n'}\left[\hat{b}_n\hat{b}_{n'}^\dagger\hat{b}_{n'} - \hat{b}_{n'}^\dagger\hat{b}_{n'}\hat{b}_n\right] \\ &= \sum_{n'} E_{n'}\left[(\hat{b}_{n'}^\dagger\hat{b}_n + \delta_{nn'})\hat{b}_{n'} - \hat{b}_{n'}^\dagger\hat{b}_{n'}\hat{b}_n\right] \\ &= \sum_{n'} E_{n'}\hat{b}_{n'}\delta_{nn'} \\ &= E_n\hat{b}_n\,. \end{aligned} \tag{3.10b}$$

Formally they are identical to the "classical" equations of motion (3.4). As in Chap. 1, where the quanta of the electromagnetic field (photons) appeared explicitly in the theory, the theory developed here describes quanta that obey *Bose–Einstein statistics.* The reason for this lies in the commutation relations (3.8), which are the same as in (1.50). These commutation relations allow us to construct states of the form

$$\begin{aligned} |N_m\rangle &= \frac{1}{\sqrt{N_m!}} \overbrace{\hat{b}_m^\dagger \hat{b}_m^\dagger \cdots \hat{b}_m^\dagger}^{N_m \text{times}} |0\rangle \\ &= \frac{1}{\sqrt{N_m!}} (\hat{b}_m^\dagger)^{N_m} |0\rangle \,, \end{aligned} \tag{3.11}$$

in which N particles of the kind m appear with the same wavefunction $\psi_m(x)$. The *number operator* $\hat{N}_n$ for particles of the kind "n" is

$$\hat{N}_n = \hat{b}_n^\dagger \hat{b}_n \,, \tag{3.12}$$

and the general state vector

$$|\cdots, n_n, \cdots, n_{n'}, \cdots\rangle \tag{3.13}$$

of the ψ field is, as previously [compare with (1.54)], a direct product of the state vectors of the field oscillators $|n_n\rangle$

$$|\cdots, n_n, \cdots, n_{n'}, \cdots\rangle = \cdots, |n_n\rangle \cdots |n_{n'}\rangle \cdots \,. \tag{3.14}$$

Here

$$\hat{N}_n |\cdots, n_n, \cdots, n_{n'}, \cdots\rangle = n_n |\cdots, n_n, \cdots, n_{n'}, \cdots\rangle \tag{3.15}$$

represents the eigenvalue equation for the particle-number operator.

The formalism developed so far holds for *bosons*, because, as we have realized, it allows several particles to have identical wavefunctions. Now, we attempt to modify the formalism, so that it can also describe *fermions.* We keep

$$\hat{H} = \sum_n E_n \hat{b}_n^\dagger \hat{b}_n$$

as the Hamiltonian. We also require that the equations of motion (3.10b) again lead to the classical equations (3.4). After some trials we observe that the result (3.10b) can be derived from (3.10a) also with commutation relations different from (3.8). For example with the *anticommutation relations*[1]

$$\begin{aligned} \left[\hat{b}_n, \hat{b}_{n'}\right]_+ &= \left[\hat{b}_n^\dagger, \hat{b}_{n'}^\dagger\right]_+ = 0 \,, \\ \left[\hat{b}_n^\dagger, \hat{b}_{n'}\right]_+ &= \delta_{nn'} \end{aligned} \tag{3.16}$$

instead of the commutation relations (3.8), where the anticommutation operation is given by

$$\left[\hat{A}, \hat{B}\right]_+ = \hat{A}\hat{B} + \hat{B}\hat{A} \,.$$

---

[1] Those go back to P. Jordan, E. Wigner: Z. Physik **47** (1928) 631.

We easily calculate with (3.9)

$$
\begin{aligned}
\mathrm{i}\hbar \frac{\mathrm{d}\hat{b}_n}{\mathrm{d}t} &= \left[\hat{b}_n, \hat{H}\right]_- \\
&= \left[\hat{b}_n, \sum_{n'} E_{n'} \hat{b}^\dagger_{n'} \hat{b}_{n'}\right] \\
&= \sum_{n'} E_{n'} \left[\hat{b}_n, \hat{b}^\dagger_{n'} \hat{b}_{n'}\right]_- \\
&= \sum_{n'} E_{n'} \left[\hat{b}_n \hat{b}^\dagger_{n'} \hat{b}_{n'} - \hat{b}^\dagger_{n'} \hat{b}_{n'} \hat{b}_n\right] \\
&= \sum_{n'} E_{n'} \left[(\delta_{nn'} - \hat{b}^\dagger_{n'} \hat{b}_n) \hat{b}_{n'} - \hat{b}^\dagger_{n'} \hat{b}_{n'} \hat{b}_n\right] \\
&= E_n \hat{b}_n + \sum_{n'} E_{n'} \left[-\hat{b}^\dagger_{n'} \hat{b}_n \hat{b}_{n'} - \hat{b}^\dagger_{n'} \hat{b}_{n'} \hat{b}_n\right] \\
&= E_n \hat{b}_n + \sum_{n'} E_{n'} \underbrace{\left[\hat{b}^\dagger_{n'} \hat{b}_{n'} \hat{b}_n - \hat{b}^\dagger_{n'} \hat{b}_{n'} \hat{b}_n\right]}_{=0} \\
&= E_n \hat{b}_n \,,
\end{aligned}
\tag{3.17}
$$

and recognize that we have derived the same equations of motions as previously in (3.10b). Notice, however, that the anticommutation relations (3.16) do not possess the same *algebraic properties* as the commutation relations do.

The commutation relations (3.8) possess the same algebra as the *classical Poisson brackets*

$$
\{A, B\}_{q,p} = \sum_{i=1}^{f} \left(\frac{\partial A}{\partial q_i} \frac{\partial B}{\partial p_i} - \frac{\partial B}{\partial q_i} \frac{\partial A}{\partial p_i} \right) . \tag{3.18}
$$

We elaborate on this point in some more detail.

The canonical momenta p_j and the corresponding coordinates q_j obviously fulfill the equations

$$
\begin{aligned}
\{q_i, p_j\}_{q,p} &= \delta_{ij} \,, \\
\{q_i, q_j\}_{q,p} &= \{p_i, p_j\}_{q,p} = 0 \,.
\end{aligned}
$$

Different *canonical variables* Q_i and P_j, once introduced, also have to obey

$$
\begin{aligned}
\{Q_i, P_j\}_{q,p} &= \delta_{ij} \,, \\
\{Q_i, Q_j\}_{q,p} &= \{P_i, P_j\}_{q,p} = 0 \,.
\end{aligned}
\tag{3.19}
$$

These are the fundamental Poisson bracket relations. They are independent of the set of coordinates and momenta q_i, p_j; therefore we can skip the index at the curly brackets. We observe from (3.8) and (3.16) that the commutators (3.8) as well as the anticommutators (3.16) formally obey the same rules. However, the *algebraic properties of the Poisson brackets*, which we will list in the following, resemble those of the commutators (3.8) only and not of the anticommutators (3.16). As a result of the definition (3.18) we calculate

$$\begin{aligned}\{A,B\} &= -\{B,A\}\\ \{A,C\} &= 0\,, \qquad \text{if } c \text{ is a number}\,,\\ \{(A_1+A_2),B\} &= \{A_1,B\}+\{A_2,B\}\,,\\ \{A_1A_2,B\} &= \{A_1,B\}A_2+A_1\{A_2,B\}\,,\\ \{A,\{B,C\}\} &+ \{B,\{C,A\}\} + \{C,\{A,B\}\} = 0 \quad \text{(Jakobi identity)}\,.\end{aligned} \tag{3.20}$$

Here we have paid attention to the ordering of the various quantities to include possibly noncommuting quantities. We will demonstrate in Exercise 3.1 that the commutators (3.8) do indeed fulfill the equations analogous to (3.20), whereas the anticommutators (3.16) do not fulfill these equations. Hence, we conclude that the *anticommutators do not have a classical analogy.* This should not lead to a misunderstanding: the number operator $\hat{N} = \sum_n \hat{N}_n$ and the Hamiltonian $\hat{H} = \sum_n E_n \hat{N}_n$ have classical limits because both are bilinear in $\hat{b}_n^\dagger$ and $\hat{b}_n$ and commute with each other. This shows that the anticommutation relations represent something typically new, which appears only in quantum mechanics. We will see soon that they incorporate the Pauli principle, which does not exist on a classical level. These conclusions are supported by the following physical observation. To become measurable classically a field amplitude has to be very strong; it must be possible to have a large number of particles in the same state so that their fields can add up coherently (see Sect. 1.5 on coherent states). This implies that particles that give rise to classical fields have to obey Bose–Einstein statistics. We conclude, for example, that light quanta have to be Bose particles because strong electromagnetic fields can be produced and measured classically. For the case of electrons in a metal, which obey Fermi–Dirac statistics, only quantities like the energy, charge, or current density are measurable classically; these quantities are bilinear combinations of the field amplitudes (operators $\hat{b}^+$ and $\hat{b}$). The amplitude of the electron field alone is linear only in $\hat{b}^+$ or $\hat{b}$ and cannot be measured classically.

EXERCISE

3.1 Do the Commutators and Anticommutators Fulfill the Poisson Bracket Algebra?

Problem. Show that the commutation relations (3.8) but not the anticommutation relations (3.16) fulfill the algebra formulated in (3.20).

Solution. For the commutators $[A,B] = AB - BA$ we calculate explicitly:

1) $[A,B] = AB - BA = -(BA - AB) = -[B,A]\,,$

2) $[A_1+A_2,B] = (A_1+A_2)B - B(A_1+A_2)$
$= A_1B - BA_1 + A_2B - BA_2$
$= [A_1,B] + [A_2,B]\,,$

Exercise 3.1.

3) $[A_1, B]\,A_2 + A_1\,[A_2, B] = (A_1B - BA_1)A_2 + A_1(A_2B - BA_2)$
$= A_1BA_2 - BA_1A_2 + A_1A_2B - A_1BA_2$
$= A_1A_2B - BA_1A_2 = [A_1A_2, B]\ ,$

4) $[A, [B, C]] + [B, [C, A]] + [C, [A, B]]$
$= A\,[B, C] - [B, C]\,A + B\,[C, A] - [C, A]\,B + C\,[A, B] - [A, B]\,C$
$= A(BC - CB) - (BC - CB)A + B(CA - AC)$
$\quad - (CA - AC)B + C(AB - BA) - (AB - BA)C$
$= ABC - ACB - BCA + CBA + BCA - BAC$
$\quad - CAB + ACB + CAB - CBA - ABC + BAC = 0\,.$

Indeed, the commutators fulfill the algebra (3.20). Now we consider the anti-commutators $\{A, B\} = AB + BA$:

1) $\{A, B\} = AB + BA \neq -\{B, A\} = -(AB + BA)\,,$
this relation does not hold!

2) $\{A_1 + A_2, B\} = (A_1 + A_2)B + B(A_1 + A_2)$
$= A_1B + BA_1 + A_2B + BA_2 = \{A_1, B\} + \{A_2, B\}\,,$

3) $\{A_1A_2, B\} = A_1A_2B + BA_1A_2\,,$ whereas
$\{A_1, B\}A_2 + A_1\{A_2, B\} = A_1BA_2 + BA_1A_2 + A_1A_2B + A_1BA_2$
$= \{A_1A_2, B\} + 2A_1BA_2 \neq \{A_1A_2, B\}\,,$
this relation does not hold either!

4) $\{A, \{B, C\}\} + \{B, \{C, A\}\} + \{C, \{A, B\}\} \neq 0\,,$
since the anticommutators yield only terms with *positive* sign, which do not cancel.

Now we determine the eigenvalues of $\hat{N}_n = \hat{b}^\dagger_n \hat{b}_n$. The following relation is helpfull for the case of the Fermi–Dirac statistics:

$$\begin{aligned}(\hat{b}^\dagger_n \hat{b}_n)(\hat{b}^\dagger_n \hat{b}_n) &= \hat{b}^\dagger_n (1 - \hat{b}^\dagger_n \hat{b}_n)\hat{b}_n \\ &= \hat{b}^\dagger_n \hat{b}_n - \hat{b}^\dagger_n \hat{b}^\dagger_n \hat{b}_n \hat{b}_n \\ &= \hat{b}^\dagger_n \hat{b}_n\,.\end{aligned} \tag{3.21}$$

The term $\hat{b}^\dagger_n \hat{b}^\dagger_n \hat{b}_n \hat{b}_n$ vanishes because (3.16) means it is antisymmetric with respect to the exchange of the first two (or the last two) operators. From (3.16) it follows that

$$\hat{b}^\dagger_n \hat{b}^\dagger_n = 0 = \hat{b}_n \hat{b}_n\,.$$

In shorthand notation, relation (3.21) can be written as $\hat{N}^2_n = \hat{N}_n$. With n_n we denote the eigenvalue of $\hat{b}^\dagger_n \hat{b}_n$; then

$$\hat{N}_n |n_n\rangle = \hat{b}^\dagger_n \hat{b}_n |n_n\rangle = n_n |n_n\rangle \tag{3.22}$$

holds and due to (3.21) it follows that

$$\hat{N}^2_n |n_n\rangle = \hat{b}^\dagger_n \hat{b}_n \hat{b}^\dagger_n \hat{b}_n |n_n\rangle = n^2_n |n_n\rangle = n_n |n_n\rangle\,. \tag{3.23}$$

Therefore $n_n^2 = n_n$, so that

$$n_n = 0 \qquad \text{or} \qquad n_n = 1 . \tag{3.24}$$

We realize that for the case of Fermi statistics [anticommutation relations (3.16)] at most one particle can appear with the wavefunction $\psi_n(x)$. We say: *only one particle can occupy the state* $|\psi_n\rangle$ *for the case of Fermi–Dirac statistics.* This is the *Pauli principle.* It stems from the introduction of the commutation relations (3.16) (for fermions). They represent the fundamental element of the theory.

We add that generalizations of this statistic exist. They were introduced by Green,[2] Greenberg and Messiah,[3] and Ohnuki and Kamefuchi[4] and are called *parastatistics.* For example, it is possible to bring in s particles into one state; we call this parastatistics of degree s. In its limiting cases, it can represent a Bose as well as a Fermi statistic. The fundamental commutation relations are not given by twofold commutators but by threefold commutators:

$$\left[\hat{a}_i, \left[\hat{a}_j^\dagger, \hat{a}_k\right]_\mp\right]_- = 2\delta_{ij}\hat{a}_k . \tag{3.25}$$

A particle that obeys parastatistics can be thought of as a composite particle with intrinsic degrees of freedom. For example, parastatistics has been applied to atomic nuclei, which have been interpreted as parabosons (with integer spin) or parafermions (with half-integer spin), respectively.[5] In this respect the transition from even–even nuclei as pure bosons to complex particles with intrinsic degrees of freedom can be described ("loss of the bosonic character"). Intrinsic degrees of freedom are not important for even–even nuclei as initial products; they are "frozen". For the final products these intrinsic degrees of freedom are "open". We shall further discuss this point in the following exercises.

EXERCISE

3.2 Threefold Commutators from an Expansion of Paraoperators

Problem. Often it is convenient to expand the paraoperators into usual operators, i.e.

$$\hat{a}_i = \sum_{\alpha=1}^{s} \hat{a}_i(\alpha) ,$$

where s determines the degree of the parastatistics. Here the operators $\hat{a}_i(\alpha)$ obey the standard commutation relations

[2] H.S. Green: Phys. Rev. **90** (1953) 270.
[3] O.W. Greenberg, A.M.L. Messiah: Phys. Rev. **138** (1965) B1155.
[4] Y. Ohnuki, S. Kamefuchi: Phys. Rev. **170** (1968) 1279; Ann. Phys. NY **51** (1969) 337; Ann. Phys. NY **78** (1973) 64; Prog. Theor. Phys. **50** (1973) 258.
[5] See for example H.J. Fink, W. Scheid, W. Greiner: J. Phys. G **3** (1977) 1119.

Exercise 3.2.

$$\left.\begin{aligned}\left[\hat{a}_i(\alpha),\hat{a}_j^\dagger(\alpha)\right] &= \delta_{ij}\,,\\ \left\{\hat{a}_i(\alpha),\hat{a}_j^\dagger(\beta)\right\} &= 0 \quad \text{for} \quad \alpha\neq\beta\end{aligned}\right\}\ \text{for the para-Bose statistics}$$

or

$$\left.\begin{aligned}\left\{\hat{a}_i(\alpha),\hat{a}_j^\dagger(\alpha)\right\} &= \delta_{ij}\,,\\ \left[\hat{a}_i(\alpha),\hat{a}_j^\dagger(\beta)\right] &= 0 \quad \text{for} \quad \alpha\neq\beta\end{aligned}\right\}\ \text{for the para-Fermi statistics.}$$

Show that the threefold commutators for the paraoperators follow from these commutation relations. For convenience we have denoted the anticommutator by curly brackets: $\{\,,\} = [\,,]_+$.

Solution. First we show for the para-Bose case that

$$\begin{aligned}
\left[\hat{a}_i,\{\hat{a}_j^\dagger,\hat{a}_k\}\right] &= \left[\sum_{\alpha=1}^{s}\hat{a}_i(\alpha),\sum_{\beta=1}^{s}\sum_{\gamma=1}^{s}\{\hat{a}_j^\dagger(\beta),\hat{a}_k(\gamma)\}\right]\\
&= \left[\sum_{\alpha}\hat{a}_i(\alpha),\sum_{\beta}\{\hat{a}_j^\dagger(\beta),\hat{a}_k(\beta)\}\right]\\
&\qquad (\text{because}\quad \{\hat{a}_i(\alpha),\hat{a}_j^\dagger(\beta)\}=0 \quad \text{for } \alpha\neq\beta)\\
&= \sum_{\alpha,\beta}\left[\hat{a}_i(\alpha),\hat{a}_j^\dagger(\beta)\hat{a}_k(\beta)+\hat{a}_k(\beta)\hat{a}_j^\dagger(\beta)\right]\\
&= \sum_{\alpha,\beta}\left\{\left[\hat{a}_i(\alpha),\hat{a}_j^\dagger(\beta)\hat{a}_k(\beta)\right]+\left[\hat{a}_i(\alpha),\hat{a}_k(\beta)\hat{a}_j^\dagger(\beta)\right]\right\}\\
&= \sum_{\alpha,\beta}\Bigg\{\hat{a}_j^\dagger(\beta)\overbrace{[\hat{a}_i(\alpha),\hat{a}_k(\beta)]}^{=0}+\left[\hat{a}_i(\alpha),\hat{a}_j^\dagger(\beta)\right]\hat{a}_k^\dagger(\beta)\\
&\qquad +\hat{a}_k(\beta)\left[\hat{a}_i(\alpha),\hat{a}_j^\dagger(\beta)\right]+\underbrace{[\hat{a}_i(\alpha),\hat{a}_k(\beta)]}_{=0}\hat{a}_j^\dagger(\beta)\Bigg\}\\
&= \sum_{\alpha,\beta}\{\delta_{ij}\delta_{\alpha\beta}\hat{a}_k(\beta)+\delta_{ij}\delta_{\alpha\beta}\hat{a}_k(\beta)\}\\
&= 2\delta_{ij}\sum_{\alpha=1}^{s}\hat{a}_k(\alpha)
\end{aligned}$$

and consequently

$$\left[\hat{a}_i,\{\hat{a}_j^\dagger,\hat{a}_k\}\right] = 2\delta_{ij}\hat{a}_k\,.$$

An analogous proof holds for the para-Fermi statistics

$$\left[\hat{a}_i,\left[\hat{a}_j^\dagger,\hat{a}_k\right]\right] = \left[\sum_{\alpha=1}^{s}\hat{a}_i(\alpha),\left[\sum_{\beta=1}^{s}\hat{a}_j^\dagger(\beta),\sum_{\gamma=1}^{s}\hat{a}_k(\gamma)\right]\right]\,.$$

Exercise 3.2.

Since

$$\left[\hat{a}_i(\alpha), \hat{a}_j^\dagger(\beta)\right] = -\left[\hat{a}_j^\dagger(\beta), \hat{a}_i(\alpha)\right] = 0$$

for $\alpha \neq \beta$ it follows that

$$\begin{aligned}
\left[\hat{a}_i, \left[\hat{a}_j^\dagger, \hat{a}_k\right]\right] &= \left[\sum_{\alpha=1}^{s} \hat{a}_i(\alpha), \sum_{\beta=1}^{s} \left[\hat{a}_j^\dagger(\beta), \hat{a}_k(\beta)\right]\right] \\
&= \sum_{\alpha,\beta=1} \left[\hat{a}_i(\alpha), \hat{a}_j^\dagger(\beta)\hat{a}_k(\beta) + \hat{a}_k(\beta)\hat{a}_j^\dagger(\beta)\right] \\
&= \sum_{\alpha,\beta} \left\{ \left[\hat{a}_i(\alpha), \hat{a}_j^\dagger(\beta)\hat{a}_k(\beta)\right] + \left[\hat{a}_i(\alpha), \hat{a}_k(\beta)\hat{a}_j^\dagger(\beta)\right] \right\} \\
&= \sum_{\alpha,\beta} \left\{ \hat{a}_j^\dagger(\beta) \overbrace{\left[\hat{a}_i(\alpha), \hat{a}_k(\beta)\right]}^{=0} + \left[\hat{a}_i(\alpha), \hat{a}_j^\dagger(\beta)\right] \hat{a}_k(\beta) \right. \\
&\qquad \left. + \hat{a}_k(\beta) \left[\hat{a}_i(\alpha), \hat{a}_j^\dagger(\beta)\right] + \underbrace{\left[\hat{a}_i(\alpha), \hat{a}_k(\beta)\right]}_{=0} a_j^\dagger(\beta) \right\} \\
&= \sum_{\alpha,\beta} \left\{ \delta_{ij}\delta_{\alpha\beta}\hat{a}_k(\beta) + \hat{a}_k(\beta)\delta_{ij}\delta_{k\beta} \right\} \\
&= 2\delta_{ij} \sum_{\alpha=1}^{s} \hat{a}_k(\alpha) ,
\end{aligned}$$

i.e.

$$\left[\hat{a}_i, \left[\hat{a}_j^\dagger, \hat{a}_k\right]\right] = 2\delta_{ij}\hat{a}_k .$$

EXERCISE

3.3 More on Paraoperators: Introduction of the Operator $\hat{C}_{jk}$

Problem. In order to simplify the threefold commutators we introduce the operator $\hat{C}_{jk}$:

$$\begin{aligned}
\left[\hat{a}_j, \hat{a}_k^\dagger\right] &= -\hat{C}_{jk} + s\delta_{jk} \qquad \text{for para-Fermi statistics,} \\
\left\{\hat{a}_j, \hat{a}_k^\dagger\right\} &= +\hat{C}_{jk} + s\delta_{jk} \qquad \text{for para-Bose statistics.}
\end{aligned}$$

Here the anticommutator is denoted by curly brackets: $\{,\} = [,]_+$. What do the threefold commutators (3.25) look like with the help of $\hat{C}_{jk}$? Use the expansion of Exercise 3.2 for $\hat{a}_i = \sum_{\alpha=1}^{s} \hat{a}_i(\alpha)$ to derive an explicit representation of $\hat{C}_{jk}$ and use this representation to discuss the action of $\hat{C}_{jk}$ on $|0\rangle$.

Exercise 3.3.

Solution. From

$$\left[\hat{a}_j, \hat{a}_k^\dagger\right] = -\hat{C}_{jk} + s\delta_{jk}$$

and (3.25) it follows for para-Fermi statistics that

$$\begin{aligned}[\hat{a}_i, \hat{C}_{jk} - s\delta_{jk}] &= [\hat{a}_i, \hat{C}_{jk}] - s\overbrace{[\hat{a}_i, \delta_{jk}]}^{=0} \\ &= [\hat{a}_i, \hat{C}_{jk}] = 2\delta_{ik}\hat{a}_j .\end{aligned}$$

Analogously we deduce for para-Bose statistics that

$$\begin{aligned}\{\hat{a}_j, \hat{a}_k^\dagger\} &= \hat{C}_{jk} + s\delta_{jk} , \\ [\hat{a}_i, \hat{C}_{jk} + s\delta_{jk}] &= [\hat{a}_i, \hat{C}_{jk}] = 2\delta_{ik}\hat{a}_j .\end{aligned}$$

With the use of the operator $\hat{C}_{jk}$ the commutators for the para-Bose and para-Fermi statistics become identical.

In order to obtain a representation for $\hat{C}_{jk}$ we transform

$$\begin{aligned}\hat{C}_{jk} &= s\delta_{jk} - \left[\hat{a}_j, \hat{a}_k^\dagger\right] = s\delta_{jk} - \sum_{\alpha,\beta}^{s} \underbrace{\left[\hat{a}_j(\alpha), \hat{a}_k^\dagger(\beta)\right]}_{=0 \text{ for } \alpha\neq\beta} \\ &= s\delta_{jk} - \sum_{\alpha} \left[\hat{a}_j(\alpha), \hat{a}_k^\dagger(\alpha)\right] \\ &= s\delta_{jk} - \sum_{\alpha=1}^{s} \Big(\hat{a}_j(\alpha)\hat{a}_k^\dagger(\alpha) - \hat{a}_k^\dagger(\alpha)\hat{a}_j(\alpha) \\ &\qquad + \hat{a}_k^\dagger(\alpha)\hat{a}_j(\alpha) - \hat{a}_k^\dagger(\alpha)\hat{a}_j(\alpha)\Big) \\ &= s\delta_{jk} - \sum_{\alpha=1}^{s} \left\{\left\{\hat{a}_j(\alpha), \hat{a}_k^\dagger(\alpha)\right\} - 2\hat{a}_k^\dagger(\alpha)\hat{a}_j(\alpha)\right\} \\ &= s\delta_{jk} - \underbrace{\sum_{\alpha=1}^{s} \delta_{jk}}_{s\delta_{jk}} + 2\sum_{\alpha=1}^{s} \hat{a}_k^\dagger(\alpha)\hat{a}_j(\alpha) ,\end{aligned}$$

i.e.

$$\hat{C}_{jk} = 2\sum_{\alpha=1}^{s} \hat{a}_k^\dagger(\alpha)\hat{a}_j(\alpha)$$

for para-Fermi statistics.

The result for para-Bose statistics is the same. From

$$\hat{a}_j(\alpha)|0\rangle = 0$$

it follows that

$$\hat{C}_{jk}|0\rangle = 0 .$$

Furthermore,

$$\begin{aligned}\langle 0|\hat{a}_k^\dagger(\alpha) &= (\hat{a}_k(\alpha)|0\rangle)^\dagger \\ &= 0 ,\end{aligned}$$

Exercise 3.3.

so that

$$\langle 0|\hat{C}_{jk} = 0 .$$

These properties become important for Exercise 3.4!

EXERCISE

3.4 Occupation Numbers of Para-Fermi States

Problem. With the results derived in Exercises 3.2 and 3.3 discuss the occupation numbers of the para-Fermi states:

$$\chi_N = (\hat{a}_k^\dagger)^N|0\rangle ;$$

the degree of the para-Fermi statistics is given by s. Hint: Examine the norm of the state!

Solution. We suppress the quantum number k because it does not turn out to be necessary. The norm of the state χ_N is given by

$$||\chi_N|| = \langle\chi_N|\chi_N\rangle ,$$

i.e.

$$||\chi_N|| = \langle 0|\hat{a}^N(\hat{a}^\dagger)^N|0\rangle .$$

First of all, we consider the simple cases with $N = 1, 2, \ldots$. Furthermore, it is convenient to pull a factor out of the matrix element:

$$||\chi_N|| = \frac{1}{s^N}\langle 0|\hat{a}^N(\hat{a}^\dagger)^N|0\rangle .$$

For $N = 1$ we get

$$||\chi_1|| = \frac{1}{s}\langle 0|\hat{a}\hat{a}^\dagger|0\rangle .$$

We employ the operator $\hat{C}$:

$$\left[\hat{a}, \hat{a}^\dagger\right] = -\hat{C} + s ,$$

i.e.

$$||\chi_1|| = \frac{1}{s}\langle 0|\hat{a}^\dagger\hat{a} - \hat{C} + s|0\rangle .$$

Since $\hat{a}|0\rangle = \hat{C}|0\rangle = 0$ and $\langle 0|0\rangle = 1$ it follows that

$$||\chi_1|| = \frac{1}{s}\langle 0|s|0\rangle = \langle 0|0\rangle = 1 .$$

Hence, we deduce

$$||\chi_1|| = 1 .$$

The $N = 2$ case yields

$$\begin{aligned} ||\chi_2|| &= \frac{1}{s^2}\langle 0|\hat{a}\hat{a}\hat{a}^\dagger\hat{a}^\dagger|0\rangle = \frac{1}{s^2}\langle 0|\hat{a}(\hat{a}^\dagger\hat{a} - \hat{C} + s)\hat{a}^\dagger|0\rangle \\ &= \frac{1}{s^2}\langle 0|\hat{a}\hat{a}^\dagger\hat{a}\hat{a}^\dagger|0\rangle = -\frac{1}{s^2}\langle 0|\hat{a}\hat{C}\hat{a}^\dagger|0\rangle + \frac{1}{s}\underbrace{\langle 0|\hat{a}\hat{a}^\dagger|0\rangle}_{=1} . \end{aligned}$$

Exercise 3.4. Because $[\hat{a},\hat{C}]=2\hat{a}$ we can make the transformation

$$\begin{aligned}
||\chi_\alpha|| &= 1+\frac{1}{s^2}\langle 0|\hat{a}\hat{a}^\dagger(\hat{a}^\dagger\hat{a}-\hat{C}+s)|0\rangle-\frac{1}{s^2}\langle 0|(\hat{C}\hat{a}+2\hat{a})\hat{a}^\dagger|0\rangle \\
&= 1+\frac{1}{s^2}\underbrace{\langle 0|\hat{a}\hat{a}^\dagger\hat{a}^\dagger\hat{a}|0\rangle}_{=0}-\frac{1}{s^2}\underbrace{\langle 0|\hat{C}\hat{a}\hat{a}^\dagger|0\rangle}_{=0}-\frac{1}{s^2}\langle 0|2\hat{a}\hat{a}^\dagger|0\rangle \\
&= -2-\frac{2}{s} \\
&= 2\left(1-\frac{1}{s}\right).
\end{aligned}$$

For $N=3$ we find

$$\begin{aligned}
||\chi_3|| &= \frac{1}{s^3}\langle 0|\hat{a}\hat{a}\hat{a}\hat{a}^\dagger\hat{a}^\dagger\hat{a}^\dagger|0\rangle = \frac{1}{s^3}\langle 0|\hat{a}\hat{a}\left[\hat{a}^\dagger\hat{a}-\hat{C}+s\right]\hat{a}^\dagger\hat{a}^\dagger|0\rangle \\
&= \frac{1}{s^3}\langle 0|\hat{a}\hat{a}\hat{a}^\dagger\hat{a}\hat{a}^\dagger\hat{a}^\dagger|0\rangle-\frac{1}{s^3}\langle 0|\hat{a}\hat{a}\hat{C}\hat{a}^\dagger\hat{a}^\dagger|0\rangle \\
&\quad \underbrace{+\frac{1}{s^2}\langle 0|\hat{a}\hat{a}\hat{a}^\dagger\hat{a}^\dagger|0\rangle}_{=||\chi_2||} \\
&= \frac{1}{s^3}\langle 0|\hat{a}\hat{a}\hat{a}^\dagger\left[\hat{a}^\dagger\hat{a}-\hat{C}+s\right]\hat{a}^\dagger|0\rangle \\
&\quad -\frac{1}{s^3}\langle 0|\hat{a}\left[\hat{C}\hat{a}+2\hat{a}\right]\hat{a}^\dagger\hat{a}^\dagger|0\rangle+||\chi_2|| \\
&= \frac{1}{s^3}\langle 0|\hat{a}\hat{a}\hat{a}^\dagger\hat{a}^\dagger\hat{a}\hat{a}|0\rangle-\frac{1}{s^3}\langle 0|\hat{a}\hat{a}\hat{a}^\dagger\hat{C}\hat{a}^\dagger|0\rangle \\
&\quad \underbrace{+\frac{1}{s^2}\langle 0|\hat{a}\hat{a}\hat{a}^\dagger\hat{a}^\dagger|0\rangle}_{=||\chi_2||}-\frac{1}{s^3}\langle 0|\hat{a}\hat{C}\hat{a}\hat{a}^\dagger\hat{a}^\dagger|0\rangle \\
&\quad -\frac{2}{s^3}\langle 0|\hat{a}\hat{a}\hat{a}^\dagger\hat{a}^\dagger|0\rangle+||\chi_2|| \\
&= \frac{1}{s^3}\langle 0|\hat{a}\hat{a}\hat{a}^\dagger\hat{a}^\dagger\left[\hat{a}^\dagger\hat{a}-\hat{C}+s\right]|0\rangle-\frac{1}{s^3}\langle 0|\hat{a}\hat{a}\hat{a}^\dagger\hat{C}\hat{a}^\dagger|0\rangle \\
&\quad +2||\chi_2||-\frac{1}{s^3}\langle 0|\left[\hat{C}\hat{a}+2\hat{a}\right]\hat{a}\hat{a}^\dagger\hat{a}^\dagger|0\rangle-\frac{2}{s}||\chi_2|| \\
&= \frac{1}{s^3}\langle 0|\hat{a}\hat{a}\hat{a}^\dagger\hat{a}^\dagger\hat{a}^\dagger\hat{a}|0\rangle-\frac{1}{s^3}\langle 0|\hat{a}\hat{a}\hat{a}^\dagger\hat{a}^\dagger\hat{C}|0\rangle \\
&\quad +\frac{1}{s^2}\langle 0|\hat{a}\hat{a}\hat{a}^\dagger\hat{a}^\dagger|0\rangle-\frac{1}{s^3}\langle 0|\hat{a}\hat{a}\hat{a}^\dagger\hat{C}\hat{a}^\dagger|0\rangle+2||\chi_2|| \\
&\quad -\frac{1}{s^3}\langle 0|\hat{C}\hat{a}\hat{a}\hat{a}^\dagger\hat{a}^\dagger|0\rangle-\frac{2}{s^3}\langle 0|\hat{a}\hat{a}\hat{a}^\dagger\hat{a}^\dagger|0\rangle-\frac{2}{s}||\chi_2|| \\
&= 3||\chi_2||-\frac{1}{s^3}\langle 0|\hat{a}\hat{a}\hat{a}^\dagger\hat{C}\hat{a}^\dagger|0\rangle-2\frac{2}{s}||\chi_2||\,.
\end{aligned}$$

Since $\hat{C}|0\rangle=0$ it holds that $(\hat{C}|0\rangle)^\dagger=\langle 0|\hat{C}^\dagger=0=\langle 0|\hat{C}$; therefore $\hat{C}=\hat{C}^\dagger$, so that $\hat{C}\hat{a}^\dagger=\hat{a}^\dagger\hat{C}+2\hat{a}^\dagger$. Hence, we derive

Exercise 3.4.

$$\begin{aligned}
||\chi_3|| &= 3||\chi_2|| - \frac{1}{s^3}\langle 0|\hat{a}\hat{a}\hat{a}^\dagger\left[\hat{a}^\dagger\hat{C} + 2\hat{a}^\dagger\right]|0\rangle - 2\frac{2}{s}||\chi_2|| \\
&= 3||\chi_2|| - \frac{1}{s^3}\underbrace{\langle 0|\hat{a}\hat{a}\hat{a}^\dagger\hat{a}^\dagger\hat{C}|0\rangle}_{=0} - \frac{2}{s^3}\langle 0|\hat{a}\hat{a}\hat{a}^\dagger\hat{a}^\dagger|0\rangle - 2\frac{2}{s}||\chi_2|| \\
&= 3||\chi_2|| - 3\frac{2}{s}||\chi_2|| = 3\left(1-\frac{2}{s}\right)||\chi_2|| ,
\end{aligned}$$

i.e.

$$||\chi_3|| = 3\left(1-\frac{2}{s}\right)2\left(1-\frac{1}{s}\right) .$$

From these three examples we discover the general formula

$$||\chi_N|| = N!\,1\left(1-\frac{1}{s}\right)\cdots\left(1-\frac{N-1}{s}\right) .$$

This can be proven by induction, but this proof is very tedious!

The norm of a state has to be positive definite, i.e. $||\chi_N|| > 0$. We recognize from the general relation for $||\chi_N||$ that

$$||\chi_N|| > 0 \iff \left(1-\frac{N-1}{s}\right) > 0 ;$$

thus,

$$N \leq s .$$

We conclude that a state of para-Fermi statistics of the degree s contains at most $N = s$ particles. For states with $N > s$ the norm vanishes or becomes negative!

We return to the Fermi–Dirac statistics and determine the matrix elements of $\hat{b}_n^\dagger$ and $\hat{b}_n$. We begin with

$$\hat{N}_n|n_n\rangle = \hat{b}_n^\dagger\hat{b}_n|n_n\rangle = n_n|n_n\rangle , \tag{3.26}$$

where $n_n = 0$ or 1 because of (3.24). We examine the vector

$$\hat{b}_n^\dagger|n_n\rangle$$

and apply the operator $\hat{N}_n = \hat{b}_n^\dagger\hat{b}_n$ to it. This gives

$$\begin{aligned}
\hat{N}_n\hat{b}_n^\dagger|n_n\rangle &= \hat{b}_n^\dagger\hat{b}_n\hat{b}_n^\dagger|n_n\rangle = \hat{b}_n^\dagger\left(1-\hat{b}_n^\dagger\hat{b}_n\right)|n_n\rangle \\
&= \hat{b}_n^\dagger\left(1-\hat{N}_n\right)|n_n\rangle \\
&= (1-n_n)\hat{b}_n^\dagger|n_n\rangle .
\end{aligned} \tag{3.27}$$

This reveals that $\hat{b}_n^\dagger|n_n\rangle$ represents an eigenvector of $\hat{b}_n^\dagger\hat{b}_n$ with eigenvalue $1-n_n$. Then we note

$$\hat{b}_n^\dagger|n_n\rangle = C_n|1-n_n\rangle . \tag{3.28}$$

The constant C_n can be determined from the scalar product

$$\begin{aligned}\left(\hat{b}_n^\dagger|n_n\rangle\right)^\dagger \hat{b}_n^\dagger|n_n\rangle &= \langle n_n|\,\hat{b}_n\hat{b}_n^\dagger|n_n\rangle \\ &= \langle n_n|\,1-\hat{b}_n^\dagger\hat{b}_n|n_n\rangle \\ &= \langle n_n|\,(1-n_n)|n_n\rangle \\ &= 1-n_n \;=\; |C_n|^2 \end{aligned} \tag{3.29}$$

and results in

$$C_n \;=\; \mathrm{e}^{\mathrm{i}\alpha_n}\sqrt{1-n_n} \;\equiv\; \Theta_n\sqrt{1-n_n}\,. \tag{3.30}$$

Here the phase factor $\Theta_n = \mathrm{e}^{\mathrm{i}\alpha_n}$ is of modulus 1. We now consider the state $\hat{b}_n|n_n\rangle$; in complete analogy to (3.27), it holds that

$$\begin{aligned}\hat{N}_n\hat{b}_n|n_n\rangle &= \hat{b}_n^\dagger\hat{b}_n\hat{b}_n|n_n\rangle \\ &= (1-\hat{b}_n\hat{b}_n^\dagger)\hat{b}_n|n_n\rangle \\ &= \hat{b}_n|n_n\rangle - \hat{b}_n\hat{N}_n|n_n\rangle \\ &= (1-n_n)\hat{b}_n|n_n\rangle\,. \end{aligned} \tag{3.31}$$

Obviously $\hat{b}_n|n_n\rangle$ is also an eigenvector of $\hat{N}_n$ with eigenvalue $1-n_n$, which means

$$\hat{b}_n|n_n\rangle \;=\; D_n|1-n_n\rangle \tag{3.32}$$

with

$$|D_n|^2 \;=\; \left(\hat{b}_n|n_n\rangle\right)^\dagger \hat{b}_n|n_n\rangle \;=\; \langle n_n|\hat{b}_n^\dagger\hat{b}_n|n_n\rangle\,.$$

Thus

$$D_n \;=\; \mathrm{e}^{\mathrm{i}\alpha_n'}\sqrt{n_n} \;\equiv\; \Theta_n'\sqrt{n_n}\,,$$

where the phase factor $\alpha_n' = \alpha_n$ can be chosen as in (3.30). Therefore we are allowed to write

$$D_n \;=\; \mathrm{e}^{\mathrm{i}\alpha_n}\sqrt{n_n} \;=\; \Theta_n\sqrt{n_n} \tag{3.33}$$

and (3.32) becomes

$$\hat{b}_n|n_n\rangle \;=\; \mathrm{e}^{\mathrm{i}\alpha_n}\sqrt{n_n}|1-n_n\rangle \;=\; \Theta_n\sqrt{n_n}|1-n_n\rangle\,. \tag{3.34}$$

Owing to (3.9), the most general fermion state $|\cdots,n_n,\cdots,n_{n'},\cdots\rangle$ factorizes into one particle fermion states $|n\rangle$; hence,

$$|\cdots,n_n,\cdots,n_{n'},\cdots\rangle \;=\; \cdots|n_n\rangle\cdots|n_{n'}\rangle\cdots\,. \tag{3.35}$$

This relation is in complete analogy to (3.14) for the case of bosons.

For the boson case (3.14) the occupation numbers are $n_n = 0,1,2,\dots$, whereas for the fermion case (3.35) the occupation numbers remain as 0 or 1 because of (3.24), i.e. $n_n = 0,1$; this is the principal difference between the two cases. In order to distinguish the differences due to the different quantization more clearly, we list again the most important relations for bosons and fermions:

(a) **bosons**

$$\begin{aligned} \hat{b}_n|\cdots n_n\cdots\rangle &= \sqrt{n_n}|\cdots n_n-1\cdots\rangle\,, \\ \hat{b}_n^\dagger|\cdots n_n\cdots\rangle &= \sqrt{n_n+1}|\cdots n_n+1\cdots\rangle\,. \end{aligned} \tag{3.36}$$

(b) **fermions**

$$\begin{aligned} \hat{b}_n|\cdots n_n\cdots\rangle &= \Theta_n\sqrt{n_n}|\cdots 1-n_n\cdots\rangle\,, \\ \hat{b}_n^\dagger|\cdots n_n\cdots\rangle &= \Theta_n\sqrt{1-n_n}|\cdots 1-n_n\cdots\rangle\,. \end{aligned} \tag{3.37}$$

For both cases $\hat{b}_n$ represents an *annhilation operator* because $\hat{b}_n$ reduces the particle number n_n by one. On the same footing $\hat{b}_n^+$ represents a *creation operator* for both cases. For the boson case the phase factor has been chosen to equal 1; this is feasable because with this the boson commutation relations

$$\begin{aligned} \hat{b}_n^\dagger\hat{b}_{n'}^\dagger - \hat{b}_{n'}^\dagger\hat{b}_n^\dagger &= 0\,, \\ \hat{b}_n\hat{b}_{n'} - \hat{b}_{n'}\hat{b}_n &= 0\,, \\ \hat{b}_n\hat{b}_{n'}^\dagger - \hat{b}_{n'}^\dagger\hat{b}_n &= \delta_{nn'} \end{aligned} \tag{3.38}$$

could be deduced from relations (3.36). In the following problem we again explain this step.

EXERCISE

3.5 On the Boson Commutation Relations

Problem. Derive the boson commutation relation (3.8) from relations (3.36) for the boson creation and annhilation operators.

Solution.
(a)

$$\begin{aligned} &(\hat{b}_n\hat{b}_{n'} - \hat{b}_{n'}\hat{b}_n)|\cdots n_n\cdots n_{n'}\cdots\rangle \\ &\quad = (\sqrt{n_{n'}}\sqrt{n_n} - \sqrt{n_{n'}}\sqrt{n_n})|\cdots n_{n-1}\cdots n_{n'-1}\cdots\rangle \;=\; 0 \end{aligned}$$

for arbitrary $|\cdots n_n\cdots n_{n'}\cdots\rangle$, i.e.

$$\left[\hat{b}_n,\hat{b}_{n'}\right]_- = 0\,.$$

(b)

$$\begin{aligned} &(\hat{b}_n^\dagger\hat{b}_{n'}^\dagger - \hat{b}_{n'}^\dagger\hat{b}_n^\dagger)|\cdots n_n\cdots n_{n'}\cdots\rangle \\ &\quad = \left(\sqrt{n_n+1}\sqrt{n_{n'}+1} - \sqrt{n_{n'}+1}\sqrt{n_n+1}\right)|\cdots n_{n+1}\cdots n_{n'+1}\cdots\rangle = 0 \end{aligned}$$

for arbitrary $|\cdots n_n\cdots n_{n'}\cdots\rangle$, i.e.

$$\left[\hat{b}_n^\dagger,\hat{b}_{n'}^\dagger\right]_- = 0\,.$$

Exercise 3.5.

(c)

$$
\begin{aligned}
&(\hat{b}_n\hat{b}^\dagger_{n'} - \hat{b}^\dagger_{n'}\hat{b}_n)|\cdots n_n \cdots n_{n'}\cdots\rangle \\
&\quad = \begin{cases} \left(\sqrt{n_n+1}\sqrt{n_n+1} - \sqrt{n_n}\sqrt{n_n}\,\right)|\cdots n_n \cdots\rangle \;=\; |\cdots n_n\cdots\rangle \\ \hfill \text{for } n' = n\,, \\ \left(\sqrt{n_{n'}+1}\sqrt{n_n} - \sqrt{n_n}\sqrt{n_{n'}+1}\,\right)|\cdots n_{n-1}\cdots n_{n'+1}\cdots\rangle \;=\; 0 \\ \hfill \text{for } n' \neq n\,, \end{cases}
\end{aligned}
$$

i.e.

$$\left[\hat{b}_n, \hat{b}^\dagger_{n'}\right] = \delta_{nn'}\,.$$

For the case of fermions the choice of phase factor is more complicated. We select $\Theta_n = \mathrm{e}^{\mathrm{i}\alpha_n}$ in such a way that fermion commutation relations

$$
\begin{aligned}
\hat{b}^\dagger_n\hat{b}^\dagger_{n'} + \hat{b}^\dagger_{n'}\hat{b}^\dagger_n &= 0\,, \\
\hat{b}_n\hat{b}_{n'} + \hat{b}_{n'}\hat{b}_n &= 0
\end{aligned} \tag{3.39}
$$

follow from (3.37) with an arbitrary state $|\cdots n_n\cdots n_{n'}\cdots\rangle$. Here it is an excellent excercise to convince ourselves that relations (3.39) do not follow from (3.37) for an arbitrary but fixed choice of phase $\Theta_n = \mathrm{e}^{\mathrm{i}\alpha_n}$. We immediately calculate that

$$
\begin{aligned}
&(\hat{b}_n\hat{b}_{n'} + \hat{b}_{n'}\hat{b}_n)|\cdots n_n\cdots n_{n'}\cdots\rangle \\
&\quad =;\Theta_n\Theta_{n'}\left(\sqrt{n_n}\sqrt{n_{n'}} + \sqrt{n_{n'}}\sqrt{n_n}\,\right)|\cdots n_n\cdots n_{n'}\cdots\rangle \\
&\quad \neq 0\,.
\end{aligned} \tag{3.40}
$$

Here it was assumed that Θ_n depends only on n_n and not on the remaining occupation numbers. Evidently this strategy leads to a contradiction with (3.39). In order to remove it we proceed in the following way: We order the states $|n_n\rangle$ of the system in an arbitrary but then fixed manner; for example,

$$|n_1 n_2\cdots n_k\cdots\rangle = |n_1\rangle|n_2\rangle\cdots|n_k\rangle\cdots\,. \tag{3.41}$$

The operation of $\hat{b}_k$ or $\hat{b}^\dagger_k$ on these states is then given by (3.37), where we choose the phase $\Theta_k = +1$ or -1 depending on the number of preceding occupied states to be even or odd. In other words: If an even number of states $n_\nu = 1$ is to be found within the states reaching from $|n_1\rangle$ to $|n_{k-1}\rangle$, then we set

$$
\begin{aligned}
\hat{b}_k|n_1\cdots n_k\cdots\rangle &= \sqrt{n_k}|n_1\cdots n_{k-1}\cdots\rangle\,, \\
\hat{b}^\dagger_k|n_1\cdots n_k\cdots\rangle &= \sqrt{1-n_k}|n_1\cdots n_{k+1}\cdots\rangle
\end{aligned} \tag{3.42a}
$$

for even $\sum_{\nu=1}^{k-1} n_\nu$. For the case when the number of occupied states within $|n_1\rangle$ to $|n_k\rangle$ is odd, we fix the phase $\Theta_k = -1$; hence,

$$\begin{aligned} \hat{b}_k|n_1\cdots n_k\cdots\rangle &= -\sqrt{n_k}|n_1\cdots n_{k-1}\cdots\rangle\,, \\ \hat{b}_k^\dagger|n_1\cdots n_k\cdots\rangle &= -\sqrt{1-n_k}|n_1\cdots n_{k+1}\cdots\rangle \end{aligned} \tag{3.42b}$$

for odd $\sum_{\nu=1}^{k-1} n_\nu$. We can combine the two relations (3.42a) and (3.42b) to give

$$\begin{aligned} \hat{b}_k|n_1\cdots n_k\cdots\rangle &= (-1)^{\sum_{\nu=1}^{k-1} n_\nu}\sqrt{n_k}|n_1\cdots n_{k-1}\cdots\rangle\,, \\ \hat{b}_k^\dagger|n_1\cdots n_k\cdots\rangle &= (-1)^{\sum_{\nu=1}^{k-1} n_\nu}\sqrt{1-n_k}|n_1\cdots n_{k+1}\cdots\rangle\,. \end{aligned} \tag{3.42}$$

The difference between the phase choice (3.42) and the choice (3.40) attempted before is that now *the phase Θ_k depends on the occupation numbers of the preceding states $|n_k\rangle$*. In (3.40) Θ_k depended on the index k only. This choice of phase builds on the construction of the state $|n_1\cdots n_k\cdots\rangle$ out of the creation operators $\hat{b}_\nu^\dagger$:

$$|n_1\cdots n_k\cdots\rangle = (\hat{b}_1^\dagger)^{n_1}(\hat{b}_2^\dagger)^{n_2}\cdots(\hat{b}_k^\dagger)^{n_k}\cdots|0\rangle\,.$$

The operation

$$\hat{b}_k|n_1\cdots n_k\cdots\rangle = \hat{b}_k(\hat{b}_1^\dagger)^{n_1}(\hat{b}_2^\dagger)^{n_2}\cdots(\hat{b}_k^\dagger)^{n_k}\cdots|0\rangle$$

makes the commutation of $\hat{b}_k$ with all operations preceding $(\hat{b}_k^\dagger)^{n_k}$ necessary. Because

$$\hat{b}_k\hat{b}_\nu^\dagger = -\hat{b}_\nu^\dagger\hat{b}_k \quad \text{for} \quad \nu \neq k\,,$$

a factor -1 comes up; it appears $\sum_{\nu=1}^{k-1} n_\nu$ times. This leads to

$$\hat{b}_k(\hat{b}_1^\dagger)^{n_1}\cdots(\hat{b}_k^\dagger)^{n_k}\cdots|0\rangle = (-1)^{\sum_{\nu=1}^{k-1} n_\nu}(\hat{b}_1^\dagger)^{n_1}\cdots\hat{b}_k(\hat{b}_k^\dagger)^{n_k}\cdots|0\rangle$$

and explains naturally the choice made for the phase factors.

EXERCISE

3.6 Consistency of the Phase Choice for Fermi States with the Fermion Commutation Relations

Problem. Show that the phase choice (3.42) is consistent with the commutation relations (3.39) for an arbitrary state $|n_1\cdots n_k\cdots\rangle$.

Solution. We compute

$$\hat{b}_l\hat{b}_k|n_1\cdots n_k\cdots n_l\cdots\rangle \tag{1}$$

and

$$\hat{b}_k\hat{b}_l|n_1\cdots n_k\cdots n_l\cdots\rangle\,, \tag{2}$$

where $|n_1\cdots n_k\cdots n_l\cdots\rangle$ represents an arbitrary state. With no loss of generality we assume $l > k$ for the ordering of the operators. So that the operations (1) and (2) are not identical to zero from the very beginning, n_k and n_l must

Exercise 3.6.

both be equal to 1 for the state $|n_1 \cdots n_k \cdots n_l \cdots\rangle$. The operation (1) then yields

$$\hat{b}_k \hat{b}_l |n_1 \cdots n_k \cdots n_l \cdots\rangle = \Theta_l \Theta_k \sqrt{n_k}\sqrt{n_l} |n_1 \cdots n_k - 1 \cdots n_l - 1 \cdots\rangle . \quad (3)$$

It empties first state l and then state k. The operation

$$\hat{b}_l \hat{b}_k |n_1 \cdots n_k \cdots n_l \cdots\rangle = \Theta'_l \Theta'_k \sqrt{n_k}\sqrt{n_l} |n_1 \cdots n_k - 1 \cdots n_l - 1 \cdots\rangle \quad (4)$$

empties first state k and then state l. Thus

$$\Theta'_k = \Theta_k = (-1)^{\sum_\nu^{k-1} n_\nu}$$

remains unaltered. The phase $\Theta'_l = -\Theta_l$ differs exactly by the factor -1 from Θ_l because state k has been emptied by $\hat{b}_k$ before, so $n'_k = n_k - 1 = 0$. Obviously the result is

$$\hat{b}_k \hat{b}_l |n_1 \cdots n_k \cdots n_l \cdots\rangle = -\hat{b}_l \hat{b}_k |n_1 \cdots n_k \cdots n_l \cdots\rangle \quad (5)$$

or

$$(\hat{b}_k \hat{b}_l + \hat{b}_l \hat{b}_k) |n_1 \cdots n_k \cdots n_l \cdots\rangle = 0 . \quad (6)$$

This is exactly the second relation in (3.39). The first relation in (3.39) can be deduced in a completely analogous manner. The same holds for

$$\hat{b}_k \hat{b}_l^\dagger + \hat{b}_l^\dagger \hat{b}_k = \delta_{lk} .$$

Since the ket vectors $|n_1 \cdots n_k \cdots\rangle$ represent all possible states of the many-particle problem, they form a complete set; consequently, the anticommutation relations (3.16) follow as operator equations from (3.42).

3.1 Spin-Statistics Theorem

We have become acquainted with two possibilities for the quantization, the one with commutators (3.8) and the other with anticommutators (3.16). With a Hamiltonian of the form (3.9) both possibilities lead to equations of motion [(3.10b) and (3.17)] for the operators, which are formally identical to the "classical equations of motion" (3.4). This reflects a criterion to accept both quantization procedures as possible quantization schemes. Immediately the question has to be raised as to when to quantize with commutators and when to quantize with anticommutators. Put in another way, when does Bose–Einstein statistics and when does Fermi–Dirac statistics apply? When should the Pauli principle hold and when not?

An answer exists in the general form of a theorem from Pauli:[6] *Particles with integer spin have to obey Bose–Einstein statistics and particles with half-integer spin have to obey Fermi–Dirac statistics.* We presume that the spin quantum number is already contained in the quantum number n of the wavefunction ψ_n for fermions.

[6] W. Pauli: Phys. Rev. **58** (1940) 716.

Later in this series on theoretical physics we will discuss and explain this theorem in much more detail.[7] For the moment we only indicate the deeper roots: the energy spectra of Hamiltonians have to be bounded from below (i.e. have to be essentially positive) and the validity of *microcausality* has to be claimed.

Remark. We assume the existence of spin-1/2 particles to be fundamental; all other particles with different spins (0,1,3/2,...) can be thought of as being built up of spin-1/2 particles. Under this aspect it seems to be plausible that spin-1/2 particles have to obey the "more fundamental" Fermi–Dirac quantization. Then all particles with integer spin have to be Bose particles because they are build up of an even number of Fermi particles. Hence, for example the exchange of two spin-1 particles would correspond to an *exchange of two pairs of spin-1/2 particles.* If the Bose quantization held for the elementary spin-1/2 particles themselves, then consequently all particles would be Bose particles. Thus, we have to demonstrate that in any case spin-1/2 particles have to obey Fermi–Dirac statistics.

3.2 Relationship Between Second Quantization and Elementary Quantum Mechanics

The formulation of quantum mechanics (Hamiltonian, equation of motion, and so on) in terms of creation and annihilation operators is called *second quantization.* It was first presented in a famous paper by P. Jordan, O. Klein, and E.P. Wigner.[8] Now, we could imagine that a quantum-mechanical equation (for example, the Schrödinger equation) could describe more (or new and additional) quantum-mechanical processes after second quantization has been applied than before. Put into other words, the conventional many-particle Schrödinger equation could yield different (more or less) results compared to a second-quantized Schrödinger equation. This is not the case. We will show that conventional (elementary) quantum mechanics follows from the second-quantized theory; the former is contained in the latter. However, the *second-quantized theory is more general.* It can be extended to describe processes such as particle creation and annhilation as occuring, for example, in β decay, quantum electrodynamics, and the theory of strong interactions.

We concentrate again on the expansion (3.3) of the wave field. We apply the quantum relations (3.8) and (3.16) to the coefficients b_n, so that the coefficients b_n turn into operators $\hat{b}_n$

$$b_n \rightarrow \hat{b}_n ;$$

the field $\psi(\boldsymbol{x},t)$ of (3.3) becomes a field operator $\hat{\psi}(\boldsymbol{x},t)$. We write for the *field operator*

[7] W. Greiner, J. Reinhardt: *Quantum Electrodynamics*, 2nd ed. (Springer, Berlin, Heidelberg 1994).

[8] P. Jordan, O. Klein: Z. Phys. **45** (1927) 751; P. Jordan, E.P. Wigner: Z. Phys. **47** (1928) 631.

$$\hat{\psi}(\boldsymbol{x},t) = \sum_n \hat{b}_n \psi_n(\boldsymbol{x}) = \sum_n \hat{b}_n \langle \boldsymbol{x}|n\rangle \, ; \tag{3.43a}$$

for the Hermitean conjugate field operator we get

$$\hat{\psi}^\dagger(\boldsymbol{x},t) = \sum_n \hat{b}_n^\dagger \psi_n^*(\boldsymbol{x}) = \sum_n \hat{b}_n^\dagger \langle n|\boldsymbol{x}\rangle \, . \tag{3.43b}$$

Here also the bracket notation for the states $\psi_n(\boldsymbol{x}) = \langle \boldsymbol{x}|n\rangle$, $\psi_n^*(\boldsymbol{x}) = \langle n|\boldsymbol{x}\rangle$ has been indicated. We shall use it from time to time when it seems suitable. Obviously, the field operator $\hat{\psi}(\boldsymbol{x},t)$ annihilates a particle at position $\boldsymbol{x}$ and at time t; it is a linear combination of annihilation operators $\hat{b}_n$ which depends on position ($\boldsymbol{x}$) and time (t). Similarly $\hat{\psi}^\dagger(\boldsymbol{x},t)$ *has to be interpreted as an operator creating a particle at position* $\boldsymbol{x}$ *and time* t. The commutation relations for the field operators $\hat{\psi}(\boldsymbol{x},t)$ and $\hat{\psi}^\dagger(\boldsymbol{x},t)$ can be derived from those between $\hat{b}_n$ and $\hat{b}_{n'}^\dagger$:

$$\begin{aligned}\left[\hat{\psi}(\boldsymbol{x},t), \hat{\psi}^\dagger(\boldsymbol{x}',t)\right]_\pm &= \sum_n \sum_{n'} \left[\hat{b}_n, \hat{b}_{n'}^\dagger\right]_\pm \psi_n(\boldsymbol{x}) \psi_{n'}^*(\boldsymbol{x}') \\ &= \sum_n \sum_{n'} \delta_{nn'} \psi_n(\boldsymbol{x}) \psi_n^*(\boldsymbol{x}') \\ &= \sum_n \psi_n(\boldsymbol{x}) \psi_n^*(\boldsymbol{x}') \\ &= \delta(\boldsymbol{x}-\boldsymbol{x}') \, .\end{aligned} \tag{3.44}$$

Here (3.8) and (3.16) and, for the last step, the completeness relation of the functions $\psi_n(\boldsymbol{x})$ have been used. Analogously we find

$$\left[\hat{\psi}(\boldsymbol{x},t), \hat{\psi}(\boldsymbol{x}',t)\right]_\pm = 0 \, , \tag{3.45a}$$

$$\left[\hat{\psi}^\dagger(\boldsymbol{x},t), \hat{\psi}^\dagger(\boldsymbol{x}',t)\right]_\pm = 0 \, . \tag{3.45b}$$

Relations (3.44) and (3.45) represent the well-known *equal-time commutation relations for field operators.* "Equal-time" indicates that the field operators are commuted at different positions ($\boldsymbol{x}, \boldsymbol{x}'$) but at the same time $t, t' = t$. In our lectures on field quantization[9] of this series we generalize the commutation relations (3.44) and (3.45) with respect to $t \neq t'$. Equations (3.44) and (3.45) can be interpreted in the following way: All over space the field operators *do not influence* each other at the same time except at an equal space–time point.

The Hamiltonian $\hat{H}$, (3.9), follows from the expectation value of the one-particle Hamiltonian $\left[-\frac{\hbar^2}{2m}\boldsymbol{\nabla}^2 + V\right]$ [compare with (3.1)] with the field operators $\hat{\psi}$:

$$\begin{aligned}\hat{H} &= \int d^3x \hat{\psi}^\dagger(\boldsymbol{x},t) \left[-\frac{\hbar^2}{2m}\boldsymbol{\nabla}^2 + V\right] \hat{\psi}(\boldsymbol{x},t) \\ &= \int d^3x \sum_n \hat{b}_n^\dagger \psi_n^*(\boldsymbol{x}) \left[-\frac{\hbar^2}{2m}\boldsymbol{\nabla}^2 + V\right] \sum_{n'} \hat{b}_{n'} \psi_{n'}(\boldsymbol{x})\end{aligned} \tag{3.46a}$$

[9] W. Greiner, J. Reinhardt: *Field Quantization* (Springer, Berlin, Heidelberg 1996).

$$
\begin{aligned}
&= \sum_{n,n'} \hat{b}_n^\dagger \hat{b}_{n'} \int d^3x \psi_n^*(\boldsymbol{x}) \left[-\frac{\hbar^2}{2m}\boldsymbol{\nabla}^2 + V\right] \psi_{n'}(\boldsymbol{x}) \\
&= \sum_{n,n'} \hat{b}_n^\dagger \hat{b}_{n'} \int d^3x \psi_n^*(\boldsymbol{x}) E_{n'} \psi_{n'}(\boldsymbol{x}) \\
&= \sum_{n,n'} E_{n'} \hat{b}_n^\dagger \hat{b}_{n'} \int d^3x \psi_n^*(\boldsymbol{x}) \psi_{n'}(\boldsymbol{x}) \\
&= \sum_{n,n'} E_{n'} \hat{b}_n^\dagger \hat{b}_{n'} \delta_{nn'} = \sum_n E_n \hat{b}_n^\dagger \hat{b}_n \, .
\end{aligned} \tag{3.46b}
$$

It is interesting that Heisenberg's equations of motion for the field operators,

$$
-\frac{\hbar}{\mathrm{i}}\frac{\partial}{\partial t}\hat{\psi}(\boldsymbol{x},t) = \left[\hat{\psi}(\boldsymbol{x},t), \hat{H}\right]_- \, , \tag{3.47}
$$

with the Hamiltonian $\hat{H}$ of (3.46) precisely result in the time-dependent Schrödinger equation; this demonstates the consistency of the theory. We find for the right-hand side of (3.47)

$$
\begin{aligned}
\left[\hat{\psi}(\boldsymbol{x},t), \hat{H}\right]_- &= \left[\hat{\psi}(\boldsymbol{x},t), \int d^3x' \hat{\psi}^\dagger(\boldsymbol{x}',t) \left[-\frac{\hbar^2}{2m}\boldsymbol{\nabla}'^2 + V(\boldsymbol{x}')\right] \hat{\psi}(\boldsymbol{x}',t)\right]_- \\
&= \int d^3x' \left\{ \hat{\psi}(\boldsymbol{x},t)\hat{\psi}^\dagger(\boldsymbol{x}',t) \left[-\frac{\hbar^2}{2m}\boldsymbol{\nabla}'^2 + V(\boldsymbol{x}')\right] \hat{\psi}(\boldsymbol{x}',t) \right. \\
&\quad \left. -\hat{\psi}^\dagger(\boldsymbol{x}',t) \left[-\frac{\hbar^2}{2m}\boldsymbol{\nabla}'^2 + V(\boldsymbol{x}')\right] \hat{\psi}(\boldsymbol{x}',t)\hat{\psi}(\boldsymbol{x},t) \right\} .
\end{aligned} \tag{3.48}
$$

Employing the commutation relation (3.44),

$$
\hat{\psi}(\boldsymbol{x},t)\hat{\psi}^\dagger(\boldsymbol{x}',t) = \delta(\boldsymbol{x}-\boldsymbol{x}') \mp \hat{\psi}^\dagger(\boldsymbol{x}',t)\hat{\psi}(\boldsymbol{x},t) \, ,
$$

we transform the second to last term in (3.48) into

$$
\begin{aligned}
&\int d^3x' \hat{\psi}(\boldsymbol{x},t)\hat{\psi}^\dagger(\boldsymbol{x}',t) \left[-\frac{\hbar^2}{2m}\boldsymbol{\nabla}'^2 + V(\boldsymbol{x}')\right] \hat{\psi}(\boldsymbol{x}',t) \\
&\quad = \int d^3x' \delta(\boldsymbol{x}-\boldsymbol{x}') \left[-\frac{\hbar^2}{2m}\boldsymbol{\nabla}'^2 + V(\boldsymbol{x}')\right] \hat{\psi}(\boldsymbol{x},t) \\
&\quad\quad \mp \int d^3x' \hat{\psi}^\dagger(\boldsymbol{x}',t)\hat{\psi}(\boldsymbol{x},t) \left[-\frac{\hbar^2}{2m}\boldsymbol{\nabla}'^2 + V(\boldsymbol{x}')\right] \hat{\psi}(\boldsymbol{x}',t) \\
&\quad = \left[-\frac{\hbar^2}{2m}\boldsymbol{\nabla}^2 + V(\boldsymbol{x})\right] \hat{\psi}(\boldsymbol{x},t) \\
&\quad\quad +(\mp 1) \int d^3x \hat{\psi}^\dagger(\boldsymbol{x}',t) \left[-\frac{\hbar^2}{2m}\boldsymbol{\nabla}'^2 + V(\boldsymbol{x}')\right] \hat{\psi}(\boldsymbol{x},t)\hat{\psi}(\boldsymbol{x}',t) \\
&\quad = \left[-\frac{\hbar^2}{2m}\boldsymbol{\nabla}^2 + V(\boldsymbol{x})\right] \hat{\psi}(\boldsymbol{x},t) \\
&\quad\quad +(\mp 1)(\mp 1) \int d^3x' \hat{\psi}^\dagger(\boldsymbol{x}',t) \left[-\frac{\hbar^2}{2m}\boldsymbol{\nabla}'^2 + V(\boldsymbol{x}')\right] \hat{\psi}(\boldsymbol{x}',t)\hat{\psi}(\boldsymbol{x},t) \, .
\end{aligned}
$$

We realize that in any case the second term is positive (and independent of the statistics) and cancels the last term in (3.48). Thus, independent of the choice of commutators or anticommutators for the quantization (Bose–Einstein or Fermi–Dirac statistics), we get for the commutator

$$\left[\hat{\psi}(\boldsymbol{x},t),\hat{H}\right]_- = \left[-\frac{\hbar^2}{2m}\boldsymbol{\nabla}^2 + V(\boldsymbol{x})\right]\hat{\psi}(\boldsymbol{x},t)\,.$$

Equation (3.47) becomes

$$-\frac{\hbar}{\mathrm{i}}\frac{\partial\hat{\psi}(\boldsymbol{x},t)}{\partial t} = \left[-\frac{\hbar^2}{2m}\boldsymbol{\nabla}^2 + V(\boldsymbol{x})\right]\hat{\psi}(\boldsymbol{x},t)\,. \tag{3.49}$$

Thus, Schrödinger's equation also holds for the field operators.

With the help of the field operators we are also able to define the *particle-number density operator*. It seems reasonable to try the expression

$$\hat{n}(\boldsymbol{x},t) = \hat{\psi}^\dagger(\boldsymbol{x},t)\hat{\psi}(\boldsymbol{x},t) \tag{3.50}$$

and to introduce the *total particle-number operator*

$$\hat{N}(t) = \int \mathrm{d}^3x\,\hat{n}(\boldsymbol{x},t) = \int \mathrm{d}^3x\,\hat{\psi}^\dagger(\boldsymbol{x},t)\hat{\psi}(\boldsymbol{x},t)\,. \tag{3.51}$$

With (3.43) we immediately calculate

$$\begin{aligned}\hat{N}(t) &= \int \mathrm{d}^3x \sum_n \hat{b}_n^\dagger\psi_n^*(\boldsymbol{x}) \sum_{n'} \hat{b}_{n'}\psi_{n'}(\boldsymbol{x}) \\ &= \sum_{n,n'} \hat{b}_n^\dagger\hat{b}_{n'} \int \mathrm{d}^3x\psi_n^*(\boldsymbol{x})\psi_{n'}(\boldsymbol{x}) \\ &= \sum_n \hat{b}_n^\dagger\hat{b}_n = \sum_n \hat{N}_n\,. \end{aligned} \tag{3.52}$$

This is in agreement with our previous result (3.12). Moreover,

$$\frac{\mathrm{d}\hat{N}}{\mathrm{d}t} = -\frac{\mathrm{i}}{\hbar}\left[\hat{N},\hat{H}\right]_- = 0\,; \tag{3.53}$$

see the following exercise. The operator of the total particle number is a constant in time; this is a very satisfying result.

EXERCISE

3.7 Constancy of the Total Particle-Number Operator

Problem. Show that $\mathrm{d}\hat{N}/\mathrm{d}t = 0$.

Solution. Since

$$\frac{\mathrm{d}\hat{N}}{\mathrm{d}t} = -\frac{\mathrm{i}}{\hbar}\left[\hat{N},\hat{H}\right]\,,$$

it is sufficient to show that

$$\left[\hat{N},\hat{H}\right] = 0$$

Exercise 3.7.

with

$$\hat{N} = \int d^3x'\hat{\psi}^\dagger(\boldsymbol{x}',t)\boldsymbol{\psi}(\boldsymbol{x}',t)\,,$$

$$\hat{H} = \int d^3x\,\hat{\psi}^\dagger(\boldsymbol{x},t)\left[-\frac{\hbar^2}{2m}\boldsymbol{\nabla}^2 + V(\boldsymbol{x})\right]\hat{\psi}(\boldsymbol{x},t)\,.$$

It holds

$$\hat{N}\hat{H} = \int d^3x'\,d^3x\,\hat{\psi}^\dagger(\boldsymbol{x}')\hat{\psi}(\boldsymbol{x}')\hat{\psi}^\dagger(\boldsymbol{x})\left[-\frac{\hbar^2}{2m}\boldsymbol{\nabla}^2 + V(\boldsymbol{x})\right]\hat{\psi}(\boldsymbol{x})$$

and

$$\hat{H}\hat{N} = \int d^3x'\,d^3x\,\hat{\psi}^\dagger(\boldsymbol{x})\left[-\frac{\hbar^2}{2m}\boldsymbol{\nabla}^2 + V(\boldsymbol{x})\right]\hat{\psi}(\boldsymbol{x})\hat{\psi}^\dagger(\boldsymbol{x}')\hat{\psi}(\boldsymbol{x}')\,,$$

where we have left out the argument t. With the help of the commutation relations

$$\hat{\psi}(\boldsymbol{x}')\hat{\psi}^\dagger(\boldsymbol{x}) = \delta(\boldsymbol{x}-\boldsymbol{x}') \pm \hat{\psi}^\dagger(\boldsymbol{x})\hat{\psi}(\boldsymbol{x}')$$

$\hat{N}\hat{H}$ is transformed into

$$\begin{aligned}\hat{N}\hat{H} &= \int d^3x'\,d^3x\,\hat{\psi}^\dagger(\boldsymbol{x}')\left[\delta(\boldsymbol{x}'-\boldsymbol{x}) \pm \hat{\psi}^\dagger(\boldsymbol{x})\hat{\psi}(\boldsymbol{x}')\right]\\ &\quad\times\left[-\frac{\hbar^2}{2m}\boldsymbol{\nabla}^2 + V(\boldsymbol{x})\right]\hat{\psi}(\boldsymbol{x})\\ &= \int d^3x'\,d^3x\,\hat{\psi}^\dagger(\boldsymbol{x}')\left[-\frac{\hbar^2}{2m}\boldsymbol{\nabla}^2 + V(\boldsymbol{x})\right]\hat{\psi}(\boldsymbol{x})\delta(\boldsymbol{x}-\boldsymbol{x}')\\ &\quad\pm\int d^3x'\,d^3x\,\hat{\psi}^\dagger(\boldsymbol{x}')\hat{\psi}^\dagger(\boldsymbol{x})\hat{\psi}(\boldsymbol{x}')\left[-\frac{\hbar^2}{2m}\boldsymbol{\nabla}^2 + V(\boldsymbol{x})\right]\hat{\psi}(\boldsymbol{x})\,.\end{aligned}$$

We concentrate on the second term. Since

$$\left[\hat{\psi}(\boldsymbol{x}'),\hat{\psi}(\boldsymbol{x})\right]_\pm = \left[\hat{\psi}^\dagger(\boldsymbol{x}'),\hat{\psi}^\dagger(\boldsymbol{x})\right]_\pm = 0\,,$$

it follows that

$$\begin{aligned}&\int d^3x'\,d^3x\,\hat{\psi}^\dagger(\boldsymbol{x}')\hat{\psi}^\dagger(\boldsymbol{x})\hat{\psi}(\boldsymbol{x}')\left[-\frac{\hbar^2}{2m}\boldsymbol{\nabla}^2 + V(\boldsymbol{x})\right]\hat{\psi}(\boldsymbol{x})\\ &= \int d^3x'\,d^3x\,\hat{\psi}^\dagger(\boldsymbol{x})\hat{\psi}^\dagger(\boldsymbol{x}')\hat{\psi}(\boldsymbol{x}')\left[-\frac{\hbar^2}{2m}\boldsymbol{\nabla}^2 + V(\boldsymbol{x})\right]\hat{\psi}(\boldsymbol{x})\\ &= \int d^3x'\,d^3x\,\hat{\psi}^\dagger(\boldsymbol{x})\left[-\frac{\hbar^2}{2m}\boldsymbol{\nabla}^2 + V(\boldsymbol{x})\right]\hat{\psi}^\dagger(\boldsymbol{x}')\hat{\psi}(\boldsymbol{x}')\hat{\psi}(\boldsymbol{x})\\ &= \pm\int d^3x'\,d^3x\,\hat{\psi}^\dagger(\boldsymbol{x})\left[-\frac{\hbar^2}{2m}\boldsymbol{\nabla}^2 + V(\boldsymbol{x})\right]\hat{\psi}^\dagger(\boldsymbol{x}')\hat{\psi}(\boldsymbol{x})\hat{\psi}(\boldsymbol{x}')\\ &= \int d^3x'\,d^3x\,\hat{\psi}^\dagger(\boldsymbol{x})\left[-\frac{\hbar^2}{2m}\boldsymbol{\nabla}^2 + V(\boldsymbol{x})\right]\\ &\quad\times\left\{\pm\{\hat{\psi}(\boldsymbol{x})\hat{\psi}^\dagger(\boldsymbol{x}') - \delta(\boldsymbol{x}'-\boldsymbol{x})\}\right\}\hat{\psi}(\boldsymbol{x}')\end{aligned}$$

Exercise 3.7.

and we deduce

$$\begin{aligned}\hat{N}\hat{H} &= \int d^3x'\, d^3x\, \hat{\psi}^\dagger(\boldsymbol{x}') \left[-\frac{\hbar^2}{2m}\boldsymbol{\nabla}^2 + V(\boldsymbol{x})\right] \hat{\psi}(\boldsymbol{x})\delta(\boldsymbol{x}-\boldsymbol{x}') \\ &\quad - \int d^3x'\, d^3x\, \hat{\psi}^\dagger(\boldsymbol{x}) \left[-\frac{\hbar^2}{2m}\boldsymbol{\nabla}^2 + V(\boldsymbol{x})\right] \hat{\psi}(\boldsymbol{x}')\delta(\boldsymbol{x}-\boldsymbol{x}') \\ &\quad + \int d^3x'\, d^3x\, \hat{\psi}^\dagger(\boldsymbol{x}) \left[-\frac{\hbar^2}{2m}\boldsymbol{\nabla}^2 + V(\boldsymbol{x})\right] \hat{\psi}(\boldsymbol{x})\hat{\psi}^\dagger(\boldsymbol{x}')\hat{\psi}(\boldsymbol{x}') \\ &= \hat{H}\hat{N}\,,\end{aligned}$$

i.e.

$$\left[\hat{N},\hat{H}\right] = 0 = \frac{d\hat{N}}{dt}\,.$$

The ket vector $|0\rangle$ designates the *vacuum state*; it shall represent the state without any real particles,[10] so that

$$\hat{b}_n|0\rangle = 0$$

for any $\hat{b}_n$. Because of (3.43) we immediately realize that

$$\hat{\psi}(\boldsymbol{x},t)|0\rangle = \sum_n \psi_n(\boldsymbol{x},t)\hat{b}_n|0\rangle = 0\,. \tag{3.54}$$

This result confirms that *the vacuum does not contain particles.* Thus, it becomes clear that the *vacuum expectation value of the field operator* vanishes:

$$\langle 0|\hat{\psi}(\boldsymbol{x},t)|0\rangle = 0\,. \tag{3.55}$$

Hence, we expect that

$$\hat{\psi}^\dagger(\boldsymbol{x},t)|0\rangle \tag{3.56}$$

describes a state where a particle stays at position $\boldsymbol{x}$. To convince ourselves we first calculate the action of the particle density operator $\hat{n}(\boldsymbol{x},t)$ of (3.50) on this state and derive

$$\begin{aligned}\hat{n}(\boldsymbol{x}',t)\hat{\psi}^\dagger(\boldsymbol{x},t)|0\rangle &= \hat{\psi}^\dagger(\boldsymbol{x}',t)\hat{\psi}(\boldsymbol{x}',t)\hat{\psi}^\dagger(\boldsymbol{x},t)|0\rangle \\ &= \hat{\psi}^\dagger(\boldsymbol{x}',t)\left[\delta(\boldsymbol{x}-\boldsymbol{x}') \pm \hat{\psi}^\dagger(\boldsymbol{x},t)\hat{\psi}(\boldsymbol{x}',t)\right]|0\rangle \\ &= \delta(\boldsymbol{x}-\boldsymbol{x}')\hat{\psi}^\dagger(\boldsymbol{x},t)|0\rangle\,.\end{aligned} \tag{3.57}$$

The last term of the second to last step vanishes because of (3.54). We realize that $\hat{\psi}^\dagger(x,t)|0\rangle$ represents an *eigenvector of the particle-number density operator with eigenvalue* $\delta(\boldsymbol{x}'-\boldsymbol{x})$. The eigenvalue is zero everywhere except at $\boldsymbol{x}'=\boldsymbol{x}$. At this point ($\boldsymbol{x}'=\boldsymbol{x}$) the particle density becomes so large that an integration over the vicinity of this position yields 1. Then the validity of the following relation becomes clear:

[10] Later on we will explain the concept of vacuum in more detail.

$$\hat{N}\hat{\psi}^\dagger(\boldsymbol{x},t)|0\rangle = \hat{\psi}^\dagger(\boldsymbol{x},t)|0\rangle . \tag{3.58}$$

It can be proven with the help of (3.57). We get

$$\begin{aligned}\int \mathrm{d}^3x'\,\hat{n}(\boldsymbol{x}',t)\hat{\psi}^\dagger(\boldsymbol{x},t)|0\rangle &= \int \mathrm{d}^3x'\delta(\boldsymbol{x}'-\boldsymbol{x})\hat{\psi}^\dagger(\boldsymbol{x},t)|0\rangle \\ &= \hat{\psi}^\dagger(\boldsymbol{x},t)|0\rangle ,\end{aligned}$$

which shows that also $\hat{\psi}^\dagger(x,t)|0\rangle$ is an *eigenvector to the total particle-number operator with eigenvalue* 1. The interpretation of $\hat{\psi}^\dagger(\boldsymbol{x},t)|0\rangle$ as a *one-particle state* has been justified! Similarily,

$$\hat{\psi}^\dagger(\boldsymbol{x}_1,t)\hat{\psi}^\dagger(\boldsymbol{x}_2,t)|0\rangle \tag{3.59}$$

represents a *two-particle state* with a particle at $\boldsymbol{x}_1$ and one at $\boldsymbol{x}_2$. *Many-particle states* can be constructed in an analogous fashion. We come back to the one-particle state (3.56) and construct a superposition of such states,

$$|\chi_1,t\rangle = \int \mathrm{d}^3x\,\chi_1(\boldsymbol{x})\hat{\psi}^\dagger(\boldsymbol{x},t)|0\rangle , \tag{3.60}$$

where the function $\chi_1(x)$ is an ordinary function and should not be interpreted as an operator! According to the rules of quantum mechanics we interpret

$$|\chi_1(\boldsymbol{x})|^2\,\mathrm{d}^3x \tag{3.61}$$

as the probability of finding a particle within the volume d^3x in state $\hat{\psi}^\dagger(x,t)|0\rangle$. Because of (3.57) this probability is identical to the probability of finding a particle in d^3x at all. Evidently $|\chi_1(\boldsymbol{x},t)|^2\,\mathrm{d}^3x$ plays the role of the probability density, which in conventional quantum mechanics is given by $|\psi(\boldsymbol{x},t)|^2$. Now we will construct $\chi_1(\boldsymbol{x})$ in such a way that $|\chi_1,t\rangle$ becomes an eigenvector to $\hat{H}$ with eigenvalue E. This leads to the eigenvalue equation

$$\hat{H}|\chi_1,t\rangle = E|\chi_1,t\rangle \tag{3.62}$$

or

$$\begin{aligned}&\int \mathrm{d}^3x\,\hat{\psi}^\dagger(\boldsymbol{x},t)\left[-\frac{\hbar^2}{2m}\boldsymbol{\nabla}^2+V(\boldsymbol{x})\right]\hat{\psi}(\boldsymbol{x},t)\int \mathrm{d}^3x'\,\chi_1(\boldsymbol{x}')\hat{\psi}^\dagger(\boldsymbol{x}',t)|0\rangle \\ &\quad = E\int \mathrm{d}^3x'\,\chi_1(\boldsymbol{x}')\hat{\psi}^\dagger(\boldsymbol{x}',t)|0\rangle .\end{aligned} \tag{3.63}$$

We transform the left-hand side into

$$\begin{aligned}&\int \mathrm{d}^3x\,\hat{\psi}^\dagger(\boldsymbol{x},t)\left[-\frac{\hbar^2}{2m}\boldsymbol{\nabla}^2+V(\boldsymbol{x})\right]\int \mathrm{d}^3x'\,\chi_1(\boldsymbol{x}')\hat{\psi}(\boldsymbol{x},t)\hat{\psi}^\dagger(\boldsymbol{x}',t)|0\rangle \\ &\quad = \int \mathrm{d}^3x\,\hat{\psi}^\dagger(\boldsymbol{x},t)\left[-\frac{\hbar^2}{2m}\boldsymbol{\nabla}^2+V(\boldsymbol{x})\right] \\ &\qquad \times \int \mathrm{d}^3x'\,\chi_1(\boldsymbol{x}')\left[\delta(\boldsymbol{x}-\boldsymbol{x}')\pm\hat{\psi}^\dagger(\boldsymbol{x}',t)\hat{\psi}(\boldsymbol{x},t)\right]|0\rangle \\ &\quad = \int \mathrm{d}^3x\int \mathrm{d}^3x'\,\hat{\psi}^\dagger(\boldsymbol{x},t)\left[-\frac{\hbar^2}{2m}\boldsymbol{\nabla}^2+V(\boldsymbol{x})\right]\delta(\boldsymbol{x}-\boldsymbol{x}')\chi_1(\boldsymbol{x}')|0\rangle \\ &\quad = \int \mathrm{d}^3x\hat{\psi}^\dagger(\boldsymbol{x},t)\left[-\frac{\hbar^2}{2m}\boldsymbol{\nabla}^2+V(\boldsymbol{x})\right]\chi_1(\boldsymbol{x})|0\rangle .\end{aligned} \tag{3.64}$$

Using (3.64), we can summarize (3.63) as

$$\int \mathrm{d}^3x\,\hat{\psi}^\dagger(\boldsymbol{x},t)\left\{\left[-\frac{\hbar^2}{2m}\boldsymbol{\nabla}^2+V(\boldsymbol{x})\right]\chi_1(\boldsymbol{x})-E\chi_1(\boldsymbol{x})\right\}|0\rangle \;=\; 0\,.$$

Since this holds for any arbitrary $\hat{\psi}^\dagger(x,t)$ it follows that

$$\left[-\frac{\hbar^2}{2m}\boldsymbol{\nabla}^2+V(\boldsymbol{x})\right]\chi_1(\boldsymbol{x}) \;=\; E\chi_1(\boldsymbol{x})\,. \tag{3.65}$$

This is the one-particle Schrödinger equation for $\chi_1(\boldsymbol{x})$. The interpretation (3.61) and the validity of the Schrödinger equation (3.65) for $\chi_1(\boldsymbol{x})$ mean the function $\chi_1(\boldsymbol{x})$ *has to be identical with the one-particle wavefunction* of elementary quantum mechanics. Hence, the formalism of the second quantization yields the conventional one-particle Schrödinger equation.

In order to derive the many-particle Schrödinger equation out of the field-theoretical formalism we introduce the many-particle state

$$|\chi_n,t\rangle = \int \mathrm{d}^3x_1\cdots\int \mathrm{d}^3x_n\,\chi_n(\boldsymbol{x}_1,\cdots,\boldsymbol{x}_n)\hat{\psi}^\dagger(\boldsymbol{x}_1,t)\cdots\hat{\psi}^\dagger(\boldsymbol{x}_n,t)|0\rangle \tag{3.66}$$

in analogy to (3.60). Then we interpret

$$|\chi_n(\boldsymbol{x}_1,\cdots,\boldsymbol{x}_n)|^2\,\mathrm{d}^3x_1\cdots\mathrm{d}^3x_n \tag{3.67}$$

as the probability of finding particle 1 in d^3x_1, particle 2 in d^3x_2, ..., particle n in d^3x_n. In analogy to (3.62) we now require that the state $|\chi_n,t\rangle$ obeys the Schrödinger equation

$$\hat{H}|\chi_n,t\rangle \;=\; E|\chi_n,t\rangle\,, \tag{3.68}$$

where $\hat{H}$ is given by (3.46a). This requirement leads to a condition for the amplitude function $\chi_n(\boldsymbol{x}_1,\ldots,\boldsymbol{x}_n)$. In order to determine it we write (3.68) in full detail as

$$\begin{aligned}
&\int \mathrm{d}^3x\hat{\psi}^\dagger(\boldsymbol{x},t)\left[-\frac{\hbar^2}{2m}\boldsymbol{\nabla}^2+V(\boldsymbol{x})\right]\hat{\psi}(\boldsymbol{x},t)\\
&\quad\times\int \mathrm{d}^3x_1'\cdots\mathrm{d}^3x_n'\,\chi_n(\boldsymbol{x}_1',\cdots,\boldsymbol{x}_n')\hat{\psi}^\dagger(\boldsymbol{x}_1',t)\cdots\hat{\psi}^\dagger(\boldsymbol{x}_n',t)|0\rangle\\
&\quad= E\int \mathrm{d}^3x_1'\cdots\mathrm{d}^3x_n'\,\chi_n(\boldsymbol{x}_1',\cdots,\boldsymbol{x}_n')\hat{\psi}^\dagger(\boldsymbol{x}_1',t)\cdots\hat{\psi}^\dagger(\boldsymbol{x}_n',t)|0\rangle\,.
\end{aligned} \tag{3.69}$$

The left-hand side becomes

$$\begin{aligned}
&\int \mathrm{d}^3x\,\mathrm{d}^3x_1'\cdots\mathrm{d}^3x_n'\hat{\psi}^\dagger(\boldsymbol{x},t)\left[-\frac{\hbar^2}{2m}\boldsymbol{\nabla}^2+V(\boldsymbol{x})\right]\hat{\psi}(\boldsymbol{x},t)\\
&\quad\times\chi_n(\boldsymbol{x}_1',\cdots,\boldsymbol{x}_n')\hat{\psi}^\dagger(\boldsymbol{x}_1',t)\cdots\hat{\psi}^\dagger(\boldsymbol{x}_n',t)|0\rangle\\
&= \int \mathrm{d}^3x\,\mathrm{d}^3x_1'\cdots\mathrm{d}^3x_n'\hat{\psi}^\dagger(\boldsymbol{x},t)\,[\cdots]\,\chi_n(\boldsymbol{x}_1',\cdots,\boldsymbol{x}_n')\\
&\quad\times\hat{\psi}(\boldsymbol{x},t)\hat{\psi}^\dagger(\boldsymbol{x}_1,t)\cdots\hat{\psi}^\dagger(\boldsymbol{x}_n',t)|0\rangle\\
&= \int \mathrm{d}^3x\,\mathrm{d}^3x_1'\cdots\mathrm{d}^3x_n'\hat{\psi}^\dagger(\boldsymbol{x},t)\,[\cdots]\,\chi_n(\boldsymbol{x}_1',\cdots,\boldsymbol{x}_n')\\
&\quad\times\left[\delta(\boldsymbol{x}-\boldsymbol{x}_1')\pm\hat{\psi}^\dagger(\boldsymbol{x}_1',t)\hat{\psi}(\boldsymbol{x},t)\right]\hat{\psi}^\dagger(\boldsymbol{x}_2',t)\cdots\hat{\psi}^\dagger(\boldsymbol{x}_n',t)|0\rangle
\end{aligned}$$

$$= \int d^3x\, d^3x_1' \cdots d^3x_n' \hat{\psi}^\dagger(\boldsymbol{x},t)\, [\cdots]\, \chi_n(\boldsymbol{x}_1',\cdots,\boldsymbol{x}_n')$$
$$\times \Big\{ \delta(\boldsymbol{x}-\boldsymbol{x}_1')\hat{\psi}^\dagger(\boldsymbol{x}_2',t)\cdots\hat{\psi}^\dagger(\boldsymbol{x}_n',t)|0\rangle$$
$$\pm\hat{\psi}^\dagger(\boldsymbol{x}_1',t)\Big[\delta(\boldsymbol{x}-\boldsymbol{x}_2')\pm\hat{\psi}^\dagger(\boldsymbol{x}_2',t)\hat{\psi}(\boldsymbol{x},t)\Big]\,\hat{\psi}^\dagger(\boldsymbol{x}_3',t)\cdots\hat{\psi}^\dagger(\boldsymbol{x}_n',t)|0\rangle\Big\}$$
$$= \int d^3x\, d^3x_1' \cdots d^3x_n' \hat{\psi}^\dagger(\boldsymbol{x},t)\, [\cdots]\, \chi_n(\boldsymbol{x}_1',\cdots,\boldsymbol{x}_n')$$
$$\times\Big\{\sum_i (\pm 1)^{i-1}\delta(\boldsymbol{x}-\boldsymbol{x}_i')\hat{\psi}^\dagger(\boldsymbol{x}_1',t)$$
$$\times\cdots\hat{\psi}^\dagger(\boldsymbol{x}_{i-1}',t)\hat{\psi}^\dagger(\boldsymbol{x}_{i+1}',t)\cdots\hat{\psi}^\dagger(\boldsymbol{x}_n',t)|0\rangle\Big\}$$
$$= \int d^3x_1'\, d^3x_2' \cdots d^3x_n' \sum_i (\pm 1)^{i-1}\hat{\psi}^\dagger(\boldsymbol{x}_i',t)\left[-\frac{\hbar^2}{2m}\boldsymbol{\nabla}_i'^2 + V(\boldsymbol{x}_i')\right]$$
$$\times\, \chi_n(\boldsymbol{x}_1',\cdots,\boldsymbol{x}_n')\hat{\psi}^\dagger(\boldsymbol{x}_1',t)\cdots\hat{\psi}^\dagger(\boldsymbol{x}_n',t)|0\rangle\,. \qquad (3.70)$$

Observe that within the last expression, all operators $\hat{\psi}^\dagger(\boldsymbol{x}_\nu',t)$ appear on the far right except for the operator $\hat{\psi}^\dagger(\boldsymbol{x}_i',t)$ with argument $\boldsymbol{x}_i'$. This field operator appears left of $[\cdots]$. All operators $\hat{\psi}^\dagger(\boldsymbol{x}_\nu',t)$ to the right of $[\cdots]\,\chi_n(x_1\cdots x_n)$ can be shifted to the left of $[\cdots]$ without any restriction. This gives

$$\int d^3x_1' \cdots d^3x_n' \sum_i (\pm 1)^{i-1}\hat{\psi}^\dagger(\boldsymbol{x}_i',t)\hat{\psi}^\dagger(\boldsymbol{x}_1',t)\cdots\hat{\psi}^\dagger(\boldsymbol{x}_{i-1}',t)\hat{\psi}^\dagger(\boldsymbol{x}_{i+1}',t)\cdots$$
$$\times\,\hat{\psi}^\dagger(\boldsymbol{x}_n',t)\left[-\frac{\hbar^2\boldsymbol{\nabla}_i'^2}{2m}+V(\boldsymbol{x}_i')\right]\chi_n(\boldsymbol{x}_1',\cdots,\boldsymbol{x}_n')|0\rangle\,.$$

If we shift the operator $\hat{\psi}^\dagger(\boldsymbol{x}_i')$ appearing on the left within this expression to its "normal" place between $\hat{\psi}^\dagger(\boldsymbol{x}_{i-1}',t)$ and $\hat{\psi}^\dagger(\boldsymbol{x}_{i+1}',t)$, we have to make $i-1$ permutations. This produces a factor $(\pm 1)^{i-1}$. Finally we get

$$\int d^3x_1' \cdots d^3x_n' \sum_i (\pm 1)^{i-1}(\pm 1)^{i-1}\hat{\psi}^\dagger(\boldsymbol{x}_1',t)\cdots\hat{\psi}^\dagger(\boldsymbol{x}_{i-1}',t)\hat{\psi}^\dagger(\boldsymbol{x}_i',t)$$
$$\times\,\hat{\psi}^\dagger(\boldsymbol{x}_{i+1}',t)\cdots\hat{\psi}^\dagger(\boldsymbol{x}_n',t)\left[-\frac{\hbar^2\boldsymbol{\nabla}_i^{2'}}{2m}+V(\boldsymbol{x}_i')\right]\chi_n(\boldsymbol{x}_1',\cdots,\boldsymbol{x}_n')|0\rangle$$
$$= \int d^3x_1' \cdots d^3x_n' \hat{\psi}^\dagger(\boldsymbol{x}_1',t)\cdots\hat{\psi}^\dagger(\boldsymbol{x}_n',t)$$
$$\times\sum_i\left[-\frac{\hbar^2\boldsymbol{\nabla}_i^{2'}}{2m}+V(\boldsymbol{x}')\right]\chi_n(\boldsymbol{x}_1',\cdots,\boldsymbol{x}_n')|0\rangle\,. \qquad (3.71)$$

It is remarkable that all distinctions have vanished as a result of the various $\pm$ signs. With the help of (3.71), we can write (3.69) in the form

$$\int d^3x_1' \cdots d^3x_n' \hat{\psi}^\dagger(\boldsymbol{x}_1',t)\cdots\hat{\psi}^\dagger(\boldsymbol{x}_n',t)$$
$$\times\left\{\left[\sum_i\left(-\frac{\hbar^2\boldsymbol{\nabla}_i^{2'}}{2m}\right)+V(\boldsymbol{x}_i')-E\right]\chi_n(\boldsymbol{x}_1',\cdots,\boldsymbol{x}_n')\right\}|0\rangle\,. \qquad (3.72)$$

This has to hold for arbitrary states $\hat{\psi}^\dagger(\boldsymbol{x}'_1, t) \cdots \hat{\psi}^\dagger(\boldsymbol{x}_n, t)|0\rangle$; hence, we deduce

$$\left(\sum_i \left\{-\frac{\hbar^2}{2m}\boldsymbol{\nabla}_i'^2 + V(\boldsymbol{x}'_i)\right\}\right) \chi_n(\boldsymbol{x}'_1, \cdots, \boldsymbol{x}'_n) = E\chi_n(\boldsymbol{x}_1, \cdots, \boldsymbol{x}_n)\,. \tag{3.73}$$

This represents an *n-particle Schrödinger equation* for the amplitude $\chi_n(\boldsymbol{x}_1, \ldots, \boldsymbol{x}_n)$. We realize that the formalism of second quantization also incorporates the elementary quantum mechanics of an arbitrary number of *noninteracting* particles. Since the field-theoretical formulation, i.e. the second quantization, clearly expresses the quantum nature of the processes (creation and annhilation of particles) as well as the wave character (particles are annihilated and created in specific states χ_i) it should be acknowledged as the true and complete formulation of quantum mechanics.

4. Quantum Fields with Interaction

We examine now the Hamiltonian of two (or more) free particle *fields* and their mutual interaction. It appears to be obvious to simply add the Hamiltonians of the free fields and to introduce their interaction suitably. We have already come across the interaction between charged particles and photons. In the following we elaborate on how this well-known system is completely described from the field-theoretical point of view, i.e. in the language of second quantization. With the introduction of creation and annhilation operators $\hat{a}^\dagger_{\boldsymbol{k}\sigma}, \hat{a}_{\boldsymbol{k}\sigma}$ for photons the radiation field itself has already been treated field-theoretically. But we now also want to express the particle processes with particle creation and annihilation operators $\hat{b}^\dagger_n$ and $\hat{b}_n$. The noninteracting particle field will be described with Hamiltonian (3.46) and the photon fields through Hamiltonians (1.40) or (1.52). The interaction (2.6) between the two fields is introduced via minimal coupling:

$$\hat{\boldsymbol{p}} \longrightarrow \hat{\boldsymbol{p}} - \frac{e}{c}\hat{\boldsymbol{A}} .$$

We insert this replacement directly into (3.46) and obtain the *total Hamiltonian* by addition of the mentioned terms:

$$\begin{aligned} \hat{H} &= \int \mathrm{d}^3x\, \hat{\psi}^\dagger(\boldsymbol{x},t) \left[\frac{1}{2m}\left(\hat{\boldsymbol{p}} - \frac{e}{c}\hat{\boldsymbol{A}}\right)^2 + V(\boldsymbol{x})\right] \hat{\psi}(\boldsymbol{x},t) \\ &\quad + \int \mathrm{d}^3x\, \frac{1}{8\pi}(\hat{\boldsymbol{E}}^2 + \hat{\boldsymbol{B}}^2) \\ &= \hat{H}_{\mathrm{mp}} + \hat{H}_{\mathrm{rad}} + \hat{H}_{\mathrm{int}} . \end{aligned} \tag{4.1}$$

Here $\hat{H}_{\mathrm{mp}}$ represents the Hamiltonian of the particle field (many-particle Hamiltonian),

$$\hat{H}_{\mathrm{mp}} = \int \mathrm{d}^3x\, \hat{\psi}^\dagger(\boldsymbol{x},t) \left[-\frac{\hbar^2}{2m}\boldsymbol{\nabla}^2 + V(\boldsymbol{x})\right] \hat{\psi}(\boldsymbol{x},t) , \tag{4.2}$$

and $\hat{H}_{\mathrm{rad}}$ represents the Hamiltonian of the radiation field,

$$\hat{H}_{\mathrm{rad}} = \int \mathrm{d}^3x\, \frac{1}{8\pi}(\hat{\boldsymbol{E}}^2 + \hat{\boldsymbol{B}}^2) = \sum_{\boldsymbol{k},\sigma} \left(\hbar\omega_k + \frac{1}{2}\right) \hat{a}^\dagger_{\boldsymbol{k}\sigma}\hat{a}_{\boldsymbol{k}\sigma} . \tag{4.3}$$

The Hamiltonian for the interaction between the particle field and the radiation field is given by

$$\hat{H}_{\mathrm{int}} = \int \mathrm{d}^3x\, \hat{\psi}^\dagger(\boldsymbol{x},t) \left[-\frac{e\hbar}{imc}\hat{\boldsymbol{A}} \cdot \boldsymbol{\nabla} + \frac{e^2}{2mc^2}\hat{\boldsymbol{A}}^2\right] \hat{\psi}(\boldsymbol{x},t) . \tag{4.4}$$

The particle-field operators $\hat{\psi}$, $\hat{\psi}^\dagger$ are given by (3.43) and are an expression of the particle annihilation operators $\hat{b}_n$ and creation operators $\hat{b}_n^\dagger$. Analogously to (2.8) and (2.9), (4.4) also splits into parts $\hat{H}'_{\text{int}}$ and $\hat{H}''_{\text{int}}$, which are proportional to $\hat{\boldsymbol{A}}$ and $\hat{\boldsymbol{A}}^2$, respectively. Substituting into (4.4) the expansion (1.37) for the vector potential field operator $\hat{\boldsymbol{A}}$ in terms of photon creation and annhilation operators,

$$\hat{\boldsymbol{A}}(\boldsymbol{x},t) = \sum_{\boldsymbol{k},\sigma} N_{\boldsymbol{k}} \boldsymbol{\epsilon}_{\boldsymbol{k},\sigma} \left(\hat{a}_{\boldsymbol{k}\sigma} \mathrm{e}^{\mathrm{i}\boldsymbol{k}\cdot\boldsymbol{x}} + \hat{a}^\dagger_{\boldsymbol{k}\sigma} \mathrm{e}^{-\mathrm{i}\boldsymbol{k}\cdot\boldsymbol{x}} \right) ,$$

and for the particle field operator the expansion (3.43) in terms of particle creation and annihilation operators,

$$\hat{\psi}(\boldsymbol{x},t) = \sum_n \hat{b}_n \psi_n(\boldsymbol{x}) ,$$

$$\hat{\psi}^\dagger(\boldsymbol{x},t) = \sum_n \hat{b}^\dagger_n \psi^*_n(\boldsymbol{x}) ,$$

we obtain the result

$$\hat{H}_{\text{int}} = \hat{H}'_{\text{int}} + \hat{H}''_{\text{int}} , \tag{4.5}$$

where

$$\hat{H}'_{\text{int}} = \sum_{\boldsymbol{k},\sigma} \sum_n \sum_{n'} \boldsymbol{\epsilon}_{\boldsymbol{k},\sigma} \cdot \left[\boldsymbol{M}(\boldsymbol{k},n,n') \hat{b}^\dagger_n \hat{b}_{n'} \hat{a}_{\boldsymbol{k}\sigma} + \boldsymbol{M}(-\boldsymbol{k},n,n') \hat{b}^\dagger_n \hat{b}_{n'} \hat{a}^\dagger_{\boldsymbol{k}\sigma} \right] \tag{4.6}$$

and

$$\begin{aligned} \hat{H}''_{\text{int}} = \sum_{\boldsymbol{k}_1,\sigma_1} \sum_{\boldsymbol{k}_1,\sigma_2} \sum_n \sum_{n'} & \hat{b}^\dagger_n \hat{b}_{n'} \boldsymbol{\epsilon}_{\boldsymbol{k}_1,\sigma_1} \cdot \boldsymbol{\epsilon}_{\boldsymbol{k}_2,\sigma_2} \\ & \times \left[M(\boldsymbol{k}_1,\boldsymbol{k}_2,n,n') \hat{a}_{\boldsymbol{k}_1,\sigma_1} \hat{a}_{\boldsymbol{k}_2,\sigma_2} \right. \\ & + M(\boldsymbol{k}_1,-\boldsymbol{k}_2,n,n') \hat{a}_{\boldsymbol{k}_1,\sigma_1} \hat{a}^\dagger_{\boldsymbol{k}_2,\sigma_2} \\ & + M(-\boldsymbol{k}_1,\boldsymbol{k}_2,n,n') \hat{a}^\dagger_{\boldsymbol{k}_1,\sigma_1} \hat{a}_{\boldsymbol{k}_2,\sigma_2} \\ & \left. + M(-\boldsymbol{k}_1,-\boldsymbol{k}_2,n,n') \hat{a}^\dagger_{\boldsymbol{k}_1,\sigma_1} \hat{a}^\dagger_{\boldsymbol{k}_2,\sigma_2} \right] . \end{aligned} \tag{4.7}$$

The matrix elements $\boldsymbol{M}(\boldsymbol{k},n,n')$ and $M(\boldsymbol{k}_1,\boldsymbol{k}_2,n,n')$ are given by

$$\boldsymbol{M}(\boldsymbol{k},n,n') = \sqrt{\frac{2\pi\hbar c^2}{L^3\omega_k}} \int \mathrm{d}V\, \psi^*_n(\boldsymbol{x}) \left(-\frac{e\hbar}{imc} e^{i\boldsymbol{k}\cdot\boldsymbol{x}} \boldsymbol{\nabla} \right) \psi_{n'}(\boldsymbol{x}) \tag{4.8a}$$

and

$$\begin{aligned} M(\boldsymbol{k}_1,\boldsymbol{k}_2,n,n') = & \frac{2\pi^2\hbar c^2}{L^3} \frac{1}{\sqrt{\omega_{k_1}\omega_{k_2}}} \\ & \times \int \mathrm{d}V\, \psi^*_n(\boldsymbol{x}) \frac{e^2}{2mc^2} \mathrm{e}^{\mathrm{i}(\boldsymbol{k}_1+\boldsymbol{k}_2)\cdot\boldsymbol{x}} \psi_{n'}(\boldsymbol{x}) , \end{aligned} \tag{4.8b}$$

respectively. The unperturbed Hamiltonian

$$\hat{H}_0 = \hat{H}_{\text{mp}} + \hat{H}_{\text{rad}} \tag{4.9}$$

has eigenvectors

$$|\cdots N_n \cdots\rangle |\cdots n_{\boldsymbol{k}\sigma} \cdots\rangle \tag{4.10}$$

with eigenvalues

$$\sum_n E_n N_n + \sum_{\boldsymbol{k},\sigma} \hbar\omega_k \left(n_{\boldsymbol{k}\sigma} + \frac{1}{2}\right) . \tag{4.11}$$

Hence, it holds that

$$\begin{aligned} &\hat{H}_0 |\cdots N_n \cdots\rangle |\cdots n_{\boldsymbol{k}\sigma} \cdots\rangle \\ &= \left(\sum_n E_n N_n + \sum_{\boldsymbol{k},\sigma} \hbar\omega_k \left(n_{\boldsymbol{k}\sigma} + \frac{1}{2}\right)\right) |\cdots N_n \cdots\rangle |\cdots n_{\boldsymbol{k}\sigma} \cdots\rangle . \end{aligned} \tag{4.12}$$

The interaction (4.5), more explicitly (4.6) and (4.7), causes transitions between states (4.10). For example, the term proportional to $\hat{b}^\dagger_n \hat{b}_{n'} \hat{a}_{\boldsymbol{k}\sigma}$ annihilates a photon with momentum $\hbar\boldsymbol{k}$ and polarization σ, annihilates a particle in state $|n'\rangle$, and at the same time creates a particle in state $|n\rangle$. We say that through the annihilation of a photon of the kind $\boldsymbol{k}\sigma$, a particle is scattered from state $|n'\rangle$ to state $|n\rangle$. The matrix element $\boldsymbol{\epsilon}_{\boldsymbol{k}\sigma} \cdot \boldsymbol{M}(\boldsymbol{k},\sigma,n,n')$ in (4.6) represents the amplitude for this process. Analogously, the term proportional to $\hat{b}^\dagger_n \hat{b}_{n'} \hat{a}^\dagger_{\boldsymbol{k}\sigma}$ describes the emission of photon $\boldsymbol{k}\sigma$ with simultaneous scattering of the particle from $|n'\rangle$ to $|n\rangle$. In the same spirit the terms contained in H''_{int} can be interpreted. We illustrate these processes in Figs. 4.1–4.5.

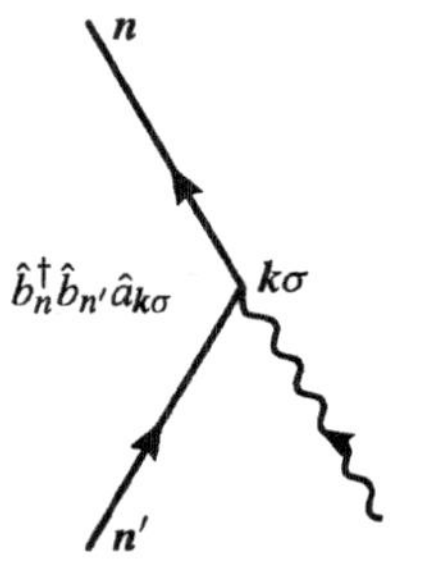

Fig. 4.1. Scattering of a particle from $|n'\rangle$ to $|n\rangle$ by absorption of a photon $\boldsymbol{k}\sigma$

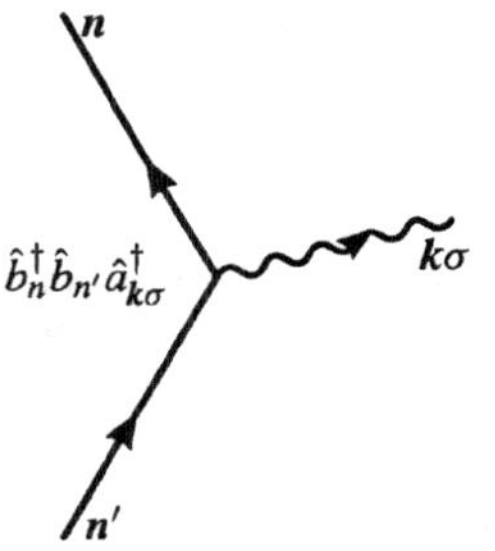

Fig. 4.2. Scattering of a particle from $|n'\rangle$ to $|n\rangle$ by emission of a photon $\boldsymbol{k}\sigma$

Often the interactions (4.6) and (4.7) are written in abbreviated form:

$$\hat{H}'_{\text{int}} = \sum_{\boldsymbol{k},\sigma} \sum_n \sum_{n'} \boldsymbol{\epsilon}_{\boldsymbol{k}\sigma} \cdot \left[\boldsymbol{M}(\boldsymbol{k},\sigma,n,n') \hat{b}^\dagger_n \hat{b}_{n'} \hat{a}_{\boldsymbol{k}\sigma} + \text{H.c.}\right] , \tag{4.13}$$

$$\begin{aligned} \hat{H}''_{\text{int}} &= \sum_{\boldsymbol{k}_1,\sigma_1} \sum_{\boldsymbol{k}_2,\sigma_2} \sum_n \sum_{n'} \boldsymbol{\epsilon}_{\boldsymbol{k}_1,\sigma_1} \cdot \boldsymbol{\epsilon}_{\boldsymbol{k}_2,\sigma_2} \\ &\quad \times \Big[M(\boldsymbol{k}_1,\sigma_1,\boldsymbol{k}_2,\sigma_2,n,n') \hat{b}^\dagger_n \hat{b}_{n'} \hat{a}_{\boldsymbol{k}_1,\sigma_1} \hat{a}_{\boldsymbol{k}_2,\sigma_2} \\ &\quad + M(\boldsymbol{k}_1,\sigma_1,-\boldsymbol{k}_2,\sigma_2,n,n') \hat{b}^\dagger_n \hat{b}_{n'} \hat{a}_{\boldsymbol{k}_1\sigma_1} \hat{a}^\dagger_{\boldsymbol{k}_2\sigma_2} + \text{H.c.}\Big] . \end{aligned} \tag{4.14}$$

This notion is indeed correct. For instance we consider the Hermitean conjugated (H.c.) part of the first expression (4.13) for H'_{int} explicitly:

$$\sum_{\boldsymbol{k},\sigma} \sum_n \sum_{n'} \boldsymbol{\epsilon}_{\boldsymbol{k}\sigma} \cdot \boldsymbol{M}^*(\boldsymbol{k},\sigma,n,n') \hat{b}^\dagger_{n'} \hat{b}_n \hat{a}^\dagger_{\boldsymbol{k}\sigma} .$$

Owing to (4.8a) we point out that

$$\begin{aligned} &\boldsymbol{\epsilon}_{\boldsymbol{k}\sigma} \cdot \boldsymbol{M}^*(\boldsymbol{k},n,n') \\ &= \sqrt{\frac{2\pi\hbar c^2}{L^3\omega_k}} \int \mathrm{d}V\, \psi_n(\boldsymbol{x}) \left(\frac{e\hbar}{\mathrm{i}mc} \mathrm{e}^{-\mathrm{i}\boldsymbol{k}\cdot\boldsymbol{x}} \boldsymbol{\epsilon}_{\boldsymbol{k}\sigma} \cdot \boldsymbol{\nabla}\right) \psi^*_{n'}(\boldsymbol{x}) \quad \left(\begin{matrix}\text{partial} \\ \text{integration}\end{matrix}\right) \\ &= -\sqrt{\frac{2\pi\hbar c^2}{L^3\omega_k}} \int \mathrm{d}V\, \psi^*_{n'}(\boldsymbol{x}) \left(\frac{e\hbar}{\mathrm{i}mc} \mathrm{e}^{-\mathrm{i}\boldsymbol{k}\cdot\boldsymbol{x}} \boldsymbol{\epsilon}_{\boldsymbol{k}\sigma} \cdot (-\mathrm{i}\boldsymbol{k} + \boldsymbol{\nabla})\right) \psi_n(\boldsymbol{x}) \\ &\qquad\qquad (\boldsymbol{\epsilon}_{\boldsymbol{k}\sigma} \cdot \boldsymbol{k} = 0) \\ &= \boldsymbol{\epsilon}_{\boldsymbol{k}\sigma} \sqrt{\frac{2\pi\hbar c^2}{L^3\omega_k}} \int \mathrm{d}V\, \psi^*_{n'}(\boldsymbol{x}) \left(-\frac{e\hbar}{\mathrm{i}mc} \mathrm{e}^{-\mathrm{i}\boldsymbol{k}\cdot\boldsymbol{x}} \cdot \boldsymbol{\nabla}\right) \psi_n(\boldsymbol{x}) \\ &= \boldsymbol{\epsilon}_{\boldsymbol{k}\sigma} \cdot \boldsymbol{M}(-\boldsymbol{k},n',n) . \end{aligned}$$

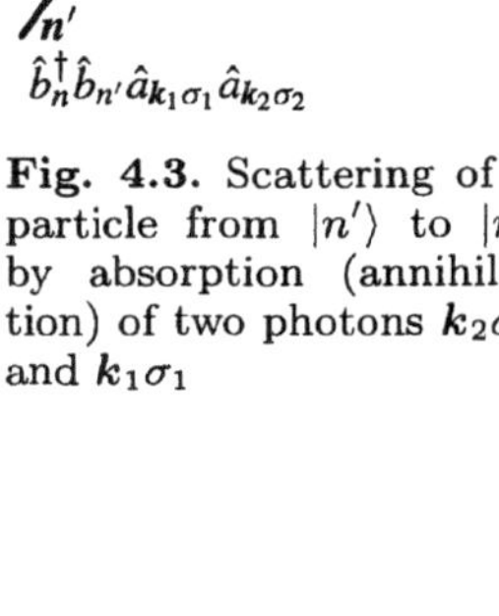

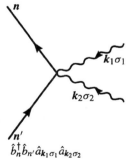

Fig. 4.3. Scattering of a particle from $|n'\rangle$ to $|n\rangle$ by absorption (annihilation) of two photons $\boldsymbol{k}_2\sigma_2$ and $\boldsymbol{k}_1\sigma_1$

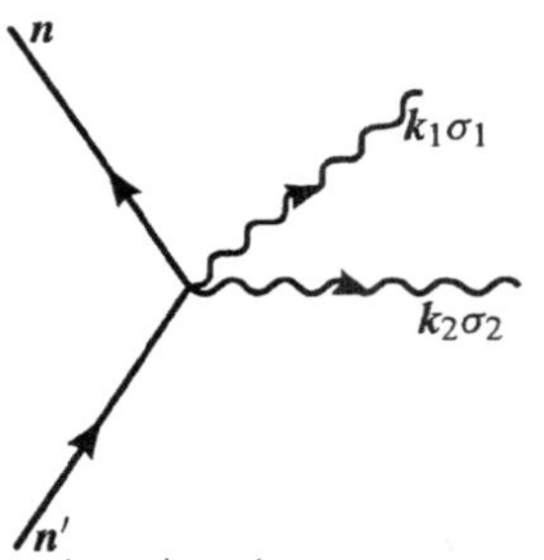

Fig. 4.4. Scattering of a particle from $|n'\rangle$ to $|n\rangle$ with simultaneous emission of two photons $\boldsymbol{k}_2\sigma_2$ and $\boldsymbol{k}_1\sigma_1$

From this we deduce

$$\begin{aligned}
&\sum_{\boldsymbol{k},\sigma}\sum_{n}\sum_{n'} \boldsymbol{\epsilon}_{\boldsymbol{k}\sigma}\cdot \boldsymbol{M}^*(\boldsymbol{k},n,n')\hat{b}_{n'}^\dagger \hat{b}_n \hat{a}_{\boldsymbol{k}\sigma}^\dagger \\
&\quad = \sum_{\boldsymbol{k},\sigma}\sum_{n}\sum_{n'} \boldsymbol{\epsilon}_{\boldsymbol{k}\sigma}\cdot \boldsymbol{M}(-\boldsymbol{k},n',n)\hat{b}_{n'}^\dagger \hat{b}_n \hat{a}_{\boldsymbol{k}\sigma}^\dagger \\
&\quad = \sum_{\boldsymbol{k},\sigma}\sum_{n}\sum_{n'} \boldsymbol{\epsilon}_{\boldsymbol{k}\sigma}\cdot \boldsymbol{M}(-\boldsymbol{k},n,n')\hat{b}_{n}^\dagger \hat{b}_{n'} \hat{a}_{\boldsymbol{k}\sigma}^\dagger \quad . \qquad (4.15)
\end{aligned}$$

We have exchanged the summation indices in the last step. Result (4.15) is identical to the second term of (4.6). On the same footing it is easy to show the identity of (4.14) and (4.7). The field-theoretical formalism (second quantization) presented here clearly expresses the quantum character of the processes. However, it does not modify the physical content of the theory; it is only the particle feature that is explicitly illustrated by second quantization.

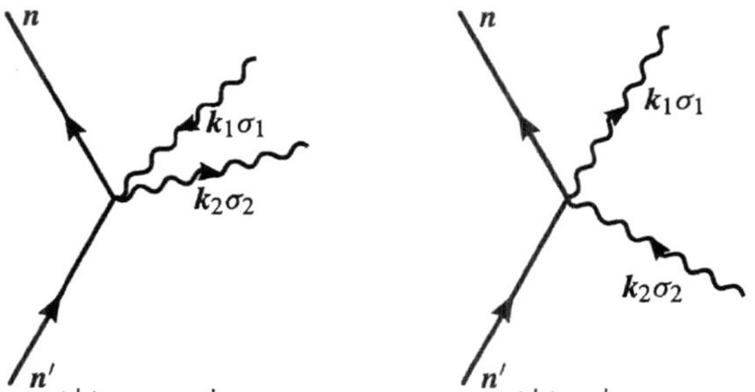

Fig. 4.5. Scattering of a particle from $|n'\rangle$ to $|n\rangle$ via emission of a photon $\boldsymbol{k}_2\sigma_2$ and simultaneous absorption of another photon $\boldsymbol{k}_1\sigma_1$, and vice versa. Here we can think of particle scattering as being caused by photon scattering

EXAMPLE

4.1 Nonrelativistic Bremsstrahlung

This example can be solved rigorously with the methods developed and presented during the course of the first three sections. We will now approach it with the tools of second quantization in order to demonstrate the advantages of this formalism.

We know from classical electrodynamics that an accelerated *charged particle radiates.* In particular such a *Bremsstrahlung* process happens in the collision of two charged particles; during the collision the particles first slow down and then accelerate again or vice versa. Bremsstrahlung occurs in various collision processes: for example, heavy-ion scattering (see Fig. 4.6), proton–proton scattering; and electron–electron scattering.

Here we discuss electron–nucleus scattering, i.e. the collision of an electron and an atomic nucleus and their bremsstrahlung. Because the mass of the atomic nucleus with A nucleons is much larger than the electron mass,

$$\frac{m_{\text{nucleon}}}{m_{\text{electron}}}A \approx 2000\,A\,,$$

recoil effects can be neglected and the nuclei can be considered as being at rest. This approximation is not valid for nucleus–nucleus bremsstrahlung. Without any loss of generality we place the nucleus at the origin $\boldsymbol{x} = 0$ of our coordinate system and denote the position vector for the electron as $\boldsymbol{x}$ (compare with Fig. 4.7). The interaction potential between the electron and nucleus is described by the Coulomb potential

$$V(\boldsymbol{x}) = -\frac{Ze^2}{|\boldsymbol{x}|} = -\frac{Ze^2}{r} \,. \tag{1}$$

We treat it as a perturbation; according to (4.2) we then write for the interaction operator

$$\hat{H}_V = \int \mathrm{d}^3x\, \hat{\psi}^\dagger(\boldsymbol{x}) V(\boldsymbol{x}) \hat{\psi}(\boldsymbol{x}) \,. \tag{2}$$

The states $|n\rangle$ of the free particle are now given by plane electron waves

$$\psi_{\boldsymbol{q}}(\boldsymbol{x}) = \langle \boldsymbol{x}|\boldsymbol{q}\rangle = \frac{1}{\sqrt{L^3}} \mathrm{e}^{\mathrm{i}\boldsymbol{q}\cdot\boldsymbol{x}} \,. \tag{3}$$

$\hbar\boldsymbol{q}$ represents the electron momentum. According to (3.43) and (3) the field operator $\hat{\psi}(\boldsymbol{x}, t)$ becomes

$$\hat{\psi}(\boldsymbol{x}, t) = \sum_{\boldsymbol{q}} \hat{b}_{\boldsymbol{q}} \psi_{\boldsymbol{q}}(\boldsymbol{x}) \,,$$

and

$$\hat{\psi}^\dagger(\boldsymbol{x}, t) = \sum_{\boldsymbol{q}} \hat{b}^\dagger_{\boldsymbol{q}} \psi^*_{\boldsymbol{q}}(\boldsymbol{x}) \,. \tag{4}$$

The Hamiltonian for the (free) particles reads [see (4.2)]

$$\begin{aligned}
\hat{H}_{\mathrm{mp}} &= \int \mathrm{d}V\, \hat{\psi}^\dagger(\boldsymbol{x}, t) \left(-\frac{\hbar^2\boldsymbol{\nabla}^2}{2m}\right) \hat{\psi}(\boldsymbol{x}, t) \\
&= \int \mathrm{d}V \sum_{\boldsymbol{q}} \hat{b}^\dagger_{\boldsymbol{q}} \psi^*_{\boldsymbol{q}}(\boldsymbol{x}) \left(-\frac{\hbar^2\boldsymbol{\nabla}^2}{2m}\right) \sum_{\boldsymbol{q}'} \hat{b}_{\boldsymbol{q}'} \psi_{\boldsymbol{q}'}(\boldsymbol{x}) \\
&= \sum_{\boldsymbol{q},\boldsymbol{q}'} \frac{(\hbar\boldsymbol{q}')^2}{2m} \hat{b}^\dagger_{\boldsymbol{q}} \hat{b}_{\boldsymbol{q}'} \int \mathrm{d}V\, \psi^*_{\boldsymbol{q}}(\boldsymbol{x}) \psi_{\boldsymbol{q}'}(\boldsymbol{x}) \\
&= \sum_{\boldsymbol{q},\boldsymbol{q}'} \frac{(\hbar\boldsymbol{q}')^2}{2m} \hat{b}^\dagger_{\boldsymbol{q}} \hat{b}_{\boldsymbol{q}'} \delta_{\boldsymbol{q},\boldsymbol{q}'} \\
&= \sum_{\boldsymbol{q}} \frac{(\hbar\boldsymbol{q})^2}{2m} \hat{b}^\dagger_{\boldsymbol{q}} \hat{b}_{\boldsymbol{q}} \,.
\end{aligned} \tag{5}$$

Here, m represents the electron mass. Expression (5) is plausible; the total kinetic energy is given as a sum of one-particle energies belonging to the various states $\psi_{\boldsymbol{q}}(\boldsymbol{x})$ multiplied by the number operator $\hat{b}^\dagger_{\boldsymbol{q}} \hat{b}_{\boldsymbol{q}}$. With (4) the (Coulomb) interaction (2) explicitly reads

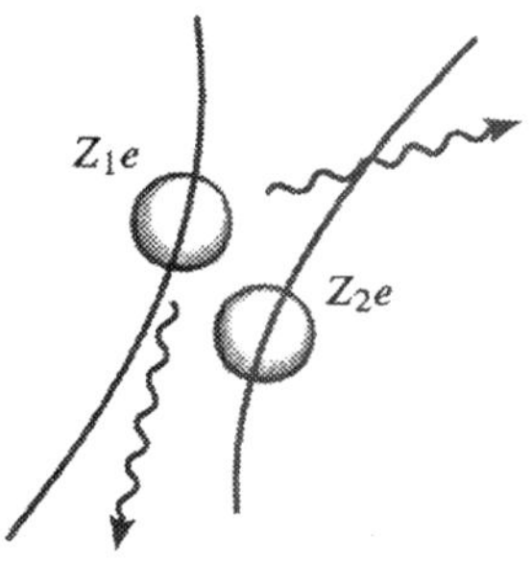

Fig. 4.6. Bremsstrahlung of atomic nuclei in heavy-ion collisions. Bremsstrahlung represents an important background for quasimolecular radiation from molecules temporarily formed during the collision. This process has been calculated by Reinhardt et al.[1] and experimentally detected by J.S. Greenberg et al.[2]

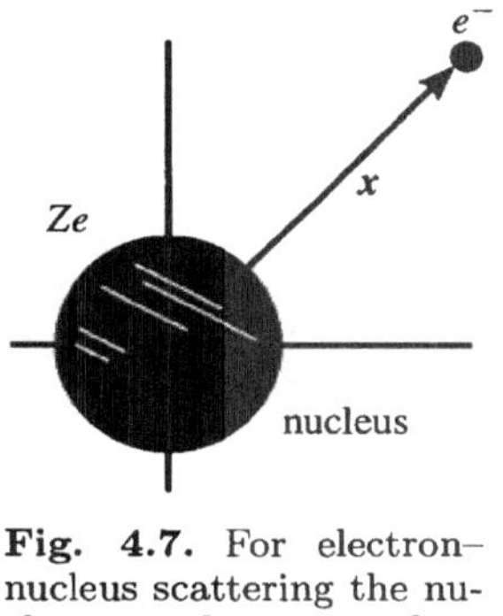

Fig. 4.7. For electron–nucleus scattering the nucleus can be assumed to rest at the origin of the coordinate system

[1] J. Reinhardt, G. Soff, W. Greiner: Z. Phys. A **276** (1976) 285.
[2] H.P. Trautvetter, J.S. Greenberg, P. Vincent: Phys. Rev. Lett. **37** (1967) 202.

Example 4.1.

$$\hat{H}_V = \sum_{q,q'} \hat{b}^\dagger_q \hat{b}_{q'} \int dV \frac{e^{-iq\cdot x}}{\sqrt{L^3}} V(x) \frac{e^{iq'\cdot x}}{\sqrt{L^3}}$$
$$= \sum_{q,q'} \hat{b}^\dagger_q \hat{b}_{q'} \overline{V}(q'-q) , \tag{6}$$

where

$$\overline{V}(q'-q) = \int dV \frac{1}{L^3} e^{i(q'-q)\cdot x} V(x) \tag{7}$$

represents the *Fourier transform of the interaction potential* $V(x)$. For the case of Coulomb interaction this expression can be computed in closed form:

$$\overline{V}(q) = \int dV \frac{1}{L^3} e^{iq\cdot x}\left(-\frac{Ze^2}{|x|}\right) = -\frac{Ze^2}{L^3}\int dV \frac{e^{iq\cdot x}}{|x|}$$
$$= -\frac{Ze^2}{L^3}\left(-\frac{1}{q^2}\right)\int \frac{dV}{|x|}\triangle e^{iq\cdot x} = \frac{Ze^2}{L^3q^2}\int\left(\triangle\frac{1}{|x|}\right)e^{iq\cdot x}\,dV$$
$$= -\frac{4\pi Ze^2}{L^3q^2}\int \delta(x)\, e^{iq\cdot x}\,dV$$
$$= -\frac{4\pi Ze^2}{L^3q^2} . \tag{8}$$

Use has been made of the relation

$$\triangle\frac{1}{|x|} = -4\pi\delta(x) , \tag{9}$$

which should be familiar from classical electrodynamics. Also, a partial integration has been employed twice. The explicit expressions of the radiation matrix elements (4.8a), and eventually also (4.8b), are needed. They can be easily calculated with the electron wavefunctions (3). We consider only the dominating interaction with the radiation field $\hat{H}'_{\rm int}$ (4.6); the relevant matrix elements (4.8a) are

$$M(k,\sigma,q,q') = \sqrt{\frac{2\pi\hbar c^2}{L^3\omega_k}}\left(-\frac{e\hbar}{mc}\right) q'\cdot\epsilon_{k,\sigma}\int dV \frac{e^{-iq\cdot x}}{\sqrt{L^3}}\, e^{ik\cdot x}\,\frac{e^{iq'\cdot x}}{\sqrt{L^3}}$$
$$= \sqrt{\frac{2\pi\hbar c^2}{L^3\omega_k}}\left(-\frac{e\hbar}{mc}\right) q'\cdot\epsilon_{k,\sigma}\int dV \frac{e^{i(q'+k-q)\cdot x}}{L^3}$$
$$= \sqrt{\frac{2\pi\hbar c^2}{L^3\omega_k}}\left(-\frac{e\hbar}{mc}\right) q'\cdot\epsilon_{k,\sigma}\,\delta_{q,q'+k} . \tag{10}$$

Hence, the interaction operator (4.6) becomes

$$H'_{\rm int} = \left(-\frac{e\hbar}{mc}\right)\sum_{k,\sigma}\sum_q \sqrt{\frac{2\pi\hbar c^2}{L^3\omega_k}}\, q\cdot\epsilon_{k,\sigma}\left(\hat{b}^\dagger_{q+k}\hat{b}_q\hat{a}_{k\sigma} + \hat{b}^\dagger_q\hat{b}_{q+k}\hat{a}^\dagger_{k\sigma}\right) . \tag{11}$$

The sum over q' breaks down because of the Kronecker delta in (10). From our previous investigations into the emission and absorption of photons from free particles (see Chap. 2, (2.48) and following) we already know that these

Example 4.1.

processes do not exist for free particles. They are only possible in a surrounding medium with an index of refraction $n > 1$ (Cherenkov radiation, Chap. 2). Therefore bremsstrahlung must occur first as a scattering of the charged particle at the nucleus from an initial to an intermediate state and a successive photon emission with simultaneous scattering of the particle into the final state. It is the presence of the atomic nucleus as a force center which makes the emission of radiation (bremsstrahlung) feasible. In other words: *bremsstrahlung is a second-order process*, for which $\hat{H}'_{\text{int}}$, (11), as well as $\hat{H}_V$, (6), contribute as a perturbation. We list the transition matrix element up to second order:

$$M_{\text{fi}} = \langle f|\hat{H}_{\text{int}} + \hat{H}_V|i\rangle + \sum_I \frac{\langle f|\hat{H}'_{\text{int}} + \hat{H}_V|I\rangle\langle I|\hat{H}'_{\text{int}} + \hat{H}_V|i\rangle}{E_{\text{i}} - E_{\text{I}} + \text{i}\eta} . \tag{12}$$

The initial and final states have to be as follows:

$$\begin{aligned} |i\rangle &= |\text{one electron with } \boldsymbol{q}_1\rangle_{\text{P}}|\text{no photon}\rangle_{\text{R}} \\ &= |\boldsymbol{q}_1\rangle_{\text{P}}|0\rangle_{\text{R}} = \hat{b}^\dagger_{\boldsymbol{q}_1}|0\rangle_{\text{P}}|0\rangle_{\text{R}} , \end{aligned} \tag{13}$$

$$\begin{aligned} |f\rangle &= |\text{one electron with } \boldsymbol{q}_3\rangle_{\text{P}}|\text{one photon } \boldsymbol{k}\sigma\rangle_{\text{R}} \\ &= |\boldsymbol{q}_3\rangle_{\text{P}}\hat{a}^\dagger_{\boldsymbol{k}\sigma}|0\rangle_{\text{R}} = \hat{b}^\dagger_{\boldsymbol{q}_3}|0\rangle_{\text{P}}\hat{a}^\dagger_{\boldsymbol{k}\sigma}|0\rangle_{\text{R}} . \end{aligned} \tag{14}$$

Indices P and R indicate particle and radiation (photon) space, respectively. It is immediately seen that the first-order term in (12) cannot contribute, i.e.

$$\langle f|\hat{H}'_{\text{int}} + \hat{H}_V|i\rangle = \langle f|\hat{H}'_{\text{int}}|i\rangle = 0 ,$$

because it leads to the problem of photon emission from free particles, as mentioned earlier, which does not exist because of the impossibility of energy and momentum conservation. $\langle f|\hat{H}_V|i\rangle$ does not contribute because with H_V alone no photon is created. Thus, only the transition matrix elements of second order,

$$M_{\text{fi}} = \sum_I \frac{\langle f|\hat{H}'_{\text{int}} + H_V|I\rangle\langle I|\hat{H}'_{\text{int}} + H_V|i\rangle}{E_{\text{i}} - E_{\text{I}} + \text{i}\eta} , \tag{15}$$

remain for the bremsstrahlung process. There are only two possibilities for the intermediate states $|I\rangle$:

$$\begin{aligned} |I_1\rangle &= |\text{one electron with } \boldsymbol{q}_2\rangle_{\text{P}}|\text{no photon}\rangle_{\text{R}} \\ &= |\boldsymbol{q}_2\rangle_{\text{P}}|0\rangle_{\text{R}} = \hat{b}_{\boldsymbol{q}_2}|0\rangle_{\text{P}}|0\rangle_{\text{R}} \end{aligned} \tag{16}$$

and

$$\begin{aligned} |I_2\rangle &= |\text{one electron with } \boldsymbol{q}'_2\rangle_{\text{P}}|\text{one photon } \boldsymbol{k}\sigma\rangle_{\text{R}} \\ &= |\boldsymbol{q}'_2\rangle_{\text{P}}\hat{a}^\dagger_{\boldsymbol{k}\sigma}|0\rangle_{\text{R}} = \hat{b}^\dagger_{\boldsymbol{q}_2'}|0\rangle_{\text{P}}\hat{a}^\dagger_{\boldsymbol{k}\sigma}|0\rangle_{\text{R}} . \end{aligned} \tag{17}$$

We illustrate these two processes in Fig. 4.8.

From our previous investigations we know that momentum conservation holds at a vertex (from the Latin expression "convertere", which means "to turn around") of the form shown in Fig. 4.9 [compare, for example, with (2.51)]. Hence we can immediately write down momentum conservation for the vertices illustrated in Fig. 4.8 for which a photon is emitted:

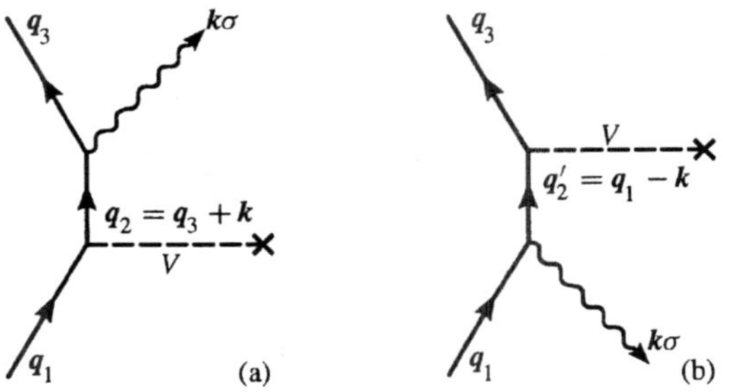

Fig. 4.8a,b. Diagrams illustrating bremsstrahlung of an electron scattered at a nucleus. For case (a), first scattering at the nucleus into the intermediate state $q_2 = q_3 + k$ takes place because of the Coulomb potential, which is indicated with $-\overset{V}{-}-$x; then the emission process of photon k, σ follows with simultaneous scattering of the particle into the final state q_3. For case (b), first the photon $k\sigma$ is emitted with simultaneous scattering of the particle into the intermediate state $q_2' = q_1 - k$; then the particle is scattered again, this time into the final state q_3 because of the Coulomb potential $V(x)$

$$
\begin{aligned}
\hbar q_2 &= \hbar q_3 + \hbar k \,, \\
\hbar q_2' &= \hbar q_1 - \hbar k \,.
\end{aligned}
\tag{18}
$$

Since only the two intermediate states $|I_\nu\rangle$ of (16) and (17) are possible, the matrix element (15) simplifies to

$$
M_{\mathrm{fi}} = \frac{\langle f|\hat{H}'_{\mathrm{int}}|I_1\rangle\langle I_1|H_V|i\rangle}{E_{\mathrm{i}} - E_{I_1}} + \frac{\langle f|\hat{H}_V|I_2\rangle\langle I_2|H'_{\mathrm{int}}|i\rangle}{E_{\mathrm{i}} - E_{I_2}} \,. \tag{19}
$$

The sum runs over only two terms, $|I_1\rangle$ and $|I_2\rangle$; as a consequence, the infinitesimal quantity η is not needed in the denominator, which usually manipulates poles occuring eventually in the infinite sum. Expression (19) exactly reflects the contributions from the two diagrams in Fig. 4.8, which have to be *added coherently.*

With (6) and (7) the relevant matrix elements are calculated straightforwardly:

$$
\begin{aligned}
\langle I_1|\hat{H}_V|i\rangle &= \left\langle \hat{b}^\dagger_{q_2}|0\rangle_{\mathrm{P}}|0\rangle_{\mathrm{R}}\right| \sum_{q,q'} \hat{b}^\dagger_q \hat{b}_{q'} \overline{V}(q'-q) \left|\hat{b}^\dagger_{q_1}|0\rangle_{\mathrm{P}}|0\rangle_{\mathrm{R}}\right\rangle \\
&= \sum_{q,q'} \overline{V}(q'-q) \left\langle \hat{b}^\dagger_{q_2}|\hat{b}^\dagger_q \hat{b}_{q'}|\hat{b}^\dagger_{q_1}\right\rangle \\
&= \sum_{q,q'} \overline{V}(q'-q) \langle 0|\, \hat{b}_{q_2}\hat{b}^\dagger_q \left(\delta_{q'q_1} - \hat{b}^\dagger_{q_1}\hat{b}_{q'}\right)|0\rangle \\
&= \sum_{q,q'} \overline{V}(q'-q)\delta_{q'q_1}\langle 0|\hat{b}_{q_2}\hat{b}^\dagger_q|0\rangle \\
&= \sum_{q,q'} \overline{V}(q'-q)\delta_{q',q_1}\langle 0|\delta_{q_2,q} - \hat{b}_{q_2}\hat{b}^\dagger_q|0\rangle \\
&= \sum_{q,q'} \overline{V}(q'-q)\delta_{q'q_1}\delta_{q_2,q} \\
&= \overline{V}(q_2 - q_1) = -\frac{4\pi Z e^2}{L^3(q_2-q_1)^2} \\
&= -\frac{4\pi Z e^2}{L^3(q_3 + k - q_1)^2} \,.
\end{aligned}
\tag{20}
$$

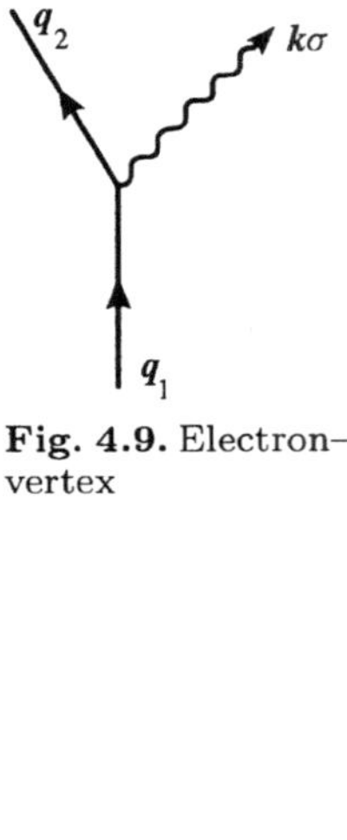

Fig. 4.9. Electron–photon vertex

Similarly follows

Example 4.1.

$$
\begin{aligned}
\langle f|\hat{H}_V|I_2\rangle &= \Big\langle \hat{b}^\dagger_{q_3}|0\rangle_{\mathrm{P}}\hat{a}^\dagger_{k\sigma}|0\rangle_{\mathrm{R}}\Big| \\
&\quad \times \sum_{q,q'} \overline{V}(q-q')\hat{b}^\dagger_q\hat{b}_{q'}\Big|\hat{b}^\dagger_{q_2}|0\rangle_{\mathrm{P}}\hat{a}^\dagger_{k\sigma}|0\rangle_{\mathrm{R}}\Big\rangle \\
&\vdots \\
&= \overline{V}(q_3-q_2') = -\frac{4\pi Ze^2}{L^3(q_3-q_2')^2} \\
&= -\frac{4\pi Ze^2}{L^3(q_3+k-q_1)^2}\,,
\end{aligned}
\tag{21}
$$

where we have used (18). With (11) the radiation matrix elements are calculated as follows:

$$
\begin{aligned}
\langle f|\hat{H}'_{\mathrm{int}}|I_1\rangle &= \Big\langle \hat{b}^\dagger_{q_3}|0\rangle_{\mathrm{P}}\hat{a}^\dagger_{k\sigma}|0\rangle_{\mathrm{R}}\Big|\left(-\frac{e\hbar}{mc}\right)\sum_{k',\sigma'}\sum_{q}\sqrt{\frac{2\pi\hbar c^2}{L^3\omega_{k'}}}\,q\cdot\epsilon_{k'\sigma'} \\
&\quad \times\left\{\hat{b}^\dagger_{q+k'}\hat{b}_q\,\hat{a}_{k'\sigma'}+\hat{b}^\dagger_q\hat{b}_{q+k'}\hat{a}^\dagger_{k'\sigma'}\right\}\Big|\hat{b}^\dagger_{q_2}|0\rangle_{\mathrm{P}}|0\rangle_{\mathrm{R}}\Big\rangle \\
&= -\frac{e\hbar}{mc}\sum_{k',\sigma'}\sum_{q}\sqrt{\frac{2\pi\hbar c^2}{L^3\omega_{k'}}}\,q\cdot\epsilon_{k'\sigma'}\Big\langle \hat{b}^\dagger_{q_3}|0\rangle_{\mathrm{P}}\hat{a}^\dagger_{k\sigma}|0\rangle_{\mathrm{R}}\Big| \\
&\quad \times\left\{\hat{b}^\dagger_{q+k'}\hat{b}_q\,\hat{a}_{k',\sigma'}+\hat{b}^\dagger_q\hat{b}_{q+k'}\hat{a}^\dagger_{k'\sigma'}\right\}\Big|\hat{b}^\dagger_{q_2}|0\rangle_{\mathrm{P}}|0\rangle_{\mathrm{R}}\Big\rangle \\
&= -\frac{e\hbar}{mc}\sum_{k',\sigma'}\sum_{q}\sqrt{\frac{2\pi\hbar c^2}{L^3\omega_{k'}}}\,q\cdot\epsilon_{k'\sigma'} \\
&\quad \times\Bigg\{\Big\langle \hat{b}^\dagger_{q_3}|0\rangle_{\mathrm{P}}\Big|\hat{b}^\dagger_{q+k'}\hat{b}_q\Big|\hat{b}^\dagger_{q_2}|0\rangle_{\mathrm{P}}\Big\rangle\overbrace{\Big\langle\hat{a}^\dagger_{k\sigma}|0\rangle_{\mathrm{R}}\Big|\hat{a}_{k'\sigma'}|0\rangle_{\mathrm{R}}\Big\rangle}^{=0} \\
&\quad +\Big\langle \hat{b}^\dagger_{q_3}|0\rangle_{\mathrm{P}}\Big|\hat{b}_q\hat{b}_{q+k'}\Big|\hat{b}^\dagger_{q_2}|0\rangle_{\mathrm{P}}\Big\rangle\overbrace{\Big\langle\hat{a}^\dagger_{k\sigma}|0\rangle_{\mathrm{R}}\Big|\hat{a}^\dagger_{k',\sigma'}|0\rangle_{\mathrm{R}}\Big\rangle}^{=\delta_{k,k'}\delta_{\sigma\sigma'}}\Bigg\} \\
&= -\frac{e\hbar}{mc}\sqrt{\frac{2\pi\hbar c^2}{L^3\omega_k}}\sum_q q\cdot\epsilon_{k\sigma}\;{}_{\mathrm{P}}\langle 0|\hat{b}_{q_3}\hat{b}^\dagger_q\hat{b}_{q+k}\hat{b}^\dagger_{q_2}|0\rangle_{\mathrm{P}} \\
&= -\frac{e\hbar}{mc}\sqrt{\frac{2\pi\hbar c^2}{L^3\omega_k}}\,q_3\cdot\epsilon_{k\sigma}\delta_{q_2,q_3+k}\,.
\end{aligned}
\tag{22}
$$

Similarly it follows that

$$
\begin{aligned}
\langle I_2|\hat{H}'_{\mathrm{int}}|i\rangle &= \Big\langle \hat{b}^\dagger_{q_2'}|0\rangle_{\mathrm{P}}\hat{a}^\dagger_{k\sigma}|0\rangle_{\mathrm{R}}\Big|\left(-\frac{e\hbar}{mc}\right)\sum_{k',\sigma'}\sum_{q}\sqrt{\frac{2\pi\hbar c^2}{L^3\omega_{k'}}}\,q\cdot\epsilon_{k'\sigma'} \\
&\quad \times\left\{\hat{b}^\dagger_{q+k'}\hat{b}_q\,\hat{a}_{k'\sigma'}+\hat{b}^\dagger_q\hat{b}_{q+k'}\hat{a}^\dagger_{k'\sigma'}\right\}\Big|\hat{b}^\dagger_{q_1}|0\rangle_{\mathrm{P}}|0\rangle_{\mathrm{R}}\Big\rangle
\end{aligned}
$$

Example 4.1.

$$\vdots$$

$$\begin{aligned}
&= -\frac{e\hbar}{mc}\sqrt{\frac{2\pi\hbar c^2}{L^3\omega_k}}(\boldsymbol{q}_1-\boldsymbol{k})\cdot\boldsymbol{\epsilon}_{k\sigma}\delta_{\boldsymbol{q}_2',\boldsymbol{q}_1-\boldsymbol{k}} \\
&= -\frac{e\hbar}{mc}\sqrt{\frac{2\pi\hbar c^2}{L^3\omega_k}}\boldsymbol{q}_1\cdot\boldsymbol{\epsilon}_{k\sigma}\delta_{\boldsymbol{q}_2',\boldsymbol{q}_1-\boldsymbol{k}}\,. \qquad (23)
\end{aligned}$$

As we have already realized earlier in the context of (18) the two δ functions appearing in (22) and (23) express momentum conservation at the photon vertices. Obviously, it also follows naturally from the formalism of second quantization. We still have to calculate the energy denominators of (19). Since we deal with real particles, this is an easy task:

$$\begin{aligned}
E_{\mathrm{i}} &= \frac{\hbar^2 q_1^2}{2m}\,, \\
E_{\mathrm{I}_1} &= \frac{\hbar^2}{2m}(\boldsymbol{q}_3+\boldsymbol{k})^2\,, \qquad E_{\mathrm{I}_2} = \frac{\hbar^2}{2m}(\boldsymbol{q}_1-\boldsymbol{k})^2+\hbar\omega_k\,, \qquad (24)\\
E_{\mathrm{f}} &= \frac{\hbar^2 q_3^2}{2m}+\hbar\omega_k\,.
\end{aligned}$$

Energy conservation for the total process is deduced from the δ function appearing in the expression for the transition probability per unit time,

$$\left(\frac{\text{trans. prob.}}{\text{time}}\right) = \frac{2\pi}{\hbar}|M_{\mathrm{fi}}|^2\delta(E_{\mathrm{f}}-E_{\mathrm{i}})\,, \qquad (25)$$

which is a result of the general formalism of time-dependent perturbation theory. For our case, energy conservation reads

$$E_{\mathrm{f}} = E_{\mathrm{i}} \quad \text{or} \quad \frac{\hbar^2 q_1^2}{2m} = \frac{\hbar^2 q_3^2}{2m}+\hbar\omega_k\,. \qquad (26)$$

With this result the energy denominators of (19) can be cast into the following form:

$$\begin{aligned}
E_{\mathrm{i}}-E_{\mathrm{I}_1} &= \frac{\hbar^2 q_1^2}{2m}-\frac{\hbar^2}{2m}(\boldsymbol{q}_3+\boldsymbol{k})^2 = \frac{\hbar^2}{2m}(q_1^2-q_3^2)-\frac{\hbar^2}{m}\boldsymbol{q}_3\cdot\boldsymbol{k}-\frac{\hbar^2}{2m}\boldsymbol{k}^2 \\
&= \hbar\omega_k-\hbar\omega_k\left(\frac{\boldsymbol{k}}{k}\cdot\frac{\boldsymbol{v}_3}{c}\right)-\frac{(\hbar\omega_k)^2}{2mc^2} \\
&= \hbar\omega_k\left\{1-\frac{\boldsymbol{k}\cdot\boldsymbol{v}_3}{kc}-\frac{\hbar k}{2mc}\right\}\,, \qquad (27)\\
E_{\mathrm{i}}-E_{\mathrm{I}_2} &= \frac{\hbar^2 q_1^2}{2m}-\frac{\hbar^2}{2m}(\boldsymbol{q}_1-\boldsymbol{k})^2-\hbar\omega_k \\
&= \frac{\hbar^2}{2m}(q_1^2-q_1^2)+\frac{\hbar^2}{m}\boldsymbol{q}_1\cdot\boldsymbol{k}-\frac{\hbar^2}{2m}\boldsymbol{k}^2-\hbar\omega_k \\
&= -\hbar\omega_k\left(1-\frac{\boldsymbol{k}\cdot\boldsymbol{v}_1}{kc}+\frac{\hbar k}{2mc}\right)\,. \qquad (28)
\end{aligned}$$

Here the particle velocities

Example 4.1.

$$v_1 = \frac{\hbar q_1}{m} \quad \text{and} \quad v_3 = \frac{\hbar q_3}{m} \tag{29}$$

have been introduced. With (20), (23), and (28) we can now write down the bremsstrahlung matrix element, (19) explicitly; we get

$$\begin{aligned} M_{\mathrm{fi}} &= \frac{-4\pi Z e^2}{L^3 (q_3 + k - q_1)^2} \left(\frac{-e\hbar}{mc}\right) \sqrt{\frac{2\pi\hbar c^2}{L^3 \omega_k}} \\ &\quad \times \epsilon_{k\sigma} \cdot \left(\frac{q_3}{\hbar\omega_k \left(1 - \frac{k\cdot v_3}{kc} - \frac{\hbar k}{2mc}\right)} - \frac{q_1}{\hbar\omega_k \left(1 - \frac{k\cdot v_1}{kc} + \frac{\hbar k}{2mc}\right)} \right) \\ &= \frac{4\pi Z e^2}{L^3 (q_3 + k - q_1)^2} \sqrt{\frac{2\pi\hbar e^2}{L^3 \omega_k^3 \hbar}} \\ &\quad \times \epsilon_{k\sigma} \cdot \left(\frac{v_3}{1 - \frac{k\cdot v_3}{kc} - \frac{\hbar k}{2mc}} - \frac{v_1}{1 - \frac{k\cdot v_1}{kc} + \frac{\hbar k}{2mc}} \right) . \end{aligned} \tag{30}$$

All the time [see (24)] we have assumed nonrelativistic electrons, i.e. $v/c \ll 1$. Moreover, the photon momentum $\hbar k$ will remain small with respect to the particle momentum in general, so that

$$\hbar k \ll \hbar q = mv .$$

Consequently, the denominator of (30) can be replaced by 1, and the denominator of the first factor of (30) can be approximated by

$$(q_3 + k - q_1)^2 \approx (q_3 - q_1)^2 .$$

Then, (30) simplifies to

$$M_{\mathrm{fi}} = \frac{4\pi Z e^2 \hbar^2}{L^3 m^2 |\Delta v|^2} \sqrt{\frac{2\pi e^2}{\hbar L^3 \omega_k^3}} \epsilon_{k\sigma} \cdot \Delta v , \tag{31}$$

where

$$\Delta v = v_3 - v_1 \tag{32}$$

represents the difference between the electron velocities in the initial and final state.

The *total cross section σ for bremsstrahlung* is defined as the sum over all transition probabilities per unit time for transition into all possible final states and divided by the incoming *electron current*

$$\frac{v_1}{L^3} . \tag{33}$$

This yields

$$\begin{aligned} \sigma &= \frac{1}{v_1/L^3} \frac{L^3}{(2\pi)^3} \int q_3^2 \, \mathrm{d}q_3 \, \mathrm{d}\Omega_{\mathrm{e}} \frac{L^3}{(2\pi)^3} \\ &\quad \times \int k^2 \, \mathrm{d}k \, \mathrm{d}\Omega_{\mathrm{k}} \frac{2\pi}{\hbar} |M_{\mathrm{fi}}|^2 \delta\left(\frac{\hbar^2 q_3^2}{2m} - \frac{\hbar^2 q_1^2}{2m}\right) . \end{aligned} \tag{34}$$

A summation over electron states on the one hand and over photon states on the other hand has been performed already; also replacement (2.18) has been applied:

Example 4.1.

$$\sum_{\boldsymbol{k}} \xrightarrow{L^3\to\infty} \frac{L^3}{(2\pi)^3}\int d^3k\,,$$
$$\sum_{\boldsymbol{q}} \xrightarrow{L^3\to\infty} \frac{L^3}{(2\pi)^3}\int d^3q_3\,. \tag{35}$$

$d\Omega_e$ is the solid angle element in which the electrons scatter and $d\Omega_k$ the one for photons (see Fig. 4.10). Inserting (31) into (34) yields

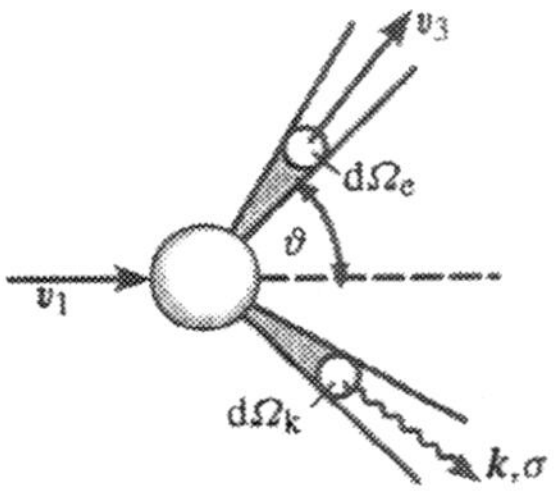

Fig. 4.10. Solid angle elements into which the electron and photon are scattered

$$\sigma = \frac{1}{v_1}\left(\frac{4\pi Ze^2\hbar^2}{m^2}\right)^2 \frac{2\pi e^2}{\hbar}\frac{2\pi}{\hbar}\frac{1}{(2\pi)^6}\int q_3^2\,dq_3\,d\Omega_e\int k^2\,dk\,d\Omega_k$$
$$\times\frac{|\boldsymbol{\epsilon}_{\boldsymbol{k}\sigma}\cdot\Delta\boldsymbol{v}\,|}{|\Delta\boldsymbol{v}\,|^4\omega^3}\frac{2m}{\hbar^2}\delta(q_3^2-q_1^2)\,. \tag{36}$$

Since

$$\delta(q_3^2-q_1^2) = \delta\,(q_3+q_1)(q_3-q_1) = \frac{1}{2q_1}\delta(q_3-q_1)$$

(only positive q_3 have to be considered in the integral $\int\ldots dq_3$) and

$$|\Delta v| = 2v_1\sin\frac{\vartheta}{2} \tag{37}$$

(see Fig. 4.11) we arrive at

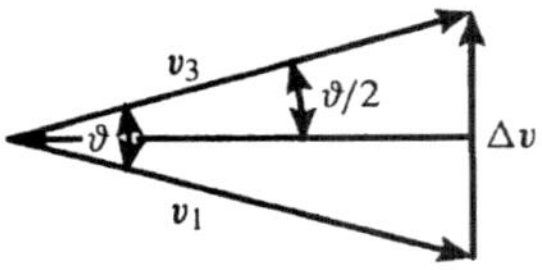

Fig. 4.11. The $\delta(q_3-q_1)$ function guarantees $q_3 = q_1$ and thus also $|\boldsymbol{v}_3| = |\boldsymbol{v}_1|$. Since we have neglected the photon energy of the final state this result describes elastic scattering of the electron. Directly from the figure we deduce $|\Delta v| = 2v_1\sin\vartheta/2$

$$\sigma = \frac{1}{v_1}\left(\frac{4\pi Ze^2\hbar^2}{m^2}\right)^2 \frac{2\pi e^2}{\hbar}\frac{2\pi}{\hbar}\frac{1}{(2\pi)^6}\frac{1}{2q_1}q_1^2\frac{2m}{\hbar^2}\frac{1}{(2v_1)^4}$$
$$\times\int d\Omega_e\int d\Omega_k\frac{k^2\,dk}{\omega^3}\frac{|\boldsymbol{\epsilon}_{\boldsymbol{k}\sigma}\cdot\Delta\boldsymbol{v}\,|^2}{\sin^4\vartheta/2}$$
$$= \frac{Z^2e^4}{m^2v_1^4}\frac{e^2}{16\pi^2\hbar}\int d\Omega_e\int d\Omega_k\frac{k^2\,dk}{\omega^3}\frac{|\boldsymbol{\epsilon}_{\boldsymbol{k}\sigma}\cdot\Delta\boldsymbol{v}\,|}{\sin^4\vartheta/2}$$
$$= \frac{Z^2e^4}{m^2v_1^4}\frac{e^2}{16\pi^2c^3\hbar}\int d\Omega_e\int d\Omega_k\frac{d\omega}{\omega}\frac{|\boldsymbol{\epsilon}_{\boldsymbol{k},\sigma}\cdot\Delta\boldsymbol{v}|}{\sin^4\vartheta/2}\,. \tag{38}$$

The differential cross section for scattering of an electron into the solid angle element $d\Omega_e$ with angle ϑ and simultaneous emission of a photon with polarization $\boldsymbol{\epsilon}_{\boldsymbol{k}\sigma}$ in the frequency domain between ω and $\omega+d\omega$ into the solid angle element $d\Omega_k$ can be read off conveniently:

$$\frac{d^3\sigma}{d\Omega_e\,d\Omega_k\,d\omega} = \frac{Z^2e^4}{m^2v_1^4\sin^4\vartheta/2}\frac{e^2|\boldsymbol{\epsilon}_{\boldsymbol{k}\sigma}\cdot\Delta\boldsymbol{v}\,|^2}{16\pi^2c^3\hbar\omega}\,. \tag{39}$$

The first factor represents *Rutherford's scattering cross section* for elastic electron scattering at the nucleus (consult Exercise 4.2). Apparently the second factor in (39) has to be the probability $dW/d\Omega_k\,d\omega$ that a photon with frequency ω will be emitted into $d\Omega_k$. The probability for two events to occur at the same time is equal to the product of the single probabilities. Hence, the differential cross section (39) takes on a very plausible form:

$$\frac{d^3\sigma}{d\Omega_e\,d\Omega_k\,d\omega} = \left(\frac{d\sigma}{d\Omega_e}\right)_{\text{Rutherford}}\frac{dW}{d\Omega_k\,d\omega}\,. \tag{40}$$

Example 4.1.

From (39) we unveil that the energy spectrum of the photon behaves like $\mathrm{d}\omega/\omega$; as a consequence the probability of a photon being emitted with energy zero appears to increase to infinity. This behavior also shows up in the relativistic theory of bremsstrahlung; it is known as the *infrared catastrophe*. In order to better understand and remove this unreasonable result a careful analysis of the actual experimental conditions is required for the observation of bremsstrahlung. The crucial point is that every counting device possesses only a finite energy resolution. *If it detects an electron scattered inelastically into a finite energy interval around $\omega = 0$*, then it will also detect an electron scattered elastically. Therefore we have to consider the elastic as well as the inelastic scattering cross section for a comparison with experiment; both processes, of course, up to the same order in e^2. *Since the bremsstrahlung part of (39) is of second order in e^2 relative to elastic scattering* the so-called radiation corrections up to order e^2 have to be taken into account for the electron scattering cross section

$$\left(\frac{\mathrm{d}\sigma}{\mathrm{d}\Omega_\mathrm{e}}\right)_{\text{elastic}} .$$

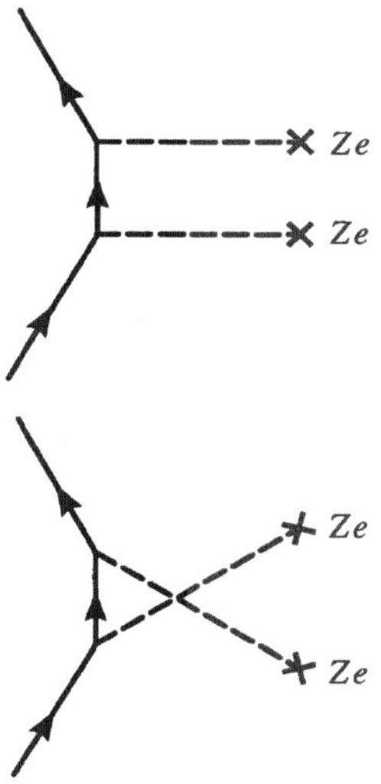

Fig. 4.12. Second-order Coulomb scattering

Two different contributions exist. On the one hand, graphs of the form given in Fig. 4.12 describe the second-order scattering of electrons in a Coulomb potential (confer Exercises 4.2 and 4.4).

On the other hand, a second-order interaction of the electron with itself exists via the radiation field; the graphs depicted in Fig. 4.13 give an illustration. They depict a virtual photon, which is first created at the electron site and then falls back again to the electron like a boomerang. This is in contrast to the graphs depicted earlier for which the second photon also runs to the source (Ze); consult again the discussion about self-energy at the end of Chap. 1. A systematic calculation of these graphs within relativistic theory shows that the contributions from radiation corrections contain a divergent term, too, which exactly cancels the divergence (39) of bremsstrahlung at $\hbar\omega = 0$. Hence, the infrared catastrophe does not exist in reality!

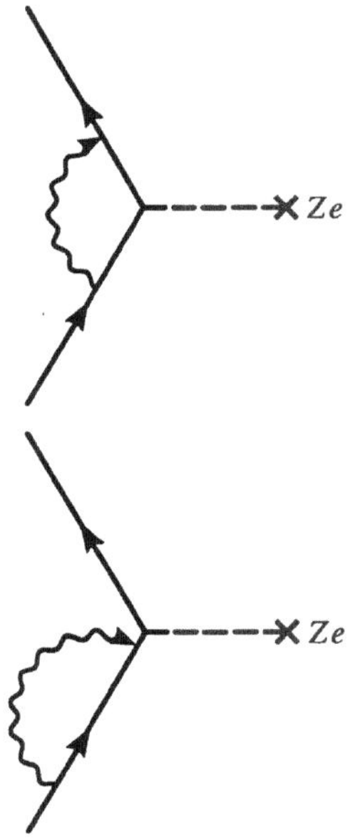

Fig. 4.13. Radiation corrections to Coulomb scattering

EXERCISE

4.2 Rutherford Scattering Cross Section

Problem. Derive Rutherford's cross section for the scattering of an electron at a fixed point nucleus with charge Ze.

Solution. In first quantization the Hamiltonian reads

$$\hat{H} = \frac{\hat{\boldsymbol{p}}^2}{2m} - \frac{Ze^2}{r} .$$

As discussed earlier, in second quantization the interaction takes on the form

$$\hat{H}_V = -\sum_{\boldsymbol{q}}\sum_{\boldsymbol{q}'} \hat{b}^\dagger_{\boldsymbol{q}} \hat{b}_{\boldsymbol{q}'} \frac{4\pi Z e^2}{L^3|\boldsymbol{q}-\boldsymbol{q}'|^2} = -\frac{4\pi Z e^2}{L^3}\sum_{\boldsymbol{q}}\sum_{\boldsymbol{k}} \hat{b}^\dagger_{\boldsymbol{q}-\boldsymbol{k}} \hat{b}_{\boldsymbol{q}} \frac{1}{\boldsymbol{k}^2} .$$

Exercise 4.2.

We consider a transition from a particle state $|i\rangle = \hat{b}_{\boldsymbol{p}}^{\dagger}|0\rangle$ to the final state $|f\rangle = \hat{b}_{\boldsymbol{p}'}^{\dagger}|0\rangle$. $\hbar\boldsymbol{p}$ and $\hbar\boldsymbol{p}'$ denote the particle momenta before and after the collision, respectively. The transition matrix element becomes

$$\begin{aligned} M_{\mathrm{fi}} &= \langle f|\hat{H}_V|i\rangle = -\frac{4\pi Ze^2}{L^3}\sum_{\boldsymbol{q},\boldsymbol{k}}\frac{1}{\boldsymbol{k}^2}\langle 0|\hat{b}_{\boldsymbol{p}'}\hat{b}_{\boldsymbol{q}-\boldsymbol{k}}^{\dagger}\hat{b}_{\boldsymbol{q}}\hat{b}_{\boldsymbol{p}}^{\dagger}|0\rangle \\ &= -\frac{4\pi Ze^2}{L^3}\sum_{\boldsymbol{q},\boldsymbol{k}}\frac{1}{\boldsymbol{k}^2}\langle 0|(\delta_{\boldsymbol{p}'\boldsymbol{q}-\boldsymbol{k}} - \hat{b}_{\boldsymbol{q}-\boldsymbol{k}}^{\dagger}\hat{b}_{\boldsymbol{p}'})\hat{b}_{\boldsymbol{q}}\hat{b}_{\boldsymbol{p}}^{\dagger}|0\rangle \\ &= -\frac{4\pi Ze^2}{L^3}\sum_{\boldsymbol{q},\boldsymbol{k}}\delta_{\boldsymbol{p}'\boldsymbol{q}-\boldsymbol{k}}\delta_{\boldsymbol{p}\boldsymbol{q}}\frac{1}{\boldsymbol{k}^2} \\ &= -\frac{4\pi Ze^2}{L^3}\frac{1}{(\boldsymbol{p}-\boldsymbol{p}')^2}\,. \end{aligned}$$

To obtain the cross section one has to sum over all final momenta $\boldsymbol{p}'$ and divide by the electron current $\hbar p/mL^3$:

$$\begin{aligned} \sigma &= \frac{L^3 m}{\hbar p}\frac{L^3}{(2\pi)^3}\int p'^2\,\mathrm{d}p'\,\mathrm{d}\Omega_{\mathrm{p}'}\frac{2\pi}{\hbar}\frac{16\pi^2(Ze^2)^2}{L^6}\frac{1}{(\boldsymbol{p}'-\boldsymbol{p})^4} \\ &\quad\times\delta\left(\frac{\hbar^2\boldsymbol{p}^2}{2m}-\frac{\hbar^2\boldsymbol{p}'^2}{2m}\right) \\ &= \frac{4mZ^2e^4}{\hbar^2 p}\frac{2m}{\hbar^2}\int p'^2\,\mathrm{d}p'\,\mathrm{d}\Omega_{\mathrm{p}'}\frac{\delta(p^2-p'^2)}{(\boldsymbol{p}'-\boldsymbol{p})^4} \\ &= \frac{4m^2Z^2e^4}{\hbar^4}\int\mathrm{d}\Omega_{\mathrm{p}'}\int\frac{p'^2}{p^2}\frac{\delta(p-p')}{(p^2+p'^2-2pp'\cos\vartheta)^2}\,\mathrm{d}p' \\ &= \frac{4m^2Z^2e^4}{\hbar^4}\int\mathrm{d}\Omega_{\mathrm{p}'}\frac{1}{(1-\cos\vartheta)^2} \\ &= \frac{Z^2e^2}{4E^2}\int\frac{\mathrm{d}\Omega}{4\sin^4\vartheta/2}\,. \end{aligned}$$

Thus, we arrive at the well-known result

$$\left(\frac{\mathrm{d}\sigma}{\mathrm{d}\Omega}\right)_{\mathrm{Rutherford}} = \frac{Z^2e^4}{16E^2}\frac{1}{\sin^4\vartheta/2}$$

for the differential cross section. Rutherford's scattering cross section diverges for small scattering angles. Performing an integration over the solid angle we realize that the total scattering cross section $\sigma = \int(\mathrm{d}\sigma/\mathrm{d}\Omega)\mathrm{d}\Omega$ diverges, too. This stems from the strong singularity $\sim 1/\vartheta^4$ at $\vartheta = 0$; this singularity has the same origin as the infrared catastrophe we have already experienced in a calculation of bremsstrahlung.

EXERCISE

4.3 Lifetime of the Hydrogen 2s State with Respect to Two-Photon Decay (in Second Quantization)

Problem. Calculate the lifetime of the 2s state of the hydrogen atom with respect to two-photon decay (compare with Exercise 2.10) with the methods of second quantization.

Solution. The relevant Hamiltonians already derived are given by

$$\hat{H}' = \sum_{k,\sigma}\sum_{n,n'} M(-k,\sigma,n,n')\hat{b}_n^\dagger \hat{b}_{n'}\hat{a}_{k\sigma}^\dagger$$

and (1)

$$\hat{H}'' = \sum_{k_1,\sigma_1}\sum_{k_2,\sigma_2}\sum_{n,n'} M(-k_1,\sigma_1,-k_2,\sigma_2,n,n')\hat{b}_n^\dagger \hat{b}_{n'}\hat{a}_{k_1,\sigma_1}^\dagger \hat{a}_{k_2,\sigma_2}^\dagger ,$$

with matrix elements

$$\begin{aligned} M(k,\sigma,n,n') &= \left(\frac{2\pi\hbar c}{L^3\omega_k}\right)^{1/2}\int \mathrm{d}^3x\,\psi_n^*\left(-\frac{e\hbar}{\mathrm{i}mc}\,\mathrm{e}^{\mathrm{i}\boldsymbol{k}\cdot\boldsymbol{x}}\boldsymbol{\epsilon}_{\boldsymbol{k}\sigma}\cdot\boldsymbol{\nabla}\right)\psi_{n'}\,, \\ M(k_1,\sigma_1,k_2,\sigma_2,n,n') &= \left(\frac{2\pi\hbar c}{L^3}\right)\frac{1}{(\omega_{k_1}\omega_{k_2})^{1/2}} \\ &\quad\times\int \mathrm{d}^3x\,\psi_n^*\left(\frac{e^2}{2mc^2}\boldsymbol{\epsilon}_{\boldsymbol{k}_1\sigma_1}\cdot\boldsymbol{\epsilon}_{\boldsymbol{k}_2\sigma_2}\,\mathrm{e}^{\mathrm{i}(\boldsymbol{k}_1+\boldsymbol{k}_2)\cdot\boldsymbol{x}}\right)\psi_{n'}\,. \end{aligned} \quad (2)$$

$\hat{H}'$ produces two photons in second order, whereas $\hat{H}''$ directly produces two photons. No other term in (4.7) contributes to the two-photon process. In the initial state the electron occupies the 2s state and no photon (radiation) is present:

$$|i\rangle = \hat{b}_{2\mathrm{s}}^\dagger|0\rangle_\mathrm{e}|0\rangle_\mathrm{rad}\,. \quad (3)$$

In the final state the electron occupies the 1s state and two photons exist with quantum numbers $(k_\mathrm{f},\sigma_\mathrm{f})$ and $(k_\mathrm{f}',\sigma_\mathrm{f}')$:

$$|f\rangle = \hat{b}_{1\mathrm{s}}^\dagger\hat{a}_{k_\mathrm{f},\sigma_\mathrm{f}}^\dagger\hat{a}_{k_\mathrm{f}',\sigma_\mathrm{f}'}^\dagger|0\rangle_\mathrm{e}|0\rangle_\mathrm{rad}\,. \quad (4)$$

With the initial and final states given we can calculate the matrix elements (2) of the Hamiltonians (1)

$$\begin{aligned} M_\mathrm{fi}^{(1)} &= \langle f|\hat{H}''|i\rangle \\ &= {}_\mathrm{rad}\langle 0|\,{}_\mathrm{e}\langle 0|\hat{a}_{k_\mathrm{f}',\sigma_\mathrm{f}'}\hat{a}_{k_\mathrm{f},\sigma_\mathrm{f}}\hat{b}_{1\mathrm{s}}\hat{H}''\hat{b}_{2\mathrm{s}}^\dagger|0\rangle_\mathrm{e}|0\rangle_\mathrm{rad} \\ &= \sum_{k_1,\sigma_1}\sum_{k_2,\sigma_2}\sum_{n,n'} M(-k_1,\sigma_1,-k_2,\sigma_2,n,n') \\ &\quad\times {}_\mathrm{rad}\langle 0|\,{}_\mathrm{e}\langle 0|\hat{a}_{k_\mathrm{f}',\sigma_\mathrm{f}'}\hat{a}_{k_\mathrm{f},\sigma_\mathrm{f}}\hat{b}_{1\mathrm{s}}\hat{b}_n^\dagger\hat{b}_{n'}\hat{a}_{k_1,\sigma_1}^\dagger\hat{a}_{k_2,\sigma_2}^\dagger\hat{b}_{2\mathrm{s}}^\dagger|0\rangle_\mathrm{e}|0\rangle_\mathrm{rad}\,. \end{aligned}$$

Exercise 4.3.

Using the commutation relations

$$\begin{aligned}\hat{a}_i\hat{a}_j^\dagger - \hat{a}_j^\dagger\hat{a}_i &= \delta_{ij} \,, \\ \hat{b}_i\hat{b}_j^\dagger + \hat{b}_j^\dagger\hat{b}_i &= \delta_{ij}\end{aligned}$$

we obtain

$$\begin{aligned}&\langle 0|\hat{a}_{k_f',\sigma_f'}\hat{a}_{k_f,\sigma_f}\hat{b}_{1s}\hat{b}_n^\dagger\hat{b}_{n'}\hat{a}^\dagger_{k_1,\sigma_1}\hat{a}^\dagger_{k_2,\sigma_2}\hat{b}^\dagger_{2s}|0\rangle \\ &= \langle 0|\hat{a}_{k_f',\sigma_f'}\hat{a}_{k_f,\sigma_f}\hat{a}^\dagger_{k_1,\sigma_1}\hat{a}^\dagger_{k_2,\sigma_2}|0\rangle\,\langle 0|\hat{b}_{1s}\hat{b}_n^\dagger\hat{b}_{n'}\hat{b}^\dagger_{2s}|0\rangle\end{aligned}$$

for the matrix element of operators. After commutation the photon part becomes:

$$\begin{aligned}&\langle 0|\hat{a}_{k_f',\sigma_f'}\left(\hat{a}^\dagger_{k_1,\sigma_1}\hat{a}_{k_f,\sigma_f} + \delta_{k_1,k_f}\delta_{\sigma_1,\sigma_f}\right)\hat{a}^\dagger_{k_2,\sigma_2}|0\rangle \\ &= \langle 0|\hat{a}_{k_f',\sigma_f'}\hat{a}^\dagger_{k_1,\sigma_1}\hat{a}_{k_f,\sigma_f}\hat{a}^\dagger_{k_2,\sigma_2}|0\rangle \\ &\quad + \delta_{k_1,k_f}\delta_{\sigma_1,\sigma_f}\langle 0|\hat{a}_{k_f',\sigma_f'}\hat{a}^\dagger_{k_2,\sigma_2}|0\rangle \\ &= \langle 0|\hat{a}_{k_f',\sigma_f'}\hat{a}^\dagger_{k_1,\sigma_1}\left(\hat{a}^\dagger_{k_2,\sigma_2}\hat{a}_{k_f,\sigma_f} + \delta_{k_2,k_f}\delta_{\sigma_2,\sigma_f}\right)|0\rangle \\ &\quad + \delta_{k_1,k_f}\delta_{\sigma_1,\sigma_f}\langle 0|\hat{a}^\dagger_{k_2,\sigma_2}\hat{a}_{k_f',\sigma_f'} + \delta_{k_2,k_f'}\delta_{\sigma_2,\sigma_f'}|0\rangle \\ &= \delta_{k_2,k_f}\delta_{\sigma_2,\sigma_f}\langle 0|\hat{a}_{k_f',\sigma_f'}\hat{a}^\dagger_{k_1,\sigma_1}|0\rangle + \delta_{k_1,k_f}\delta_{\sigma_1,\sigma_f}\delta_{k_2,k_f'}\delta_{\sigma_2,\sigma_f'} \\ &= \delta_{k_2,k_f}\delta_{\sigma_2,\sigma_f}\delta_{k_1,k_f'}\delta_{\sigma_1,\sigma_f'} + \delta_{k_2,k_f'}\delta_{\sigma_2,\sigma_f'}\delta_{k_1,k_f}\delta_{\sigma_1,\sigma_f} \,.\end{aligned}$$

For the electron part it follows:

$$\begin{aligned}\langle 0|\hat{b}_{1s}\hat{b}_n^\dagger\hat{b}_{n'}\hat{b}^\dagger_{2s}|0\rangle &= \langle 0|\hat{b}_{1s}\hat{b}_n^\dagger\left(\delta_{n',2s} - \hat{b}^\dagger_{2s}\hat{b}_{n'}\right)|0\rangle \\ &= \delta_{n',2s}\langle 0|\hat{b}_{1s}\hat{b}_n^\dagger|0\rangle \\ &= \delta_{n',2s}\langle 0|\delta_{n,1s} - \hat{b}_n^\dagger\hat{b}_{1s}|0\rangle \\ &= \delta_{n',2s}\delta_{n,1s} \,,\end{aligned}$$

i.e. the electron has to occupy the 2s state as its initial state and the 1s state as its final state. Then we get

$$\begin{aligned}M_{fi}^{(1)} &= \sum_{k_1,\sigma_1}\sum_{k_2,\sigma_2}\sum_{n,n'} M(-k_1,\sigma_1,-k_2,\sigma_2,n,n')\delta_{n',2s}\delta_{n,1s} \\ &\quad\times\left(\delta_{k_2,k_f}\delta_{\sigma_2,\sigma_f}\delta_{k_1,k_f'}\delta_{\sigma_1,\sigma_f'} + \delta_{k_2,k_f'}\delta_{\sigma_2,\sigma_f'}\delta_{k_1,k_f}\delta_{\sigma_1,\sigma_f}\right) \\ &= M(-k_f',\sigma_f',-k_f,\sigma_f,1s,2s) + M(-k_f,\sigma_f,-k_f',\sigma_f',1s,2s) \,. \qquad (5)\end{aligned}$$

In order to calculate the matrix element of second order we need the intermediate states

$$|z\rangle = \hat{a}^\dagger_{k_z\sigma_z}\hat{b}_z^\dagger|0\rangle \qquad (6)$$

for which only one photon is present ($\hat{H}'$ can produce only one photon in first order). The matrix element reads

$$M_{fi}^{(2)} = \sum_z \frac{1}{E_i - E_z}\left(\langle f|\hat{H}'|z\rangle\langle z|\hat{H}'|i\rangle\right) \,. \qquad (7)$$

We evaluate the matrix elements separately: *Exercise 4.3.*

$$
\begin{aligned}
\langle f|\hat{H}'|z\rangle &= \langle 0|\hat{a}_{k'_f,\sigma'_f}\hat{a}_{k_f,\sigma_f}\hat{b}_{1s}\hat{H}'\hat{a}^\dagger_{k_z,\sigma_z}\hat{b}^\dagger_z|0\rangle \\
&= \sum_{k,\sigma}\sum_{n,n'} M(-k,\sigma,n,n') \\
&\quad \times \langle 0|\hat{a}_{k'_f,\sigma'_f}\hat{a}_{k_f,\sigma_f}\hat{b}_{1s}\hat{b}^\dagger_n\hat{b}_{n'}\hat{a}^\dagger_{k,\sigma}\hat{a}^\dagger_{k_z,\sigma_z}\hat{b}^\dagger_z|0\rangle \\
&= \sum_{k,\sigma}\sum_{n,n'} M(-k,\sigma,n,n')\langle 0|\hat{a}_{k'_f,\sigma'_f}\hat{a}_{k_f,\sigma_f}\hat{a}^\dagger_{k,\sigma}\hat{a}^\dagger_{k_z,\sigma_z}|0\rangle \\
&\quad \times \langle 0|\hat{b}_{1s}\hat{b}^\dagger_n\hat{b}_{n'}\hat{b}^\dagger_z|0\rangle\,.
\end{aligned}
$$

Furthermore,

$$
\begin{aligned}
\langle 0|\hat{b}_{1s}\hat{b}^\dagger_n\hat{b}_{n'}\hat{b}^\dagger_z|0\rangle &= \langle 0|\hat{b}_{1s}\hat{b}^\dagger_n\left(\delta_{n',z} - \hat{b}^\dagger_z\hat{b}_{n'}\right)|0\rangle \\
&= \delta_{n',z}\langle 0|\hat{b}_{1s}\hat{b}^\dagger_n|0\rangle \\
&= \delta_{n',z}\langle 0|\delta_{n,1s} - \hat{b}^\dagger_n\hat{b}_{1s}|0\rangle \\
&= \delta_{n',z}\delta_{n,1s}
\end{aligned}
$$

and

$$
\begin{aligned}
&\langle 0|\hat{a}_{k'_f,\sigma'_f}\hat{a}_{k_f,\sigma_f}\hat{a}^\dagger_{k,\sigma}\hat{a}^\dagger_{k_z,\sigma_z}|0\rangle \\
&\quad = \langle 0|\hat{a}_{k'_f,\sigma'_f}\left(\hat{a}^\dagger_{k,\sigma}\hat{a}_{k_f,\sigma_f} + \delta_{k,k_f}\delta_{\sigma,\sigma_f}\right)\hat{a}^\dagger_{k_z,\sigma_z}|0\rangle \\
&\quad = \langle 0|\hat{a}_{k'_f,\sigma'_f}\hat{a}^\dagger_{k,\sigma}\hat{a}_{k_f,\sigma_f}\hat{a}^\dagger_{k_z,\sigma_z}|0\rangle \\
&\qquad + \delta_{k,k_f}\delta_{\sigma,\sigma_f}\langle 0|\hat{a}_{k'_f,\sigma'_f}\hat{a}^\dagger_{k_z,\sigma_z}|0\rangle \\
&\quad = \langle 0|\hat{a}_{k'_f,\sigma'_f}\hat{a}^\dagger_{k,\sigma}\left(\hat{a}^\dagger_{k_z,\sigma_z}\hat{a}_{k_f,\sigma_f} + \delta_{k_z,k_f}\delta_{\sigma_z,\sigma_f}\right)|0\rangle \\
&\qquad \delta_{k,k_f}\delta_{\sigma,\sigma_f}\langle 0|\delta_{k_z,k'_f}\delta_{\sigma_z,\sigma'_f} + \hat{a}^\dagger_{k_z,\sigma_z}\hat{a}_{k'_f,\sigma'_f}|0\rangle \\
&\quad = \delta_{k_z,k_f}\delta_{\sigma_z,\sigma_f}\langle 0|\hat{a}_{k'_f,\sigma'_f}\hat{a}^\dagger_{k,\sigma}|0\rangle + \delta_{k,k_f}\delta_{\sigma,\sigma_f}\delta_{k_z,k'_f}\delta_{\sigma_z,\sigma'_f} \\
&\quad = \delta_{k_z,k_f}\delta_{\sigma_z,\sigma_f}\langle 0|\delta_{k,k'_f}\delta_{\sigma,\sigma'_f} + \hat{a}^\dagger_{k,\sigma}\hat{a}_{k'_f,\sigma'_f}|0\rangle \\
&\qquad + \delta_{k,k_f}\delta_{\sigma,\sigma_f}\delta_{k_z,k'_f}\delta_{\sigma_z,\sigma'_f} \\
&\quad = \delta_{k_z,k_f}\delta_{\sigma_z,\sigma_f}\delta_{k,k'_f}\delta_{\sigma,\sigma'_f} + \delta_{k,k_f}\delta_{\sigma,\sigma_f}\delta_{k_z,k'_f}\delta_{\sigma_z,\sigma_f}\,,
\end{aligned}
$$

so that

$$
\begin{aligned}
\langle f|\hat{H}'|z\rangle &= \sum_{k,\sigma}\sum_{n,n'} M(-k,\sigma,n,n')\delta_{n'z}\delta_{n,1s} \\
&\quad \times \left(\delta_{k_z,k_f}\delta_{\sigma_z,\sigma_f}\delta_{k,k'_f}\delta_{\sigma,\sigma'_f} + \delta_{k,k_f}\delta_{\sigma,\sigma_f}\delta_{k_z,k'_f}\delta_{\sigma_f,\sigma'_f}\right) \\
&= \sum_{k,\sigma}\sum_{n,n'} M(-k,\sigma,n,n')\delta_{n',z}\delta_{n,1s} \\
&\quad \times \left(\delta_{k_z,k_f}\delta_{\sigma_z,\sigma_f}\delta_{k,k'_f}\delta_{\sigma,\sigma'_f} + \delta_{k,k_f}\delta_{\sigma,\sigma_f}\delta_{k_z,k'_f}\delta_{\sigma_z,\sigma'_f}\right)\,. \qquad (8)
\end{aligned}
$$

Exercise 4.3.

Analogously we derive

$$\begin{aligned}
\langle z|\hat{H}'|i\rangle &= \langle 0|\hat{b}_z\hat{a}_{k_z}\hat{H}'\hat{b}^\dagger_{2\mathrm{s}}|0\rangle \\
&= \sum_{k',\sigma'}\sum_{m,m'} M(-k,\sigma,m,m') \\
&\quad\times \langle 0|\hat{b}_z\hat{a}_{k_z,\sigma_z}\hat{b}^\dagger_m\hat{b}_{m'}\hat{a}^\dagger_{k',\sigma'}\hat{b}^\dagger_{2\mathrm{s}}|0\rangle \\
&= \sum_{k',\sigma'}\sum_{m,m'} M(-k,\sigma,m,m') \\
&\quad\times \langle 0|\hat{b}_z\hat{b}^\dagger_m\hat{b}_{m'}\hat{b}^\dagger_{2\mathrm{s}}|0\rangle\langle 0|\hat{a}_{k_z,\sigma_z}\hat{a}^\dagger_{k',\sigma'}|0\rangle\,.
\end{aligned}$$

It holds

$$\begin{aligned}
\langle 0|\hat{a}_{k_z,\sigma_z}\hat{a}^\dagger_{k',\sigma'}|0\rangle &= \langle 0|\delta_{k_z,k'}\delta_{\sigma_z,\sigma'} + \hat{a}^\dagger_{k',\sigma'}\hat{a}_{k_z,\sigma_z}|0\rangle \\
&= \delta_{k_z,k'}\delta_{\sigma_z,\sigma'}
\end{aligned}$$

and

$$\langle 0|\hat{b}_z\hat{b}^\dagger_m\hat{b}_{m'}\hat{b}^\dagger_{2\mathrm{s}}|0\rangle = \delta_{m',2\mathrm{s}}\langle 0|\hat{b}_z\hat{b}^\dagger_m|0\rangle = \delta_{m',2\mathrm{s}}\delta_{m,z}\,,$$

so that

$$\langle z|\hat{H}'|i\rangle = \sum_{k',\sigma'}\sum_{m,m'} M(-k',\sigma',m,m')\delta_{k_z,k'}\delta_{\sigma_z,\sigma'}\delta_{m',2\mathrm{s}}\delta_{m,z}\,. \tag{9}$$

Hence,

$$\begin{aligned}
M^{(2)}_{\mathrm{fi}} &= \sum_{z,k_z,\sigma_z}\frac{1}{E_\mathrm{i}-E_z}\left\{\sum_{k,\sigma}\sum_{n,n'} M(-k,\sigma,n,n')\delta_{n',z}\delta_{n,1\mathrm{s}}\right. \\
&\quad\left.\times\left[\delta_{k_z,k_\mathrm{f}}\delta_{\sigma_z,\sigma_\mathrm{f}}\delta_{k,k'_\mathrm{f}}\delta_{\sigma,\sigma'_\mathrm{f}} + \delta_{k,k_\mathrm{f}}\delta_{\sigma,\sigma_\mathrm{f}}\delta_{k_z,k'_\mathrm{f}}\delta_{\sigma_z,\sigma'_\mathrm{f}}\right]\right\} \\
&\quad\times\left[\sum_{k',\sigma'}\sum_{m,m'} M(-k',\sigma',m,m')\delta_{k_z,k'}\delta_{\sigma_z,\sigma'}\delta_{m',2\mathrm{s}}\delta_{m,z}\right] \\
&= \sum_{z,k_z,\sigma_z}\frac{1}{E_\mathrm{i}-E_z}\left[M(-k'_\mathrm{f},\sigma'_\mathrm{f},1\mathrm{s},z)\delta_{k_z,k_\mathrm{f}}\delta_{\sigma_z,\sigma_\mathrm{f}}\right. \\
&\quad\left.+M(-k_\mathrm{f},\sigma_\mathrm{f},1\mathrm{s},z)\delta_{k_z,k'_\mathrm{f}}\delta_{\sigma_z,\sigma'_\mathrm{f}}\right]M(-k_z,\sigma_z,z,2\mathrm{s})\,.
\end{aligned} \tag{10}$$

We write

$$\frac{1}{E_\mathrm{i}-E_z} = \frac{1}{E_{2\mathrm{s}}-E_z-\hbar\omega_z}\,,$$

where E_z represents the energy of the electron and $\hbar\omega_z$ the photon energy of the intermediate state z:

$$\begin{aligned}
M^{(2)}_{\mathrm{fi}} &= \sum_z\left[\frac{1}{E_{2\mathrm{s}}-E_z-\hbar\omega_\mathrm{f}}M(-k'_\mathrm{f},\sigma'_\mathrm{f},1\mathrm{s},z)M(-k_\mathrm{f},\sigma_\mathrm{f},z,2\mathrm{s})\right. \\
&\quad\left.+\frac{1}{E_{2\mathrm{s}}-E_z-\hbar\omega'_\mathrm{f}}M(-k_\mathrm{f},\sigma_\mathrm{f},1\mathrm{s},z)M(-k'_\mathrm{f},\sigma'_\mathrm{f},z,2\mathrm{s})\right]\,.
\end{aligned} \tag{11}$$

Two terms arise as a result of the different time ordering of the photon emission as can be read from the arguments of the matrix elements; see Fig. 4.14.

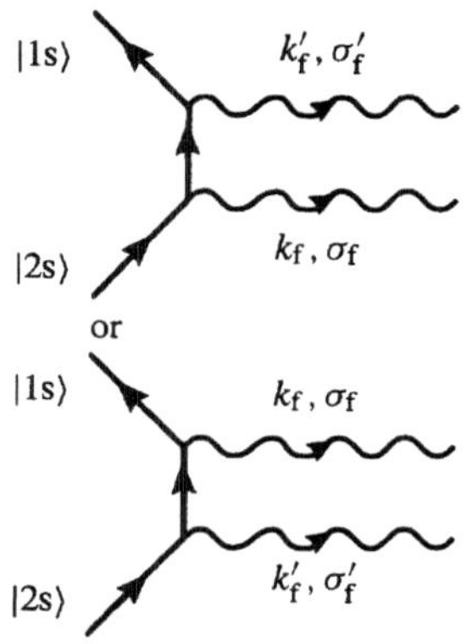

Fig. 4.14. Different time ordering of two-photon decay

Still the sum over electron states z remains in $M_{\mathrm{fi}}^{(2)}$. The total matrix element $M_{\mathrm{fi}} = M_{\mathrm{fi}}^{(1)} + M_{\mathrm{fi}}^{(2)}$ becomes

$$\begin{aligned} M_{\mathrm{fi}} &= M(-k'_{\mathrm{f}}, \sigma'_{\mathrm{f}}, -k_{\mathrm{f}}, \sigma_{\mathrm{f}}, 1\mathrm{s}, 2\mathrm{s}) + M(-k_{\mathrm{f}}, \sigma_{\mathrm{f}}, -k'_{\mathrm{f}}, \sigma'_{\mathrm{f}}, 1\mathrm{s}, 2\mathrm{s}) \\ &\quad + \sum_z \left[\frac{1}{E_{2\mathrm{s}} - E_z - \hbar\omega_{\mathrm{f}}} M(-k'_{\mathrm{f}}, \sigma'_{\mathrm{f}}, 1\mathrm{s}, z) M(-k_{\mathrm{f}}, \sigma_{\mathrm{f}}, z, 2\mathrm{s}) \right. \\ &\quad \left. + \frac{1}{E_{2\mathrm{s}} - E_z - \hbar\omega_{\mathrm{f}}} M(-k_{\mathrm{f}}, \sigma_{\mathrm{f}}, 1\mathrm{s}, z) M(-k'_{\mathrm{f}}, \sigma'_{\mathrm{f}}, z, 2\mathrm{s}) \right] . \end{aligned} \tag{12}$$

The expression for the lifetime is given by (compare with Exercise 2.10):

$$\frac{1}{\tau} = \frac{L^6}{(2\pi)^6} \frac{2\pi}{\hbar^2 c} \sum_{\sigma_{\mathrm{f}} \sigma'_{\mathrm{f}}} \int \mathrm{d}^3 k_{\mathrm{f}} \, \mathrm{d}^3 k'_{\mathrm{f}} |M_{\mathrm{fi}}|^2 \delta(k_0 - k_{\mathrm{f}} - k'_{\mathrm{f}}) \tag{13}$$

with $\hbar c k_0 = E_{2\mathrm{s}} - E_{1\mathrm{s}}$. From (2) we pull out those factors carrying dimensions:

$$\begin{aligned} M(k_1, \sigma_1, k_2, \sigma_2, n, n') &= \left(\frac{2\pi\hbar c^2}{L^3} \right) \left(\frac{1}{\omega_{k_1} \omega_{k_2}} \right)^{1/2} \frac{e^2}{2mc^2} \\ &\quad \times \int \mathrm{d}^3 x \, \psi_n^* \left(\boldsymbol{\epsilon}_{\boldsymbol{k}_1, \sigma_1} \cdot \boldsymbol{\epsilon}_{\boldsymbol{k}_2, \sigma_2} \, \mathrm{e}^{\mathrm{i}(\boldsymbol{k}_1 + \boldsymbol{k}_2) \cdot \boldsymbol{x}} \right) \psi_{n'} \end{aligned}$$

and

$$M(k, \sigma, n, n') = \left(\frac{2\pi\hbar c^2}{L^3 \omega_k} \right)^{1/2} \left(-\frac{e\hbar}{imc} \right) \int \mathrm{d}^3 x \, \psi_n^* \left(\mathrm{e}^{\mathrm{i}\boldsymbol{k} \cdot \boldsymbol{x}} \boldsymbol{\epsilon}_{\boldsymbol{k}\sigma} \cdot \boldsymbol{\nabla} \right) \psi_{n'} \, .$$

Since $M(k, \sigma, n, n')$ appears in the form $M(k, \sigma, n, n') M(k', \sigma', m, m')$ within the matrix element M_{fi} we can pull out of the total matrix element all quantities carrying dimensions:

$$M_{\mathrm{fi}} = \frac{e^2}{2mc^2} \left(\frac{2\pi\hbar c^2}{L^3} \right) \frac{1}{c\sqrt{k_{\mathrm{f}} k'_{\mathrm{f}}}} M'_{\mathrm{fi}} \, , \tag{14}$$

where M'_{fi} results from M_{fi} by substitution of the matrix elements (13) with

$$M'(k_1, \sigma_1, k_2, \sigma_2, n, n') = \int \mathrm{d}^3 x \, \psi_n^* \left(\boldsymbol{\epsilon}_{\boldsymbol{k}_1, \sigma_1} \cdot \boldsymbol{\epsilon}_{\boldsymbol{k}_2, \sigma_2} \, \mathrm{e}^{\mathrm{i}(\boldsymbol{k}_1 + \boldsymbol{k}_2) \cdot \boldsymbol{x}} \right) \psi_{n'}$$

and

$$M'(k, \sigma, n, n') = \frac{2}{m} \int \mathrm{d}^3 x \, \psi_n^* \left(\mathrm{e}^{\mathrm{i}\boldsymbol{k} \cdot \boldsymbol{x}} \boldsymbol{\epsilon}_{\boldsymbol{k}\sigma} \cdot \boldsymbol{p} \right) \psi_{n'} \, .$$

These matrix elements and in particular the sum $\sum_z$ in (12) are difficult to calculate because not only bound states but also continuum states have to be taken into account. Since M_{fi} in (14) takes on the same form as in Exercise 2.10 we can now use the same arguments as before. We set

$$\sum_{\sigma_{\mathrm{f}} \sigma'_{\mathrm{f}}} |M'_{\mathrm{fi}}|^2 \approx 1 \, .$$

The further calculation is identical to the one presented in Exercise 2.10!

EXERCISE

4.4 Second-Order Corrections to Rutherford's Scattering Cross Section

Problem. Determine the corrections of second order for Rutherford's scattering cross section.

Solution. We will calculate the contributions of the following diagrams:

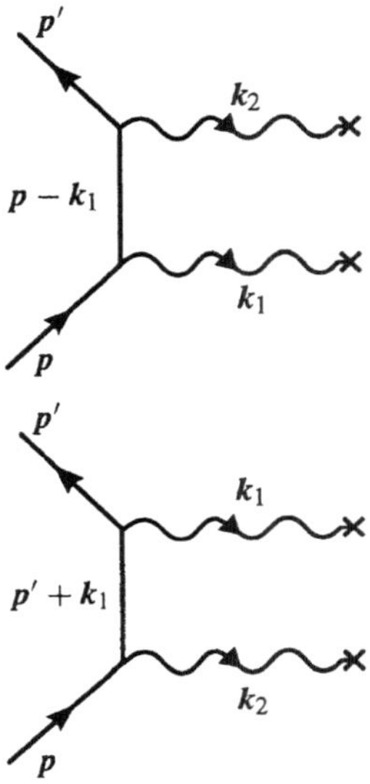

Fig. 4.15. Second-order diagrams for Rutherford scattering

We have to deal with two terms for the determination of the transition matrix element:

$$M_{\rm fi}^{(2)} = \sum_{\boldsymbol{k}_1} \frac{\langle \boldsymbol{p}'|\hat{H}_V|\boldsymbol{p}-\boldsymbol{k}_1\rangle\,\langle \boldsymbol{p}-\boldsymbol{k}_1|\hat{H}_V|\boldsymbol{p}\rangle}{(\hbar^2\boldsymbol{p}^2/2m) - (\hbar^2(\boldsymbol{p}-\boldsymbol{k}_1)^2/2m)} + \sum_{\boldsymbol{k}_1} \frac{\langle \boldsymbol{p}'|\hat{H}_V|\boldsymbol{p}'+\boldsymbol{k}_1\rangle\,\langle \boldsymbol{p}'+\boldsymbol{k}_1|\hat{H}_V|\boldsymbol{p}\rangle}{(\hbar^2\boldsymbol{p}^2/2m) - (\hbar^2(\boldsymbol{p}'+\boldsymbol{k}_1)^2/2m)} .$$

In the first sum we substitute $\boldsymbol{p}-\boldsymbol{k}_1 = \boldsymbol{q}$ and in the second sum $\boldsymbol{p}'+\boldsymbol{k}_1 = \boldsymbol{q}$; inserting $\hat{H}_V$ we derive

$$\begin{aligned} M_{\rm fi}^{(2)} &= \frac{2m}{\hbar^2}\frac{16\pi^2(Ze^2)^2}{L^6}2 \\ &\quad\times\Bigg(\sum_{\boldsymbol{q}}\frac{1}{p^2-q^2}\sum_{\boldsymbol{q}',\boldsymbol{k}'}\sum_{\boldsymbol{q}'',\boldsymbol{k}''}\langle 0|\hat{b}_{\boldsymbol{p}'}\hat{b}^\dagger_{\boldsymbol{q}'-\boldsymbol{k}'}\hat{b}_{\boldsymbol{q}'}\hat{b}^\dagger_{\boldsymbol{q}}|0\rangle\frac{1}{k'^2} \\ &\quad\times\langle 0|\hat{b}_{\boldsymbol{q}}\hat{b}^\dagger_{\boldsymbol{q}''-\boldsymbol{k}''}\hat{b}_{\boldsymbol{q}''}\hat{b}^\dagger_{\boldsymbol{p}}|0\rangle\frac{1}{k''^2}\Bigg) \\ &= \frac{2m}{\hbar^2}\frac{16\pi^2(Ze^2)^2}{L^6}2\sum_{\boldsymbol{q}}\frac{1}{p^2-q^2} \\ &\quad\times\sum_{\boldsymbol{q}',\boldsymbol{k}'\boldsymbol{q}'',\boldsymbol{k}''}\delta_{\boldsymbol{p}',\boldsymbol{q}'-\boldsymbol{k}'}\delta_{\boldsymbol{q}',\boldsymbol{q}}\delta_{\boldsymbol{q},\boldsymbol{q}''-\boldsymbol{k}''}\delta_{\boldsymbol{q}'',\boldsymbol{p}}\frac{1}{k'^2k''^2} \\ &= \frac{2m}{\hbar^2}\frac{16\pi^2(Ze^2)^2}{L^6}2\sum_{\boldsymbol{q}}\frac{1}{(p^2-q^2)(\boldsymbol{p}'-\boldsymbol{q})^2(\boldsymbol{p}-\boldsymbol{q})^2} \\ &= \frac{64\pi^2m(Ze^2)^2}{\hbar^2L^6}\frac{L^3}{(2\pi)^3}\int\frac{q^2\,\mathrm{d}q\,\mathrm{d}\Omega_q}{(p^2-q^2)(\boldsymbol{p}'-\boldsymbol{q})^2(\boldsymbol{p}-\boldsymbol{q})^2} \\ &= \frac{8m(Ze^2)^2}{\pi\hbar^2L^3}\int\frac{q^2\,\mathrm{d}q\,\mathrm{d}\Omega_q}{(p^2-q^2)(p'^2+q^2-2\boldsymbol{p}'\cdot\boldsymbol{q})(p^2+q^2-2\boldsymbol{p}\cdot\boldsymbol{q})} . \end{aligned}$$

The determination of this integral is far from easy. We follow a method presented by Dalitz.[3]

[3] R.H. Dalitz: Proc. Roy. Soc. A **206** (1950) 509.

Let us consider the integral *Exercise 4.4.*

$$\begin{aligned} I_1 &= \int \frac{\mathrm{d}^3 q}{[(\boldsymbol{p}_1 - \boldsymbol{q})^2 + \Lambda^2](p_2^2 - q^2)} \\ &= \lim_{\epsilon \to 0^+} 2\pi \int_{-1}^{+1} \mathrm{d}(\cos\vartheta) \int_0^\infty \frac{q^2\,\mathrm{d}q}{({p_1}^2 + q^2 - 2qp_1\cos\vartheta + \Lambda^2)(p_2^2 - q^2 + \mathrm{i}\epsilon)} \\ &= \lim_{\epsilon \to 0^+} \pi \int_{-1}^{+1} \mathrm{d}t \int_{-\infty}^\infty \frac{q^2\,\mathrm{d}q}{({p_1}^2 + q^2 - 2qp_1 t + \Lambda^2)(p_2^2 - q^2 + \mathrm{i}\epsilon)} \,. \end{aligned}$$

$1/(p_2^2 + \lambda^2)$ represents the Fourier transform of $1/4\pi r \exp(-r/\lambda)$, which is the exponentially screened Coulomb potential. As we will realize later on, the introduction of a finite λ is equivalent to giving the photon a finite mass λ.

The integrand becomes singular at $q = \pm(p_2 + \mathrm{i}\epsilon/2)$ and $q = p_1 t \pm \mathrm{i}\sqrt{p_2^2(1-t^2) + \Lambda^2}$. We evaluate the integral with the help of the residue theorem as we close the integration path in the upper half of the complex q plane. Here the two poles with positive imaginary part are enclosed by the integration path. Since the poles of the integrand are of first order the residues are determined by

$$\mathrm{Res}(f(q), q_i) = \lim_{q \to q_i} (q - q_i) f(q) \,.$$

For the integral we derive

$$\begin{aligned} I_1 &= -2\pi^2 \mathrm{i} \int_{-1}^{+1} \mathrm{d}t \bigg(\frac{p^2}{({p_2}^2 + {p_1}^2 - 2p_2p_1 + \Lambda^2)2p} \\ &\quad + \frac{(p_1 t + \mathrm{i}\Gamma)^2}{(p_1 t + \mathrm{i}\Gamma - p_2)(p_1 t + \mathrm{i}\Gamma + p_2)2\mathrm{i}\Gamma} \bigg) \,, \end{aligned}$$

where

$$\Gamma = \sqrt{p_1^2(1-t^2) + \Lambda^2} \,.$$

After some intricate transformations of the integrand we get

$$\begin{aligned} I_1 &= \frac{\pi^2 \mathrm{i}}{p_1} \int_{-1}^{+1} \mathrm{d}t \frac{p_1^2 t - \mathrm{i}p_1\Gamma}{\mathrm{i}\Gamma(p_2 - p_1 t + \mathrm{i}\Gamma)} = \frac{\pi^2 \mathrm{i}}{p_1} \int_{-1}^{+1} \mathrm{d}t \frac{\mathrm{d}}{\mathrm{d}t} \ln(p_2 - p_1 t + \mathrm{i}\Gamma) \\ &= \frac{\pi^2 \mathrm{i}}{p_1} \left[\ln(p_2 - p_1 + \mathrm{i}\Lambda) - \ln(p_2 + p_1 + \mathrm{i}\Lambda) \right] \,. \end{aligned}$$

Thus, we have

$$\lim_{\epsilon \to 0^+} \int \frac{\mathrm{d}^3 q}{[(\boldsymbol{p}_1 - \boldsymbol{q})^2 + \Lambda^2](p_2^2 - q^2 + \mathrm{i}\epsilon)} = \frac{2\pi^2}{p_1} \ln \frac{p_2 - p_1 + \mathrm{i}\Lambda}{p_2 + p_1 + \mathrm{i}\Lambda} \,.$$

Via partial differentiation with respect to the parameter Λ we obtain

$$\begin{aligned} I_2 &\equiv -\lim_{\epsilon \to 0^+} \int \frac{2\Lambda\,\mathrm{d}^3 q}{((q - p_1)^2 + \Lambda^2)^2(p_2^2 - q^2 + \mathrm{i}\epsilon)} \\ &= \frac{\mathrm{i}\pi^2}{p_1} \left(\frac{\mathrm{i}}{p_2 - p_1 + \mathrm{i}\Lambda} - \frac{\mathrm{i}}{p_2 + p_1 + \mathrm{i}\Lambda} \right) \,, \end{aligned}$$

from which follows

Exercise 4.4.

$$I_2 \equiv \lim_{\epsilon\to 0^+}\int \frac{d^3q}{[(q-p_1)^2+\Lambda^2]^2(p_2^2-q^2+i\epsilon)} = \frac{\pi^2}{\Lambda}\frac{1}{p_2-p_1-\Lambda^2+2p_2\Lambda i}\,.$$

Our matrix element $M_{\rm fi}^{(2)}$ takes on the form

$$\lim_{\substack{\lambda\to 0\\ \epsilon\to 0^+}}\int \frac{d^3q}{[\lambda^2+(p_2'-q)^2][\lambda^2+(p_2-q)^2](p_2^2-q^2+i\epsilon)}$$

and with a small trick it can be transformed into the form of I_2: First, we observe that a δ function appears in the formula for the cross section, which guarantees energy conservation. Hence, we only need to calculate the matrix element for the case $p_2'^2 = p_2^2$ and we write $\boldsymbol{p}_2\cdot\boldsymbol{p}_2' = p_2^2\cos\vartheta$. We make use of the identity

$$\frac{1}{ab} = \int_{-1}^{+1}\frac{dz}{2}\left(\frac{a(1+z)+b(1-z)}{2}\right)^{-2}$$

and apply it to our matrix element:

$$\begin{aligned}&\frac{1}{[\lambda^2+(\boldsymbol{p}_2'-\boldsymbol{q})^2][\lambda^2+(\boldsymbol{p}_2-\boldsymbol{q})^2]}\\ &\quad= \int_{-1}^{+1}\frac{dz}{2}\left[[\lambda^2+(\boldsymbol{p}_2'-\boldsymbol{q})^2]\frac{1+z}{2}+[\lambda^2+(\boldsymbol{p}_2-\boldsymbol{q})^2]\frac{1-z}{2}\right]^{-2}\\ &\quad= \int_{-1}^{+1}\frac{dz}{2}\left[(\boldsymbol{p}_1-\boldsymbol{q})^2+\Lambda^2\right]^{-2}\end{aligned}$$

with

$$\begin{aligned}\boldsymbol{p}_1 &= \frac{1}{2}\left[(1+z)\boldsymbol{p}_2+(1-z)\boldsymbol{p}\right]\,,\\ \Lambda^2 &= \lambda^2+\frac{1}{2}(1-z^2)\boldsymbol{p}_2{}^2+\frac{1}{2}(z^2-1)\boldsymbol{p}_2\cdot\boldsymbol{p}_2'\\ &= \lambda^2+(1-z^2)\boldsymbol{p}^2\frac{1}{2}(1-\cos\vartheta)\\ &= \lambda^2+\boldsymbol{p}^2\sin^2\frac{\vartheta}{2}(1-z^2)\,.\end{aligned}$$

Furthermore, $p_1^2 = p_2^2(\cos^2\vartheta/2+t^2\sin^2\vartheta/2)$ holds and our matrix element reads

$$M_{\rm fi}^{(2)} = \lim_{\lambda\to 0}\frac{8m(Ze^2)^2}{\hbar^2\pi L^3}\int_{-1}^{+1}\frac{dz}{2}\frac{\pi^2}{\Lambda}\frac{1}{\boldsymbol{p}_2{}^2-\boldsymbol{p}_1{}^2-\Lambda^2+i2p_2\Lambda}\,.$$

Inserting $\boldsymbol{p}_1$ and Λ, performing the integration over z, and introducing the abbreviation $\gamma = \sqrt{\lambda^4+4p_2^2(\lambda^2+p_2^2\sin^2\vartheta/2)}$ we obtain

$$\begin{aligned}M_{\rm fi}^{(2)} &= \lim_{\lambda\to 0}\frac{-8\pi m(Ze^2)^2}{\hbar^2L^3}\frac{1}{\gamma p_2\sin\vartheta/2}\\ &\quad\times\left(\arctan\frac{\lambda p_2\sin\vartheta/2}{\gamma}+\frac{i}{2}\ln\frac{\gamma+p_2^2\sin\vartheta/2}{\gamma-p_2^2\sin\vartheta/2}\right)\,.\end{aligned}$$

Exercise 4.4.

In the limit $\lambda \to 0$ the first term inside the bracket vanishes, so we finally arrive at

$$M_{\mathrm{fi}}^{(2)} \approx -\frac{4\pi m(Ze^2)^2}{\hbar^2 L^3} \frac{\mathrm{i}}{p^3 \sin^2 \vartheta/2} \ln \frac{2p_2^2 \sin \vartheta/2}{\lambda} \xrightarrow{\lambda \to 0} \infty .$$

The matrix element diverges as the introduced photon mass λ goes to zero. Here the long range of the Coulomb potential leads to a divergence which questions the result of the previous problem: How can we trust the result from first-order perturbation theory if the second order already diverges? On the other hand the experiments are in very good agreement with Rutherford's result. Later in this volume we will recognize how these and other divergences can be avoided within the framework of renormalization theory, which is widely discussed in relativistic quantum electrodynamics.

5. Infinities in Quantum Electrodynamics: Renormalization Problems

A nice feature of quantum electrodynamics is the smallness of the interaction between the charged particles and the radiation field; as a consequence, this interaction can always be treated as a perturbation. But in all this brightness a disturbing spot exists: With the use of the supposedly well-functioning perturbation theory some quantities that should remain small become infinitely large. Discussing the self-energy in Chap. 2, we have had our first encounter with this problem. Now we have to deal with these difficulties in detail in order to overcome them.

5.1 Attraction of Parallel, Conducting Plates Due to Field Quantum Fluctuations (Casimir Effect)

We begin with the infinite *zero-point energy of the radiation field* in the vacuum, which we have already encountered in (1.52):

$$W = \frac{1}{2}\sum_{\boldsymbol{k},\sigma} \hbar\omega_k = \infty \,. \tag{5.1}$$

In our previous considerations we have simply disregarded this contribution; we have concentrated on energy differences, so that (5.1) has cancelled. Thereby we have followed the argument often loosely applied that the absolute value of energy has no importance and that an arbitrary constant can be added or subtracted from it. In general this statement will not hold for sure; for example, within the general theory of relativity the absolute value of energy is physically relevant. It is the total energy that enters as a source for the gravitational field in Einstein's gravitational equations and thus determines the metric and the curvature of space.

For the quantization of the electromagnetic field as demonstrated in Chap. 1 we started with the classical expressions and then translated them into quantum mechanics. This procedure is somewhat questionable. We illustrate this point again with the classical expression (1.48) for the energy of the radiation field

$$H = \frac{1}{2}\sum_{\boldsymbol{k},\sigma} \hbar\omega_k (a_{\boldsymbol{k}\sigma}a^*_{\boldsymbol{k}\sigma} + a^*_{\boldsymbol{k}\sigma}a_{\boldsymbol{k}\sigma}) \tag{5.2a}$$

$$= \sum_{\boldsymbol{k},\sigma} \hbar\omega_k a^*_{\boldsymbol{k}\sigma}a_{\boldsymbol{k}\sigma} \, . \tag{5.2b}$$

On a classical basis both expressions are identical. From the viewpoint of quantum mechanics they are not equivalent since (1.50) means the amplitudes a and a^* become operators. Hence, expressions (5.2a) and (5.2b) become

$$\hat{H} = \frac{1}{2}\sum_{\boldsymbol{k},\sigma} \hbar\omega_k \left(\hat{a}_{\boldsymbol{k}\sigma}\hat{a}^\dagger_{\boldsymbol{k}\sigma} + \hat{a}^\dagger_{\boldsymbol{k}\sigma}\hat{a}'_{\boldsymbol{k}\sigma}\right)$$

$$= \sum_{\boldsymbol{k},\sigma} \hbar\omega_k \hat{a}^\dagger_{\boldsymbol{k}\sigma}\hat{a}_{\boldsymbol{k}\sigma} + \frac{1}{2}\sum_{\boldsymbol{k},\sigma} \hbar\omega_k \tag{5.3a}$$

and

$$\hat{H} = \sum_{\boldsymbol{k},\sigma} \hbar\omega_k \hat{a}^\dagger_{\boldsymbol{k}\sigma}\hat{a}_{\boldsymbol{k}\sigma} \, , \tag{5.3b}$$

respectively. Evidently both Hamiltonians differ by the presence of the infinite zero-point energy. Now we could argue that we have to pay careful attention to the ordering of a, a^* in the classical expression (before quantization). Then (5.3a) would be the correct Hamiltonian for the free electromagnetic field. We could also follow the reverse direction and ask what classical expression for the radiation energy would result from (5.3b) if we take into account the precise ordering. Indeed, a lengthy backward calculation[1] starting from (5.3b) yields

$$H_{\text{rad}} = \frac{1}{8\pi}\int \mathrm{d}^3x \times \left\{\hat{\boldsymbol{E}}^2 + \hat{\boldsymbol{B}}^2 + \frac{\mathrm{i}}{\sqrt{-\boldsymbol{\nabla}^2}}\left[\hat{\boldsymbol{E}}\cdot(\boldsymbol{\nabla}\times\hat{\boldsymbol{B}}) - (\boldsymbol{\nabla}\times\hat{\boldsymbol{B}})\cdot\hat{\boldsymbol{E}}\right]\right\}, \tag{5.4}$$

where the operator $1/\sqrt{-\boldsymbol{\nabla}^2}$ is defined by

$$\frac{1}{\sqrt{-\boldsymbol{\nabla}^2}}\mathrm{e}^{\mathrm{i}\boldsymbol{k}\cdot\boldsymbol{x}} = \frac{1}{k}\mathrm{e}^{\mathrm{i}\boldsymbol{k}\cdot\boldsymbol{x}}$$

for plane waves. The electromagnetic field has to be decomposed into plane wave Fourier modes in order to calculate the action of this operator properly. Note the operator character of the field operators $\hat{\boldsymbol{E}}$ and $\hat{\boldsymbol{B}}$ in (5.4); if we consider $\hat{\boldsymbol{E}}$ and $\hat{\boldsymbol{B}}$ as classical fields the term $[\hat{\boldsymbol{E}}\cdot(\boldsymbol{\nabla}\times\hat{\boldsymbol{B}}) - (\boldsymbol{\nabla}\times\hat{\boldsymbol{B}})\cdot\hat{\boldsymbol{E}}]$ vanishes and (5.4) is identical to the classical field energy (1.40). From this discussion we realize that the classical field energy can be transformed into a Hamiltonian of the wave field in many ways. We have to ask nature, i.e. experiment, for the correct Hamiltonian! The same holds for the zero-point energy of the electromagnetic field [(5.1) and (5.3a)].

[1] E.G. Harris: *A Pedestrian Approach to Quantum Theory* (Wiley, New York 1972).

Now the question arises whether somehow the zero-point energy (or parts of it) can be observed in an experiment. Casimir[2] pointed out such a possibility and Lifschitz and Fierz[3] have both investigated his argument in more detail.[4]

The line of thought is as follows. The zero-point energy of the electromagnetic field is given by

$$\sum_n \frac{1}{2}\hbar\omega_n ,$$

where n stands for $\boldsymbol{k}$. The frequencies ω_n of the electromagnetic field depend on the geometry of the volume in which the field is confined. If the geometrical form changes the frequency of the normal modes, the zero-point energy will also change. We consider a rectangular box of lenght R and basal area $A = L^2$. This box, with conducting walls, represents our volume (quantization volume), which determines the frequencies of the electromagnetic field via its geometry. We introduce a second plate at a distance d from the first one and are interested in the energy of the system depending on the position of one of the plates. Hence we subtract the energy of a reference configuration for which the mobile plate is fixed at a certain distance (for instance $1/\eta$ with $1/\eta = 1/2$; see Fig. 5.1). The difference in the energy is

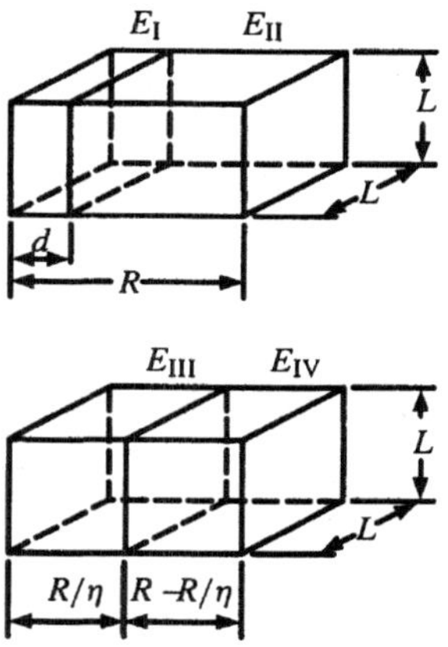

Fig. 5.1. Modification of a rectangular box leads to a modification of the zero-point energy

$$U(d,R,A) = (E_{\mathrm{I}} + E_{\mathrm{II}}) - (E_{\mathrm{III}} + E_{\mathrm{IV}}) , \tag{5.5}$$

where E_i ($i = \mathrm{I,II,III,IV}$) represents the zero-point energy of the free electromagnetic field in the corresponding region depicted in the figure. In the following, we will move the walls of our quantization volume to infinity, i.e.

$$U(d,A) = \lim_{R\to\infty} U(d,R,A) .$$

Every single term E_i is formally divergent because we consider an infinite number of normal modes with increasing frequency. Therefore it is necessary to find a physically reasonable *cut-off procedure*.

In our case, the wavelength could serve as a cut-off parameter. From classical electrodynamics we know that a good conductor turns into a bad conductor at short wavelengths (X-rays!). Hence, we can cut off the energy exponentially:

$$E_i = \sum_n \frac{1}{2}\hbar\omega_n \exp\left(-\lambda\frac{\omega_n}{c}\right) ; \tag{5.6}$$

for the final result we will again let $\lambda \to 0$.

In a rectangular waveguide with dimensions $d \times L \times L$ the frequencies of the classical and the quantum-mechanical normal modes are given by (see Chap. 1)

$$\begin{aligned} \omega_{lmn} &= c\,k_{lmn}(d,L,L) \\ &= c\sqrt{\left(\frac{l\pi}{d}\right)^2 + \left(\frac{m\pi}{L}\right)^2 + \left(\frac{n\pi}{L}\right)^2} , \end{aligned}$$

---

[2] H.B.G. Casimir: Proc. Netherlands Aka. Wetenschapen **51** (1948) 793.

[3] E.M. Lifschitz: Soviet. Phys. JETP **2** (1956) 73; M. Fierz: Hel. Phys. Acta **33** (1960) 855.

[4] For a review of the Casimir effect in all its facets see G. Plunien, B. Müller, W. Greiner: Phys. Rep. **134** (1986) 87.

where l, m, and n are positive integers. Because there are two possible polarizations, the potential energy becomes

$$
\begin{aligned}
U(d,A) &= \lim_{R\to\infty}\lim_{\lambda\to 0}\frac{1}{2}\hbar c\left\{\left[2\sum_{l,m,n}k_{lmn}\exp(-\lambda k_{lmn}) + (d\to R-d)\right]\right. \\
&\quad \left. -\left[\left(d\to\frac{R}{\eta}\right)+\left(d\to R-\frac{R}{\eta}\right)\right]\right\}, \qquad (5.7)
\end{aligned}
$$

where, for example, the bracket $(d \to R/\eta)$ is a shorthand notation of the first expression with the substitution of R/η instead of d.

We will present the further calculation in Exercise 5.1. After performing the limiting cases $R \to \infty$ and $\lambda \to 0$ we obtain

$$
U(d,A) = -\frac{\pi^2}{720}\frac{\hbar c A}{d^3}, \qquad A = L^3 . \qquad (5.8)
$$

This result is finite and does not depend on the cut-off procedure. From here we calculate the force per unit area:

$$
F = -\frac{\partial}{\partial d}\left(\frac{U(d,A)}{A}\right) = -\frac{\hbar c\pi^2}{240 d^4}; \qquad (5.9)
$$

evidently this is not only a finite result again but, more astonishing, the force depends only on the universal constants $\hbar$ and c. Obviously it is independent of e, the coupling between the electromagnetic field and matter. This force depends only on the "zero-point pressure" of the zero-point oscillations of the photon vacuum. Therefore we are allowed to say that this attracting force between the conducting plates is of pure quantum-mechanical nature. It vanishes for the limiting case $\hbar \to 0$.

We have realized that two parallel plates alter the allowed normal frequencies and thus the zero-point energy of the quantization volume (for the case $R \to \infty$ of the "universe"). From this alteration results a force that pushes the plates together.

An extension of this theory onto dielectric substances at finite temperatures was performed by Lifschitz and Fierz. An experimental test of the theory started in 1957 with studies by Abrikosova and Deriagin,[5] whereas the first real evidence was provided by Sparnaay.[6] Details of this experiment are presented in Exercise 5.2. From then on this experiment has been repeated several times with different methods. The theoretical predictions have been confirmed every time.[7]

[5] I.I. Abrikosova, B.V. Deriagin: Sov. Phys. JETP **3** (1957) 819; **4** (1957) 2.
[6] M.J. Sparnaay: Physica **24** (1958) 751.
[7] T.H. Boyer: Ann. of Physics **56** (1970) 474.

EXERCISE

5.1 Attraction of Parallel, Conducting Plates Due to the Casimir Effect

Problem. Under the assumption that the length L of the plates is large compared to the distance d, derive the relation given in the text for the potential energy of the system of plates depicted in Fig. 5.1.

Solution. To begin with we calculate the first expression in (5.7) for the potential energy; afterwards we replace d with the corresponding other quantities. We get

$$U = \hbar c \sum_{l,m,n} k_{lmn}(d,L,L) \exp[-\lambda k_{lmn}(d,L,L)] ,$$

where

$$k_{lmn} = \sqrt{\left(\frac{l\pi}{d}\right)^2 + \left(\frac{m\pi}{L}\right)^2 + \left(\frac{n\pi}{L}\right)^2} .$$

If we assume $A = L^2 \gg d^2$ then at finite energy many modes fit into the volume; this means that we can choose m/L and n/L to be very large and at the same time the energy would stay small. We are able to replace the summations over n und m by integrations:

$$U = \hbar c \sum_{l=1}^{\infty} \int_{m=0}^{\infty} \mathrm{d}m \int_{n=0}^{\infty} \mathrm{d}n\, k_{lmn}(d,L,L) \exp[-\lambda k_{lmn}(d,L,L)]$$

or

$$U = \hbar c \sum_{l=1}^{\infty} \int_{m=0}^{\infty} \mathrm{d}m \int_{n=0}^{\infty} \mathrm{d}n \sqrt{\left(\frac{l\pi}{d}\right)^2 + \left(\frac{m\pi}{L}\right)^2 + \left(\frac{n\pi}{L}\right)^2}$$
$$\times \exp\left[-\lambda\sqrt{\left(\frac{l\pi}{d}\right)^2 + \left(\frac{m\pi}{L}\right)^2 + \left(\frac{n\pi}{L}\right)^2}\right] .$$

We set

$$x = \frac{m\pi}{L} , \quad y = \frac{n\pi}{L} , \quad a = \frac{d}{\pi} ,$$

so that

$$U = \hbar c \frac{L^2}{\pi^2} \sum_{l=1}^{\infty} \int_0^{\infty} \mathrm{d}x\, \mathrm{d}y \sqrt{\left(\frac{l}{a}\right)^2 + x^2 + y^2}\, \exp\left(-\lambda\sqrt{\left(\frac{l}{a}\right)^2 + x^2 + y^2}\right) .$$

The substitution

$$z = \left(\frac{a}{l}\right)^2 (x^2 + y^2) = \left(\frac{a}{l}\right)^2 r^2$$

leads to the surface element

$$r\, \mathrm{d}r\, \mathrm{d}\phi = \frac{1}{2}\left(\frac{l}{a}\right)^2 \mathrm{d}z = \frac{1}{2}\left(\frac{\pi}{d}\right)^2 l^2\, \mathrm{d}z .$$

Exercise 5.1.

The integration over ϕ goes up to $\pi/2$ (only positive z)! We get

$$U = \frac{1}{2}\hbar c \frac{A}{\pi^2}\sum_{l=1}^{\infty}\int_0^{\pi/2} \mathrm{d}\phi \int_0^{\infty} \mathrm{d}z\,\sqrt{z+1}\left(\frac{l}{a}\right)^3 \exp\left(-\lambda\frac{l}{a}\sqrt{z+1}\right)$$

with $A = L^2$. Continuing,

$$U = \frac{1}{4}\hbar c\frac{\pi^2 A}{d^3}\sum_{l=1}^{\infty} l^3 \int_0^{\infty} \mathrm{d}z\,\sqrt{z+1}\exp\left(-\frac{\alpha l}{d}\sqrt{z+1}\right),$$

where $\alpha = \lambda\pi$. Now

$$\frac{\mathrm{d}^3}{\mathrm{d}\alpha^3}\exp\left(-\alpha\frac{l}{d}\sqrt{z+1}\right) = -\left(\frac{l}{d}\right)^3 (z+1)\sqrt{z+1}\exp\left(-\alpha\frac{l}{d}\sqrt{z+1}\right),$$

so that

$$U = -\frac{1}{4}\hbar c A\pi^2\sum_{l=1}^{\infty}\int_0^{\infty}\frac{\mathrm{d}z}{z+1}\frac{\mathrm{d}^3}{\mathrm{d}\alpha^3}\left\{\exp\left(-\frac{\alpha l}{d}\sqrt{z+1}\right)\right\}.$$

We assume uniform continuity; hence we are allowed to interchange the order of the summation, integration, and differentiation:

$$U = -\frac{1}{4}\hbar c A\pi^2\frac{\mathrm{d}^3}{\mathrm{d}\alpha^3}\int_0^{\infty}\frac{\mathrm{d}z}{z+1}\sum_{l=1}^{\infty}\exp\left(-\frac{\alpha l}{d}\sqrt{z+1}\right).$$

Since

$$\sum_{l=0}^{\infty}\mathrm{e}^{-\gamma l} = \frac{1}{1-\mathrm{e}^{-\gamma}},$$

it follows that

$$\sum_{l=1}^{\infty}\exp\left(-\frac{\alpha l}{d}\sqrt{z+1}\right) = \frac{1}{1-\exp\left(-(\alpha/d)\sqrt{z+1}\right)} - 1,$$

$$U = -\frac{1}{4}\hbar c A\pi^2\frac{\mathrm{d}^3}{\mathrm{d}\alpha^3}\int_0^{\infty}\frac{\mathrm{d}z}{z+1}\left[\frac{\exp\left(-(\alpha/d)\sqrt{z+1}\right)}{1-\exp\left(-(\alpha/d)\sqrt{z+1}\right)}\right].$$

We substitute $u = \sqrt{z+1}$ with $\mathrm{d}z = 2\sqrt{z+1}\,\mathrm{d}u$, so that

$$U = -\frac{1}{2}\hbar c A\pi^2\frac{\mathrm{d}^3}{\mathrm{d}\alpha^3}\int_1^{\infty}\frac{\mathrm{d}u}{u}\frac{\mathrm{e}^{-\gamma u}}{1-\mathrm{e}^{-\gamma u}}$$

with $\gamma = \alpha/d$, or

$$U = -\frac{1}{2}\hbar c A\pi^2\frac{\mathrm{d}^3}{\mathrm{d}\alpha^3}\int_1^{\infty}\frac{\mathrm{d}u}{u}\frac{1}{\mathrm{e}^{\gamma u}-1}.$$

Since $\mathrm{d}/\mathrm{d}\alpha = (1/d)(\mathrm{d}/\mathrm{d}\gamma)$ we obtain

$$\begin{aligned} U &= -\frac{1}{2}\hbar c A\pi^2\frac{\mathrm{d}^2}{\mathrm{d}\alpha^2}\left[\frac{1}{d}\int_1^{\infty}\frac{\mathrm{d}u}{u}\frac{\mathrm{d}}{\mathrm{d}\gamma}\left(\frac{1}{\mathrm{e}^{\gamma u}-1}\right)\right] \\ &= -\frac{1}{2}\hbar c A\pi^2\frac{\mathrm{d}^2}{\mathrm{d}\alpha^2}\left[-\frac{1}{d}\int_1^{\infty}\mathrm{d}u\,\frac{\mathrm{e}^{\gamma u}}{(\mathrm{e}^{\gamma u}-1)^2}\right].\end{aligned}$$

Exercise 5.1.

Now we substitute $x = e^{\gamma u}$ and arrive at

$$U = \frac{1}{2}\hbar c A \pi^2 \frac{d^2}{d\alpha^2}\left[\frac{1}{\alpha}\int_{e^\gamma}^{\infty}\frac{dx}{(x-1)^2}\right] .$$

With $z = x - 1$ we get

$$\begin{aligned} U &= \frac{1}{2}\hbar c A \pi^2 \frac{d^2}{d\alpha^2}\left[\frac{1}{\alpha}\int_{e^\gamma - 1}^{\infty}\frac{dz}{z^2}\right] \\ &= \frac{1}{2}\hbar c A \pi^2 \frac{d^2}{d\alpha^2}\left[\frac{1}{\alpha}\left(-\frac{1}{z}\right)_{e^\gamma - 1}^{\infty}\right] \\ &= \frac{1}{2}\hbar c A \pi^2 \frac{d^2}{d\alpha^2}\left[\frac{1}{\alpha}\frac{1}{e^{\alpha/d} - 1}\right] \\ &= \frac{\hbar c \pi^2 A}{2d}\frac{d^2}{d\alpha^2}\left[\left(\frac{d}{\alpha}\right)^2 \frac{\alpha/d}{e^{\alpha/d} - 1}\right] . \end{aligned}$$

The Taylor series expansion of $y/(e^y - 1)$ reads:

$$\frac{y}{e^y - 1} = \sum_{n=0}^{\infty}\frac{B_n}{n!}y^n ,$$

where B_n represent the Bernoulli numbers, which are given by $B_0 = 1$, $B_1 = -1/2$, $B_2 = 1/6$, $B_3 = 0$, $B_4 = -1/30$, ...[8] Hence

$$\begin{aligned} U &= \frac{\hbar c \pi^2 A}{2d}\frac{d^2}{d\alpha^2}\left[\left(\frac{d}{\alpha}\right)^2 \sum_{n=0}^{\infty}\frac{B_n}{n!}\left(\frac{\alpha}{d}\right)^n\right] \\ &= \frac{\pi^2}{2d}\hbar c A \frac{d^2}{d\alpha^2}\left[\sum_{n=0}^{\infty}\frac{B_n}{n!}\left(\frac{\alpha}{d}\right)^{n-2}\right] . \qquad (1) \end{aligned}$$

The actual difference between the potential energies of the two configurations of plates (see Fig. 5.1) is given by

$$U(d, R, A) = (E_{I} + E_{II}) - (E_{III} + E_{IV}) . \qquad (2)$$

We insert result (1) into (2) and let the distance between plates go to infinity, $R \to \infty$, as well as $\alpha \to 0$:

$$\begin{aligned} U(d, A) = \lim_{R\to\infty}\lim_{\alpha\to 0}\frac{\pi^2}{2}\hbar c A \frac{d^2}{d\alpha^2}&\left\{\left[\left(\frac{1}{d}\sum_{n=0}^{\infty}\frac{B_n}{n!}\left(\frac{\alpha}{d}\right)^{n-2}\right) + (d \to R - d)\right]\right. \\ &\left. - \left[\left(d \to \frac{R}{\eta}\right) + \left(d \to R - \frac{R}{\eta}\right)\right]\right\} . \end{aligned}$$

Here $(d \to R - d)$ means that d inside the first bracket has been replaced by $(R - d)$ and so forth.

[8] See, for example: G. Arfken, H.S. Weber: *Mathematical Methods for Physicists* (Academic Press, New York 1995).

Exercise 5.1.

We consider the first terms of the expansion:

$$U(d,A) = \lim_{R\to\infty}\lim_{\alpha\to 0}\frac{\pi^2}{2}\hbar cA\frac{\mathrm{d}^2}{\mathrm{d}\alpha^2}\left\{\left[\left(\frac{d}{\alpha^2}-\frac{1}{2\alpha}+\frac{1}{12d}-\frac{\alpha^2}{720d^3}+\dots\right)\right.\right.$$
$$\left.\left.+\left(d\to R-d\right)\right]-\left[\left(d\to\frac{R}{\eta}\right)+\left(d\to R-\frac{R}{\eta}\right)\right]\right\}.$$

As we sum over all four contributions, the terms with α^{-2} and α^{-1} cancel. The next term vanishes once differentiated with respect to α. Therefore we are left with

$$U(d,A) = \lim_{R\to\infty}\lim_{\alpha\to 0}\frac{\pi^2}{2}\hbar cA\frac{\mathrm{d}^2}{\mathrm{d}\alpha^2}\left[-\frac{\alpha^2}{720}\left(\frac{1}{(R-d)^3}\right.\right.$$
$$\left.\left.+\frac{1}{d^3}-\frac{1}{(R/\eta)^3}-\frac{1}{(R-R/\eta)^3}\right)+\dots\right]$$
$$= \lim_{R\to\infty}\lim_{\alpha\to 0}\left[-\frac{\pi^2}{720}\hbar cA\left(\frac{1}{d^3}+\frac{1}{(R-d)^3}-\frac{1}{(R/\eta)^3}-\frac{1}{(R-R/\eta)^3}\right)\right.$$
$$\left.+\text{ terms, which contain powers of }\alpha\right]$$

and get

$$U(d,A) = -\frac{\pi^2}{720}\frac{\hbar cA}{d^3}.$$

Hence, the potential stays finite and does not depend on the cut-off parameter λ nor on the coupling constant α; it depends only on the dimensions of the plates. The resulting force

$$F = -\frac{\partial}{\partial d}U(d,A) = -\frac{\pi^2}{240}\hbar c\frac{A}{d^4}$$

is attractive; both plates attract each other!

EXAMPLE

5.2 Measurement of the Casimir Effect

The first measurements of the Casimir effect were performed by Abrikosova[9] and collaborators as well as by Sparnaay.[10] Here we briefly describe the method applied by Sparnaay (see Fig. 5.2).

The attraction prevailing between the two metal plates at very short distances (d) causes a rotation of the weigh beam (b). Then the capacity of the condenser (c) is changed. Since the scale is supported at the point e, the distance d between plates of the measurement condenser decreases, i.e. the capacity increases. The measured capacity is a measure for the expansion of the spring and thus for the force to be measured. With this method it is possible to achieve a precision of $\approx 10^{-3}\,\text{dyn/cm}^2$. The procedure is as follows. A

[9] I.I. Abrikosova, B.V. Deriagin: Sov. Phys. JETP **3** (1957) 819; **4** (1957) 2.
[10] M.J. Sparnaay: Physica **24** (1958) 751.

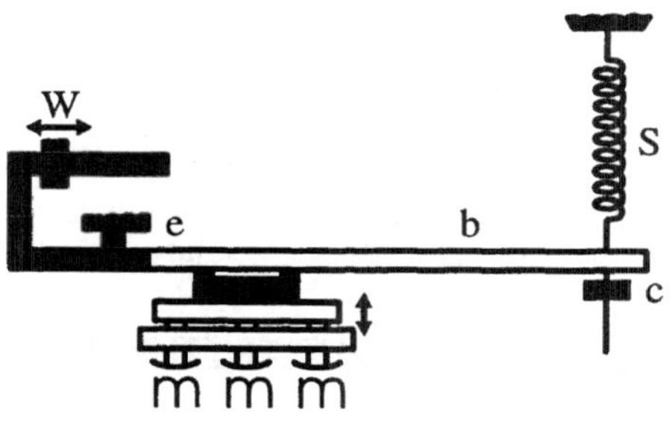

Fig. 5.2. Sparnaay's apparatus for the measurement of the Casimir effect

distance (d) between plates is chosen which is large enough so that the force acting between the plates is smaller than the measurement resolution. With the help of weight W and spring S the scale is brought into equilibrium. Now the distance between the plates is changed with the micrometer screws (m) and the corresponding force is measured via the condenser (c).

In this way the dependence of the force on the distance d is obtained which is depicted in Fig. 5.3. The width of the measured points goes back to the micrometer screws. In the region $F \ll 0.01\,\mathrm{dyn/cm}^2$ the quality of the measurement of the force becomes uncertain. The broken line represents the theoretical prediction as derived above. We see that the agreement between theory and experiment is remarkably good within the experimental uncertainty. The experimental results obtained by Abrikosova show the same agreement with the theoretical predictions.

The measured dependence of the force on the distance excludes explanations (such as van der Waals forces) other than the Casimir effect. The

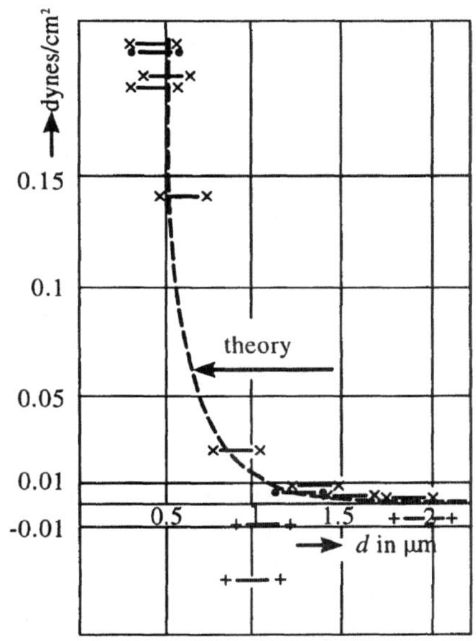

Fig. 5.3. Comparision between theoretical and experimental results for the Casimir effect

Example 5.2.

experiment described here has been presented in an oversimplified way. In real life, precision work is necessary: The plates have to be entirely free of dust, which is achieved only by very complicated procedures. Furthermore, all electrostatic charge must be removed since it would lead to additional forces. In addition, rigorous requirements have to be demanded for the plates to be really flat. These are only some points that have to be taken care of.[11]

EXAMPLE

5.3 Casimir's Approach Towards a Model for the Electron

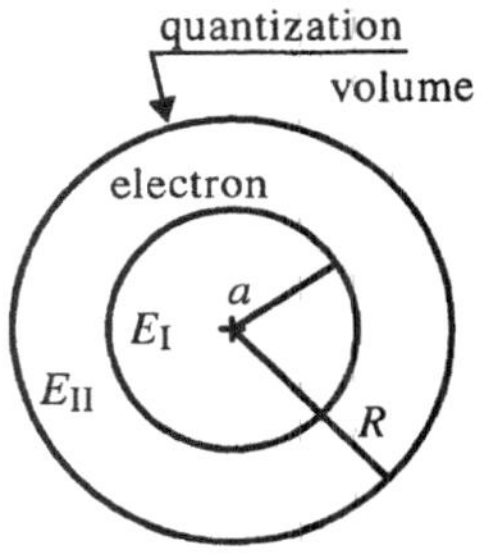

Fig. 5.4. Spherical shell model of the electron

The outcome of the zero-point energy for the plate condenser inspired Casimir[12] to develop a model for the electron. He assumed that the electron can be viewed as a spherical shell with homogenous surface charge e (see Fig. 5.4). The electrostatic energy of a spherical shell with radius a is given by

$$E_{\mathrm{e}} = \frac{e^2}{2a} .$$

The corresponding pressure is $P_{\mathrm{e}} = -\partial E/\partial V$. With $V = 4\pi a^3/3$ we obtain

$$P_{\mathrm{e}} = -\frac{1}{4\pi a^2}\frac{\partial E_{\mathrm{e}}}{\partial a} = \frac{e^2}{8\pi a^4} ,$$

i.e. the electrostatic pressure puffs up the spherical shell.

The presence of the spherical shell alters the normal modes of the electromagnetic field. The typical wave-number is $k_n \sim 1/a$, so that

$$E_{\mathrm{N}} = \sum_n \frac{1}{2}\hbar\omega_n = \frac{1}{2}\hbar c \sum_n k_n = -C\left(\frac{\hbar c}{2a}\right)$$

with constant C. The corresponding zero-point pressure reads

$$P_{\mathrm{N}} = -\frac{1}{4\pi a^2}\frac{\partial E_{\mathrm{N}}}{\partial a} = -C\frac{\hbar c}{8\pi a^4} .$$

The electron remains stable once the electrostatic pressure and the zero-point pressure cancel each other:

$$P_{\mathrm{e}} + P_{\mathrm{N}} = 0 = \frac{e^2}{8\pi a^4} - C\frac{\hbar c}{8\pi a^4} ,$$

i.e. for the case that the constant,

$$C = \frac{e^2}{\hbar c} = \alpha ,$$

is equal to Sommerfeld's fine-structure constant. Since C follows from the geometrical boundary conditions for the electromagnetic field inside the spherical

[11] G. Plunien, B. Müller, W. Greiner: Phys. Rep. **134** (1986) 87.
[12] H.B.G. Casimir: Physica **19** (1953) 846.

Example 5.3.

shell, it would be a first-time opportunity to calculate an elementary constant, i.e.

$$\alpha = \frac{e^2}{\hbar c} .$$

The simple result for the plate condenser seems to be encouraging: if we use the spherical shell as a crude model and set $A = \pi a^2$ as well as $d = a$ it follows

$$E_{\mathrm{N}} = -\frac{\pi^2 \hbar c A}{720 d^3} = -\frac{\pi^3 \hbar c}{720 a} = -\frac{\pi^3}{360}\left(\frac{\hbar c}{2a}\right) ;$$

hence, $C = \pi^3/360 \approx 0.036$ whereas $\alpha \approx 1/137 \approx 0.0073$.

However, the correct calculation for the spherical shell[13] leads to the failure of this model: The value for C turns out to be $C = -0.093$ and has the wrong sign. Thus, the resulting pressure is also positive and, like the electromagnetic pressure, puffs up the spherical shell!

5.2 Renormalization of the Electron Mass

Now we examine in more detail the self-energy of the electron, which we first encountered in Chap. 2 and which turned out to be an infinite quantity. The problem of self-energy of a charged particle already exists in classical electrodynamics. For example, a small conducting sphere of radius R and with charge e possesses an electric field

$$\boldsymbol{E} = -\frac{e}{r^2}\frac{\boldsymbol{r}}{r} , \qquad r \geq R ;$$

see Fig. 5.5. For the limit of a point charge ($R \to 0$) the electric field energy of this sphere

$$\begin{aligned} W &= \frac{1}{8\pi}\int_{r\geq R} \boldsymbol{E}^2\, \mathrm{d}^3x = \frac{1}{8\pi}\, 4\pi e^2 \int_R^\infty \frac{1}{r^4}\, r^2\, \mathrm{d}r \\ &= \frac{e^2}{2}\int_R^\infty \frac{\mathrm{d}r}{r^2} = \frac{e^2}{2R} , \end{aligned} \tag{5.10}$$

becomes infinite.

This is the longitudinal (Coulomb) energy of the field. In quantum field theory an additional energy exists as a result of the transverse electromagnetic field. In the following, we will investigate this quantity within the framework of nonrelativistic quantum electrodynamics, which we have pursued so far. In our lectures on quantum electrodynamics[14] we will become acquainted with the relativistic theory, which is more exact and correct. For our purposes

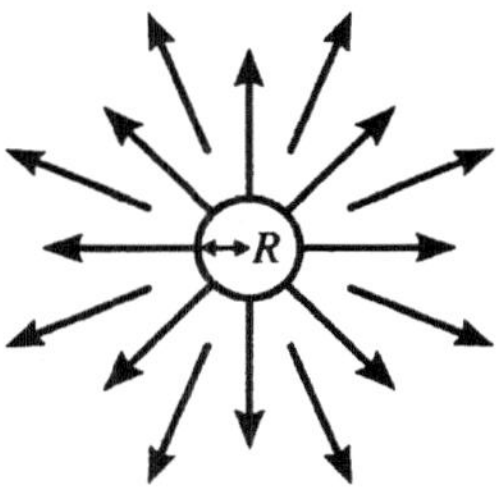

Fig. 5.5. Electric field of a charged, conducting sphere

[13] T.H. Boyer: Phys. Rev. **174** (1968) 1764.

[14] W. Greiner, J. Reinhardt: *Quantum Electrodynamics*, 2nd ed. (Springer, Berlin, Heidelberg 1994).

here it is sufficient to consider the nonrelativistic interaction operator of the electron with the transverse radiation field:

$$\begin{aligned}\hat{H}_{\text{int}} &= -\frac{e}{mc}\hat{\boldsymbol{p}}\cdot\hat{\boldsymbol{A}} + \frac{e^2}{2mc^2}\hat{\boldsymbol{A}}^2 \\ &= \hat{H}'_{\text{int}} + \hat{H}''_{\text{int}}\,;\end{aligned} \tag{5.11}$$

it already allows us to present the essential line of thought in a simple and clear fashion. As usual we treat $\hat{H}_{\text{int}}$ as a perturbation.

$$|\boldsymbol{q}\rangle \tag{5.12}$$

stands for a free electron state with momentum $\boldsymbol{p} = \hbar\boldsymbol{q}$. Its wavefunction is given by

$$\psi_{\boldsymbol{q}}(\boldsymbol{x}) = \langle\boldsymbol{x}|\boldsymbol{q}\rangle = \frac{1}{\sqrt{L^3}}\,\mathrm{e}^{\mathrm{i}\boldsymbol{q}\cdot\boldsymbol{x}}\,. \tag{5.13}$$

We designate the state with one free electron of momentum $\boldsymbol{p}$ and no photon as

$$|\boldsymbol{p}\rangle \equiv |\boldsymbol{q}\rangle|0\rangle_{\text{rad}}\,. \tag{5.14}$$

Then according to first-order perturbation calculation for the free electron we calculate the contribution

$$\begin{aligned}E^{(1)}_{\boldsymbol{p}} = \langle\boldsymbol{p}|\hat{H}_{\text{int}}|\boldsymbol{p}\rangle &= \langle\boldsymbol{p}|\hat{H}''_{\text{int}}|\boldsymbol{p}\rangle \\ &= \frac{e^2}{2mc^2}\langle\boldsymbol{p}|\hat{\boldsymbol{A}}^2|\boldsymbol{p}\rangle\,.\end{aligned} \tag{5.15}$$

EXAMPLE

5.4 Supplement: Historical Remark on the Electron Mass

According to classical electrodynamics the electron with its charge $(-e)$ produces an electrostatic field where the energy is given by

$$W = \alpha\frac{e^2}{R}\,. \tag{1}$$

The coefficient α depends on how the charge $(-e)$ is distributed over the small sphere with radius R. The order of magnitude for α is 1: for example, $\alpha = 1/2$ for a charge distribution confined to the spherical surface (compare with result (5.10) for a conducting sphere) or $\alpha = 3/5$ for a uniform charge distribution over the whole sphere. The eigenmass of the electron consists of two parts:

1) the *mechanical mass* m', which is not coupled to the field energy and
2) the *electromagnetic mass* m_{el}, which is coupled to the field energy.

Hence, we have

$$m = m' + m_{\text{el}}\,. \tag{2}$$

Example 5.4.

According to (5.10) the mass coupled to the field is determined by

$$m_{\rm el} = \frac{W}{c^2} = \alpha \frac{e^2}{Rc^2} \,. \qquad (3)$$

Strictly speaking, classical electrodynamics provides the following expression for the mass coupled to the electrostatic field:

$$m_{\rm el} = \frac{4}{3}\frac{W}{c^2} \,. \qquad (4)$$

The factor 4/3 instead of 1 results from the instability of the charge distribution as identical charge elements inside the sphere repell each other. Thus, additional forces of nonelectromagnetic origin have to be introduced in order to produce equilibrium. If we consider nonlinear electrodynamics these difficulties do not occur because they lead to stable charge distributions inside the electron. For this case the correct relation (3) between the field-coupled mass and the field energy is obtained. Lorentz succeeded in showing that the electromagnetic mass depends on the velocity v of the charge according to the relation

$$m_{\rm el} = \frac{(m_0)_{\rm el}}{\sqrt{1 - v^2/c^2}} \,. \qquad (5)$$

This relation is easy to understand because Maxwell's equations have to be Lorentz covariant. Hence, the electrostatic energy has to transform like the fourth component of a four-vector, from which (5) follows (consult the lectures on classical mechanics and classical electrodynamics of this series).[15]

According to Lorentz the mechanical mass m' should be a constant; he assumed the validity of Newtonian mechanics. However, experimental experience showed that the total electron mass changes according to the law (5). It appeared that the total electron mass possesses field character. Nowadays a value of $R \ll 10^{-16}$ cm for the electron radius, deduced from scattering experiments, is used. Lorentz used a value $R \approx 10^{-13}$ cm and calculated with (3) a reasonable mass of $m \approx 10^{-27} g = m_{\rm electron}$. Then came the theory of relativity and required that the mechanical mass m' had to obey the law (5), too.[15–16] The apparent consistency of these considerations achieved so far broke down and the nature of the electron mass again faded away into darkness.

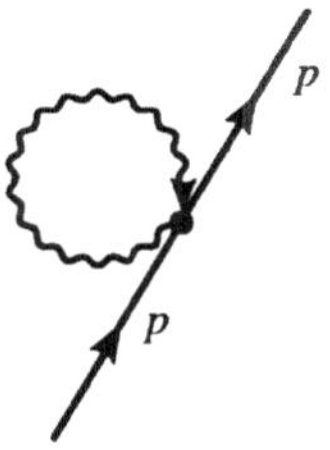

Fig. 5.6. Emission and absorption processes as contained in the term proportional to $\hat{\boldsymbol{A}}^2$ in (5.11)

In first order the term $\hat{\boldsymbol{p}}\cdot\hat{\boldsymbol{A}}$ from (5.11) does not contribute because it does not contain terms of the form $\hat{a}^\dagger\hat{a}$ or $\hat{a}\hat{a}^\dagger$, which describe the emission and reabsorption of photons. On the other hand, the term $\hat{\boldsymbol{A}}^2$ describes such simultaneous creation and annihilation processes, as shown graphically in Fig. 5.6.

With the help of (2.9) and (5.14) we deduce immediately that

---

[15] W. Greiner: *Classical Mechanics I* (Springer, New York) in preparation; W. Greiner: *Classical Electrodynamics* (Springer, New York 1996).

[16] W. Greiner, J. Rafelski: *Special Relativity* (Springer, New York) in preparation.

$$\begin{aligned}
E_p^{(1)} &= \frac{e^2}{2mc^2}\sum_{k\sigma}\sum_{k'\sigma'}\left(\frac{2\pi\hbar c^2}{L^3}\right)\frac{\epsilon_{k\sigma}\cdot\epsilon_{k'\sigma'}}{\sqrt{\omega_k\omega_{k'}}} \\
&\quad\times\underbrace{\int\frac{1}{\sqrt{L^3}}\,\mathrm{e}^{-\mathrm{i}q\cdot x}\,\mathrm{e}^{\mathrm{i}(k-k')\cdot x}\frac{1}{\sqrt{L^3}}\,\mathrm{e}^{\mathrm{i}q\cdot x}\,\mathrm{d}^3x}_{=\delta_{k,k'}}\,\delta_{\sigma\sigma'} \\
&= \frac{e^2}{2mc^2}\sum_{k\sigma}\left(\frac{2\pi\hbar c^2}{L^3}\right)\frac{\epsilon_{k\sigma}\cdot\epsilon_{k\sigma}}{\omega_k} \\
&= \frac{e^2}{2mc^2}\frac{L^3}{(2\pi)^3}2\int\frac{k^2\,\mathrm{d}k\,\mathrm{d}\Omega_k}{\omega_k}\left(\frac{2\pi\hbar c^2}{L^3}\right) \\
&= \frac{e^2\hbar}{mc}\frac{1}{2}\frac{1}{(2\pi)^2}(4\pi)^2\int_0^\infty k\,\mathrm{d}k \\
&= \frac{e^2\hbar}{\pi mc}\int_0^\infty k\,\mathrm{d}k \\
&= \infty\,.
\end{aligned} \tag{5.16}$$

Obviously this contribution to the energy of the free electron is infinite but independent of the electron momentum p. *It is equal for all electrons and, as soon as energy differences are taken into account, it drops out.* Therefore we will not consider this contribution further.

Next we turn to the second-order contribution of the term $\hat{p}\cdot\hat{A}$ appearing in the interaction (5.11), i.e. $\hat{H}'_{\mathrm{int}}$. This contribution will be much more interesting; in second order it yields the contribution

$$\begin{aligned}
E_p^{(2)} &= \sum_I\frac{\langle p|\hat{H}'_{\mathrm{int}}|I\rangle\,\langle I|\hat{H}'_{\mathrm{int}}|p\rangle}{E_p-E_I} \\
&= \frac{e^2}{m^2c^2}\sum_I\frac{|\langle I|\hat{p}\cdot\hat{A}|p\rangle|^2}{E_p-E_I}
\end{aligned} \tag{5.17}$$

to the energy. Since $\hat{A}$ is linear in the photon creation and annihilation operators $\hat{a}$, $\hat{a}^\dagger$ the intermediate states $|I\rangle$ have to contain one photon. Hence, they are of the form

$$\begin{aligned}
|I\rangle &= \hat{b}_q^\dagger|0\rangle_{\mathrm{P}}\hat{a}_{k\sigma}^\dagger|0\rangle_{\mathrm{rad}}\,, \\
|p\rangle &= \hat{b}_p^\dagger|0\rangle_{\mathrm{P}}\,,
\end{aligned} \tag{5.18}$$

and

$$\begin{aligned}
E_p &= \frac{(\hbar p)^2}{2m}\,, \\
E_I &= \frac{(\hbar q)^2}{2m}+\hbar\omega_k
\end{aligned} \tag{5.19}$$

denote the corresponding energies. With (2.8) we obtain immediately

$$\langle I|\hat{\boldsymbol{p}}\cdot\hat{\boldsymbol{A}}|\boldsymbol{p}\rangle = -\sqrt{\frac{2\pi\hbar c^2}{L^3\omega_k}}(\hbar\boldsymbol{p})\cdot\boldsymbol{\epsilon}_{\boldsymbol{k}\sigma}\underbrace{\int \mathrm{d}^3x\,\frac{\mathrm{e}^{-\mathrm{i}\boldsymbol{q}\cdot\boldsymbol{x}}}{\sqrt{L^3}}\,\mathrm{e}^{-\mathrm{i}\boldsymbol{k}\cdot\boldsymbol{x}}\,\frac{\mathrm{e}^{\mathrm{i}\boldsymbol{p}\cdot\boldsymbol{x}}}{\sqrt{L^3}}}_{=\delta_{\boldsymbol{q}+\boldsymbol{k},\boldsymbol{p}}}$$

$$= -\sqrt{\frac{2\pi\hbar c^2}{L^3\omega_k}}(\hbar\boldsymbol{p})\cdot\boldsymbol{\epsilon}_{\boldsymbol{k}\sigma}\delta_{\boldsymbol{q},\boldsymbol{p}-\boldsymbol{k}}\,. \tag{5.20}$$

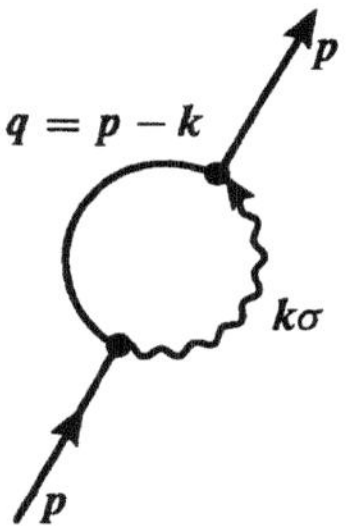

Fig. 5.7. Diagram of the self-energy. An incoming electron of momentum $\boldsymbol{p}$ scatters into the intermediate state $\boldsymbol{q}$ by emitting a photon $\boldsymbol{k}\sigma$ and then it scatters back into the old state $\boldsymbol{p}$ by reabsorbing the photon

The Kronecker delta administers momentum conservation; hence we derive for (5.17) the expression

$$E_{\boldsymbol{p}}^{(2)} = \frac{e^2}{m^2c^2}\sum_{\boldsymbol{k},\sigma}\left(\frac{2\pi\hbar c^2}{L^3\omega_k}\right)\frac{|\hbar\boldsymbol{p}\cdot\boldsymbol{\epsilon}_{\boldsymbol{k}\sigma}|^2}{\frac{\hbar^2p^2}{2m}-\left(\frac{\hbar^2(\boldsymbol{p}-\boldsymbol{k})^2}{2m}+\hbar kc\right)}\,. \tag{5.21}$$

This is exactly the real part $\Re(\Delta E_\mathrm{i})$ of the self-energy obtained earlier in (2.105a). The processes involved are exemplified in the diagram in Fig. 5.7.

We continue with (5.21) and use

$$\begin{aligned}\sum_\sigma|\boldsymbol{p}\cdot\boldsymbol{\epsilon}_{\boldsymbol{k}\sigma}|^2 &= \sum_\sigma(\boldsymbol{p}\cdot\boldsymbol{\epsilon}_{\boldsymbol{k}\sigma})(\boldsymbol{\epsilon}_{\boldsymbol{k}\sigma}\cdot\boldsymbol{p})\\ &= \boldsymbol{p}\cdot\left(\sum_\sigma\boldsymbol{\epsilon}_{\boldsymbol{k}\sigma}\boldsymbol{\epsilon}_{\boldsymbol{k}\sigma}\right)\cdot\boldsymbol{p}\\ &= \boldsymbol{p}\cdot\left(\hat{I}-\frac{\boldsymbol{k}\cdot\boldsymbol{k}}{k^2}\right)\cdot\boldsymbol{p}\\ &= p^2(1-\cos^2\vartheta)\,.\end{aligned} \tag{5.22}$$

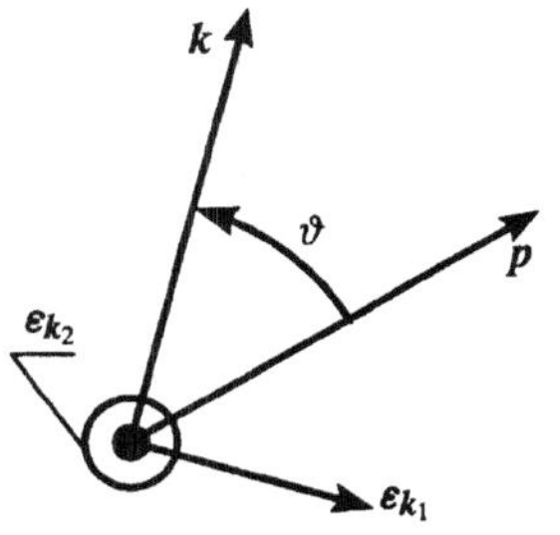

Fig. 5.8. The geometry of the emission process of a virtual photon: $\boldsymbol{p}$ represents the electron momentum and $\boldsymbol{k}$ represents the photon momentum

Here ϑ represents the angle between the photon and electron (see Fig. 5.8). Starting with

$$\sum_\sigma\boldsymbol{\epsilon}_{\boldsymbol{k}\sigma}\boldsymbol{\epsilon}_{\boldsymbol{k}\sigma} = \boldsymbol{\epsilon}_{\boldsymbol{k}1}\boldsymbol{\epsilon}_{\boldsymbol{k}1}+\boldsymbol{\epsilon}_{\boldsymbol{k}2}\boldsymbol{\epsilon}_{\boldsymbol{k}2}\,,$$

we have introduced a third unit vector $\boldsymbol{k}/k$ to construct

$$\hat{I} = \boldsymbol{\epsilon}_{\boldsymbol{k}1}\boldsymbol{\epsilon}_{\boldsymbol{k}1}+\boldsymbol{\epsilon}_{\boldsymbol{k}2}\boldsymbol{\epsilon}_{\boldsymbol{k}2}+\frac{\boldsymbol{k}}{k}\,\frac{\boldsymbol{k}}{k}\,.$$

Hence, we obtain

$$\sum_\sigma\boldsymbol{\epsilon}_{\boldsymbol{k}\sigma}\boldsymbol{\epsilon}_{\boldsymbol{k}\sigma} = \hat{I}-\frac{\boldsymbol{k}\boldsymbol{k}}{k^2}\,, \tag{5.23}$$

which we have used in (5.22). Insertion of (5.22) into (5.21) leads to

$$\begin{aligned}E_{\boldsymbol{p}}^{(2)} &= \frac{e^2}{m^2c^2}\left(\frac{2\pi\hbar c^2}{L^3}\right)\frac{L^3}{(2\pi)^3}\\ &\quad\times\int\mathrm{d}^3k\,\frac{1}{\omega_k}\,\frac{\hbar^2p^2(1-\cos^2\vartheta)}{\frac{\hbar^2p^2}{2m}-\left(\frac{\hbar^2p^2}{2m}-\frac{\hbar^2\boldsymbol{p}\cdot\boldsymbol{k}}{m}+\frac{(\hbar k)^2}{2m}+\hbar kc\right)}\\ &= \frac{e^2}{m^2c^2}\,\frac{\hbar c^2}{(2\pi)^2}\int\frac{k^2\,\mathrm{d}k\,d(\cos\vartheta)\,\mathrm{d}\phi}{kc}\,\frac{(\hbar p)^2(1-\cos^2\vartheta)}{-\hbar kc\left(1-\frac{\hbar p}{mc}\cos\vartheta+\frac{\hbar kc}{2mc^2}\right)}\end{aligned}$$

$$= -\frac{e^2}{m^2c^2}\frac{(\hbar p)^2}{(2\pi)^2}2\pi\int_0^\infty \mathrm{d}k\int_{-1}^{+1}\mathrm{d}(\cos\vartheta)\frac{1-\cos^2\vartheta}{1-\frac{v}{c}\cos\vartheta+\frac{\hbar kc}{2mc^2}}$$
$$= \frac{(\hbar p)^2}{2m}\left(-\frac{e^2}{mc^2\pi}\int_{-1}^{+1}\mathrm{d}(\cos\vartheta)(1-\cos^2\vartheta)\int_0^\infty\frac{\mathrm{d}k}{1-\frac{v}{c}\cos\vartheta+\frac{\hbar kc}{2mc^2}}\right).$$

Obviously the last integral over k diverges logarithmically because for large k it behaves like $\int \mathrm{d}k/k$. With other words: The self-energy $E_p^{(2)}$ is divergent. If we combine this contribution with the energy of zeroth order $E_p^{(0)}$ the total energy of the free particle becomes

$$\begin{aligned} E_p &= E_p^{(0)} + E_p^{(2)} \\ &= \frac{(\hbar p)^2}{2m} \\ &\quad -\frac{(\hbar p)^2}{2m}\underbrace{\left(\frac{e^2}{mc^2\pi}\int_{-1}^{+1}\mathrm{d}(\cos\vartheta)(1-\cos^2\vartheta)\int_0^\infty\frac{\mathrm{d}k}{1-\frac{v}{c}\cos\vartheta+\frac{\hbar kc}{2mc^2}}\right)}_{=D} \\ &= \frac{(\hbar p)^2}{2m}(1-D)\,. \end{aligned} \tag{5.24}$$

D stands for the divergent part. If the extreme nonrelativistic approximation $v/c \ll 1$, $\hbar\omega/mc^2 \ll 1$ is applied to D, which, of course, cannot be fulfilled in the total integration domain, then we arrive at the result

$$\begin{aligned} D &= \frac{e^2}{mc^2\pi}\int_{-1}^{+1}\mathrm{d}(\cos\vartheta)(1-\cos^2\vartheta)\int_0^\infty \mathrm{d}k \\ &= \frac{e^2}{mc^2\pi}\left(\frac{4}{3}\right)\int_0^\infty \mathrm{d}k \\ &= \frac{4}{3\pi}\frac{e^2}{mc^2}\int_0^\infty \mathrm{d}k\,, \end{aligned}$$

which is often referred to in the literature but inconsistently. Within this approximation the expression for D diverges even linearly and (5.24) becomes

$$E_p = \frac{(\hbar p)^2}{2m}\left(1-\frac{4}{3\pi}\frac{e^2}{mc^2}\int_0^\infty \mathrm{d}k\right). \tag{5.25}$$

We will pursue the following point of view: The mass m in the energy of zeroth order,

$$E_p^{(0)} = \frac{(\hbar p)^2}{2m}, \tag{5.26}$$

represents the mass of the *naked* electron, which does not interact with the electromagnetic field. It can only be fictitious because the electromagnetic interaction cannot be switched off. Only under the presence of the electromagnetic field is the mass of the electron experimentally observable. Therefore

$$\frac{1}{m_{\mathrm{exp}}} = \frac{1}{m}(1-D) \tag{5.27}$$

has to hold (in lowest order). In other words: The unknown fictitious mass m of the electron has to be conditioned (divergent) such that

$$m_{\text{exp}} = \frac{m}{1-D}$$

always has a finite value, i.e. the value for the electron mass deduced from experiment. This procedure of interpreting the divergent self-energy as a change from the fictitious mass m of the electron to the *real* mass m_{exp} is known as the *renormalization of the mass.*

5.3 The Splitting of the Hydrogen States $2s_{1/2}$–$2p_{3/2}$: The Lamb Shift

According to both Schrödinger's and Dirac's theories the $2s_{1/2}$ and $2p_{1/2}$ states of hydrogenlike point nuclei are degenerate. This is true for pointlike central nuclei, i.e. in the case of hydrogen for pointlike protons. It has been one of the greatest achievements of atomic theory that the fine structure of the atomic spectra (splitting of the states with $j = l \pm s$) could be explained quantitatively thanks to the consideration of spin–orbit coupling (which is automatically contained in the Dirac theory). However, the precise measurement of the hydrogen $2s_{1/2} - 2p_{3/2}$ splitting led to some doubts with respect to theory; small deviations from the theoretically predicted splitting have been observed in experiment.

At first (1930–40) the accuracy of the measurements did not allow the exact determination of this shift. Only the development of microwave techniques made a precise investigation possible. In 1947 Lamb and Retherford[17] investigated the energetic position of the $2s_{1/2}$ state with the help of high-frequency spectroscopy and discovered a shift of

$$\Delta\nu \approx 1058\,\text{MHz (present value)}\,.$$

This is depicted schematically in Fig. 5.9.

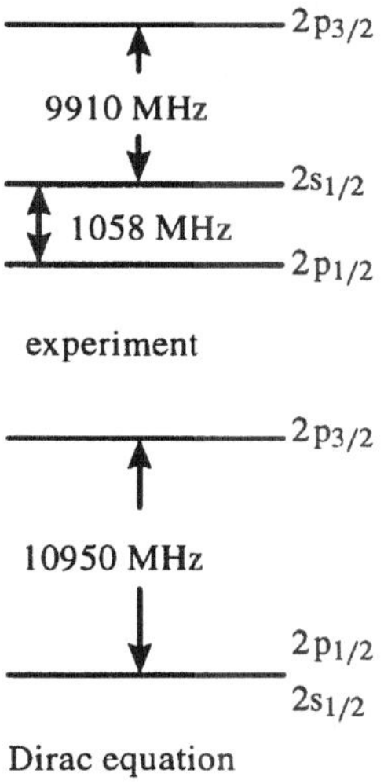

Fig. 5.9. Energy levels of the relevant hydrogen states

This so-called *Lamb shift* is a result of the interaction of the electron with its virtual radiation field. We have discussed the latter in the last section. Now we want to present the experiment, which has played a key role in the development of quantum electrodynamics, and then we concentrate on the corresponding theoretical calculations.

An atomic beam of hydrogen in the $1s_{1/2}$ ground state can be produced by dissociation of molecular hydrogen at high temperatures. An impinging electron current excites some of the atoms into the $(n = 2)$ state. Via optical transitions the two $2p_{1/2}$ and $2p_{3/2}$ levels soon decay into the 1s ground state. However, from the $2s_{1/2}$ level only the transition into the $2p_{1/2}$ level is possible, since the $l = 0$ transition into the $1s_{1/2}$ ground state is prohibited; the lowest electromagnetic multipole is the electric dipole with $l = 1$. Hence, the $2s_{1/2}$ state can be regarded as metastable. These metastable atoms are then collected in a metal target. Contrary to the case for an atomic beam with

[17] W.E. Lamb, Jr., R.C. Retherford: Phys. Rev. **72** (1947) 241.

atoms in the ground state, an intensive emission of electrons is observed from the metastable atoms; electrons can escape more easily from the metastable states because of the lower binding energy (Coulomb excitation). The intensity of the metastable beams can be altered as they have to run through a spatial region where electromagnetic radiation of the excitation frequency $2s_{1/2}$–$2p_{3/2}$ prevails. Moreover several of the magnetic substates can be studied with the help of the Zeemann effect. If a magnetic field is applied to the microwave region, the excitation into the $2p_{3/2}$ state can be studied with three different frequencies (see Fig. 5.10).

Lamb and Retherford's experiment is sketched in the next example.

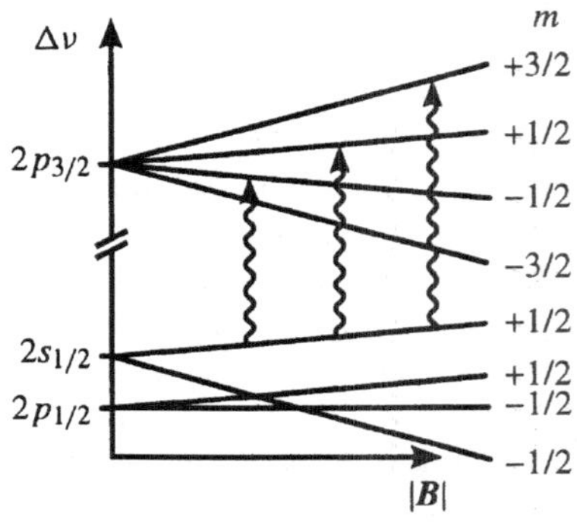

Fig. 5.10. Sketch of the splitting of the hydrogen energy levels $2p_{3/2}$, $2p_{1/2}$, $2s_{1/2}$ in a magnetic field and of the excitation frequencies of Lamb and Retherford

Since the experimental result of the hydrogen $2p_{3/2}$–$2s_{1/2}$ splitting deviates from the predictions of the Dirac equation by about 10%, this could have easily been used against the theory. However, in 1947 a strong belief in quantum electrodynamics already prevailed. The Lamb shift was thought to be a consequence of the interaction with the electromagnetic field. When the corresponding calculation for the shift was performed for the first time, it turned out to be infinite. Bethe[18] demonstrated that these difficulties can be removed with a renormalization of mass. We will follow Bethe's thinking and perform a nonrelativistic calculation.

EXAMPLE

5.5 Lamb and Retherford's Experiment

At first, H_2 molecules are dissociated into hydrogen in an oven. Then the H atoms are bombarded with electrons and are excited to transitions into states with $n = 2$. Mainly metastable excitations into the $2s_{1/2}$ states survive. Owing to the low binding energy of their electrons these excited hydrogen atoms are able to emit electrons intensively. The corresponding current is measured with a collector and an attached galvanometer.

[18] H.A. Bethe: Phys. Rev. **72** (1947) 339.

Example 5.5.

As the hydrogen atoms have to pass through a resonator which contains electromagnetic radiation with the transition frequency $2s_{1/2} \to 2p_{3/2}$ the current diminishes since the induced gamma emission causes the number of metastable atoms to be reduced. The experimental set-up is sketched in Fig. 5.11.

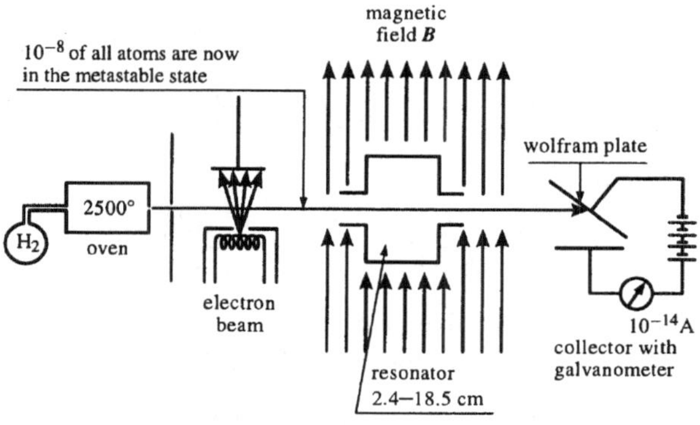

Fig. 5.11. Lamb and Retherford's experimental set-up

The splitting due to the Lamb shift is of the order of 10^{-6} eV, which lies in the microwave region. Technically it is more convenient to observe this energy difference directly with a high-frequency method instead of optically investigating the splitting of the H_α line. In particular, the latter procedure is difficult to perform because of the Doppler shift of the atoms moving with respect to the source; the resulting frequency shift,

$$\Delta\nu = 7 \times 10^{-7} \sqrt{T/A}\, \nu \,,$$

is proportional to ν (T in Kelvin; A atomic weight).

As soon as $2s \to 2p$ transitions occur in the field region, the current is reduced because metastable atoms have been removed.

The self-energy of a state $|n\rangle$ of the hydrogen atom is represented by the diagram shown in Fig. 5.12. This diagram is analogous to the one for a free electron, sketched in Fig. 5.7. The only difference is that now the electron states $|n\rangle$ are bound and not free. The correction to the energy is given by

$$E_n^{(2)} = \frac{e^2}{m^2c^2} \sum_{n'} \sum_{k\sigma} \left(\frac{2\pi\hbar c^2}{L^3 kc} \right) \frac{|\langle n'|\hat{\boldsymbol{p}} \cdot \boldsymbol{\epsilon}_{k\sigma}|n\rangle|^2}{E_n - E_{n'} - \hbar kc} \,, \tag{5.28}$$

in analogy to (5.17) and (5.21), where the dipole approximation

$$\boldsymbol{p} \cdot \boldsymbol{\epsilon}_{k\sigma}\, \mathrm{e}^{-\mathrm{i}\boldsymbol{k}\cdot\boldsymbol{x}} = \boldsymbol{p} \cdot \boldsymbol{\epsilon}_{k\sigma} \,,$$

has been used. Employing (5.23) and (2.18) we simplify this expression to

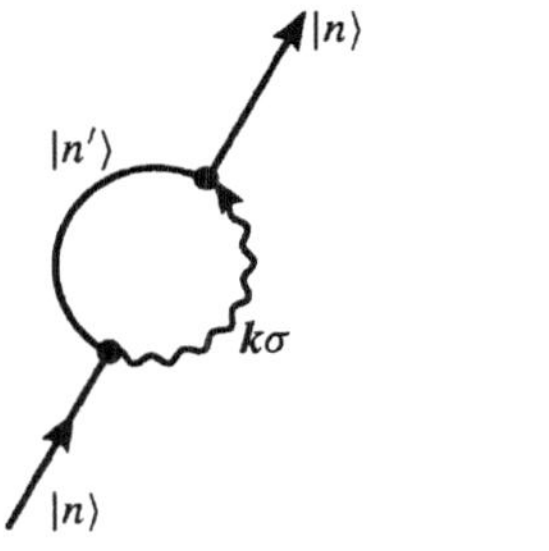

Fig. 5.12. Self-energy of a state $|n\rangle$ of the hydrogen atom

$$E_n^{(2)} = \frac{e^2}{m^2c^2}\sum_{n'}\left(\frac{2\pi\hbar c^2}{L^3 c}\right)\frac{L^3}{(2\pi)^3} \times \int_0^\infty \frac{k^2\,\mathrm{d}k}{k}\int \mathrm{d}\Omega \frac{|\langle n'|\hat{p}|n\rangle|^2}{E_n - E_{n'} - \hbar kc}(1-\cos^2\vartheta)\,. \tag{5.29}$$

Since

$$\int \mathrm{d}\Omega(1-\cos^2\vartheta) = -2\pi\int_1^{-1}\mathrm{d}(\cos\vartheta)(1-\cos^2\vartheta) = -2\pi\left(x-\frac{x^3}{3}\right)\Bigg|_1^{-1} = \frac{8\pi}{3}\,,$$

it follows that

$$E_n^{(2)} = \frac{2}{3\pi}\frac{e^2\hbar}{m^2c}\sum_{n'}\int_0^\infty k\,\mathrm{d}k\frac{|\langle n'|\hat{p}|n\rangle|^2}{E_n - E_{n'} - \hbar kc}\,. \tag{5.30}$$

This energy is as divergent as the self-energy for a free particle obtained previously in (5.25). At this point Bethe argued as follows: For a free electron the matrix elements $\langle n'|\hat{p}|n\rangle$ yield only diagonal contributions and (5.30) becomes the formerly derived second term of (5.25)

$$E_p^{(2)} = -\frac{4}{3\pi}\frac{e^2}{mc^2}\left(\frac{(\hbar p)^2}{2m}\right)\int_0^\infty \mathrm{d}k\,. \tag{5.31}$$

However, according to the mass-renormalization scheme this contribution to the kinetic energy of a free electron is interpreted as a contribution of the electromagnetic mass to the total mass m_{exp} of the electron. For a bound electron, which has no sharp momentum, a term analogous to (5.31) is obtained once the square of the momentum $(\hbar p)^2$ is replaced by the expectation value of $\langle n|\hat{\boldsymbol{p}}^2|n\rangle$. With the completeness relation

$$\langle n|\hat{\boldsymbol{p}}^2|n\rangle = \sum_{n'}|\langle n'|\hat{\boldsymbol{p}}|n\rangle|^2\,, \tag{5.32}$$

(5.31) becomes

$$E_{n(\mathrm{mass})}^{(2)} = -\frac{4}{3\pi}\frac{e^2}{mc^2}\frac{1}{2m}\sum_{n'}|\langle n'|\hat{\boldsymbol{p}}|n\rangle|^2\int_0^\infty \mathrm{d}k = -\frac{2}{3\pi}\frac{e^2\hbar}{m^2c}\sum_{n'}\int_0^\infty k\,\mathrm{d}k\frac{|\langle n'|\hat{\boldsymbol{p}}|n\rangle|^2}{\hbar kc}\,. \tag{5.33}$$

$E_{n(\mathrm{mass})}^{(2)}$ represents the correction to the kinetic energy according to the change of the electromagnetic electron mass in the state $|n\rangle$. The energy $E_{n(\mathrm{mass})}^{(2)}$ has to be subtracted from (5.30) in order to obtain the real and observable energy shift $\Delta E_n^{(2)}$:

$$\Delta E_n^{(2)} = \frac{2}{3\pi}\frac{e^2\hbar}{m^2c}\sum_{n'}\int_0^\infty k\,\mathrm{d}k|\langle n'|\hat{\boldsymbol{p}}\,|n\rangle|^2 \times\left[\frac{1}{E_n - E_{n'} - \hbar kc} + \frac{1}{\hbar kc}\right]\,. \tag{5.34}$$

Because of

$$\frac{1}{E_n - E_{n'} - \hbar kc} + \frac{1}{\hbar kc} = \frac{E_n - E_{n'}}{\hbar kc(E_n - E_{n'} - \hbar kc)} \xrightarrow{k\to\infty} -\frac{E_n - E_{n'}}{\hbar^2 k^2 c^2}$$

(5.34) is logarithmically divergent. Here Bethe assumed that the integral analogous to (5.34) has to converge in a relativistic theory. This turned out to be correct for the following reason. In a relativistic theory, states exist with positive and negative energy, which, according to the equation

$$E = \pm\sqrt{(m_0c^2)^2 + (pc)^2}\,, \tag{5.35}$$

have to lie either above $+m_0c^2$ or beneath $-m_0c^2$. According to Dirac the lower states have to be occupied with electrons; otherwise an electron would decay into even lower states and emit radiation. Thus lower states occupied with electrons prevent a *"radiation catastrophe"* because the Pauli principle prohibits such transitions.

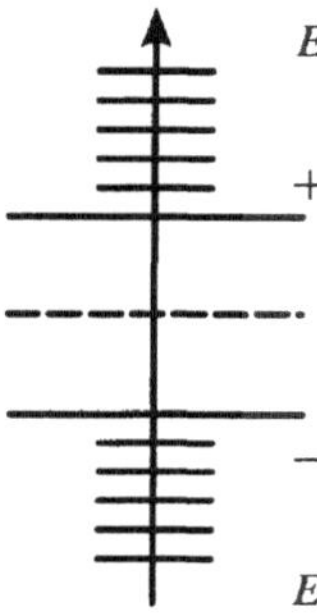

Fig. 5.13. Typical energy spectrum of a relativistic theory

If we attempt to construct a wave packet out of relativistic wavefunctions that describes an electron with positive energy, states of negative energy do not contribute as they are already occupied according to the Pauli principle. *Together* the states with positive and negative energy are complete; but no subset is complete! In other words: With states of positive energy alone we cannot construct a $\delta^3(\boldsymbol{r})$ function. Hence, a wave packet consisting of wavefunctions with positive energy cannot be compressed to arbitrary small volumes; this can only be achieved up to the Compton wavelength of the electron,

$$\lambda_{\text{Compton}} = \frac{h}{mc}\,.$$

In a way, the electron has acquired a spatial dimension; it is no longer localized as a point. For this reason a relativistic theory automatically contains this necessary restriction of the integration over k [compare with (5.34)]. This leads to the convergence of the correponding integral. For a nonrelativistic theory this convergence can be achieved by introducing a cut-off of the *integral* (5.34) *at the photon energy*

$$\hbar kc \approx mc^2 \Longrightarrow k_{\text{Compton}} = \frac{mc}{\hbar} = \frac{1}{\lambda_{\text{Compton}}}\,.$$

This yields

$$\begin{aligned}
\Delta E_n^{(2)} &= \frac{2}{3\pi}\frac{e^2\hbar}{m^2c}\sum_{n'}|\langle n'|\hat{\boldsymbol{p}}|n\rangle|^2 \int_0^{k_{\text{Compton}}} \frac{(E_n - E_{n'})k\,\mathrm{d}k}{\hbar kc(E_n - E_{n'} - \hbar kc)} \\
&= \frac{2}{3\pi}\frac{e^2}{m^2c^2}\sum_{n'}|\langle n'|\hat{\boldsymbol{p}}|n\rangle|^2 (E_n - E_{n'}) \int_0^{\infty} \frac{\mathrm{d}k}{E_n - E_{n'} - \hbar kc} \\
&= \frac{2}{3\pi}\frac{e^2}{m^2c^2}\sum_{n'}|\langle n'|\hat{\boldsymbol{p}}|n\rangle|^2 (E_n - E_{n'}) \\
&\quad\times \left[-\frac{1}{\hbar c}\ln(E_n - E_{n'} - \hbar kc)\right]_0^{k_{\text{Compton}}}
\end{aligned}$$

$$
\begin{aligned}
&= \frac{2}{3\pi}\frac{e^2}{\hbar m^2c^3}\sum_{n'}|\langle n'|\hat{\boldsymbol{p}}|n\rangle|^2(E_n-E_{n'})\ln\frac{E_n-E_{n'}-mc^2}{E_n-E_{n'}} \\
&\approx \frac{2}{3\pi}\frac{e^2}{\hbar m^2c^3}\sum_{n'}|\langle n'|\hat{\boldsymbol{p}}|n\rangle|^2(E_n-E_{n'})\ln\left|\frac{mc^2}{E_{n'}-E_n}\right| . \qquad (5.36)
\end{aligned}
$$

In the last step $(E_{n'}-E_n)$ has been neglected with respect to mc^2 in the nominator of the logarithm. In order to simplify the sum over n' even further, $(E_{n'}-E_n)$ is replaced within the argument by a suitable mean value

$$
(E_{n'}-E_n) \longrightarrow (E_{n'}-E_n)_{\text{av.}} \qquad (\text{av.} = \text{average}) . \qquad (5.37)
$$

This approximation is not as crude as it appears. The argument of the logarithm

$$
\frac{mc^2}{E_{n'}-E_n} \approx 10^5
$$

is a large number and the logarithm function is a slowly varying function. Even a deviation in $(E_{n'}-E_n)_{\text{av.}}$ by about a factor of 10 results in only a 20% incorrect value for the expression (5.36). With the approximation (5.37) the logarithm in (5.36) can be written in front of the summation; the remaining summation can be calculated as follows:

$$
\begin{aligned}
\sum_{n'}|\langle n'|\hat{\boldsymbol{p}}|n\rangle|^2(E_{n'}-E_n) &= \sum_{n'}\langle n|\hat{\boldsymbol{p}}|n'\rangle\cdot\langle n'|\hat{\boldsymbol{p}}|n\rangle(E_{n'}-E_n) \\
&= \sum_{n'}\langle n|\hat{\boldsymbol{p}}(\hat{H}_0-E_n)|n'\rangle\cdot\langle n'|\hat{\boldsymbol{p}}|n\rangle \\
&= \langle n|\hat{\boldsymbol{p}}\cdot(\hat{H}_0-E_n)\hat{\boldsymbol{p}}|n\rangle \\
&= \langle n|\hat{\boldsymbol{p}}\cdot(\hat{H}_0\hat{\boldsymbol{p}}-\hat{\boldsymbol{p}}\hat{H}_0)|n\rangle . \qquad (5.38)
\end{aligned}
$$

Here

$$
\hat{H}_0 = \frac{\hat{\boldsymbol{p}}^2}{2m}+V(\boldsymbol{x})
$$

represents the Hamiltonian of the hydrogen atom. The commutator results in

$$
\hat{H}_0\hat{\boldsymbol{p}}-\hat{\boldsymbol{p}}\hat{H}_0 = -\frac{\hbar}{\mathrm{i}}\boldsymbol{\nabla}V(\boldsymbol{x}) . \qquad (5.39)
$$

Since

$$
V(\boldsymbol{x}) = -\frac{Ze^2}{r}
$$

and

$$
\Delta V(\boldsymbol{x}) = 4\pi Ze^2\delta(\boldsymbol{x})
$$

hold, we get for (5.38)

$$
\begin{aligned}
&\langle n|\hat{\boldsymbol{p}}\cdot(\hat{H}_0\hat{\boldsymbol{p}}-\hat{\boldsymbol{p}}\hat{H}_0)|n\rangle \\
&\quad = \langle n|\frac{\hbar}{\mathrm{i}}\boldsymbol{\nabla}\cdot\left(-\frac{\hbar}{\mathrm{i}}\boldsymbol{\nabla}V(\boldsymbol{x})\right)|n\rangle \\
&\quad = \hbar^2\int\psi_n^*\boldsymbol{\nabla}\cdot(\boldsymbol{\nabla}V)\psi_n\,\mathrm{d}^3x
\end{aligned}
$$

$$= -\hbar^2 \int (\boldsymbol{\nabla}\psi_n^*) \cdot (\boldsymbol{\nabla}V)\psi_n \, d^3x$$
$$= \hbar^2 \int \psi_n^* (\boldsymbol{\nabla}^2 V)\psi_n \, d^3x + \hbar^2 \int \psi_n^* (\boldsymbol{\nabla}V) \cdot \boldsymbol{\nabla}\psi_n \, d^3x \, .$$

From the last equation it follows that

$$-\hbar^2 \int (\boldsymbol{\nabla}\psi_n^*)\psi_n \cdot (\boldsymbol{\nabla}V) \, d^3x - \hbar^2 \int \psi_n^* (\boldsymbol{\nabla}\psi_n) \cdot \boldsymbol{\nabla}V \, d^3x$$
$$= \hbar^2 \int \psi_n^* (\boldsymbol{\nabla}^2 V)\psi_n \, d^3x \, .$$

The hydrogen eigenfunctions are real and consequently the two expressions on the left-hand side are identical, so that

$$\hbar^2 \int \psi_n^* \boldsymbol{\nabla} \cdot [(\boldsymbol{\nabla}V)\psi_n] \, d^3x = -\frac{1}{2}\hbar^2 \int \psi_n^* (\boldsymbol{\nabla}^2 V)\psi_n \, d^3x \, .$$

Thus,

$$\begin{aligned} \langle n|\hat{\boldsymbol{p}} \cdot (\hat{H}_0 \hat{\boldsymbol{p}} - \hat{\boldsymbol{p}}\hat{H}_0)|n\rangle &= -\frac{1}{2}\hbar^2 \int \psi_n^* \psi_n (\boldsymbol{\nabla}^2 V) \, d^3x \\ &= \frac{\hbar^2}{2} \int \psi_n^*(\boldsymbol{x})\psi_n(\boldsymbol{x}) Z e^2 4\pi\delta(\boldsymbol{x}) \, d^3x \\ &= 2\pi\hbar^2 Z e^2 |\psi_n(0)|^2 \end{aligned} \tag{5.40}$$

and the energy shift (5.36) becomes

$$\Delta E_n^{(2)} = \frac{4}{3} Z \frac{e^4 \hbar}{m^2 c^3} |\psi_n(0)|^2 \ln \left| \frac{mc^2}{(E_{n'} - E_n)_{\text{av.}}} \right| \, . \tag{5.41}$$

We know from the theory of the hydrogen atom that states with nonzero angular momentum ($l \neq 0$) vanish at the origin:

$$\psi_{nl}(0) = 0 \qquad \text{for } l \neq 0 \, .$$

Only s states have a nonvanishing probability within the nucleus:

$$|\psi_{n,l=0}(0)|^2 = \frac{1}{\pi} \left(\frac{Z}{n a_B} \right)^3 \, . \tag{5.42}$$

Here n represents the principal quantum number and $a_B = \hbar^2/(me^2)$ is the Bohr radius. Still we need a value for $(E_{n'} - E_n)_{\text{av.}}$; Bethe calculated this value numerically to be 17.8 Ry $\approx$ 242.2 eV. Thus he arrived at

$$\frac{\Delta E_{n=2}^{(2)}}{h} \approx 1040 \, \text{MHz} \, . \tag{5.43}$$

In view of the approximations employed, this result is in excellent agreement with the experimental value for the Lamb shift of 1057 MHz. Kroll and Lamb[19] performed the relativistic calculation for the self energy. They could confirm Bethe's conjecture about the convergence of the relativistic generalisation

[19] N.M. Kroll, W.E. Lamb, Jr.: Phys. Rev. **75** (1949) 388.

of expression (5.34) and could achieve an even better agreement with the experimental result; see the volume on quantum electrodynamics.[20]

In the lectures on quantum electrodynamics we realize that no infinity (including vacuum polarization, which we will not touch here) is observable; they are included in the finite values of the observed mass and charge (*mass and charge renormalization*). Not all field theories are renormalizable. However, quantum electrodynamics is renormalizable and because of that it is possibly the most successful theory in physics. Nevertheless, the removal of infinites by renormalization is not satisfying aestheticly; we return to this point in our lectures on QED.

5.4 Is There an Inconsistency in Bethe's Approach?

The largest radius of the electron charge distribution useful for our calculations is given by the Compton wavelength $a = h/mc$. We write the interaction with the radiation field as

$$H_I = -\frac{e}{mc}\int F(\boldsymbol{x}-\boldsymbol{x}')\boldsymbol{p}\cdot\boldsymbol{A}(\boldsymbol{x}')\,\mathrm{d}^3x'\,,$$

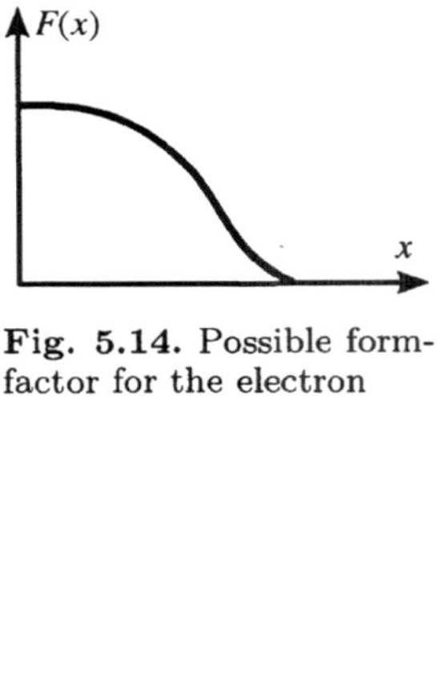

Fig. 5.14. Possible form-factor for the electron

where $F(\boldsymbol{x})$ represents the charge distribution of the electron. Hence, the electron interacts with only that part of the radiation field for which $\lambda < a$, i.e. $\hbar k < mc$, holds. The maximum photon momentum is given by $k_{\mathrm{max}} = mc/\hbar$. On the other hand, the electron is localized within the Bohr radius $a_0 = \hbar^2/me^2$. In the course of the calculation, the dipole approximation has been applied, i.e. $\exp(ikx) \approx 1$; however,

$$kx \approx k_{\mathrm{max}}a_0 = \frac{mc}{\hbar}\frac{\hbar^2}{me^2} = \frac{\hbar c}{e^2}$$

and

$$k_{\mathrm{max}}a_0 = \left(\frac{e^2}{\hbar c}\right)^{-1} = \frac{1}{\alpha} \approx 137\,,$$

so the applicability condition for the dipole approximation is *no longer* fulfilled. Furthermore, the largest photon energy is assumed to be identical to the electron rest mass mc^2. Again, this assumption makes sense only if the maximum momentum corresponds to $k_{\mathrm{max}} = mc/\hbar$, so that the electron is smeared out over the Compton wavelength. However, all measurements of the electron charge distribution clearly signal that the electron is pointlike to a very good degree ($r_{\mathrm{e}} \ll 0.1\,\mathrm{fm}$). Hence, there is no physical justification for the cut-off procedure employed by Bethe. Therefore his result has no significance; but historically it is interesting. The correct (relativistic) calculation, which does not need these dubious approximations was performed by Schwinger,

[20] W. Greiner, J. Reinhardt: *Quantum Electrodynamics*, 2nd ed. (Springer, Berlin, Heidelberg 1994).

Weisskopf, and Feynman[21] and by Kroll and Lamb. Also the so-called *vacuum polarization* is an important contribution to the Lamb shift. We discuss it in the volume on quantum electrodynamics, where the correct relativistic calculation due to Wichmann and Kroll[22] is presented. In recent years it has become possible to ionize heavy atoms completely and the investigation of, for example, lead (82 protons) with only a single electron present allows the direct measurement of the Lamb shift. This opens an interesting field of research, particularly for testing quantum electrodynamics in strong central fields.[23]

EXERCISE

5.6 The Lamb Shift

Problem. Solve the classical equations of motion for a particle in an oscillating electromagnetic field and calculate the mean quadratic displacement of the electron resulting from the action of a superposition of electromagnetic waves where the energy $\frac{1}{2}\hbar\omega$ is associated to every mode of the radiation field.

Calculate the change in the potential energy for the electron resulting from the displacement. After that, employ the additional term in the potential to calculate the shift of the hydrogen energy level (*Lamb shift*) with the help of perturbation theory.

Solution. At first we solve the classical equation of motion for the displacement $\Delta\boldsymbol{x}$ of a particle in an oscillating electric field $E = E_0 \exp(\mathrm{i}\omega t)$:

$$\frac{\mathrm{d}^2}{\mathrm{d}t^2}\Delta\boldsymbol{x} = \frac{e}{m}E = \frac{e}{m}E_0\,\mathrm{e}^{\mathrm{i}\omega t} \implies \Delta\boldsymbol{x} = -\frac{e}{m\omega^2}E = -\frac{e}{m\omega^2}E_0\,\mathrm{e}^{\mathrm{i}\omega t}\,,$$

so that

$$|\Delta\boldsymbol{x}|^2 = \frac{e^2}{m^2\omega^4}|E|^2\,.$$

We now assume that E contains the modes of the (quantized) radiation field, so that we can substitute $E \to \sum_{k\sigma} E_{k\sigma}$ and obtain the expectation value

$$\langle|\Delta\boldsymbol{x}|^2\rangle = \frac{e^2}{m^2}\sum_{k,\sigma}\frac{|E_{k\sigma}|^2}{\omega_k^4}\,.$$

Since

$$E_{k\sigma} = \frac{\mathrm{i}}{c}\left(\frac{2\pi\hbar c^2}{L^3}\right)^{1/2}\omega_k^{1/2}$$

it follows (see Exercise 1.3) that

[21] J. Schwinger, V.F. Weisskopf: Phys. Rev. **73** (1948) 1272; R.P. Feynman: Phys. Rev. **76** (1948) 939.

[22] E.H. Wichmann, N.M. Kroll: Phys. Rev. **101** (1956) 843.

[23] See e.g. S.M. Schneider, W. Greiner, G. Soff: "Vacuum polarization contribution to hyperfine-structure splitting of hydrogenlike atoms", Phys. Rev. **A50** (1994) 118; Z. Phys. **D31** (1994) 143; H. Persson, S.M. Schneider, W. Greiner, G. Soff, J. Lindgren: "Self-energy correction to the hyperfine-structure splitting of hydrogenlike atoms", Phys. Rev. Lett. **76** (1996) 1433.

Exercise 5.6.

$$|E_{k\sigma}|^2 = \frac{2\pi}{L^3}\hbar\omega_k$$

and

$$\begin{aligned}\langle|\Delta\boldsymbol{x}|^2\rangle &= \frac{2\pi\hbar}{L^3}\frac{e^2}{m^2}\sum_{k,\sigma}\frac{1}{\omega_k^3}\\ &= \frac{2\pi\hbar}{L^3}\frac{e^2}{m^2}\frac{L^3}{(2\pi)^3}2\int \mathrm{d}^3k\frac{1}{(ck)^3}\\ &= \frac{2}{(2\pi)^2}\frac{\hbar e^2}{m^2c^3}\int_0^\infty \frac{k^2\,\mathrm{d}k}{k^3}\int \mathrm{d}\Omega\\ &= \frac{2}{\pi}\frac{e^2}{\hbar c}\left(\frac{\hbar}{mc}\right)^2\int_0^\infty\frac{\mathrm{d}k}{k}\,,\end{aligned}$$

where the factor 2 keeps track of the polarizations σ. This integral diverges logarithmically; hence we introduce a cut-off for both integration boundaries:

$$\langle|\Delta\boldsymbol{x}|^2\rangle = \frac{2}{\pi}\frac{e^2}{\hbar c}\left(\frac{\hbar}{mc}\right)^2\int_{k_{\min}}^{k_{\max}}\frac{\mathrm{d}k}{k}\,.$$

The displacement $\Delta\boldsymbol{x}$ means the change in the potential given by (Taylor series expansion):

$$V(\boldsymbol{x}+\Delta\boldsymbol{x}) = \left[1+\Delta\boldsymbol{x}\cdot\boldsymbol{\nabla}+\tfrac{1}{2}(\Delta\boldsymbol{x}\cdot\boldsymbol{\nabla})^2+\cdots\right]V(\boldsymbol{x})\,.$$

We average the potential over all directions. Here we have to consider that the spherical symmetric potential $V(\boldsymbol{x})$ depends only on the modulus of $\boldsymbol{x}$:

$$\langle V(\boldsymbol{x}+\Delta\boldsymbol{x})\rangle = \frac{1}{2}\int \sin\vartheta\,\mathrm{d}\vartheta\left(1+\Delta\boldsymbol{x}\cdot\boldsymbol{\nabla}+\frac{1}{2}(\Delta\boldsymbol{x}\cdot\boldsymbol{\nabla})^2+\cdots\right)V(\boldsymbol{x})\,.$$

The term $\sim \Delta\boldsymbol{x}\cdot\boldsymbol{\nabla}$ represents an odd function. Thus, the average vanishes:

$$\langle V(\boldsymbol{x}+\Delta\boldsymbol{x})\rangle = \frac{1}{4}\int \sin\vartheta\,\mathrm{d}\vartheta(\boldsymbol{a}\cdot\boldsymbol{\nabla})(\boldsymbol{a}\cdot\boldsymbol{\nabla})V(\boldsymbol{x})+V(\boldsymbol{x})\,,$$

where $\boldsymbol{a}=\Delta\boldsymbol{x}$. Now

$$\begin{aligned}(\boldsymbol{a}\cdot\boldsymbol{\nabla})(\boldsymbol{a}\cdot\boldsymbol{\nabla})V(\boldsymbol{x}) &= a_i\partial_i a_j\partial_j V(r) \qquad \text{(sum convention!)}\\ &= a_i a_j\partial_i\partial_j V(r)\end{aligned}$$

and

$$\partial_j V = \frac{\partial r}{\partial x_j}\frac{\partial V}{\partial r} = \frac{x_j}{r}\frac{\partial V}{\partial r}\,,$$

so that

$$\begin{aligned}(\boldsymbol{a}\cdot\boldsymbol{\nabla})(\boldsymbol{a}\cdot\boldsymbol{\nabla})V(r) &= a_i a_j\partial_i\frac{x_j}{r}\frac{\partial V}{\partial r}\\ &= a_i a_j\left[\left(\frac{\delta_{ij}}{r}-\frac{x_i x_j}{r^3}\right)\frac{\partial V}{\partial r}+\frac{x_j x_i}{r^2}\frac{\partial^2 V}{\partial r^2}\right]\\ &= \left(\frac{\boldsymbol{a}^2}{r}-\frac{(\boldsymbol{a}\cdot\boldsymbol{r})(\boldsymbol{a}\cdot\boldsymbol{r})}{r^3}\right)\frac{\partial V}{\partial r}+\left(\boldsymbol{a}\cdot\frac{\boldsymbol{r}}{r}\right)^2\frac{\partial^2 V}{\partial r^2}\\ &= \frac{a^2}{r}\left(1-\cos^2\vartheta\right)\frac{\partial V}{\partial r}+a^2\cos^2\vartheta\frac{\partial^2 V}{\partial r^2}\end{aligned}$$

Exercise 5.6.

and

$$\int \sin\vartheta\,\mathrm{d}\vartheta\,(\boldsymbol{a}\cdot\boldsymbol{\nabla})(\boldsymbol{a}\cdot\boldsymbol{\nabla})V(r)$$
$$= \int_0^\pi \sin\vartheta\,\mathrm{d}\vartheta\left[\frac{a^2}{r}\left(1-\cos^2\vartheta\right)\frac{\partial V}{\partial r}+a^2\cos^2\vartheta\frac{\partial^2 V}{\partial r^2}\right]$$
$$= \frac{a^2}{r}\frac{2}{3}\frac{\partial V}{\partial r}+a^2\frac{1}{3}\frac{\partial^2 V}{\partial r^2}$$
$$= \frac{a^2}{3}\left[\frac{2}{r}\frac{\partial V}{\partial r}+\frac{\partial^2 V}{\partial r^2}\right].$$

The expression within the last square brackets is identical to $\nabla^2 V$ (because of the spherical symmetry!); hence,

$$\langle V(\boldsymbol{x}+\Delta\boldsymbol{x})\rangle = \left(1+\tfrac{1}{6}\langle|\Delta\boldsymbol{x}|^2\rangle\nabla^2\right)V(\boldsymbol{x}).$$

With

$$V = -\frac{e^2}{r}$$

and

$$\nabla^2 V = -e^2\nabla^2\left(\frac{1}{r}\right) = 4\pi e^2\delta(r)$$

it follows that

$$V_{\mathrm{pert}} = \frac{1}{6}\langle|\Delta\boldsymbol{x}|^2\rangle\nabla^2 V$$
$$= \frac{2\pi e^2}{3}\langle|\Delta\boldsymbol{x}|^2\rangle\delta^3(\boldsymbol{x})$$
$$= \frac{4}{3}\frac{e^4}{\hbar c}\left(\frac{\hbar}{mc}\right)^2\ln\left(\frac{k_{\mathrm{max}}}{k_{\mathrm{min}}}\right)\delta^3(\boldsymbol{x}).$$

This is a perturbation potential. Thus, in lowest-order perturbation theory the energy shift becomes

$$\Delta E = \langle\psi|V_{\mathrm{pert}}|\psi\rangle$$
$$= \frac{4}{3}\frac{e^4}{\hbar c}\left(\frac{\hbar}{mc}\right)^2\ln\left(\frac{k_{\mathrm{max}}}{k_{\mathrm{min}}}\right)\langle\psi|\delta^3(\boldsymbol{x})|\psi\rangle$$
$$= \frac{4}{3}\frac{e^4}{\hbar c}\left(\frac{\hbar}{mc}\right)^2\ln\left(\frac{k_{\mathrm{max}}}{k_{\mathrm{min}}}\right)|\psi(0)|^2,$$

where ψ is the unperturbed ground-state wavefunction.

This expression has exactly the form of (5.41), from which we have calculated the Lamb shift. However, the cut-off momenta $\hbar k_{\mathrm{min}}$ and $\hbar k_{\mathrm{max}}$ remain to be determined. The infrared divergency is not physical because a bound and localized electron cannot be put into oscillations by photons with long wavelenghts. Thus, $k_{\mathrm{min}} \approx 1/a_{\mathrm{B}} = \alpha/\lambda\!\!\!^{-}$ should hold. Here $a_{\mathrm{B}} = \hbar^2/(me^2)$ denotes the Bohr radius and $\lambda\!\!\!^{-} = \lambda_{\mathrm{Compton}}/(2\pi) = h/(2\pi mc) = \hbar/(mc)$ the Compton wavelength divided by 2π. As the upper boundary we choose again

Exercise 5.6.

the Compton wavelength $k_{\text{max}} = 1/\lambda\!\!\!^{-}$. In the framework of the classical theory we cannot give a satisfactory derivation for the latter. With (5.42) we arrive at

$$\Delta E = \frac{4}{3\pi}\frac{\alpha^5}{n^3}mc^2\, Z^3 \ln\left(\frac{1}{Z\alpha}\right)\delta_{nl}\,.$$

With this result the Lamb shift of the hydrogen atom ($Z = 1$) becomes $\Delta E \approx$ 660 MHz, which in view of the many crude approximations is in surprisingly good agreement with the experimental result. This illustrative explanation of the Lamb shift as a consequence of the vacuum fluctuations was given by Welton in 1948.[24]

[24] T. Welton: Phys. Rev. **74** (1948) 1157.

6. Nonrelativistic Quantum Field Theory of Interacting Particles and Its Applications

From Chap. 3 we know how an arbitrary number of particles, which *do not interact* with each other but which are subject to an external potential $V(\boldsymbol{x})$, are to be described within the framework of second quantization. Now we will extend this formalism with respect to particles interacting with each other. First, we construct an additional term in the Hamiltonian, which describes the potential interaction energy $\mathcal{V}(\boldsymbol{x},\boldsymbol{x}')$ between two particles at positions $\boldsymbol{x}$ and $\boldsymbol{x}'$, respectively. As a generalisation of (3.63) we make the ansatz

$$\begin{aligned}\hat{H} &= \int d^3x\,\hat{\Psi}^\dagger(\boldsymbol{x},t)\hat{H}_0'\hat{\Psi}(\boldsymbol{x},t)\\ &\quad+\frac{1}{2}\int d^3x\int d^3x'\,\hat{\Psi}^\dagger(\boldsymbol{x},t)\hat{\Psi}^\dagger(\boldsymbol{x}',t)\,\mathcal{V}(\boldsymbol{x},\boldsymbol{x}')\,\hat{\Psi}(\boldsymbol{x},t)\hat{\Psi}(\boldsymbol{x}',t)\ ,\end{aligned}\tag{6.1}$$

where

$$\hat{H}_0' = -\frac{\hbar^2\boldsymbol{\nabla}^2}{2m}+V(\boldsymbol{x})\tag{6.2}$$

represents the unperturbed one-particle Hamiltonian for a particle in an external potential $V(\boldsymbol{x})$. $\hat{\Psi}(\boldsymbol{x},t)$ represents the *field operators* (3.43). $\hat{H}$ itself is the Hamiltonian for the many-particle system in second quantization or, more informatively, the *field-theoretical Hamiltonian*. We have to find its solutions, i.e. its eigenstates. In order to see the connection between the field-theoretical problem (6.1) and the "classical" many-particle problem of elementary quantum mechanics we introduce the *n-particle state*

$$|\chi_n,t\rangle = \int d^3x_1\dots\int d^3x_n\chi_n(\boldsymbol{x}_1,\dots,\boldsymbol{x}_n)\,\hat{\Psi}^\dagger(\boldsymbol{x}_1,t)\dots\hat{\Psi}^\dagger(\boldsymbol{x}_n,t)\,|0\rangle\tag{6.3}$$

in analogy to our previous considerations (3.66). $\chi_n(\boldsymbol{x}_1,\dots,\boldsymbol{x}_n)$ represents an ordinary function of the n-particle coordinates $\boldsymbol{x}_1,\dots,\boldsymbol{x}_n$. According to the conventional rules of quantum mechanics, we interpret

$$|\,\chi_n(\boldsymbol{x}_1,\dots,\boldsymbol{x}_n)\,|^2\,d^3x_1\dots d^3x_n\tag{6.4}$$

as the probability of finding particle 1 in d^3x_1, particle 2 in $d^3x_2,\dots$, and particle n in d^3x_n. We determine the amplitude function $\chi_n(\boldsymbol{x}_1,\dots,\boldsymbol{x}_n)$ in such a way that $|\chi_n,t\rangle$ is an eigenstate of the field-theoretical Hamiltonian (6.1):

$$\hat{H}\,|\chi_n,t\rangle = E_n\,|\chi_n,t\rangle\ .\tag{6.5}$$

Indeed, it will be shown in Exercise 6.1 that (6.5) provides the following defining equation for the amplitude function $\chi_n(\boldsymbol{x}_1, \ldots, \boldsymbol{x}_n)$:

$$\left[\sum_{i=1}^{n}\left(-\frac{\hbar^2}{2m}\boldsymbol{\nabla}_i^2 + V(\boldsymbol{x}_i)\right) + \frac{1}{2}\sum_{i=1}^{n}\sum_{j=1}^{n}\mathcal{V}(\boldsymbol{x}_i, \boldsymbol{x}_j)\right]\chi_n(\boldsymbol{x}_1, \ldots, \boldsymbol{x}_n)$$
$$= E_n\chi_n(\boldsymbol{x}_1, \ldots, \boldsymbol{x}_n) . \tag{6.6}$$

This implies that $\chi_n(\boldsymbol{x}_1, \ldots, \boldsymbol{x}_n)$ obeys the conventional many-particle Schrödinger equation. Hence, the field-theoretical problem (6.5) is equivalent to the conventional many-particle problem (6.6). This is the same result as obtained in Chap. 3 [see (3.73)]; now this equivalence also holds for interacting particles.

EXERCISE

6.1 The Field-Theoretical Many-Particle Problem

Problem. Show the equivalence between the field-theoretical many-particle problem $\hat{H}\,|\chi_n, t\rangle = E_n\,|\chi_n, t\rangle$ and the conventional many-particle problem

$$\left[\sum_{i=1}^{n}\left(-\frac{\hbar^2}{2m}\boldsymbol{\nabla}_i^2 + V(\boldsymbol{x}_i)\right) + \frac{1}{2}\sum_{i,j}\mathcal{V}(\boldsymbol{x}_i, \boldsymbol{x}_j)\right]\chi_n(\boldsymbol{x}_1, \ldots, \boldsymbol{x}_n)$$
$$= E_n\chi_n(\boldsymbol{x}_1, \ldots, \boldsymbol{x}_n) ,$$

where

$$|\chi_n, t\rangle = \int \mathrm{d}^3x_1 \ldots \int \mathrm{d}^3x_n \chi_n(\boldsymbol{x}_1, \ldots, \boldsymbol{x}_n)\hat{\Psi}^\dagger(\boldsymbol{x}_1, t)\ldots\hat{\Psi}^\dagger(\boldsymbol{x}_n, t)\,|0\rangle .$$

Solution. We write $\hat{H}|\chi_n, t\rangle = E_n|\chi_n, t\rangle$ as

$$(\hat{H}_0 + \hat{\mathcal{V}})|\chi_n, t\rangle = E_n|\chi_n, t\rangle ,$$

where

$$\hat{H}_0 = \int \mathrm{d}^3x\,\hat{\Psi}^\dagger(\boldsymbol{x}, t)\hat{H}_0'\hat{\Psi}(\boldsymbol{x}, t)$$

with

$$\hat{H}_0' = -\frac{\hbar^2\boldsymbol{\nabla}^2}{2m} + V(\boldsymbol{x})$$

and

$$\hat{\mathcal{V}} = \frac{1}{2}\int \mathrm{d}^3x \int \mathrm{d}^3x'\,\hat{\Psi}^\dagger(\boldsymbol{x}, t)\hat{\Psi}^\dagger(\boldsymbol{x}', t)\,\mathcal{V}(\boldsymbol{x}, \boldsymbol{x}')\,\hat{\Psi}(\boldsymbol{x}, t)\hat{\Psi}(\boldsymbol{x}', t) .$$

We have already shown in (3.73) the equivalence between $\hat{H}_0|\chi_n, t\rangle = E_n|\chi_n, t\rangle$ and the "conventional" many-particle problem. Thus, it remains to show that

$$\hat{\mathcal{V}}|\,\chi_n, t\rangle$$

is equivalent to

$$\frac{1}{2}\sum_{i,j} \mathcal{V}(\boldsymbol{x}_i, \boldsymbol{x}_j)\, \chi_n(\boldsymbol{x}_1, \dots, \boldsymbol{x}_n)\ .$$

Exercise 6.1.

We write

$$\begin{aligned}
&\hat{\mathcal{V}}\,|\chi_n, t\rangle \\
&\quad = \frac{1}{2}\int \mathrm{d}^3x \int \mathrm{d}^3x'\, \hat{\Psi}^\dagger(\boldsymbol{x},t)\hat{\Psi}^\dagger(\boldsymbol{x}',t)\, \mathcal{V}(\boldsymbol{x},\boldsymbol{x}')\, \hat{\Psi}(\boldsymbol{x},t)\hat{\Psi}(\boldsymbol{x}',t) \\
&\qquad \times \int \mathrm{d}^3x_1' \dots \int \mathrm{d}^3x_n'\, \chi_n(\boldsymbol{x}_1',\dots,\boldsymbol{x}_n')\hat{\Psi}^\dagger(\boldsymbol{x}_1',t)\dots\hat{\Psi}^\dagger(\boldsymbol{x}_n',t)\,|0\rangle \\
&\quad = \frac{1}{2}\int \mathrm{d}^3x\, \mathrm{d}^3x'\, \mathrm{d}^3x_1' \dots \mathrm{d}^3x_n'\, \hat{\Psi}^\dagger(\boldsymbol{x},t)\hat{\Psi}^\dagger(\boldsymbol{x}',t)\, \mathcal{V}(\boldsymbol{x},\boldsymbol{x}') \\
&\qquad \times \hat{\Psi}(\boldsymbol{x},t)\hat{\Psi}(\boldsymbol{x}',t)\chi_n(\boldsymbol{x}_1',\dots,\boldsymbol{x}_n')\hat{\Psi}^\dagger(\boldsymbol{x}_1',t)\dots\hat{\Psi}^\dagger(\boldsymbol{x}_n',t)\,|0\rangle\ .
\end{aligned}$$

We note the commutation relation

$$\hat{\Psi}(\boldsymbol{x})\hat{\Psi}^\dagger(\boldsymbol{x}') = \pm\hat{\Psi}^\dagger(\boldsymbol{x}')\hat{\Psi}(\boldsymbol{x}) + \delta(\boldsymbol{x}-\boldsymbol{x}')\ ,$$

$$\begin{aligned}
&\hat{\mathcal{V}}\,|\chi_n, t\rangle \\
&\quad = \frac{1}{2}\int \mathrm{d}^3x\, \mathrm{d}^3x'\, \mathrm{d}^3x_1' \dots \mathrm{d}^3x_n'\, \hat{\Psi}^\dagger(\boldsymbol{x},t)\hat{\Psi}^\dagger(\boldsymbol{x}',t)\mathcal{V}(\boldsymbol{x},\boldsymbol{x}') \\
&\qquad \times \chi_n(\boldsymbol{x}_1',\dots,\boldsymbol{x}_n')\hat{\Psi}(\boldsymbol{x},t)\hat{\Psi}(\boldsymbol{x}',t)\hat{\Psi}^\dagger(\boldsymbol{x}_1',t)\dots\hat{\Psi}^\dagger(\boldsymbol{x}_n',t)\,|0\rangle \\
&\quad = \frac{1}{2}\int \mathrm{d}^3x\, \mathrm{d}^3x'\, \mathrm{d}^3x_1' \dots \mathrm{d}^3x_n'\, \hat{\Psi}^\dagger(\boldsymbol{x},t)\hat{\Psi}^\dagger(\boldsymbol{x}',t)\mathcal{V}(\boldsymbol{x},\boldsymbol{x}') \\
&\qquad \times \chi_n(\boldsymbol{x}_1',\dots,\boldsymbol{x}_n')\hat{\Psi}(\boldsymbol{x},t)\left[\delta(\boldsymbol{x}'-\boldsymbol{x}_1') \pm \hat{\Psi}^\dagger(\boldsymbol{x}_1',t)\hat{\Psi}(\boldsymbol{x}',t)\right] \\
&\qquad \times \hat{\Psi}^\dagger(\boldsymbol{x}_2',t)\dots\hat{\Psi}^\dagger(\boldsymbol{x}_n',t)\,|0\rangle\ .
\end{aligned}$$

The successive commutations are performed analogously to the method for (3.69) and the equations following it:

$$\begin{aligned}
&\hat{\mathcal{V}}\,|\chi_n, t\rangle \\
&\quad = \frac{1}{2}\int \mathrm{d}^3x\, \mathrm{d}^3x_1' \dots \mathrm{d}^3x_n'\, \hat{\Psi}^\dagger(\boldsymbol{x},t)\sum_i v(\boldsymbol{x},\boldsymbol{x}_i')\,\chi_n(\boldsymbol{x}_1',\dots,\boldsymbol{x}_n') \\
&\qquad \times \hat{\Psi}(\boldsymbol{x},t)\hat{\Psi}^\dagger(\boldsymbol{x}_1',t)\dots\hat{\Psi}^\dagger(\boldsymbol{x}_n',t)\,|0\rangle\ .
\end{aligned}$$

In the same manner the second annihilation operator goes through the commutations; it produces a second sum

$$\begin{aligned}
\hat{\mathcal{V}}\,|\chi_n, t\rangle &= \frac{1}{2}\int \mathrm{d}^3x\, \mathrm{d}^3x_1' \dots \mathrm{d}^3x_n'\, \hat{\Psi}^\dagger(\boldsymbol{x}',t)\dots\hat{\Psi}^\dagger(\boldsymbol{x}_n',t)\sum_{i,j}\mathcal{V}(\boldsymbol{x}_j',\boldsymbol{x}_i') \\
&\quad \times\chi_n(\boldsymbol{x}_1'\dots\boldsymbol{x}_n')\,|0\rangle\ .
\end{aligned}$$

Since this holds for arbitrary states

$$\hat{\Psi}^\dagger(\boldsymbol{x}_1',t)\cdots\hat{\Psi}^\dagger(\boldsymbol{x}_n',t)\,|0\rangle\ ,$$

it follows that

$$\frac{1}{2}\sum_{i,j}\mathcal{V}(\boldsymbol{x}_i',\boldsymbol{x}_j')\,\chi_n(\boldsymbol{x}_1',\dots,\boldsymbol{x}_n') \sim \hat{\mathcal{V}}\,|\chi_n,t\rangle\ .$$

For the case when the external potential $V(\boldsymbol{x})$ vanishes the n particles move only as a result of their mutual interactions. Then the field operator $\hat{\Psi}$ can be expanded into plane waves:

$$\hat{\Psi}(\boldsymbol{x},t) = \sum_{\boldsymbol{k}} \hat{b}_{\boldsymbol{k}}(t) \frac{\mathrm{e}^{\mathrm{i}\boldsymbol{k}\cdot\boldsymbol{x}}}{\sqrt{L^3}} . \tag{6.7}$$

If we further assume that the particle–particle interaction depends only on the relative distance between the particles, i.e.

$$\mathcal{V}(\boldsymbol{x},\boldsymbol{x}') = \mathcal{V}(\boldsymbol{x}-\boldsymbol{x}') , \tag{6.8}$$

(6.1) becomes

$$\begin{aligned} \hat{H} &= \sum_{\boldsymbol{k}_1}\sum_{\boldsymbol{k}_2} \hat{b}^\dagger_{\boldsymbol{k}_1}\hat{b}_{\boldsymbol{k}_2} \frac{\hbar^2 k_2^2}{2m} \int \frac{\mathrm{d}^3x}{L^3} \exp\left[\mathrm{i}(\boldsymbol{k}_2-\boldsymbol{k}_1)\cdot\boldsymbol{x}\right] \\ &\quad + \sum_{\boldsymbol{k}_1}\sum_{\boldsymbol{k}_2}\sum_{\boldsymbol{k}_3}\sum_{\boldsymbol{k}_4} \hat{b}^\dagger_{\boldsymbol{k}_1}\hat{b}^\dagger_{\boldsymbol{k}_2}\hat{b}_{\boldsymbol{k}_3}\hat{b}_{\boldsymbol{k}_4} \int\frac{\mathrm{d}^3x}{L^3}\int\frac{\mathrm{d}^3x'}{L^3} \\ &\quad \times \exp\left[\mathrm{i}(\boldsymbol{k}_3-\boldsymbol{k}_1)\cdot\boldsymbol{x}\right]\exp\left[\mathrm{i}(\boldsymbol{k}_4-\boldsymbol{k}_2)\cdot\boldsymbol{x}'\right]\mathcal{V}(\boldsymbol{x}-\boldsymbol{x}') . \end{aligned} \tag{6.9}$$

We employ

$$\int \frac{\mathrm{d}^3x}{L^3}\mathrm{e}^{\mathrm{i}(\boldsymbol{k}_2-\boldsymbol{k}_1)\cdot\boldsymbol{x}} = \delta_{\boldsymbol{k}_2,\boldsymbol{k}_1} \qquad \text{(box normalization)}$$

and use the Fourier transform of the interaction potential:

$$\bar{\mathcal{V}}(\boldsymbol{q}) = \int \frac{\mathrm{d}^3x}{L^3}\mathcal{V}(\boldsymbol{x})\,\mathrm{e}^{\mathrm{i}\boldsymbol{q}\cdot\boldsymbol{x}} ; \tag{6.10}$$

then (6.9) simplifies to

$$\hat{H} = \sum_{\boldsymbol{k}} \frac{\hbar^2k^2}{2m}\hat{b}^\dagger_{\boldsymbol{k}}\hat{b}_{\boldsymbol{k}} + \sum_{\boldsymbol{k}_1}\sum_{\boldsymbol{k}_2}\sum_{\boldsymbol{q}} \bar{\mathcal{V}}(\boldsymbol{q})\,\hat{b}^\dagger_{\boldsymbol{k}_1+\boldsymbol{q}}\hat{b}^\dagger_{\boldsymbol{k}_2-\boldsymbol{q}}\hat{b}_{\boldsymbol{k}_2}\hat{b}_{\boldsymbol{k}_1} . \tag{6.11}$$

The last term becomes more plausible once we calculate in more detail:

$$\begin{aligned} &\int\frac{\mathrm{d}^3x}{L^3}\int\frac{\mathrm{d}^3x'}{L^3}\exp\left[\mathrm{i}(\boldsymbol{k}_3-\boldsymbol{k}_1)\cdot\boldsymbol{x}\right]\exp\left[\mathrm{i}(\boldsymbol{k}_4-\boldsymbol{k}_2)\cdot\boldsymbol{x}'\right]\mathcal{V}(\boldsymbol{x}-\boldsymbol{x}') \\ &= \int\frac{\mathrm{d}^3z}{L^3}\int\frac{\mathrm{d}^3x'}{L^3}\exp\left[\mathrm{i}(\boldsymbol{k}_3-\boldsymbol{k}_1)\cdot\boldsymbol{z}\right]\mathcal{V}(\boldsymbol{z})\exp\left[\mathrm{i}(\boldsymbol{k}_4-\boldsymbol{k}_2+\boldsymbol{k}_3-\boldsymbol{k}_1)\cdot\boldsymbol{x}'\right] \\ &= \bar{\mathcal{V}}(\boldsymbol{k}_3-\boldsymbol{k}_1)\int\frac{\mathrm{d}^3x'}{L^3}\exp\left[\mathrm{i}(\boldsymbol{k}_3+\boldsymbol{k}_4-(\boldsymbol{k}_1+\boldsymbol{k}_2))\cdot\boldsymbol{x}'\right] \\ &= \bar{\mathcal{V}}(\boldsymbol{k}_3-\boldsymbol{k}_1)\,\delta_{\boldsymbol{k}_3+\boldsymbol{k}_4,\boldsymbol{k}_1+\boldsymbol{k}_2} \end{aligned} \tag{6.12}$$

with the substitution $\boldsymbol{z} = \boldsymbol{x} - \boldsymbol{x}'$. We set $\boldsymbol{q} = \boldsymbol{k}_3 - \boldsymbol{k}_1$, so that

$$\boldsymbol{k}_3 + \boldsymbol{k}_4 = \boldsymbol{k}_1 + \boldsymbol{k}_2$$

or

$$\boldsymbol{q} + \boldsymbol{k}_1 + \boldsymbol{k}_4 = \boldsymbol{k}_1 + \boldsymbol{k}_2 \implies \boldsymbol{k}_4 = \boldsymbol{k}_2 - \boldsymbol{q} , \qquad \boldsymbol{k}_3 = \boldsymbol{k}_1 + \boldsymbol{q}$$

follows from the Kronecker delta symbol. Therefore the last term of (6.9) becomes

$$
\begin{aligned}
&\sum_{k_1}\sum_{k_2}\sum_{q}\hat{b}^{\dagger}_{k_1}\hat{b}^{\dagger}_{k_2}\hat{b}_{k_1+q}\hat{b}_{k_2-q}\,\bar{V}(q)\\
&\quad=\sum_{k_1'}\sum_{k_2'}\sum_{q}\hat{b}^{\dagger}_{k_1'-q}\hat{b}^{\dagger}_{k_2'+q}\hat{b}_{k_1'}\hat{b}_{k_2'}\,\bar{V}(q)\\
&\quad=\sum_{k_1'}\sum_{k_2'}\sum_{q}\hat{b}^{\dagger}_{k_2'+q}\hat{b}^{\dagger}_{k_1'-q}\hat{b}_{k_1'}\hat{b}_{k_2'}\,\bar{V}(q)\\
&\quad=\sum_{k_1'}\sum_{k_2'}\sum_{q}\hat{b}^{\dagger}_{k_1'+q}\hat{b}^{\dagger}_{k_2'-q}\hat{b}_{k_2'}\hat{b}_{k_1'}\,\bar{V}(q)\ ,
\end{aligned}
\tag{6.13}
$$

which is identical to the last term in (6.11). The interaction term arising in (6.11) is illustrated in Fig. 6.1.

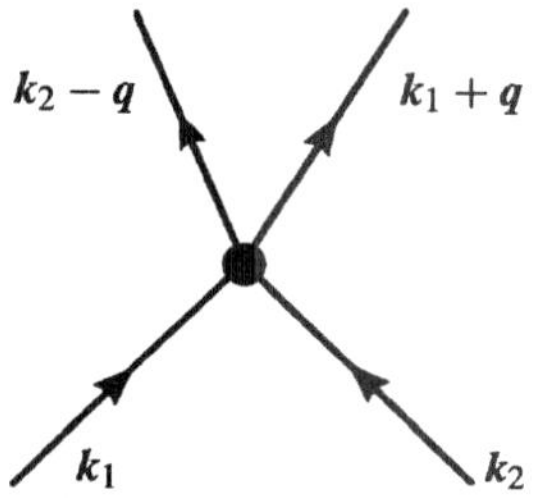

Fig. 6.1. Momentum conservation at a vertex for two-particle collisions

The operators $\hat{b}_{k_1}$ and $\hat{b}_{k_2}$ annhilate two particles with corresponding momenta $\hbar k_1$ and $\hbar k_2$ at the vertex. In the same interaction (at the same vertex) the operators $\hat{b}^{\dagger}_{k_1+q}$ and $\hat{b}^{\dagger}_{k_2-q}$ create two new particles with momenta $\hbar(k_1+q)$ and $\hbar(k_2-q)$. For this process, momentum conservation is guaranteed since

$$
\begin{aligned}
\hbar(k_1+q)\quad+\quad\hbar(k_2-q) &= \hbar k_1 \quad+\quad \hbar k_2\,, \\
\text{(momentum after collision)} &= \text{(momentum before collision)}\,.
\end{aligned}
\tag{6.14}
$$

This momentum conservation stems from the assumption (6.8) that the interaction between particles depends only on $x-x'$, which reflects translational invariance. We can rephrase this result: During the scattering the two particles exchange the momentum $\hbar q$ (at the vertex); the corresponding amplitude is determined by $\bar{v}(q)$.

6.1 Quantum Gases

A quantum gas is defined as a large number of particles that collide (interact) with each other according to the interaction potential $v(x-x')$. The particles arrive with a variety of different momenta. Within this ensemble we denote the number of particles with momentum $\hbar k$ by $N(k)$; $N(k)$ will change with time. Mutual interactions mean that particles with momentum $\hbar k$ are scattered to another momentum; on the other hand particles with a different momentum can adopt the momentum $\hbar k$ after the collision. The change in $N(k)$ is described symbolically by the following equation:

$$
\frac{\partial N(k)}{\partial t}=\sum_{k'}\sum_{q}\left\{\ \cdots\ \right\}. \tag{6.15}
$$

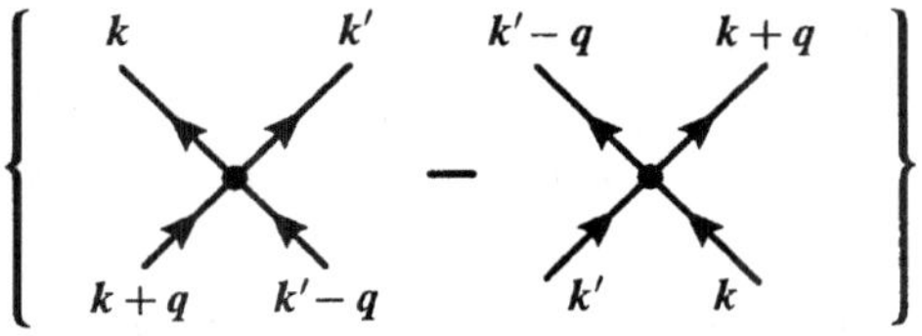

Evidently the  represents a binary scattering process. We sum over all binary collisions that lead to a particle with momentum $\hbar k$ after the collision; they increase the number $N(k)$. We subtract all processes that annihilate a particle with momentum $\hbar \boldsymbol{k}$; they decrease the number $N(\boldsymbol{k})$. In order to translate the symbolic relation (6.15) into a quantitative equation we have to introduce the *transition probability per unit time* for each diagram. We calculate the latter in first-order perturbation theory, where the interaction term is treated as a perturbation potential. We use

$$\begin{aligned} \hat{b}_n | \cdots N_n \cdots \rangle &= \sqrt{N_n} | \cdots N_n - 1 \cdots \rangle , \\ \hat{b}_n^\dagger | \cdots N_n \cdots \rangle &= \sqrt{N_n + 1} | \cdots N_n + 1 \cdots \rangle \end{aligned} \tag{6.16a}$$

for *bosons* and

$$\begin{aligned} \hat{b}_n | \cdots N_n \cdots \rangle &= \Theta_n \sqrt{N_n} | \cdots 1 - N_n \cdots \rangle , \\ \hat{b}_n^\dagger | \cdots N_n \cdots \rangle &= \Theta_n \sqrt{1 - N_n} | \cdots 1 - N_n \cdots \rangle \end{aligned} \tag{6.16b}$$

for *fermions*, where the phase Θ_n has been explained previously in (3.42). With this, the transition probability per unit time for the diagram depicted in (6.15) becomes, according to Fermi's golden rule,

$$\left(\frac{\text{trans. prob.}}{\text{time}} \right)$$

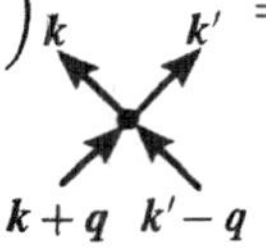

$$\begin{aligned} &= \frac{2\pi}{\hbar} \left| \bar{V}(\boldsymbol{q}) \right|^2 N(\boldsymbol{k}+\boldsymbol{q}) N(\boldsymbol{k}'-\boldsymbol{q}) \\ &\quad \times [1 \pm N(\boldsymbol{k})][1 \pm N(\boldsymbol{k}')] \\ &\quad \times \delta\left(\frac{\hbar^2}{2m} \left(|\boldsymbol{k}+\boldsymbol{q}|^2 + |\boldsymbol{k}'-\boldsymbol{q}|^2 - \boldsymbol{k}^2 - \boldsymbol{k}'^2 \right) \right) . \end{aligned} \tag{6.17}$$

The plus or minus sign within the factors $(1 \pm N)$ has to be chosen for bosons or fermions, respectively. Since the matrix element enters quadratically into Fermi's golden rule the square-root factors appearing in (6.16) are carried into N and $(N \pm 1)$. The argument of the δ function expresses energy conservation for the scattering process; note that because of elastic collisions only kinetic energy is considered. Analogously to the process for (6.17), the transition probability per unit time for the second scattering process appearing in (6.15) can be calculated. We then arrive at the following *master equation* for (6.15):

$$\begin{aligned} \frac{\partial N(\boldsymbol{k})}{\partial t} &= \sum_{\boldsymbol{k}'} \sum_{\boldsymbol{q}} \frac{2\pi}{\hbar} |\bar{V}(\boldsymbol{q})|^2 \delta\left(\frac{\hbar^2}{2m} \left(|\boldsymbol{k}+\boldsymbol{q}|^2 + |\boldsymbol{k}'-\boldsymbol{q}|^2 - \boldsymbol{k}^2 - \boldsymbol{k}'^2 \right) \right) \\ &\quad \times \{ N(\boldsymbol{k}+\boldsymbol{q}) N(\boldsymbol{k}'-\boldsymbol{q}) [1 \pm N(\boldsymbol{k})] [1 \pm N(\boldsymbol{k}')] \\ &\quad - N(\boldsymbol{k}) N(\boldsymbol{k}') [1 \pm N(\boldsymbol{k}+\boldsymbol{q})] [1 \pm N(\boldsymbol{k}'-\boldsymbol{q})] \} . \end{aligned} \tag{6.18}$$

We realize that because of factors $1 - N(\boldsymbol{k})$, transitions into already occupied states ($N(\boldsymbol{k}) = 1$) vanish for fermions; this reflects the *Pauli principle.* For bosons on the other hand, transitions into already occupied states are amplified; this is reflected in the factors of the type $1 + N(\boldsymbol{k})$. *The master equation* (6.18) *represents the quantum-mechanical generalization of the Boltzmann equation.* The scattering probability therein is proportional to $|\bar{V}(\boldsymbol{q})|^2$;

this goes back to the fact that outside the interaction region the particles here move freely without interactions, so the matrix element (6.10) is calculated with plane waves (Born approximation). In order to find an *equilibrium* solution of the Boltzmann equation (6.18) we have to require that

$$\frac{\partial N(\boldsymbol{k})}{\partial t} = 0\,. \tag{6.19}$$

This leads to the sufficient condition

$$\begin{aligned} &N(\boldsymbol{k}+\boldsymbol{q})N(\boldsymbol{k}'-\boldsymbol{q})\,[1\pm N(\boldsymbol{k})]\,[1\pm N(\boldsymbol{k}')] \\ &\quad = N(\boldsymbol{k})N(\boldsymbol{k}')\,[1\pm N(\boldsymbol{k}+\boldsymbol{q})]\,[1\pm N(\boldsymbol{k}'-\boldsymbol{q})]\;. \end{aligned} \tag{6.20}$$

We will solve this equation in Exercise 6.2; the result is

$$N(\boldsymbol{k}) = \frac{1}{A\exp\left[E(k)/k_{\mathrm{B}}T\right]\mp 1}\,. \tag{6.21}$$

Here, A is a normalization constant,

$$E\,(k) = \frac{\hbar^2k^2}{2m} \tag{6.22}$$

is the kinetic energy, and $k_{\mathrm{B}}T$ represents an energy related to temperature. The plus and minus signs in (6.21) hold for fermions and bosons, respectively; only with this assignment (6.21) is a solution of (6.20) possible.

EXERCISE

6.2 Equilibrium Solution of the Quantum-Mechanical Boltzmann Equation

Problem. Find the equilibrium solution of the Boltzmann equation. Proceed from the sufficient condition (6.20).

Solution. We start from

$$\begin{aligned} &N(\boldsymbol{k}+\boldsymbol{q})N(\boldsymbol{k}'-\boldsymbol{q})\,[1\pm N(\boldsymbol{k})]\,[1\pm N(\boldsymbol{k}')] \\ &\quad = N(\boldsymbol{k})N(\boldsymbol{k}')\,[1\pm N(\boldsymbol{k}+\boldsymbol{q})]\,[1\pm N(\boldsymbol{k}'-\boldsymbol{q})] \end{aligned}$$

and

$$\frac{N(\boldsymbol{k}+\boldsymbol{q})N(\boldsymbol{k}'-\boldsymbol{q})}{[1\pm N(\boldsymbol{k}+\boldsymbol{q})]\,[1\pm N(\boldsymbol{k}'-\boldsymbol{q})]} = \frac{N(\boldsymbol{k})N(\boldsymbol{k}')}{[1\pm N(\boldsymbol{k})]\,[1\pm N(\boldsymbol{k}')]}$$

or

$$\begin{aligned} &\ln\left(\frac{N(\boldsymbol{k}+\boldsymbol{q})}{1\pm N(\boldsymbol{k}+\boldsymbol{q})}\right) + \ln\left(\frac{N(\boldsymbol{k}'-\boldsymbol{q})}{1\pm N(\boldsymbol{k}'-\boldsymbol{q})}\right) \\ &\quad = \ln\left(\frac{N(\boldsymbol{k})}{1\pm N(\boldsymbol{k})}\right) + \ln\left(\frac{N(\boldsymbol{k}')}{1\pm N(\boldsymbol{k}')}\right)\,. \end{aligned}$$

So $\ln[N/(1\pm N)]$ is a collision invariant. Functions $a(\boldsymbol{k})$ that fulfill this balance for all possible collisions are given by

Exercise 6.2.

$$
\begin{aligned}
a_0 &= 1 \qquad \text{(particle-number conservation)}\,, \\
a_1 &= E(k) \qquad \text{(energy conservation)}
\end{aligned}
$$

owing to energy and particle-number conservation. Therefore every collision invariant is a linear combination of these elementary invariants, i.e.

$$
\ln\left(\frac{N(\boldsymbol{k})}{1 \pm N(\boldsymbol{k})}\right) = \alpha_0 + \alpha(\boldsymbol{r},t) - \beta(\boldsymbol{r},t)E(\boldsymbol{k})
$$

or

$$
N(\boldsymbol{k}) = \frac{1}{A\exp\left[\beta(\boldsymbol{r},t)E(\boldsymbol{k}) - \alpha(\boldsymbol{r},t)\right] \mp 1}\,,
$$

where $A = \mathrm{e}^{-\alpha_0}$ is a normalization constant. We define the *local temperature* as

$$
\beta(\boldsymbol{r},t) = \frac{A}{k_{\mathrm{B}}T} \qquad \text{or} \qquad T(\boldsymbol{r},t) = \frac{1}{k_{\mathrm{B}}\beta(\boldsymbol{r},t)}\,,
$$

k_{B} being the Boltzmann constant, and the *chemical potential* as

$$
\mu = \frac{\alpha(\boldsymbol{r},t)}{\beta(\boldsymbol{r},t)}\,.
$$

Then we deduce

$$
\begin{aligned}
N(\boldsymbol{k}) &= \left[A\exp\left\{\beta(\boldsymbol{r},t)\left[E(\boldsymbol{k}) - \mu(\boldsymbol{r},t)\right]\right\} \mp 1\right]^{-1}\,, \\
&\text{for } - \quad \text{Bose–Einstein} \longrightarrow 1 + N \\
&\text{for } + \quad \text{Fermi–Dirac} \longrightarrow 1 - N\,.
\end{aligned}
$$

If the chemical potential vanishes we obtain (6.21).

The normalization constant A is determined from the total number N_{P} of particles in the system according to

$$
\begin{aligned}
N_{\mathrm{P}} &= \sum_{\boldsymbol{k}} N(\boldsymbol{k}) = L^3 \int \mathrm{d}^3k\, N(\boldsymbol{k}) \\
&= L^3 \int \mathrm{d}^3k \frac{1}{A\exp\left(\hbar^2k^2/2mk_{\mathrm{B}}T\right) \pm 1}\,.
\end{aligned} \tag{6.23}
$$

The distribution functions calculated with (6.21) are called the *Fermi–Dirac* (for the plus sign) and the Bose–Einstein (for the minus sign) *distributions.*[1] In order to proceed further we need the notion of the *entropy* of a quantum gas, which characterizes the disorder of the gas. We will explain some basics about entropy in Exercise 6.6, where we will also derive the following expression:

$$
S = \pm k_{\mathrm{B}} \sum_{\boldsymbol{k}} \left[(1 \pm N(\boldsymbol{k}))\,\ln(1 \pm N(\boldsymbol{k})) \mp N(\boldsymbol{k})\,\ln N(\boldsymbol{k})\right]\,. \tag{6.24}
$$

[1] Historically it is interesting to know that Einstein was the referee of the manuscript of the Indian physicist Bose, unknown at that time, which had been submitted to Zeitschrift für Physik. He realized its deep implications and improved the paper by suggesting some valuable comments to Bose.

k_B represents the *Boltzmann constant*. Again, the upper sign holds for bosons and the lower sign for fermions. We will now show that entropy will increase permanently, i.e.

$$\frac{\mathrm{d}S}{\mathrm{d}t} \geq 0 . \tag{6.25}$$

This follows from the Boltzmann equation (6.18). For the proof we restrict ourselves to the case of bosons; for fermions the arguments are the same. We start from (6.24), which we differentiate with respect to time; then we employ again the Boltzmann equation (6.18). This gives

$$\begin{aligned}
\frac{\mathrm{d}S}{\mathrm{d}t} &= k_B \sum_{\boldsymbol{k}} \left\{ \frac{\partial N(\boldsymbol{k})}{\partial t} \{\ln[1+N(\boldsymbol{k})] - \ln N(\boldsymbol{k})\} \right. \\
&\quad \left. + [1+N(\boldsymbol{k})] \frac{\partial N(\boldsymbol{k})}{\partial t} \frac{1}{[1+N(\boldsymbol{k})]} - N(\boldsymbol{k}) \frac{\partial N(\boldsymbol{k})}{\partial t} \frac{1}{N(\boldsymbol{k})} \right\} \\
&= k_B \sum_{\boldsymbol{k}} \frac{\partial N(\boldsymbol{k})}{\partial t} \{\ln[1+N(\boldsymbol{k})] - \ln N(\boldsymbol{k})\} \\
&= k_B \sum_{\boldsymbol{k}} \sum_{\boldsymbol{k}'} \sum_{\boldsymbol{q}} \frac{2\pi}{\hbar} |\bar{\mathcal{V}}(\boldsymbol{q})|^2 \\
&\quad \times \delta \left[\frac{\hbar^2}{2m} (|\boldsymbol{k}+\boldsymbol{q}|^2 + |\boldsymbol{k}-\boldsymbol{q}|^2 - \boldsymbol{k}^2 - \boldsymbol{k}'^2) \right] \\
&\quad \times \{N(\boldsymbol{k}+\boldsymbol{q}) N(\boldsymbol{k}'-\boldsymbol{q}) [1+N(\boldsymbol{k})][1+N(\boldsymbol{k}')] \\
&\quad - N(\boldsymbol{k}) N(\boldsymbol{k}') [1+N(\boldsymbol{k}+\boldsymbol{q})][1+N(\boldsymbol{k}'-\boldsymbol{q})]\} \\
&\quad \times \{\ln[1+N(\boldsymbol{k})] - \ln N(\boldsymbol{k})\} .
\end{aligned} \tag{6.26a}$$

From (6.26a) three different, but similar, equations can be derived. The first one follows by making the replacements $\boldsymbol{q} \to -\boldsymbol{q}$, $\boldsymbol{k} \to \boldsymbol{k}+\boldsymbol{q}$, $\boldsymbol{k}' \to \boldsymbol{k}'-\boldsymbol{q}$. With the fact that $|\bar{\mathcal{V}}(-\boldsymbol{q})| = |\bar{\mathcal{V}}(\boldsymbol{q})|$ taken into account this equation reads

$$\begin{aligned}
\frac{\mathrm{d}S}{\mathrm{d}t} &= k_B \sum_{\boldsymbol{k}} \sum_{\boldsymbol{k}'} \sum_{\boldsymbol{q}} \frac{2\pi}{\hbar} |\bar{\mathcal{V}}(\boldsymbol{q})|^2 \\
&\quad \times \delta \left\{ \frac{\hbar^2}{2m} \left[|\boldsymbol{k}|^2 + |\boldsymbol{k}'|^2 - (\boldsymbol{k}+\boldsymbol{q})^2 - (\boldsymbol{k}'-\boldsymbol{q})^2 \right] \right\} \\
&\quad \times \{N(\boldsymbol{k}) N(\boldsymbol{k}') [1+N(\boldsymbol{k}+\boldsymbol{q})][1+N(\boldsymbol{k}'-\boldsymbol{q})] \\
&\quad - N(\boldsymbol{k}+\boldsymbol{q}) N(\boldsymbol{k}'-\boldsymbol{q}) [1+N(\boldsymbol{k})][1+N(\boldsymbol{k}')]\} \\
&\quad \times \{\ln[1+N(\boldsymbol{k}+\boldsymbol{q})] - \ln N(\boldsymbol{k}')\} .
\end{aligned} \tag{6.26b}$$

We obtain the third equation by making the replacements $\boldsymbol{q} \to -\boldsymbol{q}$ and $\boldsymbol{k} \to \boldsymbol{k}'$ in (6.26a):

$$\begin{aligned}
\frac{\mathrm{d}S}{\mathrm{d}t} &= k_B \sum_{\boldsymbol{k}} \sum_{\boldsymbol{k}'} \sum_{\boldsymbol{q}} \frac{2\pi}{\hbar} |\bar{\mathcal{V}}(\boldsymbol{q})|^2 \\
&\quad \times \delta \left[\frac{\hbar^2}{2m} \left(|\boldsymbol{k}'-\boldsymbol{q}|^2 + |\boldsymbol{k}+\boldsymbol{q}|^2 - \boldsymbol{k}'^2 - \boldsymbol{k}^2 \right) \right] \\
&\quad \times \{N(\boldsymbol{k}'-\boldsymbol{q}) N(\boldsymbol{k}+\boldsymbol{q}) [1+N(\boldsymbol{k}')][1+N(\boldsymbol{k})]
\end{aligned}$$

$$- N(\boldsymbol{k})N(\boldsymbol{k}')[1 + N(\boldsymbol{k}' - \boldsymbol{q})][1 + N(\boldsymbol{k} + \boldsymbol{q})]\}$$
$$\times \{\ln[1 + N(\boldsymbol{k}')] - \ln N(\boldsymbol{k}')\} \ . \tag{6.26c}$$

The fourth equation results from the substitutions $\boldsymbol{q} \to -\boldsymbol{q}$ and $\boldsymbol{k} \to \boldsymbol{k}'$ in (6.26b):

$$\begin{aligned}\frac{\mathrm{d}S}{\mathrm{d}t} &= k_{\mathrm{B}} \sum_{\boldsymbol{k}} \sum_{\boldsymbol{k}'} \sum_{\boldsymbol{q}} \frac{2\pi}{\hbar} | \bar{\mathcal{V}}(\boldsymbol{q}) |^2 \\ &\times \delta \left\{ \frac{\hbar^2}{2m} \left[\boldsymbol{k}'^2 + \boldsymbol{k}^2 - (\boldsymbol{k}' - \boldsymbol{q})^2 - (\boldsymbol{k} + \boldsymbol{q})^2 \right] \right\} \\ &\times \{N(\boldsymbol{k}')N(\boldsymbol{k})[1 + N(\boldsymbol{k}' - \boldsymbol{q})][1 + N(\boldsymbol{k} + \boldsymbol{q})] \\ &- N(\boldsymbol{k}' - \boldsymbol{q})N(\boldsymbol{k} + \boldsymbol{q})[1 + N(\boldsymbol{k}')][1 + N(\boldsymbol{k})]\} \\ &\times \{\ln[1 + N(\boldsymbol{k}' - \boldsymbol{q})] - \ln N(\boldsymbol{k}')\} \ . \end{aligned} \tag{6.26d}$$

Addition of all four equations (6.26a)–(6.26d) yields

$$\begin{aligned}4\frac{\mathrm{d}S}{\mathrm{d}t} &= k_{\mathrm{B}} \sum_{\boldsymbol{k}} \sum_{\boldsymbol{k}'} \sum_{\boldsymbol{q}} \frac{2\pi}{\hbar} | \bar{\mathcal{V}}(\boldsymbol{q}) |^2 \\ &\times \delta \left[\frac{\hbar^2}{2m} \left(|\boldsymbol{k} + \boldsymbol{q}|^2 + |\boldsymbol{k}' - \boldsymbol{q}|^2 - \boldsymbol{k}'^2 - \boldsymbol{k}^2 \right) \right] \\ &\times \{N(\boldsymbol{k} + \boldsymbol{q})N(\boldsymbol{k}' - \boldsymbol{q}')[1 + N(\boldsymbol{k})][1 + N(\boldsymbol{k}')] \\ &- N(\boldsymbol{k})N(\boldsymbol{k}')[1 + N(\boldsymbol{k} + \boldsymbol{q})][1 + N(\boldsymbol{k}' - \boldsymbol{q})]\} \\ &\times \{\ln \{N(\boldsymbol{k} + \boldsymbol{q})N(\boldsymbol{k}' - \boldsymbol{q})[1 + N(\boldsymbol{k})][1 + N(\boldsymbol{k}')]\} \\ &- \ln \{N(\boldsymbol{k})N(\boldsymbol{k}')[1 + N(\boldsymbol{k} + \boldsymbol{q})][1 + N(\boldsymbol{k}' - \boldsymbol{q})]\}\} \ . \end{aligned} \tag{6.27}$$

With the abbreviations

$$\tilde{u} = N(\boldsymbol{k} + \boldsymbol{q})N(\boldsymbol{k}' - \boldsymbol{q})[1 + N(\boldsymbol{k})][1 + N(\boldsymbol{k}')] \tag{6.28a}$$

and

$$\tilde{v} = N(\boldsymbol{k})N(\boldsymbol{k}')[1 + N(\boldsymbol{k} + \boldsymbol{q})][1 + N(\boldsymbol{k}' - \boldsymbol{q})] \ , \tag{6.28b}$$

the two last factors appearing in (6.27) can be written as

$$(\tilde{u} - \tilde{v})(\ln \tilde{u} - \ln \tilde{v}) \ . \tag{6.29}$$

This quantity is positive for $\tilde{u} > \tilde{v}$ and also for $\tilde{u} < \tilde{v}$; it vanishes for $\tilde{u} = \tilde{v}$. The condition $\tilde{u} = \tilde{v}$ is just the equilibrium condition (6.20) or (6.19). Hence, we conclude that

$$\frac{\mathrm{d}S}{\mathrm{d}t} > 0 \tag{6.30}$$

always holds except for equilibrium. Entropy increases steadily and reaches its maximum once the system is in equilibrium. This important conclusion contains, on the one hand, the *second law of thermodynamics* (increase of entropy) and, on the other hand, the so-called *H theorem from Boltzmann* (entropy is at a maximum in equilibrium). We consider now the *classical Boltzmann equation*, which we obtain from (6.18) once we perform the following limiting cases:

$$\begin{aligned}
&\text{(a)} \quad \hbar \to 0 && \text{(classical limit)}\,, \\
&\text{(b)} \quad L^3 \to \infty && \text{(large volume)}\,, \\
&\text{(c)} \quad 1 \pm N(\boldsymbol{k}) \approx 1 && \text{(gas is far from degeneracy)}\,.
\end{aligned} \tag{6.31}$$

The first two conditions do not need further explanation. Condition (c) means that every state $\boldsymbol{k}$ is occupied according to a distribution function $N(\boldsymbol{k})$. However, the latter should be close to zero so that the average occupation numbers for a momentum interval are equal for the classical and quantum-mechanical case.

We then say that *the gas is far from degeneracy*. Now we introduce further the classical *velocity distribution function* $f(v)$ via

$$\sum_{\boldsymbol{k}} N(\boldsymbol{k}) \Longrightarrow L^3 \int \mathrm{d}^3 v f(v) \tag{6.32}$$

and take notice of the substitutions

$$\hbar \boldsymbol{k} \to m\boldsymbol{v} \qquad \text{and} \qquad \hbar \boldsymbol{q} \to m\boldsymbol{u}\,. \tag{6.33}$$

Thus, the quantum-mechanical momentum $\hbar\boldsymbol{k}$ and momentum transfer $\hbar\boldsymbol{q}$ of the collision are identified with the classical particle momentum $m\boldsymbol{v}$ and classical momentum transfer $m\boldsymbol{u}$, respectively. Then

$$\sum_{\boldsymbol{q}} \Rightarrow L^3 \int \frac{\mathrm{d}^3 q}{(2\pi)^3} = \frac{L^3 m^3}{\hbar^3 (2\pi)^3} \int \mathrm{d}^3 u \tag{6.34}$$

and (6.18) becomes

$$\begin{aligned}
\frac{\partial f(\boldsymbol{v})}{\partial t} &= \int \mathrm{d}^3 v' \int \mathrm{d}^3 u \frac{L^6 m^3}{\hbar^3 (2\pi)^3} \frac{2\pi}{\hbar} \left| \bar{\mathcal{V}}\left(\frac{m\boldsymbol{u}}{\hbar}\right)\right|^2 \\
&\quad \times \delta\left(\frac{m}{2}|\boldsymbol{v}+\boldsymbol{u}|^2 + \frac{m}{2}|\boldsymbol{v}'-\boldsymbol{u}|^2 - \frac{m}{2}\boldsymbol{v}^2 - \frac{m}{2}\boldsymbol{v}'^2\right) \\
&\quad \times \left[f(\boldsymbol{v}+\boldsymbol{u}) f(\boldsymbol{v}'-\boldsymbol{u}) - f(\boldsymbol{v}) f(\boldsymbol{v}')\right]\,.
\end{aligned} \tag{6.35}$$

Rewriting the Fourier transform (6.10) as

$$\begin{aligned}
\bar{\mathcal{V}}(\boldsymbol{q}) &= \frac{(2\pi)^{3/2}}{L^3} \int \mathrm{d}^3 x\, \mathcal{V}(\boldsymbol{x}) \frac{\mathrm{e}^{\mathrm{i}\boldsymbol{q}\cdot\boldsymbol{x}}}{(2\pi)^{3/2}} \\
&= \frac{(2\pi\hbar/m)^{3/2}}{L^3} \int \mathrm{d}^3 x\, \mathcal{V}(\boldsymbol{x}) \frac{\mathrm{e}^{\mathrm{i}m\boldsymbol{u}\cdot\boldsymbol{x}/\hbar}}{(2\pi\hbar/m)^{3/2}} \equiv \left(\frac{2\pi\hbar}{m}\right)^{3/2} \frac{\bar{\bar{\mathcal{V}}}(\boldsymbol{u})}{L^3}
\end{aligned} \tag{6.36}$$

and setting the transition probability as

$$\frac{2\pi}{\hbar} \left|\bar{\bar{\mathcal{V}}}(\boldsymbol{u})\right|^2 = w(\boldsymbol{u})\,, \tag{6.37}$$

we find that (6.35) becomes

$$\begin{aligned}
\frac{\partial f(\boldsymbol{v})}{\partial t} &= \int \mathrm{d}^3 v' \int \mathrm{d}^3 u\, w(\boldsymbol{u}) \\
&\quad \times \delta\left[\frac{m}{2}(\boldsymbol{v}+\boldsymbol{u})^2 + \frac{m}{2}(\boldsymbol{v}'-\boldsymbol{u})^2 - \frac{m}{2}\boldsymbol{v}^2 - \frac{m}{2}\boldsymbol{u}^2\right] \\
&\quad \times \left[f(\boldsymbol{v}+\boldsymbol{u}) f(\boldsymbol{v}'-\boldsymbol{u}) - f(\boldsymbol{v}) f(\boldsymbol{v}')\right]\,.
\end{aligned} \tag{6.38}$$

This is the *classical Boltzmann equation*; it corresponds to the master equation (6.15) and the quantum-mechanical Boltzmann equation (6.18). The quantity $\hbar$ does not occur anymore. Note, however, that we tacitly made the descision in (6.37) to replace the quantum-mechanical transition probability by a (classical) expression $w(\boldsymbol{u})$ that is more or less independent of $\hbar$. This makes sense because for both the quantum-mechanical and for the classical case a probability exists for the scattering of two particles with momentum transfer $m\boldsymbol{u}$. Each single term in (6.38) may be interpreted and understood according to the master equation (6.15). The δ function expresses energy conservation in each single collision. We underline these points further with the following exercises.

EXERCISE

6.3 Equilibrium Solution of the Classical Boltzmann Equation

Problem. Show that

$$f(v) = \exp\left(-mv^2/2k_{\mathrm{B}}T\right)$$

represents an equilibrium solution of the classical Boltzmann equation. Is this solution unique?

Solution. We denote

$$\begin{aligned} f_1 &= f(\boldsymbol{v}+\boldsymbol{u}) , & f_2 &= f(\boldsymbol{v}'-\boldsymbol{u}) , \\ f_1' &= f(\boldsymbol{v}) , \quad \text{and} & f_2' &= f(\boldsymbol{v}') . \end{aligned} \tag{1}$$

The integrand of (6.38) vanishes if

$$f_1 f_2 = f_1' f_2' \quad \text{and} \quad \ln f_1 + \ln f_2 = \ln f_1' + \ln f_2' . \tag{2}$$

Functions $a(\boldsymbol{v})$ are called *collision invariants* if they fulfill this balance for all possible collisions. In fact, (2) states that the value of the quantity $\ln f_1 + \ln f_2$ remains the same before and after the collision. As a result of energy and momentum conservation

$$\begin{aligned} a_i(\boldsymbol{v}) &= m v_i \quad (i = 1,2,3) , \\ a_4(\boldsymbol{v}) &= \frac{m}{2} v^2 \end{aligned}$$

represent collision invariants. Conservation of particle number yields another collision invariant $a_0(\boldsymbol{v}) = 1$. In addition to these five functions a_i, only the angular momentum may be thought of as a linear independent collision invariant. Since usually "infinitely extended" gases are considered, the definition of an angular momentum $\boldsymbol{L} = \boldsymbol{r} \times \boldsymbol{p}$ makes no sense since no reference point exists for $\boldsymbol{r}$. Consequently, $\ln f(\boldsymbol{v},\boldsymbol{r},t)$ has to be a linear combination of the a_i:

$$\ln f(\boldsymbol{v},\boldsymbol{r},t) = \alpha(\boldsymbol{r},t) + \boldsymbol{\lambda}(\boldsymbol{r},t) m\boldsymbol{v} - \beta(\boldsymbol{r},t)\frac{m}{2}\boldsymbol{v}^2 .$$

Exercise 6.3.

We define the temperature $T = 1/(k_B\beta)$, the chemical potential $\mu(\boldsymbol{r},t) = \alpha(\boldsymbol{r},t)/\beta(\boldsymbol{r},t)$ and the *average flow velocity* $\boldsymbol{u} = \boldsymbol{\lambda}/\beta(\boldsymbol{r},t)$ and obtain

$$f(\boldsymbol{v},\boldsymbol{r},t) = \frac{N}{V}\left(\frac{m}{2\pi T}\right)^{3/2} \exp\left\{\beta(\boldsymbol{r},t)\left(\mu(\boldsymbol{r},t) - \frac{m}{2}[\boldsymbol{v}-\boldsymbol{u}(\boldsymbol{r},t)]^2\right)\right\}, \quad (3)$$

where the particle density $\rho = N/V$ has been introduced so that

$$\int d^3v f(\boldsymbol{v},\boldsymbol{r},t) = \rho .$$

If the average velocity $\boldsymbol{u} = 0$ and $\mu = 0$, we obtain the simplified form

$$f(\boldsymbol{v},\boldsymbol{r},t) = \rho\left(\frac{m}{2\pi k_B T}\right)^{3/2} \exp\left(-m\boldsymbol{v}^2/2k_B T\right) . \quad (4)$$

EXERCISE

6.4 From the Entropy Formula for the Bose (Fermi) Gas to the Classical Entropy Formula

Problem. Show that expression (6.24) for the entropy of a quantum gas becomes the classical formula for the entropy

$$S = -k_B \sum_{\boldsymbol{k}} N(\boldsymbol{k}) \ln N(\boldsymbol{k}) ,$$

when the gas is far from degeneracy.

Solution. The expression for entropy is given by

$$S = -k_B \sum_{\boldsymbol{k}} \{N(\boldsymbol{k}) \ln N(\boldsymbol{k}) \mp [1 \pm N(\boldsymbol{k})] \ln[1 \pm N(\boldsymbol{k})]\} ,$$

where the upper sign holds for bosons and the lower sign for fermions. As we have already stressed a classical ideal gas is characterized such that the number of possible states for the particles is large and that the average occupation number $N(\boldsymbol{k})$ of each single state remains small, i.e. the particles occupy a large number of states independently. The effects of quantum statistics, which for the case of fermions allows up to only one fermion per state but for bosons allows the average occupation number to become arbitrary large, do not play a role in the classical limit. We denote gases for which the average occupation number $N(\boldsymbol{k}) \ll 1$ remains small as *"far from degeneracy"*. Since then $N(\boldsymbol{k}) \ll 1$, we get $1 \pm N(\boldsymbol{k}) \approx 1$, and the expression for the entropy of the Bose (Fermi) gas becomes

$$S_{\text{class.}} = -k_B \sum_{\boldsymbol{k}} N(\boldsymbol{k}) \ln N(\boldsymbol{k}) .$$

EXERCISE

6.5 Proof of the H Theorem

Problem. Use the classical expression for the entropy derived in the previous problem to prove the H theorem (pronounced 'eta theorem', because H is meant to represent the greak letter eta).

Solution. The classical expression for entropy is given by

$$S = -k_{\mathrm{B}} \sum_{\boldsymbol{k}} N(\boldsymbol{k}) \ln N(\boldsymbol{k}) ,$$

i.e.

$$\frac{\mathrm{d}S}{\mathrm{d}t} = -k_{\mathrm{B}} \sum_{\boldsymbol{k}} \frac{\partial N(\boldsymbol{k})}{\partial t} \left[\ln N(\boldsymbol{k}) + 1\right] .$$

We insert $\partial N(\boldsymbol{k})/\partial t$ and use immediately the classical distribution function (6.35); it follows that

$$\begin{aligned}\frac{\mathrm{d}S}{\mathrm{d}t} &= -k_{\mathrm{B}} \int \mathrm{d}^3 v' \, \mathrm{d}^3 u \frac{m^3 L^6}{(2\pi)^2 \hbar^4} \left| \bar{\mathcal{V}}\left(\frac{m\boldsymbol{u}}{\hbar}\right)\right|^2 \\ &\quad \times \delta\left(\frac{m}{2}|\boldsymbol{v}+\boldsymbol{u}|^2 + \frac{m}{2}|\boldsymbol{v}'-\boldsymbol{u}|^2 - \frac{m}{2}\boldsymbol{v}^2 - \frac{m}{2}\boldsymbol{v}'^2\right) \\ &\quad \times \left[f(\boldsymbol{v}+\boldsymbol{u}) f(\boldsymbol{v}'-\boldsymbol{u}) - f(\boldsymbol{v}) f(\boldsymbol{v}')\right]\left[\ln f(\boldsymbol{v}) + 1\right] .\end{aligned}$$

The equilibrium solution of the Boltzmann equation reads

$$f(\boldsymbol{v}) = C \exp\left(-mv^2/2k_{\mathrm{B}}T\right) .$$

Inserting this solution into $\mathrm{d}S/\mathrm{d}t$, we immediately realize that the bracket

$$[f(\boldsymbol{v}+\boldsymbol{u})f(\boldsymbol{v}'-\boldsymbol{u}) - f(\boldsymbol{v})f(\boldsymbol{v}')]$$

vanishes, i.e. $\mathrm{d}S/\mathrm{d}t = 0$ in equilibrium.

On the other hand,

$$f(\boldsymbol{v}+\boldsymbol{u})f(\boldsymbol{v}'-\boldsymbol{u}) - f(\boldsymbol{v})f(\boldsymbol{v}') = 0$$

follows from the requirement $\mathrm{d}S/\mathrm{d}t = 0$; hence, we conclude that the Maxwell–Boltzmann distribution

$$f(\boldsymbol{v}) = C \exp\left[-\left(\frac{mv^2}{2k_{\mathrm{B}}T}\right)\right]$$

represents a unique solution.

6.2 Nearly Ideal, Degenerate Bose–Einstein Gases

We again focus on (6.23), which we will now study in more detail for the case of a Bose–Einstein gas. Considering the Bose–Einstein distribution law we realize that the number of molecules or atoms in the state with energy $E = 0$ is given by

$$N_0 = \frac{1}{A-1} = \frac{1}{e^{x_0}-1}, \tag{6.39a}$$

where the normalization constant A has been written as

$$A = e^{x_0}. \tag{6.39b}$$

Rearranging (6.39a) we get:

$$e^{-x_0} = \frac{N_0}{N_0+1}. \tag{6.40}$$

Since N_0 cannot become negative and since (6.40) approaches 1 for large N_0, we deduce that

$$0 \le e^{-x_0} < 1. \tag{6.41}$$

The value for the normalization constant x_0 is determined from the requirement to conserve the particle number:

$$N = \sum_k \frac{1}{e^{x_0}\exp(E_k/k_BT)-1}. \tag{6.42}$$

The ratio between the total number of particles and the number of particles N_0 in the state with $E=0$ is given by

$$\frac{N_0}{N} = \frac{1}{N}\left(\frac{1}{e^{x_0}-1}\right). \tag{6.43}$$

If $e^{x_0} > 1$ it holds that $N_0/N < 1$; if, on the other hand, $e^{x_0} \approx 1$ we get $(N_0/N) \to 1$. Hence, the state with $E=0$ is strongly populated only for $e^{x_0} \approx 1$. Particles in such a state are called a *condensate*. In order to analyse the expression for N further, we separate the state with $E=0$ from (6.42) and change the sum over $\boldsymbol{k}$ into an integral:

$$N = \frac{1}{e^{x_0}-1} + \frac{1}{4}\left(\frac{8m}{\hbar^2}\right)^{3/2}\pi L^3 \int_0^\infty dE \frac{E^{1/2}}{e^{x_0}\exp(E/k_BT)-1}, \tag{6.44}$$

where we have also performed a substitution from momentum to energy $E = (\hbar k)^2/2m$. With a further substitution the integral can be written as

$$\frac{1}{4}\left(\frac{8mk_BT}{\hbar^2}\right)^{3/2}\pi L^3 F' \tag{6.45}$$

with

$$F' = \int_0^\infty \frac{x^{1/2}\,dx}{e^{x_0}e^x-1} \tag{6.46}$$

and

$$x = E/k_BT.$$

We expand the integral:

$$\begin{aligned} F' &= \int_0^\infty \frac{x^{1/2}}{e^{x_0}e^x}\left(\frac{1}{1-e^{-x}e^{-x_0}}\right)dx \\ &= \int_0^\infty x^{1/2}\left(e^{-x}e^{-x_0} - e^{-2x}e^{-2x_0} + e^{-3x}e^{-3x_0} \pm \ldots\right)dx. \end{aligned} \tag{6.47}$$

We integrate term by term and use

$$\int_0^\infty \mathrm{e}^{-ax} x^{1/2}\,\mathrm{d}x \;=\; \frac{1}{2}\pi^{1/2}a^{-3/2}\,; \tag{6.48}$$

then

$$\begin{aligned} F' &= \frac{\pi^{1/2}}{2}\left(\mathrm{e}^{-x_0} + 2^{-3/2}\,\mathrm{e}^{-2x_0} + 3^{-3/2}\,\mathrm{e}^{-3x_0} + \ldots\right) \\ &= \frac{\pi^{1/2}}{2}F\,. \end{aligned} \tag{6.49}$$

Hence, we obtain

$$N \;=\; \frac{(2\pi m k_\mathrm{B} T)^{3/2}}{\hbar^3} L^3 F + \frac{1}{\mathrm{e}^{x_0} - 1} \tag{6.50}$$

or

$$\mathrm{e}^{-x_0} + 2^{-3/2}\,\mathrm{e}^{-2x_0} + 3^{-3/2}\,\mathrm{e}^{-3x_0} + \ldots \;=\; \frac{N\hbar^3}{(2\pi m k_\mathrm{B} T)^{3/2} L^3}\,, \tag{6.51}$$

if e^{x_0} is not close to 1. We can rewrite (6.51) as

$$T \;=\; \frac{\hbar^2}{2\pi m k_\mathrm{B}}\left[\frac{1}{L^3 F}\left(N - \frac{1}{\mathrm{e}^{x_0} - 1}\right)\right]^{2/3}. \tag{6.52}$$

Owing to (6.41) the largest value of e^{-x_0} becomes ≈ 1, so the largest value of F' [in (6.50)] is

$$F_0 \;=\; 1 + 2^{-3/2} + 3^{-3/2} + \ldots \;=\; \zeta(3/2) \approx 2.612\,, \tag{6.53}$$

where $\zeta(x) = \sum_{p=1}^{\infty} p^{-x}$ represents Riemann's ζ function. In the region where e^{x_0} becomes so large that we can neglect $1/(\mathrm{e}^{x_0} - 1)$ with respect to N we derive with (6.53) that

$$T_{x_0 \gg 1} \;=\; \frac{\hbar^2}{2\pi m k_\mathrm{B}}\left(\frac{N}{L^3 F}\right)^{2/3}. \tag{6.54}$$

In the second region, e^{x_0} is close to 1, so we can approximate F by F_0. However, e^{x_0} should still be sufficiently different from 1 so that $1/(\mathrm{e}^{x_0} - 1)$ can be neglected:

$$T_{x_0 \approx 1} \;=\; \frac{\hbar^2}{2\pi m k_\mathrm{B}}\left(\frac{N}{L^3 F_0}\right)^{2/3} \;=\; T_0 \tag{6.55}$$

In the third region, $F \approx F_0$ and e^{x_0} become so close to 1 that $1/(\mathrm{e}^{x_0} - 1)$ cannot be neglected anymore; instead, we get

$$\frac{1}{\mathrm{e}^{x_0} - 1} \;\approx\; \frac{1}{x_0}\,. \tag{6.56}$$

Thus,

$$N \;=\; \frac{(2\pi m k_\mathrm{B} T)^{3/2} L^3 F_0}{\hbar^3} + \frac{1}{x_0} \tag{6.57}$$

or

$$\frac{1}{x_0} = N\left[1-\left(\frac{T}{T_0}\right)^{3/2}\right] . \tag{6.58}$$

Since $x_0 > 1$ this region corresponds to $T < T_0$. We arrive at the result that a critical temperature

$$T_0 = \frac{\hbar^2}{2\pi m k_B}\left(\frac{\rho}{2.612}\right)^{2/3} \tag{6.59}$$

exists, which depends on the density $\rho = N/L^3$ of the Bose–Einstein gas; beneath this critical temperature nearly all particles are in the state with energy $E = 0$. This phenomenon is known as *Bose–Einstein condensation.* It should be clear that for $T = 0$ all particles occupy the state with $E = 0$.

Only one fluid occurs in nature for which this quantum effect becomes observable macroscopically: liquid helium, ^{4}He. We insert its density into relation (6.59) and obtain the critical temperature $T_0 = 3.13\,\mathrm{K}$. Below this temperature ^{4}He represents a macroscopic Bose–Einstein condensate.[2]

After these introductory statistical considerations we now focus on the quantum features of the Bose–Einstein gas. First we consider the noninteracting ground state of a Bose gas with N particles

$$|\Phi_0(N)\rangle = |N, 0, 0, \cdots\rangle , \tag{6.60}$$

i.e. all N particles occupy the state with $E = 0$. In the following, we will consider a box of volume $V = L^3$ and periodic boundary conditions. The creation and annihilation operators for the state with $E = \boldsymbol{k} = 0$ are denoted as $\hat{a}_0^\dagger$ and $\hat{a}_0$. The usual relations

$$\begin{aligned} \hat{a}_0|\Phi_0(N)\rangle &= N^{1/2}|\Phi_0(N-1)\rangle , \\ \hat{a}_0^\dagger|\Phi_0(N)\rangle &= (N+1)^{1/2}|\Phi_0(N+1)\rangle \end{aligned} \tag{6.61}$$

hold. Neither $\hat{a}_0$ nor $\hat{a}_0^\dagger$ annihilate the ground state; this is in contrast to the well-known property $\hat{a}_0|0\rangle = 0$ of a vacuum state, in which no particles are present. However, the Bose operators multiply the ground state with large numbers of the order $\sqrt{N}$. Therefore it is convenient to switch to new operators $\hat{b}_0$ and $\hat{b}_0^\dagger$ which we define as

$$\hat{b}_0 = V^{-1/2}\hat{a}_0 \qquad \text{and} \qquad \hat{b}_0^\dagger = V^{-1/2}\hat{a}_0^\dagger . \tag{6.62}$$

Since

$$[\hat{a}_0, \hat{a}_0^\dagger] = 1 ,$$

we get

$$[\hat{b}_0, \hat{b}_0^\dagger] = V^{-1} , \tag{6.63}$$

and

[2] Recently Bose–Einstein condensates of various total spin $F = 0$ atoms have been observed: see e.g. M.H. Anderson, J.R. Ensher, M.R. Matthews, C.E. Wiemann, E.A. Cornell: Science **269** (1995) 198; K.B. Davis, M.O. Mewes, M.R. Andrews, N.J. van Druten, D.S. Durfee, D.M. Kurn, W. Ketterle: Phys. Rev. Lett. **75** (1995) 3969; C.C. Bradley, C.A. Sackett, J.J. Tollett, R.G. Hulet: Phys. Rev. Lett. **75** (1995) 1687.

$$\begin{aligned}\hat{b}_0|\Phi_0(N)\rangle &= \left(\frac{N}{V}\right)^{1/2}|\Phi_0(N-1)\rangle\,,\\ \hat{b}_0^\dagger|\Phi_0(N)\rangle &= \left(\frac{N+1}{V}\right)^{1/2}|\Phi_0(N+1)\rangle\,.\end{aligned} \tag{6.64}$$

We are interested in the behavior of a gas, i.e. in large particle numbers and volumes. Hence, we consider the so-called *thermodynamic limit*: $N \to \infty$, $V \to \infty$, but constant density $\rho = N/V \to$ const. Then $\hat{b}_0$ and $\hat{b}_0^\dagger$ multiply the ground state with a constant whereas the commutator becomes

$$[\hat{b}_0, \hat{b}_0^\dagger] \to 0 \tag{6.65}$$

in the thermodynamic limit, i.e. we obtain the classical limit for which $\hat{b}_0$ and $\hat{b}_0^\dagger$ are ordinary C numbers.[3] For the case of photons (see Chap. 1) we have already seen that a field for which the quanta are particles behaves classically once a large number of particles occupy the same state.

So far we have considered a noninteracting ground state; now we investigate a system of bosons at $T = 0$, which is slightly nonideal. We keep the interaction term from (6.11) but, for simplicity, we replace $\overline{v}(\boldsymbol{q})$ by the factor $g/2V$:

$$\hat{H} = \sum_{\boldsymbol{k}} \hbar\omega_{\boldsymbol{k}} \hat{a}_{\boldsymbol{k}}^\dagger \hat{a}_{\boldsymbol{k}} + \frac{g}{2V} \sum_{\boldsymbol{k}_1\boldsymbol{k}_2\boldsymbol{k}_3\boldsymbol{k}_4} \hat{a}_{\boldsymbol{k}_1}^\dagger \hat{a}_{\boldsymbol{k}_2}^\dagger \hat{a}_{\boldsymbol{k}_3} \hat{a}_{\boldsymbol{k}_4} \delta_{\boldsymbol{k}_1+\boldsymbol{k}_2,\boldsymbol{k}_3+\boldsymbol{k}_4}\,. \tag{6.66}$$

g is often called the *pseudopotential*. Owing to momentum conservation we manipulate the wave-number vectors $\boldsymbol{k}_1, \dots, \boldsymbol{k}_4$ appearing in (6.66). Furthermore, we then separate the contribution for $\boldsymbol{k} = 0$ as an interaction term of the Hamiltonian

$$\begin{aligned}\hat{H}_{\text{int}} \approx \frac{g}{2V}\Big[&\hat{a}_0^\dagger \hat{a}_0^\dagger \hat{a}_0 \hat{a}_0\\ &+ {\sum_{\boldsymbol{k}}}' \left(4\hat{a}_{\boldsymbol{k}}^\dagger \hat{a}_{\boldsymbol{k}} \hat{a}_0^\dagger \hat{a}_0 + \hat{a}_{\boldsymbol{k}}^\dagger \hat{a}_{-\boldsymbol{k}}^\dagger \hat{a}_0 \hat{a}_0 + \hat{a}_0^\dagger \hat{a}_0^\dagger \hat{a}_{\boldsymbol{k}} \hat{a}_{-\boldsymbol{k}}\right)\Big]\,;\end{aligned} \tag{6.67}$$

here $\sum'$ signals that terms with $\boldsymbol{k} = 0$ have to be omitted from the summation. Also, we have kept only terms up to second order in $\hat{a}_{\boldsymbol{k}}$ and $\hat{a}_{\boldsymbol{k}}^\dagger$ and have neglected fourth-order terms; momentum conservation means that terms with an odd number of these operators do not exist.

As a further approximation we replace $\hat{a}_0$ and $\hat{a}_0^\dagger$ in (6.61) and (6.67) by C numbers:

$$\hat{a}_0, \hat{a}_0^\dagger \to \sqrt{N_0}\,, \tag{6.68}$$

where N_0 represents the number of particles in the state with $\boldsymbol{k} = 0$. It follows that

$$\begin{aligned}\hat{H}_{\text{int}} \approx \frac{g}{2V}\Big(&N_0^2 + 2N_0 {\sum_{\boldsymbol{k}}}' (\hat{a}_{\boldsymbol{k}}^\dagger \hat{a}_{\boldsymbol{k}} + \hat{a}_{-\boldsymbol{k}}^\dagger \hat{a}_{-\boldsymbol{k}})\\ &+ N_0 {\sum_{\boldsymbol{k}}}' (\hat{a}_{\boldsymbol{k}}^\dagger \hat{a}_{-\boldsymbol{k}}^\dagger + \hat{a}_{\boldsymbol{k}} \hat{a}_{-\boldsymbol{k}})\Big)\,.\end{aligned} \tag{6.69}$$

[3] The phrase *C numbers* means "commuting" numbers, i.e. ordinary real or complex numbers.

Analogously we get for the particle number operator

$$\hat{N} = \sum \hat{a}_k^\dagger \hat{a}_k = N_0 + \frac{1}{2}\sum_k{}' (\hat{a}_k^\dagger \hat{a}_k + \hat{a}_{-k}^\dagger \hat{a}_{-k}) . \tag{6.70}$$

We set $\rho = N/V$, insert (6.70) into (6.69), and obtain

$$\begin{aligned}\hat{H} &= \frac{1}{2}Vg\rho^2 + \frac{1}{2}\sum_k{}' \left[(E_k^0 + \rho g)(\hat{a}_k^\dagger \hat{a}_k + \hat{a}_{-k}^\dagger \hat{a}_{-k})\right. \\ &\quad \left. + \rho g(\hat{a}_k^\dagger \hat{a}_{-k}^\dagger + \hat{a}_k \hat{a}_{-k})\right] , \qquad \left(E_k^0 = \frac{\hbar^2 k^2}{2m}\right) . \end{aligned} \tag{6.71}$$

Here we have assumed that $N - N_0 \ll N$ and, consequently, that terms like $(\sum_k{}' \hat{a}_k^\dagger \hat{a}_k)^2$ can be neglected. This also means that most particles are in the condensate. We look for a solution of our problem (6.71), i.e. the energy eigenvalues of the Hamiltonian (6.71). For this special case the problem can be solved exactly with a canonical transformation *(Bogoliubov transformation)*:

We introduce new operators[4]

$$\begin{aligned}\hat{a}_k &= u_k \hat{A}_k + v_k \hat{A}_{-k}^\dagger , \quad & \hat{A}_k &= u_k \hat{a}_k - v_k \hat{a}_{-k}^\dagger , \\ \hat{a}_k^\dagger &= u_k \hat{A}_k^\dagger + v_k \hat{A}_{-k} , \quad & \hat{A}_k^\dagger &= v_k \hat{a}_k - u_k \hat{a}_{-k}^\dagger . \end{aligned} \tag{6.72}$$

The coefficients should be real and spherically symmetric. The transformation is canonical, once the new operators obey the canonical commutation rules:

$$\begin{aligned}\left[\hat{A}_k, \hat{A}_{k'}^\dagger\right] &= \delta_{k,k'} , \\ \left[\hat{A}_k, \hat{A}_{k'}\right] &= \left[\hat{A}_k^\dagger, \hat{A}_{k'}^\dagger\right] = 0 . \end{aligned} \tag{6.73}$$

By inserting (6.72) into (6.73) we realize that this is only fulfilled if

$$u_k^2 - v_k^2 = 1 \tag{6.74}$$

holds for every $\boldsymbol{k}$. For reasons that will become clear later on we call the new operators *quasiparticle operators*. We insert (6.72) into (6.71) and obtain

$$\begin{aligned}\hat{H} &= \frac{1}{2}Vg\rho^2 + \sum_k{}' \left[(E_k^0 + \rho g)v_k^2 + \rho g u_k v_k\right] \\ &\quad + \frac{1}{2}\sum_k{}' \left\{\left[(E_k^0 + \rho g)(u_k^2 + v_k^2) + 2u_k v_k \rho g\right] (\hat{A}_k^\dagger \hat{A}_k + \hat{A}_{-k}^\dagger \hat{A}_{-k})\right\} \\ &\quad + \frac{1}{2}\sum_k{}' \left\{\left[\rho g(u_k^2 + v_k^2) + 2u_k v_k (E_k^0 + \rho g)\right]\right. \\ &\quad \left. \times (\hat{A}_k^\dagger \hat{A}_{-k}^\dagger + \hat{A}_k \hat{A}_{-k})\right\} , \end{aligned} \tag{6.75}$$

where u_k and v_k are related via (6.74); however, the corresponding ratio may be chosen freely. In order to make $\hat{H}$ diagonal in $\hat{A}_k^\dagger$, $\hat{A}_k$ we use this freedom to eliminate the last line in (6.75); we require that

$$\rho g(u_k^2 + v_k^2) + 2u_k v_k (E_k^0 + \rho g) = 0 . \tag{6.76}$$

[4] For the following, see also Sect. 7.1, where we follow the same steps in a slightly different way.

We choose

$$u_{\boldsymbol{k}} = \cosh\varphi_{\boldsymbol{k}} , \qquad v_{\boldsymbol{k}} = -\sinh\varphi_{\boldsymbol{k}} \tag{6.77}$$

such that (6.74) is fulfilled; then (6.76) becomes

$$-\frac{2u_{\boldsymbol{k}}v_{\boldsymbol{k}}}{u_{\boldsymbol{k}}^2+v_{\boldsymbol{k}}^2} = \frac{\rho g}{E_{\boldsymbol{k}}^0+\rho g} \tag{6.78}$$

or

$$\frac{2\sinh\varphi_{\boldsymbol{k}}\cosh\varphi_{\boldsymbol{k}}}{\sinh^2\varphi_{\boldsymbol{k}}+\cosh^2\varphi_{\boldsymbol{k}}} = \frac{\sinh 2\varphi_{\boldsymbol{k}}}{\cosh 2\varphi_{\boldsymbol{k}}} = \frac{\rho g}{E_{\boldsymbol{k}}^0+\rho g} . \tag{6.79}$$

Use of the known trigonometric relations then yields

$$\tanh 2\varphi_{\boldsymbol{k}} = \frac{\rho g}{E_{\boldsymbol{k}}^0+\rho g} . \tag{6.80}$$

Since $-1 \leq \tanh 2\varphi_{\boldsymbol{k}} \leq 1$ this last equation can be fulfilled for all $\boldsymbol{k}$ only once $g > 0$, i.e. the pseudopotential has to be repulsive! It follows that

$$v_{\boldsymbol{k}}^2 = u_{\boldsymbol{k}}^2 - 1 = \tfrac{1}{2}\left[E_{\boldsymbol{k}}^{-1}(E_{\boldsymbol{k}}^0+\rho g) - 1\right] , \tag{6.81}$$

where we have defined

$$E_{\boldsymbol{k}} = \left[(E_{\boldsymbol{k}}^0+\rho g)^2 + (\rho g)^2\right]^{1/2} . \tag{6.82}$$

With the use of (6.74) to (6.82) we are able to write the Hamiltonian (6.75) in its final form:

$$\begin{aligned} \hat{H} &= \frac{1}{2}Vg\rho^2 + \frac{1}{2}{\sum_{\boldsymbol{k}}}' \left[E_{\boldsymbol{k}} - (E_{\boldsymbol{k}}^0+\rho g)\right] \\ &\quad + \frac{1}{2}{\sum_{\boldsymbol{k}}}' E_{\boldsymbol{k}}(\hat{A}_{\boldsymbol{k}}^\dagger\hat{A}_{\boldsymbol{k}} + \hat{A}_{-\boldsymbol{k}}^\dagger\hat{A}_{-\boldsymbol{k}}) . \end{aligned} \tag{6.83}$$

Here $E_{\boldsymbol{k}}^0$ stands for the free particle energies, which are obtained from (6.82) if the interaction $g = 0$, i.e. $E_{\boldsymbol{k}}^0 = (\hbar\boldsymbol{k})^2/2m$. The operator $\hat{A}_{\boldsymbol{k}}^\dagger\hat{A}_{\boldsymbol{k}}$ resembles a particle-number operator and has eigenvalues $0, 1, 2, \ldots$. Hence, the ground state of $\hat{H}$ is determined by the requirement that

$$\hat{A}_{\boldsymbol{k}}|0\rangle = 0 \qquad \text{for all } \boldsymbol{k} \neq 0 . \tag{6.84}$$

Be aware that $|0\rangle$ represents a complicated combination of unperturbed eigenfunctions since neither $\hat{a}_{\boldsymbol{k}}$ nor $\hat{a}_{\boldsymbol{k}}^\dagger$ annihilate the groundstate $|0\rangle$. Because of (6.84) we obtain for the *ground-state energy*

$$E = \langle 0|\hat{H}|0\rangle = \frac{1}{2}Vg\rho^2 + \frac{1}{2}{\sum_{\boldsymbol{k}}}'(E_{\boldsymbol{k}} - E_{\boldsymbol{k}}^0 - \rho g) . \tag{6.85}$$

Furthermore, all excited states correspond to different numbers of noninteracting bosons where each boson possesses the excitation energy $E_{\boldsymbol{k}}$:

$$E_{\boldsymbol{k}} = \left[\left(\frac{\hbar^2\boldsymbol{k}^2}{2m} + \rho g\right)^2 - (\rho g)^2\right]^{1/2} . \tag{6.86a}$$

For $\boldsymbol{k} \to 0$ this becomes

$$E_{\boldsymbol{k}}(\boldsymbol{k}\to 0) \approx \left(\frac{\rho g}{m}\right)^{1/2}\hbar k = c_s\hbar k . \tag{6.86b}$$

We realize that (6.86a) and (6.86b) describe a system of quanta with momentum $\hbar k$. The operators $\hat{A}_k$, $\hat{A}_k^\dagger$ annihilate and create those quanta we have called *quasiparticles*. The quantity c_s in (6.86b) represents a velocity, which we can interprete as the *speed of sound* in the nearly ideal degenerate Bose gas. The long-wave excitations (quasiparticles) corresponding thereto are called *phonons*. In the limit of large momenta (6.86a) becomes

$$E_k \approx \frac{\hbar^2 k^2}{2m} , \tag{6.87}$$

i.e. for large momenta the quasiparticles do not interact with each other. Hence, we obtain the qualitative result presented in Fig. 6.2.

In this section we have established that particles of an interacting gas can be treated as a gas consisting of *noninteracting quasiparticles*. Phonons represent a first example for quasiparticles. Later on we will encounter further examples.

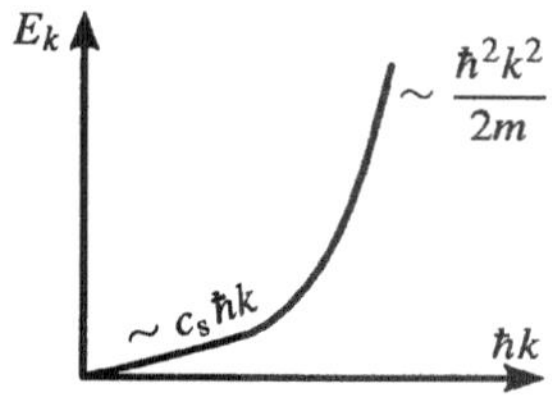

Fig. 6.2. Quasiparticle energies. For low momentum the energy grows linearly with k, for large momentum quadratically. These two domains correspond to the phonon and particle sectors, respectively

EXAMPLE

6.6 Entropy of a Quantum Gas

In order to understand the physical significance of the quantity "entropy" we consider a closed macroscopic *system*. If we separate a small part of the system from the rest, this *subsystem* is no longer closed and can interact with other parts of the system. We now investigate one of these subsystems; its distribution function $w(E)$ is a function of energy alone. For a better understanding we imagine that our system is confined to a fixed box. We use a coordinate system in which the box does not move. Under these circumstances momentum and angular momentum no longer represent constants of motion. Energy remains the only constant of motion that determines the distribution function of the system. With $\Gamma(E)$ we denote the number of quantum states for which the energy is smaller than or equal to E. Thus, the number of states of the subsystem with energies between E and $E + \mathrm{d}E$ is given by

$$\frac{\mathrm{d}\Gamma(E)}{\mathrm{d}E}\mathrm{d}E \tag{1}$$

and the probability distribution for the energies reads

$$W(E) = \frac{\mathrm{d}\Gamma(E)}{\mathrm{d}E} w(E) \tag{2}$$

with the normalization condition

$$\int W(E)\,\mathrm{d}E = 1 . \tag{3}$$

We proceed with the assumption that $W(E)$ possesses a sharp maximum at $E = \overline{E}$ (average) and that it differs significantly from zero only in the vicinity of $\overline{E}$. Therefore we can introduce a width ΔE such that

$$W(\overline{E})\,\Delta E = 1 . \tag{4}$$

Example 6.6.

The number of quantum states that correspond to the energy interval ΔE is given by

$$\Delta\Gamma = \frac{\mathrm{d}\Gamma(\overline{E})}{\mathrm{d}E}\,\Delta E\,. \tag{5}$$

The normalization condition (4) can then be written as

$$w(\overline{E})\,\Delta\Gamma = 1\,. \tag{6}$$

The interval ΔE corresponds approximately to the average fluctuation of the energy of the subsystem; hence, $\Delta\Gamma$ represents the "order of smearing" of the macroscopical state of the subsystem over its microscopical states. For a classical system, $w(E)$ is replaced by its classical distribution function $\rho(E)$ and $\Delta\Gamma$ corresponds to the phase-space element $\Delta p\,\Delta q$ in which the system prevails for nearly all the time. In order to find the number of states we consider the *Bohr–Sommerfeld quantization* rule

$$\frac{1}{2\pi\hbar}\oint p\,\mathrm{d}x = n + \frac{1}{2}$$

for a particle with one degree of freedom. The integral $\oint p\,\mathrm{d}x$ represents the area confined by the closed classical phase trajectory of the particle, i.e. the trajectory in the p-x plane (phase space) of the particle. By dividing this area into cells of area $2\pi\hbar$ each we obtain n cells in total. On the other hand, n represents the number of quantum-mechanical states with energies that are not larger than the corresponding value of the phase trajectory considered. We realize that in the quasiclassical limit every quantum-mechanical state corresponds to a phase-space cell with area $2\pi\hbar$. Put into other words, the number of states in the volume element $\Delta p\,\Delta x$ of phase space is given by

$$\frac{\Delta p\,\Delta x}{2\pi\hbar}\,.$$

If we consider s degrees of freedom this number turns out to be

$$\frac{\Delta p_1\,\Delta x_1}{2\pi\hbar}\,\frac{\Delta p_2\,\Delta x_2}{2\pi\hbar}\cdots\frac{\Delta p_s\,\Delta x_s}{2\pi\hbar}\,,$$

i.e.

$$\frac{\Delta p_s\cdots\Delta p_s\,\Delta x_1\cdots\Delta x_s}{(2\pi\hbar)^s}$$

in total. We introduce the abbreviations $\Delta p = \Delta p_1\cdots\Delta p_s$ and $\Delta q = \Delta x_1\cdots\Delta x_s$ and obtain

$$\frac{\Delta p\,\Delta q}{(2\pi\hbar)^s}\,.$$

This relation yields well known results; for example, for one particle

$$\frac{\mathrm{d}^3p}{(2\pi\hbar)^3}$$

represents the number of states within the momentum region d^3p and the unit volume of the configuration space.

Hence, we arrive at the "weight" of a phase-space cell: *Example 6.6.*

$$\Delta\Gamma = \frac{\Delta p\,\Delta q}{(2\pi\hbar)^s} . \tag{7}$$

Another approach to the derivation of the number of states in the phase-space volume is as follows. The number N of plane waves in a box of length L with periodic boundary conditions for the walls is given by

$$N = \frac{L^3}{(2\pi)^3}\,\mathrm{d}k_x\,\mathrm{d}k_y\,\mathrm{d}k_z \; ;$$

see (2.22). For a small cube $\mathrm{d}V = \mathrm{d}^3x$ we can write

$$\mathrm{d}N = \mathrm{d}V\frac{\mathrm{d}k_x}{2\pi}\frac{\mathrm{d}k_y}{2\pi}\frac{\mathrm{d}k_x}{2\pi} = \mathrm{d}V\frac{\mathrm{d}p_x}{2\pi\hbar}\frac{\mathrm{d}p_y}{2\pi\hbar}\frac{\mathrm{d}p_z}{2\pi\hbar} = \frac{\mathrm{d}^3x\,\mathrm{d}^3p}{(2\pi\hbar)^3} .$$

In addition, if every particle in the plane-wave state possesses s spin degrees of freedom $1, 2, \dots, s$ we get

$$\begin{aligned}\mathrm{d}N &= \mathrm{d}N_1\,\mathrm{d}N_2\dots\mathrm{d}N_s \\ &= \frac{\mathrm{d}^3x_1\,\mathrm{d}^3p_1}{(2\pi\hbar)^3}\frac{\mathrm{d}^3x_2\,\mathrm{d}^3p_2}{(2\pi\hbar)^3}\cdots\frac{\mathrm{d}^3x_s\,\mathrm{d}^3p_s}{(2\pi\hbar)^3} = \frac{\mathrm{d}q\,\mathrm{d}p}{(2\pi\hbar)^{3s}} ,\end{aligned}$$

where $\mathrm{d}q = \mathrm{d}^3x_1\dots\mathrm{d}^3x_s$ and $\mathrm{d}p = \mathrm{d}^3p_1\dots\mathrm{d}^3p_s$. This is exactly the former result (7); here we have counted only the three space and three momentum degrees of freedom for each of the particles separately, whereas in (7) we counted all degrees of freedom from 1 to 5. Note that we have here 2×35 degrees of freedom, whereas in (7) we have 2×5.

$\Delta\Gamma$ is called the *statistical weight* of the macroscopic state of the subsystem, and its logarithm

$$S = \ln\Delta\Gamma \tag{8}$$

is called *entropy*. Hence, in the classical limit we have

$$S = \ln\frac{\Delta p\,\Delta q}{(2\pi\hbar)^s} . \tag{9}$$

Since the number of states $\Delta\Gamma$ always has to be $\Delta\Gamma \gg 1$ we conclude $S \geq 0$! In statisitical mechanics it is customary to set

$$S = k_\mathrm{B}\ln\Delta\Gamma \tag{10}$$

instead of the dimensionless expression $S = \ln\Delta\Gamma$. Since entropy has to fulfill the thermodynamic relation $T\,\mathrm{d}S = \mathrm{d}E$ the proportionality constant k_B has to have the dimension

$$\left[\frac{\text{energy}}{\text{temperature}}\right] .$$

k_B is called the Boltzmann constant. With this relation we are able to derive the entropy for an ideal, a Bose–Einstein, and a Fermi–Dirac gas.

(a) Ideal Gas. We consider a group of N_j particles as an independent system. Its statistical weight is given by $\Delta\Gamma_j$, and the statistical weight of all particle groups is given by

Example 6.6.

$$\Delta\Gamma = \prod_j \Delta\Gamma_j . \tag{11}$$

Within the classical limit the average occupation numbers are small compared to unity, i.e. the number of particles N_j is small compared to the number G_j of states ($N_j \ll G_j$); however, N_j should still be a large number. Therefore we assume that the particles are distributed completely independently of each other over the various states. Once we assign to each of the N_j particles one of the G_j states we arrive at $(G_j)^{N_j}$ possible distributions. Some of them are identical because $N_j!$ permutations between the N_j undistinguishable particles are possible without changing the distribution. Hence, the statistical weight is given by

$$\Delta\Gamma_j = \frac{(G_j)^{N_j}}{N_j!} \tag{12}$$

and we get

$$S = k_{\mathrm{B}} \ln \Delta\Gamma = k_{\mathrm{B}} \sum_j \ln \Delta\Gamma_j \Rightarrow$$

$$S = k_{\mathrm{B}} \sum_j (N_j \ln G_j - \ln N_j!) . \tag{13}$$

Since N_j is large we can use Stirling's formula (see Exercise 6.8) and write $\ln N_j! \approx N_j \ln N_j/\mathrm{e}$, i.e.

$$S = k_{\mathrm{B}} \sum_j N_j \ln \frac{\mathrm{e}G_j}{N_j} \tag{14}$$

or, introducing the average particle number per state $\overline{n} = N_j/G_j$,

$$S = k_{\mathrm{B}} \sum_j G_j \overline{n}_j \ln \frac{\mathrm{e}}{\overline{n}_j} \tag{15}$$

or

$$S = k_{\mathrm{B}} \sum_j G_j \overline{n}_j - k_{\mathrm{B}} \sum_j G_j \overline{n}_j \ln \overline{n}_j . \tag{16}$$

The first part becomes

$$S_1 = k_{\mathrm{B}} \sum_j G_j \overline{n}_j = k_{\mathrm{B}} \sum_j N_j \overline{n}_j = k_{\mathrm{B}} N = \text{const} .$$

This constant is often dropped. Then the entropy of the ideal gas reads

$$S_I = -k_{\mathrm{B}} \sum_j G_j \overline{n}_j \ln \overline{n}_j ; \tag{17}$$

this corresponds to the expression derived earlier.

(b) Fermi Gas. Now $N_j \approx G_j$. The number of ways to distribute N_j identical particles over G_j states in such a way that a state is occupied with one particle at most results from the number of combinations of N_j out of G_j elements. This reflects a standard task of combinatorics.

Example 6.6.

Combinatorics can be illustrated with playing cards. For most of the hands of cards it does not matter in which order the cards are handed out since afterwards it is possible to rearrange them arbitrarily. Instead of the possible k-tupel only the subsets consisting of k elements of the given set are of interest. They are called *combinations of order k of n elements without repetition.* For an ideal gas we have seen that the number of *permutations* of n elements is given by $n!$. If only k elements are picked from the n elements then n elements are available to occupy the first place of the k-tupel; for the second place only $n-1$ elements are available and for place k only $n-(k-1)$ elements remain. Hence, the number of *variations* of order k consisting of n elements without repetition is given by

$$n \cdot (n-1) \cdot (n-2) \cdot \ldots \cdot [n-(k-1)] = \frac{n!}{(n-k)!} \,. \tag{18}$$

(Example: For six people to take a seat from ten chairs $10!/4! = 151\,200$ possibilities exist.)

If we further allow the possibility of rearranging k elements, i.e. k permutations, then we obtain exactly

$$\frac{n!}{k!(n-k)!} = \binom{n}{k} \tag{19}$$

combinations. (Example: In a lottery, 6 numbers are picked out of 49 numbers. The number of possibilities is $\binom{49}{6} = 13\,983\,816$.)

After these considerations we obtain for the statistical weight of N_j fermions in G_j states

$$\Delta\Gamma_j = \frac{G_j!}{N_j!(G_j-N_j)!} = \binom{G_j}{N_j} \,. \tag{20}$$

We take the logarithm, make use again of Stirling's formula $\ln N! \approx N\ln(N/\mathrm{e})$, and obtain

$$S_{\mathrm{F}} = k_{\mathrm{B}} \sum_j G_j \ln G_j - N_j \ln N_j - (G_j - N_j)\ln(G_j - N_j) \,. \tag{21}$$

Again we introduce the average number of particles per state $\overline{n}_j = N_j/G_j$ and it follows that

$$S_{\mathrm{F}} = k_{\mathrm{B}} \sum_j G_j \left[\overline{n}_j \ln \overline{n}_j + (1-\overline{n}_j)\ln(1-\overline{n}_j)\right] \,. \tag{22}$$

For the case that the G_j is independent of j we can take them out of the summation and arrive at the entropy relation for the Fermi gas already mentioned.

(c) Bose Gas. For this case an arbitrary number of particles can occupy each quantum state. As a consequence the statistical weight $\Delta\Gamma_j$ is given as the number of ways to distribute N_j particles over G_j states; all particles are allowed to occupy the same state. Using arguments analogously to the case of fermions we obtain

$$\Delta\Gamma_i = \frac{(G_j + N_j - 1)!}{(G_j - 1)!N_j!} \,; \tag{23}$$

Example 6.6.

see Exercise 6.7. We neglect the -1 because it is small compared to the G_j and N_j, take the logarithm, and again use Stirling's formula. This results in

$$S_B = k_B \sum_j (G_j + N_j)\ln(G_j + N_j) - N_j \ln N_j - G_j \ln G_j \tag{24}$$

or with $\overline{n}_j$

$$S_B = k_B \sum_j G_j \left[(1 + \overline{n}_j)\ln(1 + \overline{n}_j) - \overline{n}_j \ln \overline{n}_j\right] . \tag{25}$$

Again, if we assume G_j to be independent of j we arrive at the entropy expression for the Bose gas given earlier.

In the limit $N_j \ll G_j$, i.e. $\overline{n}_j \ll 1$, the entropy for both quantum gases becomes identical to the one for an ideal Maxwell–Boltzmann gas.

EXERCISE

6.7 Distribution of N Particles over G States (Number of Combinations)

Problem. Show that the number of ways of distributing, without restriction, N particles over G states is given by

$$M = \frac{(G + N - 1)!}{(G - 1)!N!}$$

Solution. We imagine N balls as points lying on a line:

• • • | | • • • • | • | • •

We numerate the G cells and separate them by $G - 1$ vertical lines. In the sketch shown we have: 3 balls in the first cell, 0 in the second, 4 in the third, 1 in the fourth, and 2 in the fifth. The total number of sites that are either occupied by points or lines is $G + N - 1$. Hence the number of possibilities is identical to the number of ways of choosing $G - 1$ sites for the vertical lines, i.e. the number of *combinations* of $G - 1$ elements out of $N + G - 1$; thus,

$$M = \frac{(N + G - 1)!}{(G - 1)!\,[N + G - 1 - (G - 1)]!} = \frac{(N + G - 1)!}{(G - 1)!N!} .$$

EXERCISE

6.8 Stirling's Formula

Problem. Prove Stirling's formula

$$\ln N! = N \ln\left(\frac{N}{\mathrm{e}}\right)$$

for *large* N.

Solution. It holds that

$$\begin{aligned}\ln N! &= \ln(1 \times 2 \times 3 \times \cdots \times N) \\ &= \ln 1 + \ln 2 + \ln 3 + \cdots + \ln N \\ &= \sum_{n=2}^{N} \ln(n)\,.\end{aligned}$$

For large N we replace the sum by an integral:

$$\int_1^N \ln(x)\,\mathrm{d}x = [x\ln(x) - x]_1^N = N\ln(N) - N + 1\,.$$

Since N is very large we neglect the 1 and write $N = N\ln \mathrm{e}$; we obtain

$$\ln N! \approx N \ln\left(\frac{N}{\mathrm{e}}\right)$$

for $N \gg 1$! According to Fig. 6.3 we realize that the sum $\sum_{n=2}^{N}\ln(n)$ (illustrated by the step function), i.e. $\ln N!$, is approximated by the two integrals:

$$\int_1^N \ln(x)\,\mathrm{d}x < \ln N! < \int_1^N \ln(1+x)\,\mathrm{d}x\,;$$

hence, Stirling's formula represents a lower estimate for $\ln N!$. Since

$$\frac{\int_1^N \ln(1+x)\,\mathrm{d}x}{\int_1^N \ln(x)\,\mathrm{d}x} = \frac{(N+1)\ln(N+1) - N + 1 - 2\ln 2}{N\ln N - N + 1} \longrightarrow 1\,,$$

for $N \to \infty$, the relative error of Stirling's approximation vanishes as $N \to \infty$.

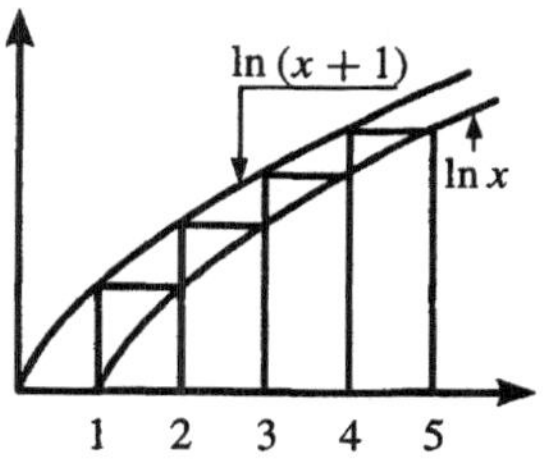

Fig. 6.3. Approximation leading to Stirling's formula

EXAMPLE

6.9 Entropy and Information

In this example we will study briefly the connection between entropy and the information about a system.

First, we have to find a statistical definition of information. We consider a situation for which P_0 different successive situations may arise, all with the same probability. We denote the information as I. For the initial situation $I_0 = 0$ since all processes are equally probable; we have no information about the system.

We consider two independent processes, which come with probabilities P_{01} and P_{02}, respectively; the total probability is then given by

$$P_0 = P_{01}P_{02}\,. \tag{1}$$

The information should be additive, i.e.

$$I = I_1 + I_2\,. \tag{2}$$

This property leads to the following, plausible definition of information:

$$I = K\ln P \tag{3}$$

Example 6.9.

with a constant K. Hence, we obtain

$$I = K \ln(P_{01}P_{02}) = I_1 + I_2 . \tag{4}$$

Furthermore it is customary to think of the information I as a dimensionless quantity, so the constant K has to be a number. In order to find a reasonable unit for information we consider the following example.

A problem exists with n different and independent choices; each single choice corresponds to either 0 or 1. The total number of possible choices is then $P = 2^n$; the information becomes $I = K \ln P = Kn \ln 2$. It is desirable to identify I with n and to take into account the binary structure of the decision between 0 and 1. Hence, we write

$$K = \frac{1}{\ln 2} = \log_2 \mathrm{e} .$$

This choice defines the unit of information: The bit (from "binary digit"). That is,

$$I = \log_2 P \tag{5}$$

measures the information in bits.

Example: We choose one from 32 cards and the information we have is

$$I = \log_2 32 = \log_2(2^5) \mathrel{\hat{=}} 5\,\mathrm{bits} . \tag{6}$$

If we choose one card each from two sets of cards, consisting each of 32 cards, then the information is given by

$$I = \log_2(2^{10}) = 10\,\mathrm{bits}$$

since $P = P_1P_2$ with $P_1 = P_2 = 32 = 2^5$. The property of addition of the information is realized.

So far we have considered events that all come with equal probability. However, if we consider a set of G letters then our familiar formula would be

$$I = (G \log_2 27)\,\mathrm{bits}$$

since 26 letters constitute our alphabet and the spacing between the words is considered the 27th letter. The information per letter would then be

$$\frac{I}{G} = \log_2 27 \approx 4.76\,\mathrm{bits} .$$

However, this result is not conclusive; we know that different letters do not appear with the same frequency within a sentence. For example, the frequency of the letter "E" is approximately $p \approx 0.105$, whereas "Q" appears with only $p \approx 0.001$.

In order to find the correct relation we consider an "alphabet" consisting of only two "letters": 0 and 1. In total we have G sites: N_0 sites have a "0" and N_1 sites have a "1".

It holds that $N_0 + N_1 = G$ and the probabilities that one of the G sites contains a "0" or "1" are

$$p_0 = \frac{N_0}{G} \quad \text{and} \quad p_1 = \frac{N_1}{G} , \tag{7}$$

Example 6.9.

i.e.

$$p_0 + p_1 = 1 . \qquad (8)$$

Now we can fill the G sites with numbers 0 and 1, one number for each site only! This automatically leads to the number of ways to occupy N_0 of the sites with zeros since the remaining sites are then filled with ones. The number of ways to fill N_0 sites with zeros is equal to the number of *combinations* of G objects from which N_0 are picked, i.e.

$$p = \frac{G!}{N_0!N_1!} . \qquad (9)$$

Hence, the information becomes

$$I = K \ln P = K (\ln G! - \ln N_0! - \ln N_1!) .$$

If the "sentence" we inspect is very long we can make use of Stirling's formula $\ln N! \approx N(\ln N - 1)$, i.e. if $G \gg 1$, $N_0 \gg 1$, and $N_1 \gg 1$, then

$$I \approx K\{G[\ln(G) - 1] - N_0[\ln(N_0) - 1] - N_1[\ln(N_1) - 1]\} . \qquad (10)$$

Since $G = N_0 + N_1$ it follows that

$$I \approx K(G \ln G - N_0 \ln N_0 - N_1 \ln N_1) \qquad (11)$$

or

$$I \approx -KG\left(\frac{N_0}{G} \ln \frac{N_0}{G} + \frac{N_1}{G} \ln \frac{N_1}{G}\right) .$$

With (7) we obtain the information per "letter" of the message:

$$\frac{I}{G} \approx -K(p_0 \ln p_0 + p_1 \ln p_1) .$$

This generalized to M different symbols $0, 1, \dots$ is called *Shannon's formula*:

$$I = -KG \sum_{j=1} p_j \ln p_j . \qquad (12)$$

If we consider the realistic frequencies of the various letters we obtain for our example of a 27-letter alphabet information of $I/G \approx 4.03$ bit per letter, i.e. smaller than calculated (4.76 bit). Our definition of the information (3) is completely equivalent to the definition of entropy

$$S = k_B \ln P , \qquad (13)$$

where k_B represents the Boltzmann constant. Hence, it makes sense to choose $K = k_B$ for the constant K and to define a new, thermodynamic unit for the information. In order to characterize the relationship between entropy and information in more detail we differentiate between two different forms of information:

(1) **free information**, which occurs if the possible cases can be viewed as abstract and have no assigned physical meaning,
(2) **bound information** I_g, which occurs if the possible cases are interpreted as a manifestation of a physical system. (Planck has denoted these forms as *complexions*).

Example 6.9.

It is this bound information which is connected to the physically relevant situations and which we take to be related to entropy. We identify the possible cases with the complexions; then it holds that

	bound information	number of complexions	entropy
initial state	$I_{g0} = 0$	P_0	$S_0 = k_B \ln P_0$
final state	$I_{g1} \neq 0$	$P_1 < P_0$	$S_1 = k_B \ln P_1$

because I_{g1} reduces the number of possible states from P_0 to P_1.

It is obvious that the system considered cannot be isolated, because when information is obtained, the entropy decreases; the number of possibilities, i.e. complexions, is reduced. This information must come from an outer system for which entropy increases.

We obtain:

$$I_0 = 0 \quad \text{and} \quad P_0 \text{ complexions}\,,$$
$$I_1 > 0 \quad \text{and} \quad P_0 \text{ complexions}\,,$$

where $I_1 = K \ln(P_0/P_1)$ and thus $(K = k_B)$

$$I_{g1} = k_B(\ln P_0 - \ln P_1) = S_0 - S_1 \tag{14}$$

or

$$S_1 = S_0 - I_{g1}\,, \tag{15}$$

respectively. The bound information appears as a negative term in the total entropy of the system:

$$\text{bound information} = \text{entropy decrease} = \text{increase of } \textit{negentropy } N\,,$$

where negentropy stands for the negative of entropy.

Let us consider a closed system. Then the law on the increase of entropy gives

$$\Delta S_1 \geq 0 \quad \text{or} \quad \Delta(S_0 - I_{g1}) \geq 0\,, \tag{16}$$

i.e. an increase in entropy S_1 can cause an increase in the entropy S_0 as well as a decrease in the information I_{g1}: *As the entropy of a system increases the information about this system decreases at the same time.* An increase of negentropy results in a gain of information, i.e. negentropy corresponds directly to the information about a system. Let us now elucidate the significance of these conceptions with several examples. At first we point out the difference between free and bound information.

(1) A person possesses information (free).
(2) This person passes the information to a friend via acoustic waves (voice) or electric waves (telephone). The information is realized in a physical process and, hence, becomes bound information.

Example 6.9.

(3) The friend is partly deaf and does not understand some words; thus, the bound information gets lost.
(4) After some time the original person forgets the information and loses the free information since it was only present within thoughts (brain).

In the following we will consider a further example which clarifies the relationship between information and entropy.

EXAMPLE

6.10 Maxwell's Demon

We consider a container divided into two parts A and B. A small hole is present in the partitioning wall between A and B. Both volumes are filled with the same gas under identical pressure. We now assume a living creature, i.e. the demon, exists who is able to "see" every single molecule. He opens and shuts the hole in such a way that the fast molecules from B go into A and that the slow molecules from A pass into B. Hence, the temperature rises in A without the input of work from outside the system. This is a contradiction of the second law of thermodynamics.

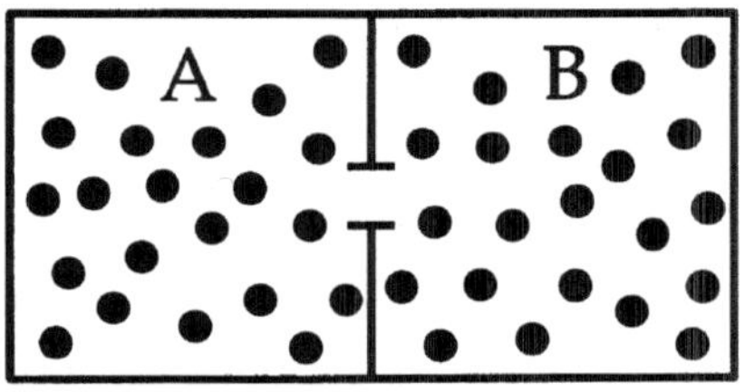

Fig. 6.4. Typical set-up for Maxwell's demon

In order to exorcize the demon we have to raise the question of whether he can really see each single molecule?[5] We have assumed the system to be isolated; hence, the demon stays in a container with constant temperature, which is filled with the corresponding black-body radiation. Here comes the solution: It is impossible to see anything in the interior of a black body. The demon can see only the thermal radiation but no molecules. Thus, he cannot use the shutter and cannot violate the second law.

What other possibilities may come to his mind? He might get a lamp so that he can see the molecules. However, the lamp represents a radiation source which is not in equilibrium with the black body; it produces negative entropy within the system. The demon is able to see the molecules, i.e. the negentropy is transformed into his information. Now he can fulfill his demonic destiny: he produces a higher temperature in A, i.e. negative entropy; he transforms his information into negentropy. We arrive at the cycle

$$\text{negentropy} \to \text{information} \to \text{negentropy}\,.$$

[5] P. Demers: Can. J. Research **22** (1944) 27; L. Brillouin: J. Appl. Phys. **22** (1951) 334.

Example 6.10.

Since the transformation process is at best complete, the demon will also allow some "wrong" molecules to pass from time to time and, thus, it holds again that

$$\Delta S \geq 0 \,, \qquad \text{because } \Delta N \leq 0 \,,$$

i.e. negentropy N gets lost and entropy is produced; as a consequence the second law of thermodynamics is not violated!

7. Superfluidity

In the preceding chapter we introduced the concept of a "quasiparticle". We now intend to elucidate this notion with respect to the interpretation of experimental facts and refer to the properties of liquid helium.

Two stable isotopes of the element Helium occur in nature: ^{3}He (two protons and one neutron: a fermion) and ^{4}He (two protons and two neutrons: a boson). Both isotopes remain liquid down to temperatures $T \approx 0\,\mathrm{K}$ at low pressures; only at high pressures (of about 30 atmospheres) they do solidify. This is in contrast to the remaining elements, which become solid easily. Since these properties are related to quantum effects, liquid helium is denoted as a *quantum liquid.* Fermi statistics state that every quantum state of ^{3}He can be occupied by only one particle. On the other hand ^{4}He obeys Bose statistics; thus it has the properties of a nearly ideal Bose gas with a weak interaction between the particles. We will examine ^{4}He in the following. If ^{4}He, which

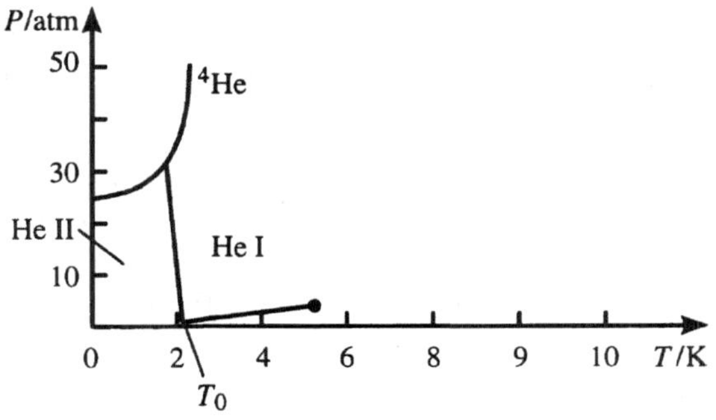

Fig. 7.1. Phase diagram of ^{4}He at low temperature

is in equilibrium with its vapour, is cooled below $T_0 = 2.17\,\mathrm{K}$, it develops into a new phase denoted as He II. This new phase has remarkable properties: He II flows through tiny capillary tubes with no friction; evidently, its viscosity is zero. On the other hand, direct measurements of the viscosity (with a cylindrical viscosimeter) show that the viscosity of He II corresponds to the viscosity of He I above T_0. This peculiar behavior has been explained with a two-fluid model by Landau[1] and Tisza[2]: He II consists of two fluids penetrating each other. One component is *superfluid* and therefore shows no viscosity;

[1] L.D. Landau: J. Phys. (Moscow) **5** (1941) 71; **11** (1947) 91.
[2] L. Tisza: Nature **141** (1938) 913.

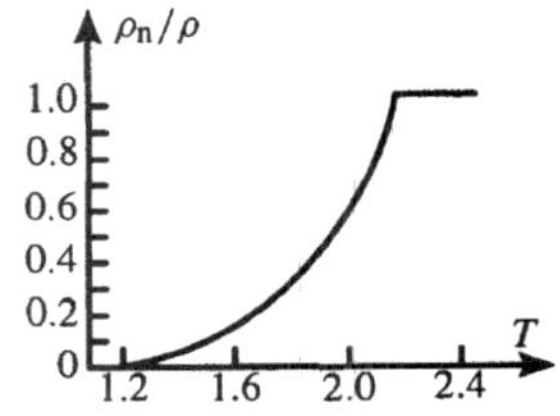

Fig. 7.2. The ratio of normal fluid density ρ_n to total density $\rho = \rho_n + \rho_s$ as a function of temperature (qualitatively). ρ_s indicates the density of the superfluid phase

the flow is curl free, i.e. $\mathrm{curl}\,\boldsymbol{v}_s = 0$, where $\boldsymbol{v}_s$ represents the velocity of the superfluid component. Its density is ρ_s. The second component with density ρ_n and velocity $\boldsymbol{v}_n$ is represented by a "normal" fluid, which has viscosity. This picture from Landau has been confirmed experimentally.[3]

The presence of two components in He II allows two different excitation modes, which differ in the relative orientation between $\boldsymbol{v}_s$ and $\boldsymbol{v}_n$: Once $\boldsymbol{v}_s$ and $\boldsymbol{v}_n$ are parallel a perturbation in He II leads to a change in density and pressure (for constant T and constant entropy); it is denoted as the *first* (or ordinary) *sound*.

If $\boldsymbol{v}_n$ and $\boldsymbol{v}_s$ point to opposite directions a wave is obtained, leading to periodic changes in ρ_s/ρ_n. This ratio depends on temperature and $\rho = \rho_s + \rho_n$ is constant (also the pressure is constant), so we obtain a temperature wave, the so-called *second sound*. A detailed investigation of this problem shows that the corresponding speed of sound is proportional to $(\rho_s/\rho_n)^{1/2}$. The second sound was observed for the first time by Peshkov.[4]

7.1 Basics of a Microscopic Theory of Superfluidity

If liquid helium (^{4}He) is cooled down to almost absolute zero its viscosity vanishes (to the degree of measureability) and it flows without friction through tiny capillaries and tiny openings. This was observed for the first time by Kapitza.[5] Landau's explanation of this phenomenon uses the concept of a "quantum liquid" (bosonic liquid), as we have discussed in the previous section. Now we will investigate in detail the microscopic theory of superfluid helium developed by Bogoliubov.[6] The method developed is of general interest and often finds applications once perturbation theory becomes useless, for example for the theory of superconductivity (see Chap. 8).

^{4}He atoms represent neutral bosons (spin 0) and interact only weakly with each other because of polarization forces (van der Waals forces). The Hamiltonian for N ^{4}He atoms can be cast into the form

$$\hat{H} = \sum_{i=1}^{N} \hat{H}(\boldsymbol{r}_i) + \sum_{i<j} \mathcal{V}(|\boldsymbol{r}_i - \boldsymbol{r}_j|) . \tag{7.1}$$

$\hat{H}(\boldsymbol{r}_i)$ represents the operator for the kinetic energy of the ith He atom and $\mathcal{V}(|\boldsymbol{r}_i - \boldsymbol{r}_j|)$ describes the interaction between the ith and jth atoms. For a complete and orthogonal set of eigenfunctions we choose the solutions of $\hat{H}(\boldsymbol{r}_i)$; these are given by plane waves

$$\varphi_k(\boldsymbol{r}) = \frac{1}{\sqrt{L^3}} \mathrm{e}^{i\boldsymbol{k}\cdot\boldsymbol{r}} , \tag{7.2}$$

which are normalized with the conventional periodic boundary conditions for a large volume L^3 [compare for example with (1.23)]. Hence, it holds that

[3] E.L. Andronikashvili: J. Phys. (Moscow) **10** (1946) 201.
[4] V.P. Peshkov: J. Phys. (Moscow) **10** (1946) 389.
[5] P.L. Kapitza: Nature **141** (1938) 74.
[6] N.N. Bogoliubov: J. Phys. (Moscow) **9** (1947) 23.

$$\begin{aligned}\hat{H}(\boldsymbol{r})\varphi_k(\boldsymbol{r}) &= \varepsilon_k\varphi_k(\boldsymbol{r}) \,, \\ \varepsilon_k &= \frac{\hbar^2k^2}{2m} \,.\end{aligned} \tag{7.3}$$

According to Chaps. 3 and 6 [(6.7) to (6.13) in particular], the total Hamiltonian can be rewritten in second quantization:

$$\hat{H} = \sum_{\boldsymbol{k}} \varepsilon_k \hat{b}^\dagger_{\boldsymbol{k}}\hat{b}_{\boldsymbol{k}} + \frac{1}{2}\sum \langle \boldsymbol{k}_1\boldsymbol{k}_2|\mathcal{V}|\boldsymbol{k}'_1\boldsymbol{k}'_2\rangle \hat{b}^\dagger_{\boldsymbol{k}_1}\hat{b}^\dagger_{\boldsymbol{k}_2}\hat{b}_{\boldsymbol{k}'_1}\hat{b}_{\boldsymbol{k}'_2} \,. \tag{7.4}$$

From (6.11) and (6.13) the matrix elements become

$$\begin{aligned} M &= \langle \boldsymbol{k}_1\boldsymbol{k}_2|\mathcal{V}|\boldsymbol{k}'_1\boldsymbol{k}'_2\rangle \\ &= \int \mathrm{d}^3r_1\,\mathrm{d}^3r_2\,\mathcal{V}(|\boldsymbol{r}_1-\boldsymbol{r}_2|)\frac{\exp\left[-\mathrm{i}(\boldsymbol{k}_1-\boldsymbol{k}'_1)\cdot\boldsymbol{r}_1\right]}{\sqrt{L^3}\sqrt{L^3}}\frac{\exp\left[-\mathrm{i}(\boldsymbol{k}_2-\boldsymbol{k}'_2)\cdot\boldsymbol{r}_2\right]}{\sqrt{L^3}\sqrt{L^3}} \\ &= \frac{1}{L^6}\int \mathrm{d}^3(\boldsymbol{r}_1-\boldsymbol{r}_2)\,\mathrm{d}^3r_2\,\mathcal{V}(|\boldsymbol{r}_1-\boldsymbol{r}_2|)\exp\left[-\mathrm{i}(\boldsymbol{k}_1-\boldsymbol{k}'_1)\cdot(\boldsymbol{r}_1-\boldsymbol{r}_2)\right] \\ &\quad\times \exp\left[-\mathrm{i}(\boldsymbol{k}_1-\boldsymbol{k}'_1+\boldsymbol{k}_2-\boldsymbol{k}'_2)\cdot\boldsymbol{r}_2\right] \\ &= \frac{1}{L^3}\int \mathrm{d}^3\rho\,\mathcal{V}(|\boldsymbol{\rho}\,|)\exp\left[-\mathrm{i}(\boldsymbol{k}_1-\boldsymbol{k}'_1)\cdot\boldsymbol{\rho}\right] \\ &\quad\times\frac{1}{L^3}\int \mathrm{d}^3r_2\exp\left[-\mathrm{i}(\boldsymbol{k}_1+\boldsymbol{k}_2-\boldsymbol{k}'_1-\boldsymbol{k}'_2)\cdot\boldsymbol{r}_2\right] \\ &= \frac{\mathcal{V}(q)}{L^3}\delta_{\boldsymbol{k}_1+\boldsymbol{k}_2,\boldsymbol{k}'_1+\boldsymbol{k}'_2} \end{aligned} \tag{7.5}$$

with

$$\begin{aligned} \mathcal{V}(q) &= \int \mathrm{d}^3r\,\mathcal{V}(|\boldsymbol{r}|)\,\mathrm{e}^{+\mathrm{i}\boldsymbol{q}\cdot\boldsymbol{r}} \\ &= -\int \mathrm{d}r\,\mathrm{d}(\cos\vartheta)\,\mathrm{d}\varphi\,r^2\,\mathcal{V}(r)\mathrm{e}^{+\mathrm{i}qr\cos\vartheta} \\ &= -2\pi\int \mathrm{d}r\,r^2\,\mathcal{V}(r)\int_1^{-1}\mathrm{d}x\,\mathrm{e}^{-\mathrm{i}qrx} \\ &= 2\pi\int \mathrm{d}r\,r^2\,\mathcal{V}(r)\frac{\mathrm{e}^{\mathrm{i}qr}-\mathrm{e}^{-\mathrm{i}qr}}{\mathrm{i}qr} \\ &= \frac{4\pi}{|\boldsymbol{q}|}\int \mathrm{d}r\,r\mathcal{V}(r)\sin(|\boldsymbol{q}|r) \,. \end{aligned} \tag{7.6}$$

Here

$$\hbar\boldsymbol{q} = \hbar(\boldsymbol{k}'_1-\boldsymbol{k}_1) = \hbar(\boldsymbol{k}'_2-\boldsymbol{k}_2) \tag{7.7}$$

represents the momentum transfer of particle 1 onto particle 2. The creation and annihilation operators for the ^{4}He atoms have been denoted as $\hat{b}^\dagger$ and $\hat{b}$, respectively. Conservation of total momentum is expressed by the Kronecker delta symbol

$$\delta_{\boldsymbol{k}_1+\boldsymbol{k}_2,\,\boldsymbol{k}'_1+\boldsymbol{k}'_2} \tag{7.8}$$

appearing in (7.5). We obtain this symbol because of the box normalization. In quantum electrodynamics, where we use a normalization with respect to δ-functions (infinitly large box), (7.8) becomes a $\delta(\boldsymbol{k}_1+\boldsymbol{k}_2-\boldsymbol{k}'_1-\boldsymbol{k}'_2)$ function.

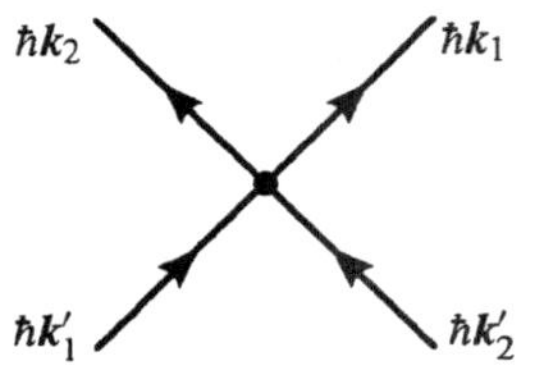

Fig. 7.3. Momentum conservation at an interaction vertex

The second term in (7.4) describes a two-particle interaction or, to be more precise, a two-boson interaction. Every single term within this summation describes the annihilation of a pair of ^{4}He atoms with momenta $\hbar \boldsymbol{k}_1'$ and $\hbar \boldsymbol{k}_2'$ and the creation of another pair with momenta $\hbar \boldsymbol{k}_1$ and $\hbar \boldsymbol{k}_2$. According to (7.7) and (7.8) conservation of the total momentum

$$\hbar \boldsymbol{k}_1' + \hbar \boldsymbol{k}_2' = \hbar \boldsymbol{k}_1 + \hbar \boldsymbol{k}_2 \tag{7.9}$$

holds for each such scattering process (see Fig. 7.3). This stems from the momentum conserving potential $\mathcal{V}(|\boldsymbol{r}_i - \boldsymbol{r}_j|)$, which only depends on the distance of the interacting particles and is therefore translationally invariant.

Inserting (7.5) into (7.4) we get

$$\hat{H} = \sum_{\boldsymbol{k}} \varepsilon_k \hat{b}_{\boldsymbol{k}}^\dagger \hat{b}_{\boldsymbol{k}} + \frac{1}{2L^3} \sum_{\boldsymbol{k}_1+\boldsymbol{k}_2=\boldsymbol{k}_1'+\boldsymbol{k}_2'} \mathcal{V}(|\boldsymbol{k}_1' - \boldsymbol{k}_1|) \hat{b}_{\boldsymbol{k}_1}^\dagger \hat{b}_{\boldsymbol{k}_2}^\dagger \hat{b}_{\boldsymbol{k}_1'} \hat{b}_{\boldsymbol{k}_2'} , \tag{7.10}$$

where the summation runs over only the momentum-conserving combinations of $\boldsymbol{k}_1, \boldsymbol{k}_2, \boldsymbol{k}_1', \boldsymbol{k}_2'$.

For the ground state of a system of bosons all particles occupy the lowest one-particle state. Contrary to fermions, bosons do not have to obey the Pauli principle; this leads to the *condensation of bosons.*

The lowest one-particle state is characterized by $\boldsymbol{k} = 0$, which is equivalent to $\varepsilon_{\boldsymbol{k}=0} = 0$. For the case when a small repulsion exists between the atoms, most of the atoms will still remain in the lowest state with $\boldsymbol{k} = 0$ and remain in the condensate. In other words, even if the interaction $\mathcal{V}$ between the ^{4}He atoms stays small, n_0 atoms will remain in the condensate, where n_0 deviates only slightly from the total number N of ^{4}He atoms.

It should be clear that at the very end of our calculations the volume size has to go to infinity, i.e. $L^3 \to \infty$; in this way we remove the dependence on the periodic boundary conditions for a normalization box with finite length L. Then, of course, $N \to \infty$ increases, too. We require for this limit that the density

$$\frac{N}{L^3} = \text{const.} \tag{7.11}$$

remains constant. Now

$$\hat{b}_0^\dagger \hat{b}_0 = \hat{n}_0 \qquad \text{and} \qquad \hat{b}_0 \hat{b}_0^\dagger = \hat{n}_0 + 1 , \tag{7.12}$$

where $\hat{n}_0$ represents the number operator for particles in the ground state. Because of $\langle 0|\hat{n}_0|0\rangle = n_0 \approx N$ it follows for their difference that

$$\frac{\hat{b}_0 \hat{b}_0^\dagger}{L^3} - \frac{\hat{b}_0^\dagger \hat{b}_0}{L^3} = \frac{1}{L^3} \to 0 . \tag{7.13}$$

In the limit $L^3 \to \infty$ we are allowed to forget the operator character of $\hat{b}_0^\dagger$ and $\hat{b}_0$; within this limit they commute and can be replaced by numbers. Bogoliubov suggested the introduction of new boson operators

$$\begin{aligned}\hat{a}_k &= \frac{1}{\sqrt{n_0}}\hat{b}_0^\dagger \hat{b}_k\,,\\ \hat{a}_k^\dagger &= \frac{1}{\sqrt{n_0}}\hat{b}_k^\dagger \hat{b}_0\end{aligned} \tag{7.14}$$

for $k \neq 0$. They represent boson–boson–hole pairs. Equation (7.10) then becomes

$$\begin{aligned}\hat{H} &= \frac{1}{2L^3}N^2\mathcal{V}(0) + {\sum_k}' \frac{(\hbar k)^2}{2m}\hat{a}_k^\dagger \hat{a}_k\\ &\quad + \frac{n_0}{2L^3}{\sum_k}' \mathcal{V}(k)[\hat{a}_k^\dagger \hat{a}_{-k}^\dagger + \hat{a}_k \hat{a}_{-k} + 2\hat{a}_k^\dagger \hat{a}_k] + \hat{H}'\\ &\equiv \hat{H}_1 + \hat{H}'\,.\end{aligned} \tag{7.15}$$

The prime (′) attached to the summation sign indicates that values $k = 0$ are excluded from the summation. Those terms that contain three or four boson operators $\hat{a}, \hat{a}^\dagger$ are collected in $\hat{H}'$ and can be neglected because the boson operators (7.14) are of the order $1/N$. This can be understood from a small excercise:

$$\begin{aligned}{\sum_k}' \hat{a}_k^\dagger \hat{a}_k &= {\sum_k}' \frac{1}{n_0}\hat{b}_k^\dagger \hat{b}_0 \hat{b}_0^\dagger \hat{b}_k\\ &= {\sum_k}' \hat{b}_k^\dagger \frac{\hat{n}_0 + 1}{n_0}\hat{b}_k\\ &= {\sum_k}' \hat{b}_k^\dagger \hat{b}_k \frac{\hat{n}_0 + 1}{n_0}\\ &= \sum_k \hat{b}_k^\dagger \hat{b}_k \frac{\hat{n}_0 + 1}{n_0} - \hat{b}_0^\dagger \hat{b}_0 \frac{\hat{n}_0 + 1}{n_0}\\ &= \hat{N}\left(\frac{\hat{n}_0 + 1}{n_0} - \frac{\hat{n}_0(\hat{n}_0 + 1)}{n_0}\right)\\ &= (\hat{N} - \hat{n}_0)\frac{\hat{n}_0 + 1}{n_0}\\ &\approx N - n_0 \to 0 \qquad \text{for} \qquad L \to \infty\,,\end{aligned} \tag{7.16}$$

where (7.14) has been used. It follows that $\hat{a}_k^\dagger \hat{a}_k \sim 1/N^2$ so

$${\sum_k}' \hat{a}_k^\dagger \hat{a}_k \sim {\sum_k}' \frac{1}{N^2} \sim \frac{N}{N^2} = \frac{1}{N} \to 0\,.$$

Thus, in the limit $N \to \infty$ we are allowed to neglect the term $\hat{H}'$ in (7.15), which contains four operators of $\hat{a}$ and $\hat{a}^\dagger$. We focus on the interesting part, i.e.

$$\begin{aligned}\hat{H}_1 &= \frac{N^2}{2L^3}\mathcal{V}(0) + {\sum_k}' \frac{(\hbar k)^2}{2m}\hat{a}_k^\dagger \hat{a}_k\\ &\quad + \frac{n_0}{2L^3}{\sum_k}' \mathcal{V}(k)\left(\hat{a}_k^\dagger \hat{a}_{-k}^\dagger + \hat{a}_k \hat{a}_{-k} + 2\hat{a}_k^\dagger \hat{a}_k\right)\,.\end{aligned} \tag{7.17}$$

No doubt the interaction matrix elements $\mathcal{V}(k)$ are small, but perturbation theory is not allowed to run the show. Infinitely many terms exist in the interaction sum of (7.17) for pairs $\boldsymbol{k}$ and $-\boldsymbol{k}$ once $|\boldsymbol{k}| \to 0$. In other words, scattering out of the condensate with one-particle states $\boldsymbol{k} = 0$ into infinitely many states $\boldsymbol{k}$, $-\boldsymbol{k}$ might take place which are energetically all close to $|\boldsymbol{k}| = 0$; a perturbation calculation then becomes impractical.

Bogoliubov suggested the following approach: Take the Hamiltonian $\hat{H}_0$ and diagonalize it with a *canonical transformation in the boson operators*:

$$\begin{aligned} \hat{a}_{\boldsymbol{k}} &= u(\boldsymbol{k})\hat{A}_{\boldsymbol{k}} + v(\boldsymbol{k})\hat{A}^{\dagger}_{-\boldsymbol{k}} , \\ \hat{a}^{\dagger}_{\boldsymbol{k}} &= u(\boldsymbol{k})\hat{A}^{\dagger}_{\boldsymbol{k}} + v(\boldsymbol{k})\hat{A}_{-\boldsymbol{k}} . \end{aligned} \tag{7.18}$$

This is exactly the same as in (6.72). We will repeat now some of the steps presented in Sect. 6 in more detail and with a different stress. $u(k)$ and $v(k)$ represent real functions with the property

$$\begin{aligned} u(-\boldsymbol{k}) &= u(\boldsymbol{k}) , \\ v(-\boldsymbol{k}) &= v(\boldsymbol{k}) . \end{aligned} \tag{7.19}$$

This guarantees that no asymmetry occurs between the bosons $\hat{a}^{\dagger}_{\boldsymbol{k}}$ and $\hat{a}^{\dagger}_{-\boldsymbol{k}}$, or $\hat{A}_{\boldsymbol{k}}$ and $\hat{A}^{\dagger}_{-\boldsymbol{k}}$, respectively. Also, (7.18) can now easily be transformed into

$$\begin{aligned} \hat{A}_{\boldsymbol{k}} &= \frac{u(\boldsymbol{k})\hat{a}_{\boldsymbol{k}} - v(\boldsymbol{k})\hat{a}^{\dagger}_{-\boldsymbol{k}}}{u^2(\boldsymbol{k}) - v^2(\boldsymbol{k})} , \\ \hat{A}^{\dagger}_{\boldsymbol{k}} &= \frac{u(\boldsymbol{k})\hat{a}^{\dagger}_{\boldsymbol{k}} - v(\boldsymbol{k})\hat{a}_{-\boldsymbol{k}}}{u^2(\boldsymbol{k}) - v^2(\boldsymbol{k})} . \end{aligned} \tag{7.20}$$

This suggests the normalization

$$u^2(\boldsymbol{k}) - v^2(\boldsymbol{k}) = 1 . \tag{7.21}$$

The latter follows from the requirement that the $\hat{A}_{\boldsymbol{k}}$ should again fulfill the boson commutation relations

$$\begin{aligned} \left[\hat{A}_{\boldsymbol{k}}, \hat{A}^{\dagger}_{\boldsymbol{k}'}\right]_- &= \delta_{\boldsymbol{k}\boldsymbol{k}'} , \\ \left[\hat{A}_{\boldsymbol{k}}, \hat{A}_{\boldsymbol{k}'}\right]_- &= 0 , \\ \left[\hat{A}^{\dagger}_{\boldsymbol{k}}, \hat{A}^{\dagger}_{\boldsymbol{k}'}\right]_- &= 0 . \end{aligned} \tag{7.22}$$

Of course, these relations must also hold for $\hat{a}_{\boldsymbol{k}}, \hat{a}^{\dagger}_{\boldsymbol{k}}$. This implies

$$\begin{aligned} \left[\hat{a}_{\boldsymbol{k}}, \hat{a}^{\dagger}_{\boldsymbol{k}'}\right]_- &= \delta_{\boldsymbol{k}\boldsymbol{k}'} \\ &= \left[u(\boldsymbol{k})\hat{A}_{\boldsymbol{k}} + v(\boldsymbol{k})\hat{A}^{\dagger}_{-\boldsymbol{k}}, u(\boldsymbol{k}')\hat{A}^{\dagger}_{\boldsymbol{k}'} + v(\boldsymbol{k}')\hat{A}_{-\boldsymbol{k}'}\right]_- \\ &= u(\boldsymbol{k})u(\boldsymbol{k}')\left[\hat{A}_{\boldsymbol{k}}, \hat{A}^{\dagger}_{\boldsymbol{k}'}\right]_- + v(\boldsymbol{k})v(\boldsymbol{k}')\left[\hat{A}^{\dagger}_{-\boldsymbol{k}}, \hat{A}_{-\boldsymbol{k}'}\right]_- \\ &= u^2(\boldsymbol{k})\delta_{\boldsymbol{k}\boldsymbol{k}'} - v^2(\boldsymbol{k})\delta_{\boldsymbol{k}\boldsymbol{k}'} \\ &= \left(u^2(\boldsymbol{k}) - v^2(\boldsymbol{k})\right)\delta_{\boldsymbol{k}\boldsymbol{k}'} \end{aligned} \tag{7.23}$$

and we arrive at the stated normalization relation (7.21). The normalization (7.21) and the reality of $u(\boldsymbol{k}), v(\boldsymbol{k})$ are fulfilled with the ansatz

$$\begin{aligned} u(\boldsymbol{k}) &= \frac{1}{\sqrt{1-D_k^2}} , \\ v(\boldsymbol{k}) &= \frac{D_k}{\sqrt{1-D_k^2}} . \end{aligned} \tag{7.24}$$

For the moment the values of the coefficients D_k remain unknown. Inserting (7.24) into (7.18) we deduce

$$\begin{aligned} \hat{a}_{\boldsymbol{k}} &= \frac{1}{\sqrt{1-D_k^2}} (\hat{A}_{\boldsymbol{k}} + D_k \hat{A}^\dagger_{-\boldsymbol{k}}) , \\ \hat{a}^\dagger_{\boldsymbol{k}} &= \frac{1}{\sqrt{1-D_k^2}} (\hat{A}^\dagger_{\boldsymbol{k}} + D_k \hat{A}_{-\boldsymbol{k}}) . \end{aligned} \tag{7.25}$$

Using this result for $\hat{H}_1$ in (7.17), we finally express $\hat{H}_1$ in terms of the new boson operators $\hat{A}_{\boldsymbol{k}}, \hat{A}^\dagger_{\boldsymbol{k}}$. This calculation will be presented in detail in Exercise 7.1. Here, we state the outcome:

$$\hat{H} = \hat{H}_0 + \sum_{\boldsymbol{k}} E(k) \hat{A}^\dagger_{\boldsymbol{k}} \hat{A}_{\boldsymbol{k}} , \tag{7.26}$$

where

$$\hat{H}_0 = \frac{N^2}{2L^3} \mathcal{V}(0) + \frac{1}{2} \sum_{\boldsymbol{k}} \left[E(k) - \left(\frac{(\hbar \boldsymbol{k})^2}{2m} + \frac{n_0}{L^3} \mathcal{V}(k) \right) \right] \tag{7.27}$$

and

$$E(k) = \sqrt{\frac{(\hbar k)^4}{4m^2} + \frac{n_0 (\hbar k)^2}{mL^3} \mathcal{V}(k)} , \tag{7.28a}$$

$$D_k = \frac{L^3}{n_0 \mathcal{V}(k)} \left(E(k) - \frac{(\hbar k)^2}{2m} - \frac{n_0}{L^3} \mathcal{V}(k) \right) . \tag{7.28b}$$

Note that $n_0/L^3 = \rho$, the density, and compare these results with (6.75)–(6.86a). As to be expected, they are the same, except that the interaction matrix element $\mathcal{V}(k)$ of (7.5) has in our former treatment been simplified to a uniform constant g, i.e. $\mathcal{V}(k) \leftrightarrow g$.

EXERCISE

7.1 Choice of Coefficients for the Bogoliubov Transformation

Problem. Determine the coefficients D_k of the canonical transformation (7.25) in such a way that the Hamiltonian $\hat{H}_1$ diagonalizes in the boson operators $\hat{A}_{\boldsymbol{k}}, \hat{A}^\dagger_{\boldsymbol{k}}$.

Exercise 7.1.

Solution. We have

$$\hat{H}_1 = \frac{N^2}{2L^3}\mathcal{V}(0) + \sum_{\boldsymbol{k}}{}' \frac{(\hbar\boldsymbol{k})^2}{2m}\hat{a}^\dagger_{\boldsymbol{k}}\hat{a}_{\boldsymbol{k}} + \frac{n_0}{2L^3}\sum_{\boldsymbol{k}}{}' \mathcal{V}(k)\left(\hat{a}^\dagger_{\boldsymbol{k}}\hat{a}^\dagger_{-\boldsymbol{k}} + \hat{a}_{\boldsymbol{k}}\hat{a}_{-\boldsymbol{k}} + 2\hat{a}^\dagger_{\boldsymbol{k}}\hat{a}_{\boldsymbol{k}}\right) . \quad (1)$$

The prime (′) at the sum indicates that the terms with $\boldsymbol{k} = 0$ have to be omitted from the summation. We introduce new boson operators $\hat{A}_{\boldsymbol{k}}, \hat{A}^\dagger_{\boldsymbol{k}}$ [see (7.18) and (7.24)]:

$$\begin{aligned}\hat{a}_{\boldsymbol{k}} &= u(k)\hat{A}_{\boldsymbol{k}} + v(k)\hat{A}^\dagger_{-\boldsymbol{k}} ,\\ \hat{a}^\dagger_{\boldsymbol{k}} &= u(k)\hat{A}^\dagger_{\boldsymbol{k}} + v(k)\hat{A}_{-\boldsymbol{k}} ,\end{aligned} \quad (2)$$

where

$$u(k) = \frac{1}{\sqrt{1-D_k^2}} , \qquad v(k) = \frac{D_k}{\sqrt{1-D_k^2}} . \quad (3)$$

Inserting (2) into (1) gives

$$\begin{aligned}\hat{H} &= \frac{N^2}{2L^3}\mathcal{V}(0)\\ &+\sum_{\boldsymbol{k}}{}'\left(\frac{(\hbar\boldsymbol{k})^2}{2m} + \frac{n_0}{L^3}\mathcal{V}(k)\right)\Big[u^2(k)\hat{A}^\dagger_{\boldsymbol{k}}\hat{A}_{\boldsymbol{k}} + v^2(k)\hat{A}_{-\boldsymbol{k}}\hat{A}^\dagger_{-\boldsymbol{k}}\\ &+ u(k)v(k)(\hat{A}^\dagger_{\boldsymbol{k}}\hat{A}^\dagger_{-\boldsymbol{k}} + \hat{A}_{-\boldsymbol{k}}\hat{A}_{\boldsymbol{k}})\Big]\\ &+\sum_{\boldsymbol{k}}{}'\frac{n_0}{2L^3}\mathcal{V}(k)\Big[u(k)v(k)(\hat{A}^\dagger_{\boldsymbol{k}}\hat{A}_{\boldsymbol{k}} + \hat{A}_{\boldsymbol{k}}\hat{A}^\dagger_{\boldsymbol{k}} + \hat{A}^\dagger_{-\boldsymbol{k}}\hat{A}_{-\boldsymbol{k}} + \hat{A}_{-\boldsymbol{k}}\hat{A}^\dagger_{-\boldsymbol{k}})\\ &+ u^2(k)(\hat{A}^\dagger_{\boldsymbol{k}}\hat{A}^\dagger_{-\boldsymbol{k}} + \hat{A}_{\boldsymbol{k}}\hat{A}_{-\boldsymbol{k}}) + v^2(k)(\hat{A}_{-\boldsymbol{k}}\hat{A}_{\boldsymbol{k}} + \hat{A}^\dagger_{-\boldsymbol{k}}\hat{A}^\dagger_{\boldsymbol{k}})\Big] .\end{aligned} \quad (4)$$

With the help of the commutation relations (7.22) and (7.19) we obtain

$$\begin{aligned}\hat{H} &= \frac{N^2}{2L^3}\mathcal{V}(0)\\ &+\sum_{\boldsymbol{k}}{}'\left[\left(\frac{(\hbar\boldsymbol{k})^2}{2m} + \frac{n_0}{L^3}\mathcal{V}(k)\right)v^2(k) + \frac{n_0}{L^3}\mathcal{V}(k)u(k)v(k)\right]\\ &+\sum_{\boldsymbol{k}}{}'\left\{\left(\frac{(\hbar\boldsymbol{k})^2}{2m} + \frac{n_0}{L^3}\mathcal{V}(k)\right)[u^2(k)+v^2(k)]\right.\\ &\left.+\frac{n_0}{L^3}\mathcal{V}(k)u(k)v(k)\right\}\hat{A}^\dagger_{\boldsymbol{k}}\hat{A}_{\boldsymbol{k}}\\ &+\sum_{\boldsymbol{k}}{}'\left[\left(\frac{(\hbar\boldsymbol{k})^2}{2m} + \frac{n_0}{L^3}\mathcal{V}(k)\right)u(k)v(k)\right.\\ &\left.+\frac{n_0}{L^3}\mathcal{V}(k)(u(k)^2+v(k)^2)\right]\left(\hat{A}^\dagger_{\boldsymbol{k}}\hat{A}^\dagger_{-\boldsymbol{k}} + \hat{A}_{\boldsymbol{k}}\hat{A}_{-\boldsymbol{k}}\right) .\end{aligned} \quad (5)$$

To bring the Hamiltonian into diagonal form, the relation

Exercise 7.1.

$$\left(\frac{(\hbar \boldsymbol{k})^2}{2m} + \frac{n_0}{L^3}\mathcal{V}(k)\right) u(k)v(k) + \frac{n_0}{L^3}\mathcal{V}(k)\left[u^2(k) + v^2(k)\right] = 0 \tag{6}$$

has to hold. With (3) this one becomes

$$\left(\frac{(\hbar \boldsymbol{k})^2}{2m} + \frac{n_0}{L^3}\mathcal{V}(k)\right) \frac{D_k}{1 - D_k^2} + \frac{n_0}{2L^3}\mathcal{V}(k)\frac{1 + D_k^2}{1 - D_k^2} = 0\,, \tag{7}$$

and we deduce for the coefficients D_k

$$D_k = \frac{L^3}{n_0 \mathcal{V}(k)}\left(\sqrt{\left(\frac{\hbar \boldsymbol{k}}{2m}\right)^2 + \frac{(\hbar \boldsymbol{k})^2}{m}\frac{n_0\mathcal{V}(k)}{L^3}} - \frac{(\hbar \boldsymbol{k})^2}{2m} - \frac{n_0}{L^3}\mathcal{V}(k)\right). \tag{8}$$

Hence, the Hamiltonian (5) takes on the form

$$\hat{H} = \hat{H}_0 + \sum_{\boldsymbol{k}}{}' E(k)\hat{A}_{\boldsymbol{k}}^\dagger \hat{A}_{\boldsymbol{k}} \tag{9}$$

with

$$\hat{H}_0 = \frac{N^2}{2L^3}\mathcal{V}(0) + \frac{1}{2}\sum_{\boldsymbol{k}}{}' \left(\mathcal{E}(k) - \frac{(\hbar \boldsymbol{k})^2}{2m} - \frac{n_0}{L^3}\mathcal{V}(k)\right) \tag{10}$$

and

$$E(k) = \sqrt{\left(\frac{(\hbar \boldsymbol{k})^2}{2m}\right)^2 + \frac{(\hbar \boldsymbol{k})^2}{m}\frac{n_0}{L^3}\mathcal{V}(k)}\,. \tag{11}$$

Equation (9) describes a system of free bosons – quasiparticles – with one-particle energy $\mathcal{E}(k)$. $\hat{H}_0$ describes the energy of the system in the ground state $|\phi_0\rangle$; for the state $|\phi_0\rangle$, free of quasiparticles, it holds that

$$\hat{A}_{\boldsymbol{k}}|\phi_0\rangle = 0\,, \quad \text{for all } k \neq 0 \tag{12}$$

and the energy becomes

$$\frac{\langle\phi_0|\hat{H}|\phi_0\rangle}{\langle\phi_0|\phi_0\rangle} = \frac{\langle\phi_0|\hat{H}_0|\phi_0\rangle}{\langle\phi_0|\phi_0\rangle} = H_0\,. \tag{13}$$

The coefficients D_k can also be determined by the requirement that the energy of the ground state $|\phi_0\rangle$ of the system, (12), takes on a minimum. According to (5) the ground-state energy is given by

$$\begin{aligned}\hat{H}_0 = {} & \frac{N^2}{2L^3}\mathcal{V}(0) \\ & + \sum_{\boldsymbol{k}}{}' \left[\left(\frac{(\hbar \boldsymbol{k})^2}{2m} + \frac{n_0}{L^3}\mathcal{V}(k)\right) v^2(k) + \frac{n_0}{L^3}\mathcal{V}(k)u(k)v(k)\right]\end{aligned} \tag{14}$$

or

$$\begin{aligned}H_0 = {} & \frac{N^2}{2L^3}\mathcal{V}(0) \\ & + \sum_{\boldsymbol{k}'}{}' \left[\left(\frac{(\hbar \boldsymbol{k}')^2}{2m} + \frac{n_0}{L^3}\mathcal{V}(k')\right) \frac{D_{k'}^2}{1 - D_{k'}^2} + \frac{n_0}{L^3}\mathcal{V}(k')\frac{D_{k'}}{1 - D_{k'}^2}\right].\end{aligned} \tag{15}$$

Exercise 7.1.

From the requirement that the variation of H_0 with respect to D_k should vanish, i.e.

$$\frac{\delta H_0}{\delta D_k} = 0\,, \tag{16}$$

the condition for D_k is obtained:

$$\left(\frac{(\hbar \boldsymbol{k})^2}{m} + \frac{2n_0}{L^3}\mathcal{V}(k)\right) D_k + \frac{N}{L^3}\mathcal{V}(k)\left(D_k^2 + 1\right) = 0\,. \tag{17}$$

This is the same result as stated in (7).

The operators $\hat{A}_{\boldsymbol{k}}^{\dagger}$, $\hat{A}_{\boldsymbol{k}}$ from (7.26) and (7.22) describe new boson operators. According to (7.26) the excitations of the system can be described as single-particle excitations

$$\hat{A}_{\boldsymbol{k}}^{\dagger}|\phi_0\rangle\,, \tag{7.29}$$

where $|\phi_0\rangle$ characterizes the ground state with

$$\hat{A}_{\boldsymbol{k}}|\phi_0\rangle = 0\,. \tag{7.30}$$

These new one-particle states (7.29) are called *quasiparticles*. According to (7.20) they consist of a linear combination of creation and annihilation operators $\hat{a}_{-\boldsymbol{k}}^{\dagger}$ and $\hat{a}_{\boldsymbol{k}}$ of the customary boson operators. The latter themselves represent particle–hole pairs, as is clearly expressed in the defining equations (7.14). As a result of the transition to quasiparticles, an important part of the interaction between the original boson particles (described by $\hat{b}_{\boldsymbol{k}}^{\dagger}$,$\hat{b}_{\boldsymbol{k}}$) has already been taken care of, i.e. "diagonalized away". Hence, quasiparticles can be viewed as freely moving structures consisting of a superposition of ordinary particles and holes. An illustrative example of a quasiparticle is given, for example, by the motion of a charged particle in a strong electrolyte, which consists of positive and negative charge carriers.

The particle ⊕, which is here positively charged, attracts negative charge carriers around it (Debye cloud) and is thus screened (see Fig. 7.4). Out of the originally charged particle a neutral quasiparticle has developed: A part of the Coulomb interaction with the electrolytic medium has been "transformed away" with the Debye cloud. It is evident that a quasiparticle, evolved in this way, possesses a different mass as compared to the original particle, which in general also depends on the velocity. This is in complete analogy with the nontrivial k dependence of $\mathcal{E}(k)$ from (7.28a). We refer to and already recommend Exercise 9.1, where the screening potential of the Debye cloud is calculated.

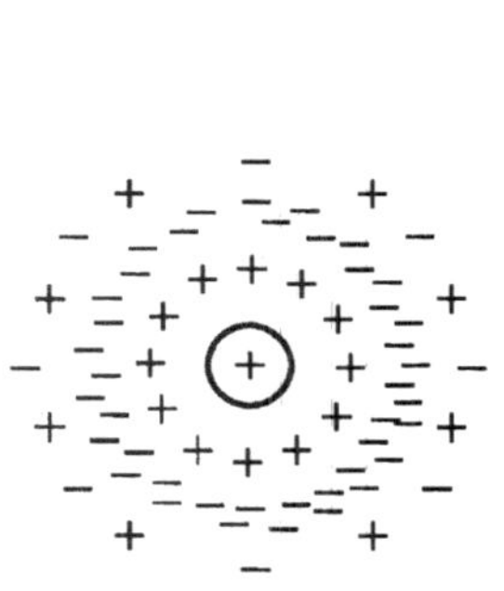

Fig. 7.4. Illustration of a quasiparticle by a Debye cloud around a charged particle ⊕ in a strong electrolyte

The part $\hat{H}_0$ is constant and independent of the quasiparticle operators $\hat{A}_{\boldsymbol{k}}^{\dagger}$,$\hat{A}_{\boldsymbol{k}}$ [see (7.27)]. Obviously, it represents the energy of the ground state of the system; we have shown this already in Exercise 7.1.

The operator $\hat{\boldsymbol{P}}$ of the total momentum of the system can be expressed in terms of quasiparticle operators, too. With (7.14) and (7.18) we obtain

$$
\begin{aligned}
\hat{\boldsymbol{P}} &= \sum_{\boldsymbol{k}} \hbar \boldsymbol{k} \hat{b}^{\dagger}_{\boldsymbol{k}} \hat{b}_{\boldsymbol{k}} \\
&= \sum_{\boldsymbol{k}} \hbar \boldsymbol{k} \hat{a}^{\dagger}_{\boldsymbol{k}} \hat{a}_{\boldsymbol{k}} \\
&= \sum_{\boldsymbol{k}} \hbar \boldsymbol{k} \left[u(\boldsymbol{k})\hat{A}^{\dagger}_{\boldsymbol{k}} + v(\boldsymbol{k})\hat{A}_{-\boldsymbol{k}}\right]\left[u(\boldsymbol{k})\hat{A}_{\boldsymbol{k}} + v(\boldsymbol{k})\hat{A}^{\dagger}_{-\boldsymbol{k}}\right] \\
&= \sum_{\boldsymbol{k}} \hbar \boldsymbol{k} \left\{\left[u^2(\boldsymbol{k}) - v^2(\boldsymbol{k})\right] \hat{A}^{\dagger}_{\boldsymbol{k}}\hat{A}_{\boldsymbol{k}} \right. \\
&\quad \left. + u(\boldsymbol{k})v(\boldsymbol{k})\left(\hat{A}^{\dagger}_{\boldsymbol{k}}\hat{A}^{\dagger}_{-\boldsymbol{k}} + \hat{A}_{-\boldsymbol{k}}\hat{A}_{\boldsymbol{k}}\right)\right\} . \qquad (7.31)
\end{aligned}
$$

The second term inside the curly bracket vanishes as it is summed over $\boldsymbol{k}$, because positive and negative $\boldsymbol{k}$ contributions cancel pairwise. With the normalization (7.21) we get

$$
\hat{\boldsymbol{P}} = \sum_{\boldsymbol{k}} \hbar \boldsymbol{k} \hat{A}^{\dagger}_{\boldsymbol{k}} \hat{A}_{\boldsymbol{k}} . \qquad (7.32)
$$

Thus, it now becomes clear that the low-lying excitations of the system containing Helium atoms at low temperatures are given by elementary (quasiparticle) excitations of the energy $E_{\boldsymbol{k}}$ (which we now write also $E(k)$ to indicate its k dependence more clearly) and the momentum $(\hbar\boldsymbol{k})$. For small excitations $n_0 \approx N$, $E(k)$ from (7.28a) can be rewritten as

$$
E_{\boldsymbol{k}} = E(k) = \sqrt{\frac{(\hbar k)^4}{4m^2} + \frac{N(\hbar k)^2}{mL^3}\mathcal{V}(k)} . \qquad (7.33)
$$

For small momenta this becomes

$$
E(k) = \sqrt{\frac{N}{mL^3}\mathcal{V}(0)}\, \hbar|\boldsymbol{k}|\,(1 + \cdots) . \qquad (7.34)
$$

The velocity of the quasiparticles is given by

$$
c_{\mathrm{s}} = \left(\frac{\partial E}{\partial \hbar k}\right)_{k=0} = \sqrt{\frac{N\mathcal{V}(0)}{mL^3}} = \sqrt{\rho\frac{\mathcal{V}(0)}{m}} . \qquad (7.35)
$$

Here $\rho = N/L^3$ denotes the particle density. It represents the velocity with which the elementary excitations are propagating, i.e. the *sound velocity*. Hence, (7.34) can also be written as

$$
E(k) \approx c_{\mathrm{s}}\hbar|\boldsymbol{k}| . \qquad (7.36)
$$

The sound velocity (7.35) has to be positive and real; otherwise the ground state would decay. This leads to the condition

$$
\mathcal{V}(0) > 0 , \qquad (7.37)
$$

which can be transformed into

$$
\begin{aligned}
\mathcal{V}(0) &= \lim_{|\boldsymbol{q}|\to 0} \int \mathrm{d}^3r\, \mathcal{V}(|\boldsymbol{r}|)\, \mathrm{e}^{\mathrm{i}\boldsymbol{q}\cdot\boldsymbol{r}} \\
&= \int \mathrm{d}^3r\, \mathcal{V}(|\boldsymbol{r}|) > 0 , \qquad (7.38)
\end{aligned}
$$

where (7.6) has been used. This implies that the interaction energy $\mathcal{V}(|\boldsymbol{r}|)$ has to be essentially positive, i.e. *repulsive.*

Now we will look at the quasiparticle energies $E(k)$ from (7.33) for large momenta. For this purpose we write $(\hbar k)^2/2m$ in front of the square root and arrive at

$$\begin{aligned} E(k) &= \frac{(\hbar k)^2}{2m}\sqrt{1+\frac{N4m}{(\hbar k)^2L^3}\mathcal{V}(k)} \\ &\approx \frac{(\hbar k)^2}{2m}\left(1+\frac{1}{2}\frac{N4m}{(\hbar k)^2L^3}\mathcal{V}(k)+\cdots\right) \\ &\approx \frac{(\hbar k)^2}{2m}+\frac{N\mathcal{V}(k)}{L^3}\,. \end{aligned} \tag{7.39}$$

As k increases we get

$$\lim_{k\to\infty}\mathcal{V}(k) \;\rightarrow\; 0 \tag{7.40}$$

according to (7.6). Then the quasiparticle energy (7.39) for larger momenta becomes

$$\lim_{k\to\infty}E(k) \;=\; \frac{(\hbar k)^2}{2m}\,, \tag{7.41}$$

which is just the kinetic energy of a free single atom. Intuitively this is correct; think of the quasiparticle model of the Debye cloud around the particle. For large velocities of the $\oplus$ particle we expect the Debye cloud to be stripped off so that we are dealing more or less with a free particle.

Hence, the momentum dependence of the quasiparticle energies (7.33) in boson systems with weak and repelling interactions has the qualitative form illustrated in Fig. 7.5.

Let us now understand how superfluidity comes about, at least qualitatively. Consider a small particle (which we here call a *cluster*, it could be one or several dirt atoms or a piece of the surrounding wall, etc.) moving through a quantum liquid with energy–momentum characteristics of the type discussed here (see Fig. 7.5), i.e. its quasiparticle energies rise linearly with momentum at low $\hbar k$ and quadratically at higher $\hbar k$. The cluster can lose energy only by causing excitations in the fluid. At temperature $\neq 0$ there are already excitations in the fluid at which the cluster may scatter and thus lose energy, but

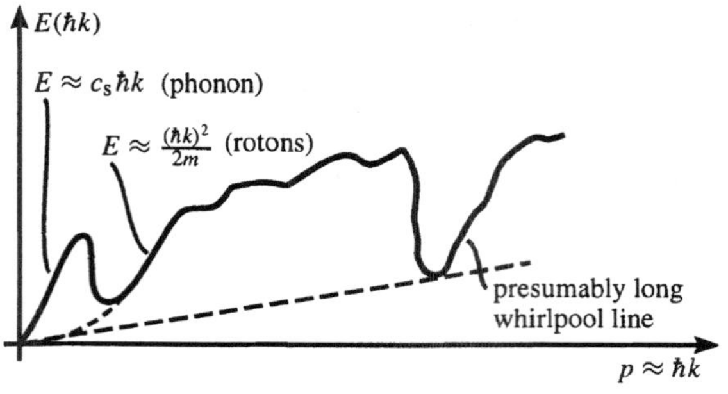

Fig. 7.5. The *full curve* illustrates the qualitative behavior of the quasiparticle energies in a superfluid medium. The free particle energy is indicated as the *dashed curve*

at zero temperature this is not the case. The initial momentum of the cluster shall be $\hbar \boldsymbol{q}$, the momentum of the excitations $\hbar \boldsymbol{k}$. In a scattering event of the cluster with the fluid, energy and momentum are conserved.

$$\begin{aligned} \hbar \boldsymbol{q} &= \hbar \boldsymbol{q}' + \hbar \boldsymbol{k} , \\ \frac{\hbar^2 q^2}{2m} &= \frac{\hbar^2 (q')^2}{2m} + E(k) . \end{aligned} \tag{7.42}$$

Here $\hbar \boldsymbol{q}'$ denotes the momentum of the cluster after the scattering. The elementary excitations of the fluid (quasiparticles) $E(k)$ are given by (7.39). The equations (7.42) cannot be satisfied. To see this let us follow up on energy conservation, which reads

$$\frac{\hbar^2 q^2}{2m} = \frac{\hbar^2}{2m}(\boldsymbol{q} - \boldsymbol{k})^2 + E(k) \tag{7.43}$$

or

$$0 = -\frac{\hbar^2}{m}\boldsymbol{q} \cdot \boldsymbol{k} + \frac{\hbar^2}{2m}\boldsymbol{k} + E(k) , \tag{7.44}$$

and it follows for the angle α between $\boldsymbol{q}$ and $\boldsymbol{k}$ that

$$\cos \alpha = \frac{E_k/\hbar k}{v} + \frac{\hbar k/2m}{v} = \frac{c_s}{v} + \frac{\hbar k/2m}{v} , \tag{7.45}$$

where $v = \hbar \boldsymbol{q}/m$ is the initial velocity of the cluster. We learned that the low-energy excitations (quasiparticles) of the fluid are phonons, for which $E_k/\hbar k > c_s$. Hence for the excitation (emission) of a phonon the cluster velocity v must be larger than c_s, i.e. $v > c_s$. This clearly follows from (7.45) for $k \to 0$, otherwise the angle α becomes imaginary. We can state that a cluster moving with velocity $v < v_{\text{crit}}$ (where $v_{\text{crit}} = c_s$ in this case) cannot lose energy to the fluid; thus there is no friction and one has superfluidity.

It does not matter, of course, in which coordinate system the analysis is carried out. Here we have choosen the fluid at rest and the cluster moving in the fluid. The same holds if the cluster is at rest and the fluid moves along the cluster. Our arguments also hold if the cluster is a dislocation of the surface surrounding the fluid (rough surface).

For liquid Helium the critical velocity is $v_{\text{crit}} \ll c_s$. The relation between momentum and energy is believed to appear as indicated in Fig. 7.5, where the critical velocity is given by the slope of the dashed line. Also indicated are the rotons and curls of high momentum; the latter are believed to be responsible for the critical velocity.

7.2 Landau's Theory of Superfluidity

Let us investigate here in more detail the ability of He II to flow through a capillary without viscosity below the critical temperature T_0. For that purpose we consider the Helium at $T = 0$; the superfluid liquid is in its ground state and is not excited.

We examine a liquid that flows with constant velocity $\boldsymbol{v}$ through a capillary. If friction were present the liquid would lose kinetic energy and the current would slow down.

It is appropriate to consider the problem in the rest frame of the liquid, i.e. the liquid is at rest and the walls are moving with velocity $-\boldsymbol{v}$. If viscosity were present then the resting Helium would start to move. This motion starts locally and gradually because of *elementary excitations*. Once an elementary excitation (quasiparticle) with momentum $\boldsymbol{p} = \hbar\boldsymbol{k}$ and energy $E(k) = E(p)$ is present the energy E_0 of the liquid (in the system for which it was initially at rest) is equal to $E(k)$ and its momentum is $\boldsymbol{P}_0 = \boldsymbol{p}$. We transform back into the rest frame of the capillary with the help of a Galilei transformation, and we arrive at

$$\mathcal{E} = E_0 + \boldsymbol{P}_0 \cdot \boldsymbol{v} + \frac{Mv^2}{2} \qquad \text{and} \qquad \boldsymbol{P} = \boldsymbol{P}_0 + M\boldsymbol{v}\,, \tag{7.46}$$

where M represents the mass of the liquid.

If only one excitation is present, i.e. $E_0 = E(p)$, $\boldsymbol{k}_0 = \boldsymbol{p}$, it follows that

$$\mathcal{E} = E(p) + \boldsymbol{p} \cdot \boldsymbol{v} + \frac{M}{2}v^2\,. \tag{7.47}$$

$Mv^2/2$ is the kinetic energy of the liquid and $E(p) + \boldsymbol{p} \cdot \boldsymbol{v}$ represents the change of energy due to the elementary excitation of the quasiparticle. Since for the case of an intrinsic friction (viscosity) the energy of the fluid, which is moving with velocity $\boldsymbol{v}$, should decrease, it has to hold that $E(p) + \boldsymbol{p} \cdot \boldsymbol{v} < 0$. If $\boldsymbol{p}$ is given, $E(p) + \boldsymbol{p} \cdot \boldsymbol{v}$ becomes minimal once $\boldsymbol{p}$ and $\boldsymbol{v}$ are antiparallel, i.e. $E(p) - |\boldsymbol{p}||\boldsymbol{v}| < 0$! Thus,

$$v > \frac{E(k)}{k}\,, \quad \hbar k = p\,. \tag{7.48}$$

Only if this condition is fulfilled can elementary excitations take place, i.e. quasiparticles can be excited, and the liquid slows down. Then it possesses viscosity because energy is lost for the quasiparticle excitations.

Condition (7.48) has to be fulfilled at least for some values of p. As already outlined before, this implies that elementary excitations can occur only once v is larger than the minimum of $E(p)/p$. We are thus led to a critical velocity:

$$v > \left(\frac{E(p)}{p}\right)_{\min} = v_{\text{crit}} \tag{7.49}$$

for quasiparticles to be excited. If $v < v_{\text{crit}}$ then condition (7.48) cannot be fulfilled: no quasiparticles can be excited and no energy is lost for intrinsic degrees of freedom. *Then the liquid is superfluid.*

The minimum value of $E(p)/p$ corresponds to the point on the curve $E(p)(p)$ for which

$$\frac{\mathrm{d}E(p)}{\mathrm{d}p} = \frac{E(p)}{p}\,. \tag{7.50}$$

Figure 7.6 sketches the situation.

Geometrically this corresponds to the point for which a straight line through the origin is at the same time a tangent to $E(p)$ [see (7.34)!]. For all other values of p, which are not solutions of (7.50), $E(p)/p$ is larger than the minimal value described by (7.50). *Superfluidity can occur only if the velocity of the liquid is smaller than the velocity of an elementary excitation;*

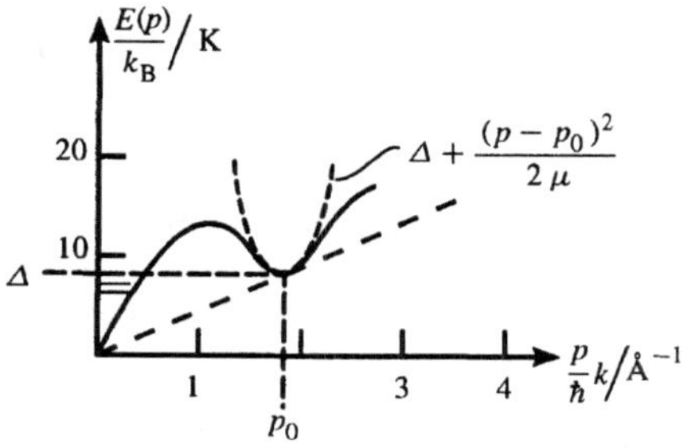

Fig. 7.6. The quasiparticle energy in K (k_B represents Boltzmann's constant) and its expansion around the minimum

otherwise superfluidity would be destroyed because the fluid would lose energy as a result of the excitation of quasiparticles. $\mathrm{d}E/\mathrm{d}p$ is the velocity of an elementary excitation. For every Bose liquid this required condition is fulfilled for one point, i.e. for $p = 0$. Near $p = 0$ the elementary excitations move with the sound velocity. These are the phonons with $E(p) = c_s p$! The condition for superfluidity is not fulfilled for sure once the velocity of the liquid is $v > c_s$. For the case of Helium, however, more than one point exists that fulfills this condition. For small p the experimental spectrum $E(p)$ (see Fig. 7.6) shows the expected behavior for phonons, i.e. $E(p) = c_s p = c_s \hbar k$. However, another minimum exists at p_0. In thermal equilibrium most of the elementary excitations have energies very close to the minimum of $E(p)$, i.e. for $p = 0$ and $p = p_0$!

In such a case, we may expand around p_0 (approximation with a parabola) and obtain

$$E(p) = k_B \Delta + \frac{(p - p_0)^2}{2\mu} , \tag{7.51}$$

where $\Delta = E(p_0)$ and μ are constants characterizing the minimum. Quasiparticles that obey this dispersion relation are called *rotons*. For ^{4}He these parameters are

$$\begin{aligned} \Delta &\equiv T_0 \approx 8.5\,\mathrm{K} , \\ k_0 = p_0/\hbar &= 1.92 \times 10^8\,\mathrm{cm}^{-1} , \\ \mu &= 0.16\, m_{(^4\mathrm{He})} . \end{aligned} \tag{7.52}$$

The second point [i.e. second solution of (7.50)] we were looking for lies right of the *minimum for rotons* at p_0 as can be identified from Fig. 7.6. With (7.51) we calculate the critical velocity v_{crit} from (7.49). It follows from condition (7.50), which now reads

$$\frac{p - p_0}{\mu} = \frac{k_B \Delta + (p - p_0)^2/2\mu}{\Delta} .$$

It has the solution

$$p = \sqrt{2\mu k_B \Delta + p_0^2}$$

and, hence,

$$v_{\text{crit}} = \frac{1}{\mu}\left(\sqrt{p_0^2 + 2\mu k_{\text{B}}\Delta} - p_0\right) .$$

Thus, superfluidity exists only if

$$v < v_{\text{crit}} = \frac{1}{\mu}\left(\sqrt{p_0^2 + 2\mu k_{\text{B}}\Delta} - p_0\right) \tag{7.53}$$

or, once we consider the experimental values (7.52) for ^{4}He (they show that $p_0^2 \gg 2\mu\Delta$),

$$v < \frac{k_{\text{B}}\Delta}{p_0} .$$

We can express this in another way: There is no superfluid flow in ^{4}He if the flow velocity is $v > k_{\text{B}}\Delta/p_0 = k_{\text{B}}T_0/p_0$! Remember that $\Delta = T_0$ is measured as a temperature – see (7.52). Consider now the Bose liquid for $T \ll T_0$ but $T \neq 0$. In this case excitations may be present. Think of a "quasiparticle gas" that moves with $\boldsymbol{v}$ relative to the liquid. This quasiparticle gas can be at rest relative to the wall, the latter moving with $-\boldsymbol{v}$ with respect to the liquid: it crawls, so to speak, out of the vessel walls because of the temperature, which causes the quasiparticle excitations. The distribution function of the gas at rest is denoted as $\rho(E)$; $\rho(E - \boldsymbol{p}\cdot\boldsymbol{v})$ represents the distribution of the moving gas. Hence, the total momentum of the system is given by

$$\boldsymbol{P} = \frac{1}{(2\pi)^3}\int \mathrm{d}^3p\,\boldsymbol{p}\,\rho(E - \boldsymbol{p}\cdot\boldsymbol{v}) . \tag{7.54}$$

For small velocities we expand with respect to $\boldsymbol{p}\cdot\boldsymbol{v}$. The first term is linear in $\boldsymbol{p}$. Because of the isotropic orientation of $\boldsymbol{p}$, this term vanishes with an integration over all directions. The second term is

$$\boldsymbol{P} = -\frac{1}{(2\pi)^3}\int \mathrm{d}^3p\,\boldsymbol{p}(\boldsymbol{p}\cdot\boldsymbol{v})\frac{\partial\rho}{\partial E} . \tag{7.55}$$

The integration over angles produces a factor $4\pi/3$ and we arrive at

$$\boldsymbol{P} = -\frac{\boldsymbol{v}}{3}4\pi\int_0^\infty \mathrm{d}p\,p^4\frac{\mathrm{d}\rho(E)}{\mathrm{d}E} . \tag{7.56}$$

For phonons the relation $E = c_{\text{s}}p$ holds. Via partial integration we obtain

$$\begin{aligned} \boldsymbol{P} &= -\boldsymbol{v}\frac{4\pi}{3c_{\text{s}}}\int_0^\infty \mathrm{d}p\,p^4\frac{\mathrm{d}\rho(p)}{\mathrm{d}p} \\ &= \boldsymbol{v}\frac{16\pi}{3c_{\text{s}}}\int_0^\infty \mathrm{d}p\,p^3\rho(p) . \end{aligned} \tag{7.57}$$

The integral

$$4\pi\int_0^\infty \mathrm{d}p\,c_{\text{s}}p\,\rho(p)p^2 = \int \mathrm{d}^3p\,E\,\rho(E)$$

gives the energy E_{ph} of a unit volume of the phonon gas. We arrive at

$$\boldsymbol{P} = \boldsymbol{v}\frac{4E_{\text{ph}}}{3c_{\text{s}}^2} = \boldsymbol{v}m^* . \tag{7.58}$$

Obviously we can interpret this in the following way. The phonon gas has an effective mass m^*. Like all the liquid, this part of it also represents mass transport. The excitations are allowed to collide with the walls and to exchange momentum; this leads to viscosity. Evidently, viscous flow can occur in a Bose gas that contains elementary excitations and does not violate the condition for the existence of superfluidity. The total transported mass m^* is not identical to the total mass of the liquid. We obtain the following picture. For $T = 0$ no excitations exist, i.e. the motion is superfluid. For $T \neq 0$ elementary excitations exist in the system, which cause a part of the gas (mass m^*) to be viscous and converts it into a "normal" liquid. A superfluid rest remains, so that

$$\rho = \rho_n + \rho_s .$$

This is the two-fluid picture we mentioned earlier.

The portions ρ_n and ρ_s depend on temperature: For $T = 0$ we have $\rho_n = 0$ and $\rho_s = \rho$, for $0 < T < T_0$ we expect $\rho_s \neq 0$ and $\rho_n \neq 0$, and for $T > T_0$ we expect $\rho_s = 0$ and $\rho_n = \rho$ (compare again with Fig. 7.2). Notice that two different kinds of elementary excitations "contribute" to the normal fluid: phonons with $E \sim p$ and rotons with $E \sim p^2$. Their relevance depends on temperature. For $T \approx 0.6\,\mathrm{K}$ the contributions from phonons and rotons are comparable whereas for larger temperatures the rotons dominate. However, except for the dispersion relation, no difference exists between phonons and rotons.

EXERCISE

7.2 An Analogy to Superfluidity in Hydrodynamics

Problem. Explain the following Gedanken experiment. A basin is filled with water. If a thin object, such as a rasor-blade, is moved through the water, a laminar current is observed for small velocities of the object. Once the velocity is increased above a critical value (approximately 23 cm/s) the current changes its character.

Solution. The critical velocity v_{crit} ($\approx$ 23 cm/s) represents the velocity of surface waves on the water. If $v > v_{\text{crit}}$ the object produces waves in such a way that the latter run away from the object, interfere according to Huygen's principle, and form a cone (with a fixed opening angle $\approx 20°$). In contrast to the Mach cone phenomenon appearing at supersonic flight velocities, this angle does not depend on the shape of the object, the velocity or any other exterior factor. It is a consequence only of the dispersion relation for water waves. In the case of deep water, where the depth of the water is much greater than the wavelength λ of the surface wave, the dispersion relation of the wave reads

$$c^2 = \frac{g\lambda}{2\pi} . \tag{1}$$

Exercise 7.2.

c denotes the speed of propagation and $g = 9.8\,\mathrm{m/s^2}$, the gravitational acceleration. Thus, the angular velocity $\omega = 2\pi c/\lambda$ is

$$\omega = g/c\,. \tag{2}$$

Now, consider a radially symmetric surface wave centered around $r = 0$,

$$\sin\left(\frac{2\pi r}{\lambda} - \omega t\right)\,, \tag{3}$$

i.e. the phase ϕ at radial distance r and time t is

$$\phi(r,t) = \frac{2\pi r}{\lambda} - \omega t = \frac{2\pi r}{\lambda} - t\sqrt{\frac{2\pi g}{\lambda}}\,. \tag{4}$$

The characteristic wavelength is obtained from the condition $\mathrm{d}\phi/\mathrm{d}\lambda = 0$,

$$\lambda_{\mathrm{ch}} = \frac{8\pi r^2}{g t^2}\,. \tag{5}$$

In the case of an object moving with velocity v, waves centered around each point of its trajectory superimpose (cf. Fig. 7.7).The characteristic wavelength of the waves centered around x_2 is

$$\lambda_{\mathrm{ch}} = \frac{8\pi r'^2}{g t^2}\,. \tag{6}$$

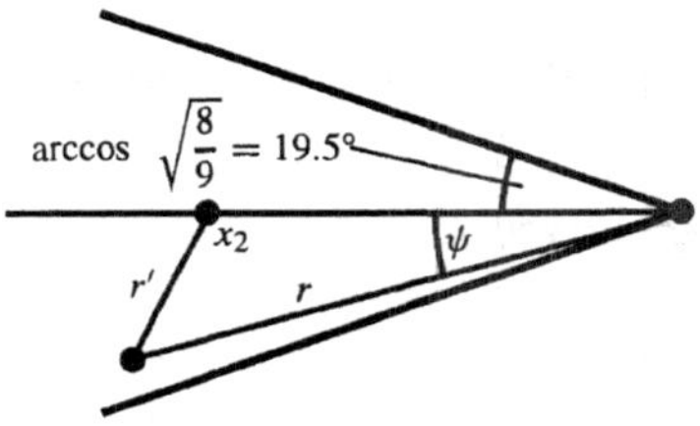

Fig. 7.7. If the object moves faster than v_{crit}, a cone with fixed angle, independent of the object's velocity, emerges

Using (5) and (6) we find

$$\phi = -\frac{g t^2}{4r'}\,. \tag{7}$$

Employing $r' = \sqrt{r^2 + v^2t^2 - 2rvt\cos\psi}$, the condition $\mathrm{d}\phi/\mathrm{d}t = 0$ (stationary phase) leads to

$$t = \frac{3}{2}\frac{r}{v}\left(\cos\psi \pm \sqrt{\cos^2\psi - 8/9}\right)\,. \tag{8}$$

For $\cos^2\psi < 8/9$, i.e. $|\psi| > 19.5°$, no real solution exists. In this case the phase has no stationary point, i.e. all phases contribute equally, and destructive interference leads to a vanishing amplitude outside this cone. If we consider this problem quantum mechanically, we could say that the moving object emitts quasiparticles (we could call them "hydrons")[7] and in this way could obtain

[7] J.L. Synge: Science 138 (1962) 13.

momentum and energy as far as $v > v_{\mathrm{crit}}$. We have the same situation as in the case of superfluidity: The laminar current corresponds to the superfluid state; no excitations do occur! The current cone determined by the quasiparticles corresponds to the normal, nonsuperfluid state.

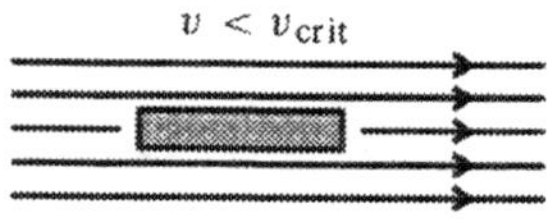

Fig. 7.8. Laminar current of a fluid around a thin object. If the object moves faster than v_{crit} "hydrons" are emitted and a Mach-cone-type wave emerges with well-defined cone angle, independent of the object's velocity

8. Pair Correlations Among Fermions and Bosons

We will make use of the formalism of second quantization developed so far in order to deepen our insights into the properties of quantum-mechanical many-body systems.

So-called correlations coin very fundamental and characteristic many-body effects. In general, correlations of a many-particle system (bosons or fermions) describe the influence of the quantum numbers of a single particle on all remaining particles of the system and their quantum numbers. We accentuate in advance: Correlations are not only produced by special interactions (forces) between particles but are also a consequence of the corresponding fundamental statistics of fermions and bosons (commutation relations between fermionic and bosonic operators). In the following, we will concentrate on correlations that result from pure exchange effects.

The lowest correlations that can be studied are the so-called *pair correlations*. In the following, we will focus on these correlations for fermions and bosons.

8.1 Pair-Correlation Function for Fermions

We consider a gas of N free, noninteracting fermions with spin 1/2 in a volume V. The ground state $|\Phi_0\rangle$ of this N-particle system is characterized such that all states with spin projection $s = \pm 1/2$ and momentum $\boldsymbol{p}$ are occupied up to the Fermi momentum p_{F}, i.e.

$$n_s(\boldsymbol{p}) \;=\; \langle\phi_0|\hat{b}_s^\dagger(\boldsymbol{p})\hat{b}_s(\boldsymbol{p})|\phi_0\rangle \;=\; \begin{cases} 1 & \text{for } |\boldsymbol{p}| \le p_{\mathrm{F}} \,, \\ 0 & \text{for } |\boldsymbol{p}| > p_{\mathrm{F}} \,. \end{cases} \tag{8.1}$$

The validity of this relation becomes instantly clear once the representation

$$|\phi_0\rangle \;=\; \prod_{s,\boldsymbol{p}}^{p_{\mathrm{F}}} \hat{b}_s^\dagger(\boldsymbol{p})|0\rangle \tag{8.2}$$

of the ground state is inserted. The Fermi momentum is fixed by the condition that the average particle-number density n of the system is constant:

$$n \;=\; \frac{N}{V} \;=\; \sum_{s,\boldsymbol{p}} n_s(\boldsymbol{p}) \;=\; \frac{2}{V}\sum_{|\boldsymbol{p}|\le p_{\mathrm{F}}} 1 \,. \tag{8.3}$$

In the thermodynamic limit ($\lim_{N\to\infty}$, $\lim_{V\to\infty}$, $N/V = \text{const}$) this leads to the relation

$$n = 2\int^{p_F} \frac{d^3p}{(2\pi)^3} = \frac{p_F^3}{3\pi^2}\,. \tag{8.4}$$

Before we concentrate directly on the two-particle correlation function in a Fermi gas, we first calculate the expectation value of the one-particle density matrix operator and of the one-particle density operator.

The operator of the *one-particle density matrix* is defined via field operators:

$$\hat{\rho}_s(\boldsymbol{r},\boldsymbol{r}') = \hat{\psi}_s^\dagger(\boldsymbol{r})\hat{\psi}_s(\boldsymbol{r}')\,. \tag{8.5a}$$

Its form is motivated by the well-known expression $\psi^*(\boldsymbol{r})\psi(\boldsymbol{r})$ for the probability density. It fulfills the symmetric relation

$$\hat{\rho}_s(\boldsymbol{r},\boldsymbol{r}') = \hat{\rho}_s^\dagger(\boldsymbol{r}',\boldsymbol{r})\,. \tag{8.5b}$$

The one-particle density operator for particles with spin projection s is the diagonal element ($\boldsymbol{r}=\boldsymbol{r}'$) of the one-particle density matrix:

$$\hat{\rho}_s(\boldsymbol{r}) = \hat{\rho}_s(\boldsymbol{r},\boldsymbol{r}) = \hat{\psi}_s^\dagger(\boldsymbol{r})\hat{\psi}_s(\boldsymbol{r})\,. \tag{8.6a}$$

Hence, the total *density operator* is given by

$$\hat{\rho}(\boldsymbol{r}) = \sum_s \hat{\rho}_s(\boldsymbol{r})\,. \tag{8.6b}$$

In order to evaluate the ground-state expectation value of $\hat{\rho}_s(\boldsymbol{r},\boldsymbol{r}')$ we employ the expansion of the field operators into free momentum states:

$$\hat{\psi}_s(\boldsymbol{r}) = \int \frac{d^3p}{(2\pi)^3}\, e^{i\boldsymbol{p}\cdot\boldsymbol{r}}\, \hat{b}_s(\boldsymbol{p})\,; \tag{8.7}$$

we obtain

$$\begin{aligned}\rho_s(\boldsymbol{r},\boldsymbol{r}') &= \langle\phi_0|\hat{\rho}_s(\boldsymbol{r},\boldsymbol{r}')|\phi_0\rangle \\ &= \int \frac{d^3p}{(2\pi)^3}\frac{d^3p'}{(2\pi)^3}\, e^{-i\boldsymbol{p}\cdot\boldsymbol{r}}\, e^{i\boldsymbol{p}'\cdot\boldsymbol{r}'}\langle\phi_0|\hat{b}_s^\dagger(\boldsymbol{p})\hat{b}_s(\boldsymbol{p}')|\phi_0\rangle\,. \end{aligned} \tag{8.8}$$

The matrix element appearing in (8.8) is nonvanishing only if $\boldsymbol{p}=\boldsymbol{p}'$, i.e.

$$\langle\phi_0|\hat{b}_s^\dagger(\boldsymbol{p})\hat{b}_s(\boldsymbol{p}')|\phi_0\rangle = (2\pi)^3\,\delta(\boldsymbol{p}-\boldsymbol{p}')\langle\phi_0|\hat{b}_s^\dagger(\boldsymbol{p})\hat{b}_s(\boldsymbol{p})|\phi_0\rangle\,, \tag{8.9}$$

and if momenta $\boldsymbol{p}$ lie inside the Fermi sphere. This yields

$$\begin{aligned}\rho_s(\boldsymbol{r},\boldsymbol{r}') &= \int^{p_F} \frac{d^3p}{(2\pi)^3}\, e^{-i\boldsymbol{p}\cdot(\boldsymbol{r}-\boldsymbol{r}')} \\ &= \frac{p_F^3}{3\pi^2}\frac{1}{\xi^3}(\sin\xi - \xi\cos\xi) \\ &= \frac{3n}{2}\frac{j_1(\xi)}{\xi}\,. \end{aligned} \tag{8.10}$$

Here, the momentum integration has been performed in spherical coordinates and the substitution $\xi = p_F|\boldsymbol{r}-\boldsymbol{r}'|$ and (8.4) have been used.

j_1 denotes a spherical Bessel function. It is related to the regular Bessel functions $J_{n+\frac{1}{2}}(x)$ via the relation $j_n(x) = \sqrt{\pi/2x}\, J_{n+\frac{1}{2}}(x)$. In particular, $j_1(x) = (\sin x/x^2) - (\cos x/x)$. The one-particle density follows directly from the one-particle density matrix (8.10):

$$\langle\phi_0|\hat{\rho}(\boldsymbol{r})|\phi_0\rangle = 2\lim_{\xi\to 0}\frac{3n}{2}\frac{j_1(\xi)}{\xi} = n\,. \tag{8.11}$$

After these preliminary considerations we now calculate the pair-correlation function for a free Fermi gas.

We are well aware of the Pauli principle for fermions from elementary quantum mechanics. As an immediate consequence particles with an identical spin projection tend to avoid each other. To be more precise: Two particles with the same spin quantum number are not allowed at the same space point. We will now inquire the probability amplitude of finding a particle with spin projection s at place $\boldsymbol{r}$ inside a Fermi gas ($s = 1/2$) given another particle with s' at $\boldsymbol{r}'$. That part $\Pi_{ss'}$ of this probability amplitude normalized to unity and independent of the average particle-number density is denoted as the *pair-correlation function*. For a free and homogenous Fermi gas it is exclusively a function of the relative distance between the two particles. Hence, the following properties of $\Pi_{ss'}$ become plausible: Particles with opposite spin, i.e. $s \neq s'$, are allowed to approach each other closely; the correlation function has to be independent of the relative distance, i.e.

$$\Pi_{ss'}(|\boldsymbol{r}-\boldsymbol{r}'|) = 1 \qquad \text{for } s \neq s'\,. \tag{8.12a}$$

Such particles are *uncorrelated*. On the other hand, for particles with identical spin it has to hold that

$$\Pi_{ss'}(|\boldsymbol{r}-\boldsymbol{r}'|) = \begin{cases} 0 & \text{for } |\boldsymbol{r}-\boldsymbol{r}'| \to 0 \\ 1 & \text{for } |\boldsymbol{r}-\boldsymbol{r}'| \to \infty \end{cases}, \qquad s = s'\,. \tag{8.12b}$$

This represents the so-called *Pauli blocking* for small distances; for large distances the correlations vanish, too. We will now determine the probability amplitude of finding a particle at place $\boldsymbol{r}'$ given another one at place $\boldsymbol{r}$ at the same time. A mathematical formulation of this problem is provided in the following way: With the help of the field operators $\hat{\psi}_s(\boldsymbol{r})$ and $\hat{\psi}_{s'}(\boldsymbol{r}')$ a particle with spin s at $\boldsymbol{r}$ and another one with spin s' at $\boldsymbol{r}'$ are annihilated from the (N-particle) ground state $|\phi_0\rangle$. The amplitude of transition into an (N-2)-particle state $|\phi_\nu^{N-2}\rangle$ is given via $\langle\phi_\nu^{N-2}|\hat{\psi}_{s'}(\boldsymbol{r}')\hat{\psi}_s(\boldsymbol{r})|\phi_0\rangle$. After a summation over all possible final states we arrive at the total probability amplitude for the annihilation of two particles:

$$\begin{aligned}
&\sum_\nu |\langle\phi_\nu^{N-2}|\hat{\psi}_{s'}(\boldsymbol{r}')\hat{\psi}_s(\boldsymbol{r})|\phi_0\rangle|^2 \\
&\quad = \langle\phi_0|\hat{\psi}_s^\dagger(\boldsymbol{r})\hat{\psi}_{s'}^\dagger(\boldsymbol{r}')\sum_\nu|\phi_\nu^{N-2}\rangle\langle\phi_\nu^{N-2}|\hat{\psi}_{s'}(\boldsymbol{r}')\hat{\psi}_s(\boldsymbol{r})|\phi_0\rangle \\
&\quad = \langle\phi_0|\hat{\psi}_s^\dagger(\boldsymbol{r})\hat{\psi}_{s'}^\dagger(\boldsymbol{r}')\hat{\psi}_{s'}(\boldsymbol{r}')\hat{\psi}_s(\boldsymbol{r})|\phi_0\rangle \equiv \left(\frac{n}{2}\right)^2\Pi_{ss'}(|\boldsymbol{r}-\boldsymbol{r}'|)\,,
\end{aligned} \tag{8.13a}$$

where use has been made of the completeness relation for two-particle states. From the form of the probability amplitude (8.13a) it becomes obvious that an equivalent formulation of the problem is provided by first annihilating a particle with spin s at position $\boldsymbol{r}$ and then calculating the expectation value with the remaining (N-1)-particle state $|\phi^{N-1}\rangle = \hat{\psi}_s(\boldsymbol{r})|\phi_0\rangle$ to find a particle with spin s' at position $\boldsymbol{r}'$:

$$\begin{aligned}&\langle\phi^{N-1}|\hat{\psi}^\dagger_{s'}(\boldsymbol{r}')\hat{\psi}_{s'}(\boldsymbol{r}')|\phi^{N-1}\rangle \\ &\quad= \langle\phi_0|\hat{\psi}^\dagger_s(\boldsymbol{r})\hat{\psi}^\dagger_{s'}(\boldsymbol{r}')\hat{\psi}_{s'}(\boldsymbol{r}')\hat{\psi}_s(\boldsymbol{r})|\phi_0\rangle\,. \qquad (8.13\text{b})\end{aligned}$$

For the further manipulation of the amplitude (8.13b) we insert an expansion into momentum states (8.7) for each field operator and obtain

$$\begin{aligned}&\left(\frac{n}{2}\right)^2 \Pi_{ss'}(|\boldsymbol{r}-\boldsymbol{r}'|) \\ &\quad= \int \frac{\mathrm{d}^3p}{(2\pi)^3}\frac{\mathrm{d}^3q}{(2\pi)^3}\frac{\mathrm{d}^3p'}{(2\pi)^3}\frac{\mathrm{d}^3q'}{(2\pi)^3}\,\mathrm{e}^{-\mathrm{i}(\boldsymbol{p}-\boldsymbol{p}')\cdot\boldsymbol{r}}\,\mathrm{e}^{-\mathrm{i}(\boldsymbol{q}-\boldsymbol{q}')\cdot\boldsymbol{r}'} \\ &\qquad\times\langle\phi_0|\hat{b}^\dagger_s(\boldsymbol{p})\hat{b}^\dagger_{s'}(\boldsymbol{q})\hat{b}_{s'}(\boldsymbol{q}')\hat{b}_s(\boldsymbol{p}')|\phi_0\rangle\,. \qquad (8.14)\end{aligned}$$

In order to obtain nonvanishing values for the appearing matrix element, the momentum integrations should run over only the Fermi sphere. Furthermore, we use the commutation relations for the operators $\hat{b}^\dagger_s(\boldsymbol{p})$ and $\hat{b}_{s'}(\boldsymbol{p}')$, which follow directly from the corresponding relations for the field operators presented earlier:

$$\begin{aligned}\left[\hat{b}^\dagger_s(\boldsymbol{p}),\hat{b}^\dagger_{s'}(\boldsymbol{p}')\right]_+ &= \left[\hat{b}_s(\boldsymbol{p}),\hat{b}_{s'}(\boldsymbol{p}')\right]_+ = 0\,, \\ \left[\hat{b}^\dagger_s(\boldsymbol{p}),\hat{b}_{s'}(\boldsymbol{p}')\right]_+ &= (2\pi)^3\,\delta_{ss'}\,\delta(\boldsymbol{p}-\boldsymbol{p}')\,. \end{aligned} \qquad (8.15)$$

The factor $(2\pi)^3$ on the right-hand side of the last equation is a normalization factor which corresponds to the $\mathrm{d}p^3/(2\pi)^3$ volume element in momentum space. Indeed, taking the vacuum expectation value and integrating over momenta yields

$$\int\frac{\mathrm{d}^3p}{(2\pi)^3}\langle 0|[\hat{b}^\dagger_s(\boldsymbol{p}),\hat{b}_{s'}(\boldsymbol{p}')]_+|0\rangle \qquad (8.16)$$

$$= \int\frac{\mathrm{d}^3p}{(2\pi)^3}\langle 0|\hat{b}^\dagger_s(\boldsymbol{p}),\hat{b}_{s'}(\boldsymbol{p}') \qquad (8.17)$$

$$+\,\hat{b}_{s'}(\boldsymbol{p}')\hat{b}^\dagger_s(\boldsymbol{p}),|0\rangle \qquad (8.18)$$

$$= \int\frac{\mathrm{d}^3p}{(2\pi)^3}\langle 0|\hat{b}_{s'}(\boldsymbol{p}')\hat{b}^\dagger_s(\boldsymbol{p}),|0\rangle \qquad (8.19)$$

$$= \int\frac{\mathrm{d}^3p}{(2\pi)^3}\langle 0|(2\pi)^3\delta_{ss'}\delta(\boldsymbol{q}-\boldsymbol{p}') \qquad (8.20)$$

$$-\,\hat{b}^\dagger_s(\boldsymbol{p}),\hat{b}_{s'}(\boldsymbol{p}')|0\rangle \qquad (8.21)$$

$$= \int\frac{\mathrm{d}^3p}{(2\pi)^3}(2\pi)^3\delta^3(\boldsymbol{q}-\boldsymbol{p}') = 1\,. \qquad (8.22)$$

We transform further:

$$\begin{aligned}
&\langle\phi_0|\hat{b}_s^\dagger(\boldsymbol{p})\hat{b}_{s'}^\dagger(\boldsymbol{q})\hat{b}_{s'}(\boldsymbol{q}')\hat{b}_s(\boldsymbol{p}')|\phi_0\rangle \\
&\quad= -(2\pi)^3\delta_{ss'}\delta(\boldsymbol{p}'-\boldsymbol{q})\langle\phi_0|\hat{b}_s^\dagger(\boldsymbol{p})\hat{b}_{s'}(\boldsymbol{q}')|\phi_0\rangle \\
&\qquad+\langle\phi_0|\hat{b}_s^\dagger(\boldsymbol{p})\hat{b}_s(\boldsymbol{p}')\hat{b}_{s'}^\dagger(\boldsymbol{q})\hat{b}_{s'}(\boldsymbol{q}')|\phi_0\rangle \\
&\quad= (2\pi)^6\delta(\boldsymbol{p}-\boldsymbol{p}')\delta(\boldsymbol{q}-\boldsymbol{q}') \\
&\qquad\times\langle\phi_0|\hat{b}_s^\dagger(\boldsymbol{p})\hat{b}_s(\boldsymbol{p}')|\phi_0\rangle\langle\phi_0|\hat{b}_{s'}^\dagger(\boldsymbol{q})\hat{b}_{s'}(\boldsymbol{q}')|\phi_0\rangle \\
&\qquad-(2\pi)^6\delta_{ss'}\delta(\boldsymbol{p}'-\boldsymbol{q})\delta(\boldsymbol{q}'-\boldsymbol{p})\,\langle\phi_0|\hat{b}_s^\dagger(\boldsymbol{p})\hat{b}_s(\boldsymbol{p})|\phi_0\rangle\,.
\end{aligned}$$

Inside the integration domain the remaining matrix elements are all equal to unity. Inserting (8.14) provides

$$\begin{aligned}
\left(\frac{n}{2}\right)^2\Pi_{ss'}(|\boldsymbol{r}-\boldsymbol{r}'|) &= \int^{p_F}\frac{\mathrm{d}^3p}{(2\pi)^3}\,1\int^{q_F}\frac{\mathrm{d}^3q}{(2\pi)^3}\,1 \\
&\quad-\delta_{ss'}\int^{p_F}\frac{\mathrm{d}^3p}{(2\pi)^3}\,\mathrm{e}^{-\mathrm{i}\boldsymbol{p}\cdot(\boldsymbol{r}-\boldsymbol{r}')}\int^{q_F}\frac{\mathrm{d}^3q}{(2\pi)^3}\,\mathrm{e}^{-\mathrm{i}\boldsymbol{q}\cdot(\boldsymbol{r}-\boldsymbol{r}')}\,.
\end{aligned}$$

Taking into account (8.3) and (8.10), we deduce that

$$\left(\frac{n}{2}\right)^2\Pi_{ss'}(|\boldsymbol{r}-\boldsymbol{r}'|) = \left(\frac{n}{2}\right)^2\delta_{ss'}[\rho_s(\boldsymbol{r},\boldsymbol{r}')]^2 \tag{8.23}$$

and we find the pair-correlation function

$$\begin{aligned}
\Pi_{ss'}(|\boldsymbol{r}-\boldsymbol{r}'|) &= 1-\delta_{ss'}\left(\frac{3j_1(p_\mathrm{F}|\boldsymbol{r}-\boldsymbol{r}'|)}{p_\mathrm{F}|\boldsymbol{r}-\boldsymbol{r}'|}\right)^2 \qquad (8.24)\\
&= 1-\delta_{ss'}\frac{\{3[\sin(p_\mathrm{F}|\boldsymbol{r}-\boldsymbol{r}'|)-p_\mathrm{F}|\boldsymbol{r}-\boldsymbol{r}'|\cos(p_\mathrm{F}|\boldsymbol{r}-\boldsymbol{r}'|)]\}^2}{(p_\mathrm{F}|\boldsymbol{r}-\boldsymbol{r}'|)^6}\,.
\end{aligned}$$

This correlation function possesses the required properties discussed before; it is sketched in Fig. 8.1.

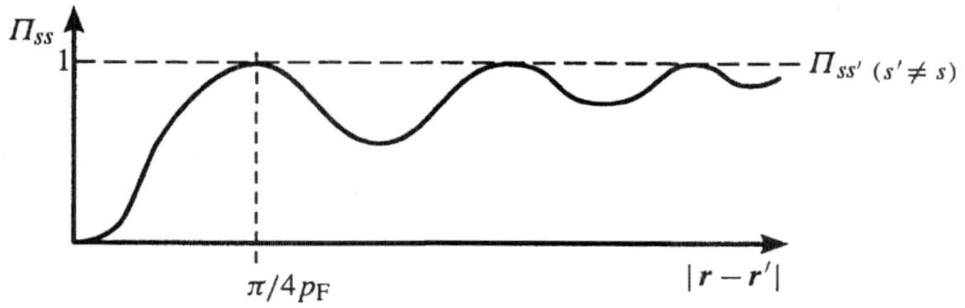

Fig. 8.1. Pair-correlation function of noninteracting spin-1/2 particles with identical spin projection

We realize that the probability of finding two fermions with parallel spins for relative distances smaller than the inverse Fermi momentum is drastically reduced. The Pauli principle induces long-range correlations for the motion of fermions with equal spin; they imply that these fermions cannot approach each other too closely. This effective repulsion results explicitly from the exchange symmetry of the many-particle wavefunction and not from the real, external interactions. This observation is formally expressed in the following rewriting of the amplitude (8.13b):

$$
\begin{aligned}
&\langle\phi_0|\hat{\psi}_s^\dagger(\boldsymbol{r})\hat{\psi}_{s'}^\dagger(\boldsymbol{r}')\hat{\psi}_{s'}(\boldsymbol{r}')\hat{\psi}_s(\boldsymbol{r})|\phi_0\rangle \\
&\quad= \langle\phi_0|\hat{\psi}_s^\dagger(\boldsymbol{r})\hat{\psi}_s(\boldsymbol{r})|\phi_0\rangle\langle\phi_0|\hat{\psi}_{s'}^\dagger(\boldsymbol{r}')\hat{\psi}_{s'}(\boldsymbol{r}')|\phi_0\rangle \\
&\qquad- \langle\phi_0|\hat{\psi}_s^\dagger(\boldsymbol{r})\hat{\psi}_{s'}(\boldsymbol{r}')|\phi_0\rangle\langle\phi_0|\hat{\psi}_{s'}^\dagger(\boldsymbol{r}')\hat{\psi}_s(\boldsymbol{r})|\phi_0\rangle \\
&\quad= \langle\phi_0|\hat{\rho}_s(\boldsymbol{r})|\phi_0\rangle\langle\phi_0|\hat{\rho}_{s'}(\boldsymbol{r}')|\phi_0\rangle \\
&\qquad-\langle\phi_0|\hat{\rho}_s(\boldsymbol{r},\boldsymbol{r}')|\phi_0\rangle\langle\phi_0|\hat{\rho}_s(\boldsymbol{r},\boldsymbol{r}')|\phi_0\rangle\delta_{ss'} \\
&\quad= \left(\frac{n}{2}\right)^2 - \delta_{ss'}(\rho_s(\boldsymbol{r},\boldsymbol{r}'))^2 .
\end{aligned}
\tag{8.25}
$$

The first term represents the direct term, whereas the second one stands for the exchange term. The relative minus sign reflects the antisymmetry.

8.2 Pair-Correlation Function for Bosons

Here we will calculate the pair-correlation function for a free, noninteracting Bose gas. We will come back to the mathematical formulation of the correlation function as presented in (8.13a); this equation is also valid for a system of N bosons. However, changes do occur because bosons do not have to obey Pauli's principle, i.e. arbitrarily many bosons are allowed to occupy the same momentum state.

We consider an N-particle state of the form

$$
|\phi\rangle = \frac{1}{\mathcal{N}}\prod_{\boldsymbol{p}_i}\left[\hat{a}^\dagger(\boldsymbol{p}_i)\right]^{n(\boldsymbol{p}_i)}|0\rangle , \tag{8.26}
$$

where $\mathcal{N}$ represents a normalization factor. Furthermore,

$$
\begin{aligned}
\hat{a}^\dagger(\boldsymbol{p})|\phi\rangle &= \sqrt{n(\boldsymbol{p})+1}\,|\phi^{[n(\boldsymbol{p})+1]}\rangle , \\
\hat{a}(\boldsymbol{p})|\phi\rangle &= \sqrt{n(\boldsymbol{p})}\,|\phi^{[n(\boldsymbol{p})-1]}\rangle .
\end{aligned}
$$

The particle-number density is given via

$$
\langle\phi|\hat{\varphi}^\dagger(\boldsymbol{r})\hat{\varphi}(\boldsymbol{r})|\phi\rangle = \int\frac{\mathrm{d}^3p}{(2\pi)^3}\,n(\boldsymbol{p}) = n . \tag{8.27}
$$

Again, we introduce an expansion into free momentum states for the field operators, this time for the boson operators:

$$
\hat{\varphi}(\boldsymbol{r}) = \int\frac{\mathrm{d}^3p}{(2\pi)^3}\,\mathrm{e}^{\mathrm{i}\boldsymbol{p}\cdot\boldsymbol{r}}\,\hat{a}(\boldsymbol{p}) . \tag{8.28}
$$

An analogous calculation of the expectation value

$$
\langle\phi|\hat{\varphi}^\dagger(\boldsymbol{r})\hat{\varphi}^\dagger(\boldsymbol{r}')\hat{\varphi}(\boldsymbol{r}')\hat{\varphi}(\boldsymbol{r})|\phi\rangle
$$

leads to the matrix element

$$
\langle\phi|\hat{a}^\dagger(\boldsymbol{p})\hat{a}^\dagger(\boldsymbol{q})\hat{a}(\boldsymbol{q}')\hat{a}(\boldsymbol{p}')|\phi\rangle .
$$

The latter is only nonvanishing if $\boldsymbol{p}=\boldsymbol{p}'$ and $\boldsymbol{q}=\boldsymbol{q}'$, or if $\boldsymbol{p}=\boldsymbol{q}'$ and $\boldsymbol{q}=\boldsymbol{p}'$. Both cases are identical once $\boldsymbol{p}=\boldsymbol{q}$. Hence, we can write the following:

$$
\begin{aligned}
&\langle\phi|\hat{a}^\dagger(\boldsymbol{p})\hat{a}^\dagger(\boldsymbol{q})\hat{a}(\boldsymbol{q}')\hat{a}(\boldsymbol{p}')|\phi\rangle \\
&\quad= [1-(2\pi)^3\delta(\boldsymbol{p}-\boldsymbol{q})](2\pi)^6\delta(\boldsymbol{p}-\boldsymbol{p}')\delta(\boldsymbol{q}-\boldsymbol{q}') \\
&\qquad\times\langle\phi|\hat{a}^\dagger(\boldsymbol{p})\hat{a}^\dagger(\boldsymbol{q})\hat{a}(\boldsymbol{q})\hat{a}(\boldsymbol{p})|\phi\rangle \\
&\qquad+[1-(2\pi)^3\delta(\boldsymbol{p}-\boldsymbol{q})](2\pi)^6\delta(\boldsymbol{p}-\boldsymbol{q}')\delta(\boldsymbol{q}-\boldsymbol{p}') \\
&\qquad\times\langle\phi|\hat{a}^\dagger(\boldsymbol{p})\hat{a}^\dagger(\boldsymbol{q})\hat{a}(\boldsymbol{p})\hat{a}(\boldsymbol{q})|\phi\rangle \\
&\qquad+(2\pi)^9\delta(\boldsymbol{p}-\boldsymbol{q})\delta(\boldsymbol{p}-\boldsymbol{p}')\delta(\boldsymbol{q}-\boldsymbol{q}') \\
&\qquad\times\langle\phi|\hat{a}^\dagger(\boldsymbol{p})\hat{a}^\dagger(\boldsymbol{p})\hat{a}(\boldsymbol{p})\hat{a}(\boldsymbol{p})|\phi\rangle\,.
\end{aligned}
\tag{8.29}
$$

We determine the appearing matrix elements one after the other:

$$
\begin{aligned}
&\langle\phi|\hat{a}^\dagger(\boldsymbol{p})\hat{a}^\dagger(\boldsymbol{q})\hat{a}(\boldsymbol{q})\hat{a}(\boldsymbol{p})|\phi\rangle \\
&\quad= n(\boldsymbol{p})\big\langle\phi^{[n(\boldsymbol{p})-1]}|\hat{a}^\dagger(\boldsymbol{q})\hat{a}(\boldsymbol{q})|\phi^{[n(\boldsymbol{p})-1]}\big\rangle \\
&\quad= n(\boldsymbol{p})\,n(\boldsymbol{q})\,, \\
&\langle\phi|\hat{a}^\dagger(\boldsymbol{p})\hat{a}^\dagger(\boldsymbol{q})\hat{a}(\boldsymbol{p})\hat{a}(\boldsymbol{q})|\phi\rangle = n(\boldsymbol{p})\,n(\boldsymbol{q})\,, \\
&\langle\phi|\hat{a}^\dagger(\boldsymbol{p})\hat{a}^\dagger(\boldsymbol{p})\hat{a}(\boldsymbol{p})\hat{a}(\boldsymbol{p})|\phi\rangle \\
&\quad= n(\boldsymbol{p})\big\langle\phi^{[n(\boldsymbol{p})-1]}|\hat{a}^\dagger(\boldsymbol{p})\hat{a}(\boldsymbol{p})|\phi^{[n(\boldsymbol{p})-1]}\big\rangle \\
&\quad= n(\boldsymbol{p})\,(n(\boldsymbol{p})-1)\,.
\end{aligned}
$$

Thus, we obtain for (8.29)

$$
\begin{aligned}
&\langle\phi|\hat{a}^\dagger(\boldsymbol{p})\hat{a}^\dagger(\boldsymbol{q})\hat{a}(\boldsymbol{q}')\hat{a}(\boldsymbol{p}')|\phi\rangle \\
&\quad= (2\pi)^6[\delta(\boldsymbol{p}-\boldsymbol{p}')\delta(\boldsymbol{q}-\boldsymbol{q}')+\delta(\boldsymbol{p}-\boldsymbol{q}')\delta(\boldsymbol{q}-\boldsymbol{p}')]n(\boldsymbol{p})\,n(\boldsymbol{q}) \\
&\qquad-(2\pi)^9\delta(\boldsymbol{p}-\boldsymbol{q})\delta(\boldsymbol{p}-\boldsymbol{p}')\delta(\boldsymbol{q}-\boldsymbol{q}')\,n(\boldsymbol{p})\,(n(\boldsymbol{p})+1)\,.
\end{aligned}
\tag{8.30}
$$

We deduce the probability amplitude of finding a particle at position $\boldsymbol{r}$ and another one at $\boldsymbol{r}'$:

$$
\begin{aligned}
&\langle\phi|\hat{\varphi}^\dagger(\boldsymbol{r})\hat{\varphi}^\dagger(\boldsymbol{r}')\hat{\varphi}(\boldsymbol{r}')\hat{\varphi}(\boldsymbol{r})|\phi\rangle \\
&\quad= \int\frac{\mathrm{d}^3p}{(2\pi)^3}\,n(\boldsymbol{p})\int\frac{\mathrm{d}^3q}{(2\pi)^3}\,n(\boldsymbol{q}) \\
&\qquad+\int\frac{\mathrm{d}^3p}{(2\pi)^3}\,\mathrm{e}^{\mathrm{i}\boldsymbol{p}\cdot(\boldsymbol{r}-\boldsymbol{r}')}\,n(\boldsymbol{p})\int\frac{\mathrm{d}^3q}{(2\pi)^3}\,\mathrm{e}^{\mathrm{i}\boldsymbol{q}\cdot(\boldsymbol{r}-\boldsymbol{r}')}\,n(\boldsymbol{q}) \\
&\qquad-\int\frac{\mathrm{d}^3p}{(2\pi)^3}\,n(\boldsymbol{p})\big(n(\boldsymbol{p})+1\big) \qquad\qquad (8.31)\\
&\quad= n^2+\left|\int\frac{\mathrm{d}^3p}{(2\pi)^3}\,\mathrm{e}^{\mathrm{i}\boldsymbol{p}\cdot(\boldsymbol{r}-\boldsymbol{r}')}\,n(\boldsymbol{p})\right|^2-\int\frac{\mathrm{d}^3p}{(2\pi)^3}\,n(\boldsymbol{p})\big(n(\boldsymbol{p})+1\big)\,.
\end{aligned}
$$

Let's inspect this result closer. In two ways it is different from the result obtained for fermions. First, the sign of the second term is positive which is a direct consequence of the exchange symmetry of bosons. Second, an additional term shows up which is a result of the fact that many bosons are allowed to occupy the same state. Furthermore we notice that this latter term is independent of the positions and, according to dimensions, is smaller than the first two terms by an order $1/V$; this will be rectified in Exercise 8.2. For a homogenous Bose gas that is extended to infinity, the pair correlation function $\Pi(\boldsymbol{r}-\boldsymbol{r}')$ is defined via

$$\langle\phi|\hat{\varphi}^\dagger(\boldsymbol{r})\hat{\varphi}^\dagger(\boldsymbol{r}')\hat{\varphi}(\boldsymbol{r}')\hat{\varphi}(\boldsymbol{r})|\phi\rangle = n^2\,\Pi(\boldsymbol{r}-\boldsymbol{r}') = n^2 + \left|\int \frac{\mathrm{d}^3p}{(2\pi)^3}\,\mathrm{e}^{\mathrm{i}\boldsymbol{p}\cdot(\boldsymbol{r}-\boldsymbol{r}')}\,n(\boldsymbol{p})\right|^2 . \tag{8.32}$$

The first term appearing in this equation emerges from the second one for the case $\boldsymbol{r}=\boldsymbol{r}'$. In other words: the exchange symmetry causes bosons to cluster. According to (8.32) the probability for finding two particles in the same space point at the same time doubles.[1] The following example provides a further illustration of this point.

EXAMPLE

8.1 Pair-Correlation Function for a Beam of Bosons

We calculate the pair-correlation function for a particle beam of non-interacting bosons, a photon beam for example. The average particle-number density $n(\boldsymbol{p})$ is characterized by a Gaussian momentum distribution

$$n(\boldsymbol{p}) = \alpha\,\mathrm{e}^{-\beta(\boldsymbol{p}-\boldsymbol{p}_0)^2/2} , \tag{1}$$

centered around momentum $\boldsymbol{p}_0$. At first we determine the Fourier integral of the second term in (8.32) with $(\boldsymbol{R}=\boldsymbol{r}-\boldsymbol{r}')$:

$$I(\boldsymbol{R}) = \alpha\int\frac{\mathrm{d}^3p}{(2\pi)^3}\,\mathrm{e}^{\mathrm{i}\boldsymbol{p}\cdot\boldsymbol{R}}\,\mathrm{e}^{-\beta(\boldsymbol{p}-\boldsymbol{p}_0)^2/2} . \tag{2}$$

We substitute $\boldsymbol{k}=\boldsymbol{p}-\boldsymbol{p}_0$ and calculate the $\boldsymbol{k}$ integral in polar coordinates; this yields:

$$I(\boldsymbol{R}) = \frac{\alpha\,\mathrm{e}^{\mathrm{i}\boldsymbol{p}_0\cdot\boldsymbol{R}}}{4\pi^2}\int_0^\infty \mathrm{d}k\,k^2\,\mathrm{e}^{-\beta k^2/2}\int_{-1}^{1}\mathrm{d}(\cos\alpha)\,\mathrm{e}^{\mathrm{i}kR\cos\alpha} = \frac{\alpha\,\mathrm{e}^{\mathrm{i}\boldsymbol{p}_0\cdot\boldsymbol{R}}}{2\pi^2 R}\,F(R) \tag{3}$$

with

$$F(R) = \int_0^\infty \mathrm{d}k\,k\,\mathrm{e}^{-\beta k^2/2}\sin(kR) . \tag{4}$$

The remaining parameter-dependent integral can be found, for example, in the book by Gradshteyn and Ryzhik.[2] However, we prefer to determine it explicitly with a favorite standard trick: The determination of the integral is reduced to the solution of a simple, ordinary differential equation.

The integral $F(R)$ can be expressed as a differentiation of another integral with respect to the same parameter:

$$F(R) = -\frac{\mathrm{d}}{\mathrm{d}R}H(R) , \qquad H(R) = \int_0^\infty \mathrm{d}k\,\mathrm{e}^{-\beta k^2/2}\cos(Rk) . \tag{5}$$

[1] Such effects are of some importance in e.g. high-energy heavy-ion collisions, where in one encounter up to 10 000 pions (bosons!) are produced.

[2] I.M. Ryzhik, A. Seffrey, I.S. Gradshteyn: *Table of Integrals, Series, and Products*, 5th ed. (Academic Press, New York 1994).

Example 8.1.

On the other hand, a partial integration of (4) yields

$$F(R) = \frac{R}{\beta}H(R)\,. \tag{6}$$

Hence, we arrive at an ordinary differential equation for the integral $H(R)$:

$$\frac{\mathrm{d}H}{\mathrm{d}R} = -\frac{R}{\beta}H\,. \tag{7}$$

With the initial condition

$$H(R=0) = \int_0^\infty \mathrm{d}k\,\mathrm{e}^{-\beta k^2/2} = \sqrt{\frac{\pi}{2\beta}} \tag{8}$$

we find the solution via a separation of variables:

$$\int_{H(0)}^{H(R)} \frac{\mathrm{d}H'}{H'} = -\frac{1}{\beta}\int_0^R \mathrm{d}R'\,R'\,,$$

$$H(R) = \sqrt{\frac{\pi}{2\beta^3}}\,\mathrm{e}^{-R^2/2\beta}\,. \tag{9}$$

An insertion into (6) yields $F(R)$ and according to (3) we get the result for the wanted Fourier integral:

$$I(\boldsymbol{R}) = \frac{\alpha\,\mathrm{e}^{\mathrm{i}\boldsymbol{p}_0\cdot\boldsymbol{R}}}{(2\pi\beta)^{3/2}}\,\mathrm{e}^{-R^2/2\beta}\,. \tag{10}$$

If we take the square of the modulus, the phase drops out and we get

$$|I(\boldsymbol{R})|^2 = \frac{\alpha^2}{8\pi^3\beta^3}\,\mathrm{e}^{-R^2/\beta}\,. \tag{11}$$

This was the method used for the second term of (8.32) in the case of the photon beam. The first term, i.e. n^2, immediately follows once we take the limit $R=0$ for the second term:

$$n^2 = \frac{\alpha^2}{8\pi^3\beta^3}\,. \tag{12}$$

In total we obtain

$$n^2\Pi(\boldsymbol{r}-\boldsymbol{r}') = n^2\left(1+\mathrm{e}^{-|\boldsymbol{r}-\boldsymbol{r}'|^2/\beta}\right)\,. \tag{13}$$

A qualitative sketch of the pair-correlation function is illustrated in Fig. 8.2.

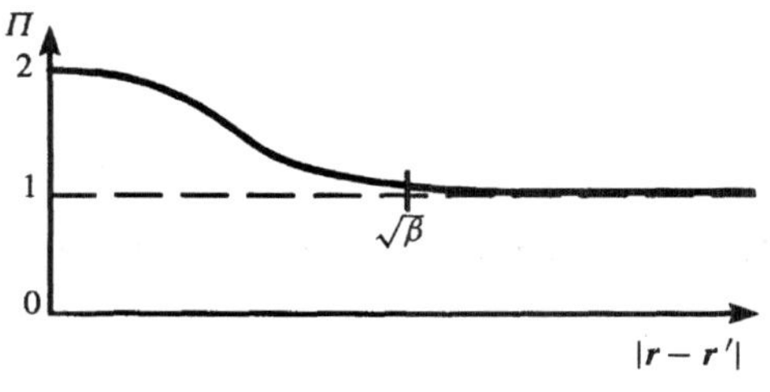

Fig. 8.2. Pair-correlation function of noninteracting bosons

Example 8.1.

Obviously, a significant increase can be observed for the probability of finding two bosons at small relative distances at the same time. In particular, the probability of finding a pair at the same position is twice as large as the asymptotic value for large relative distances.

EXERCISE

8.2 Boson Pair-Correlation Function as a Function of the Quantization Volume

Problem. For the derivation of (8.31) we have assumed, without mentioning it, that the quantization volume is equal to the unit volume $V = 1$. Show that the last term in (8.31) is of the order $1/V$ with respect to the first two terms.

Solution. We indicate again the derivation of (8.31), but this time we use an expansion of the field operators into discrete momentum states, i.e.

$$\hat{\varphi}(\boldsymbol{r}) = \sum_{\boldsymbol{p}} \frac{\mathrm{e}^{\mathrm{i}\boldsymbol{p}\cdot\boldsymbol{r}}}{\sqrt{V}} \hat{a}_{\boldsymbol{k}} \,. \tag{1}$$

Then, we obtain for the matrix element in (8.31):

$$\begin{aligned} &\langle\phi|\hat{\varphi}^\dagger(\boldsymbol{r})\hat{\varphi}^\dagger(\boldsymbol{r}')\hat{\varphi}(\boldsymbol{r}')\hat{\varphi}(\boldsymbol{r})|\phi\rangle \\ &\quad = \frac{1}{V^2}\sum_{\boldsymbol{p},\boldsymbol{q},\boldsymbol{p}',\boldsymbol{q}'} \mathrm{e}^{\mathrm{i}(\boldsymbol{p}-\boldsymbol{p}')\cdot\boldsymbol{r}}\,\mathrm{e}^{\mathrm{i}(\boldsymbol{q}-\boldsymbol{q}')\cdot\boldsymbol{r}'}\langle\phi|\hat{a}^\dagger_{\boldsymbol{p}}\,\hat{a}^\dagger_{\boldsymbol{q}}\,\hat{a}_{\boldsymbol{q}'}\hat{a}_{\boldsymbol{p}'}|\phi\rangle \,. \end{aligned} \tag{2}$$

The treatment of the matrix element appearing in (2) is straightforward and completly analogous to the previous case. Instead of δ functions for continuous momenta we have to deal with the corresponding Kronecker symbols:

$$\begin{aligned} \langle\phi|\hat{a}^\dagger_{\boldsymbol{p}}\hat{a}^\dagger_{\boldsymbol{q}}\,\hat{a}_{\boldsymbol{q}'}\hat{a}_{\boldsymbol{p}'}|\phi\rangle &= (\delta_{\boldsymbol{p}\boldsymbol{p}'}\delta_{\boldsymbol{q}\boldsymbol{q}'} + \delta_{\boldsymbol{p}\boldsymbol{q}'}\delta_{\boldsymbol{q}\boldsymbol{p}'})\,n_{\boldsymbol{p}}\,n_{\boldsymbol{q}} \\ &\quad - \delta_{\boldsymbol{p}\boldsymbol{q}}\,\delta_{\boldsymbol{p}\boldsymbol{p}'}\delta_{\boldsymbol{q}\boldsymbol{q}'}n_{\boldsymbol{p}}\,(n_{\boldsymbol{p}}+1)\,. \end{aligned} \tag{3}$$

Inserting (3) into (2) and carrying out the summations yields

$$\begin{aligned} &\langle\phi|\hat{\varphi}^\dagger(\boldsymbol{r})\hat{\varphi}^\dagger(\boldsymbol{r}')\hat{\varphi}(\boldsymbol{r}')\hat{\varphi}(\boldsymbol{r})|\phi\rangle \\ &\quad = \frac{1}{V^2}\sum_{\boldsymbol{p},\boldsymbol{q}} n_{\boldsymbol{p}}\,n_{\boldsymbol{q}} + \frac{1}{V^2}\sum_{\boldsymbol{p},\boldsymbol{q}} \mathrm{e}^{-\mathrm{i}\boldsymbol{p}\cdot(\boldsymbol{r}-\boldsymbol{r}')}\,\mathrm{e}^{\mathrm{i}\boldsymbol{q}\cdot(\boldsymbol{r}-\boldsymbol{r}')}\,n_{\boldsymbol{p}}\,n_{\boldsymbol{q}} \\ &\qquad - \frac{1}{V^2}\sum_{\boldsymbol{p}} n_{\boldsymbol{p}}\,(n_{\boldsymbol{p}}+1) \\ &\quad = n^2 + \left|\frac{1}{V}\sum_{\boldsymbol{p}} \mathrm{e}^{-\mathrm{i}\boldsymbol{p}\cdot(\boldsymbol{r}-\boldsymbol{r}')}n_{\boldsymbol{p}}\right|^2 - \frac{1}{V^2}\sum_{\boldsymbol{p}} n_{\boldsymbol{p}}\,(n_{\boldsymbol{p}}+1)\,. \end{aligned} \tag{4}$$

Once the continuum limit is performed, i.e. replacement of $\sum_{\boldsymbol{p}}$ with $V\int \mathrm{d}^3p/(2\pi)^3$ etc., we realize that the third term in (4) is of the order $1/V$ with respect to the first term.

8.3 The Hanbury-Brown and Twiss Effect

Hanbury-Brown and Twiss[3] have carried out an experiment to measure the pair-correlation function of noninteracting bosons. Here we will discuss only the basic setup of the experiment; for most of the theoretical details we refer to the original publications.

In this experiment, the probability amplitude is measured to detect two photons in different space points at neighboring time; the two photons belong to a coherent photon beam which is produced by a source Q. At first, the primary beam is divided into two identical coherent beams via a semipermeable mirror S. In this way we avoid placing two detectors one behind the other. Coming from the mirror, the two partial beams hit the two detectors D_1 and D_2, which are at different distances from the mirror. In Fig. 8.3 the experimental setup is sketched schematically.

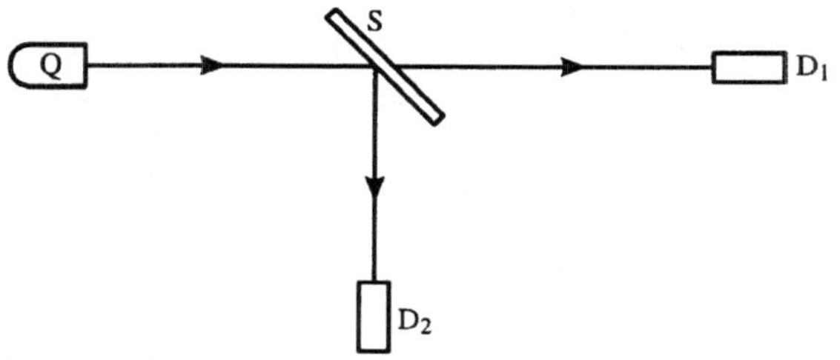

Fig. 8.3. Schematic setup of the Hanbury-Brown and Twiss experiment

The light intensities, which are a measure for the average photon numbers, are measured: $J_1(t)$ in detector D_1 at time t and $J_2(t+T)$ in detector D_2 at a later time $t+T$. The difference in time T defines the relative distance $r = cT$ between two points in the primary beam. The product of the two intensities is averaged with respect to t over the observation window τ. This measurement prescription is equivalent to the measurement of two photons at two different points of the beam with a relative distance cT. For fixed T the observed, correlated average value

$$\langle J_1(t)\, J_2(t+T)\rangle_\tau \; = \; \frac{1}{\tau}\int^{\tau} \mathrm{d}t\, J_1(t) J_2(t+T)$$

shows the same characteristic trend as a function of T as has been derived for the pair-correlation function $\Pi(r)$ in the previous example.

This experiment appears to reveal a typical quantum effect in Bose gases. It is quite remarkable that it can also be explained completely within the tools of classical electrodynamics. This should not be too surprising; we remember that a light beam (macroscopic electromagnetic field) contains very many photons and, thus, also contains many photons with the same momentum. An analysis of correlation functions in terms of Glauber states, which we have briefly introduced in Chap. 1 in order to calculate, for example, the classical value of the $\boldsymbol{E}$ field, would lead to classical quantities such as the

[3] R. Hanbury-Brown, R.Q. Twiss: Nature, **177** (1956) 27; **178** (1956) 1447.

intensity. As squares of electric and magnetic field amplitudes they represent a measure for the number of photons. Hence, the Hanbury-Brown and Twiss experiment teaches us that the bosonic character of photons is already contained in the superposition principle for classical electromagnetic fields. The following oversimplified consideration shall illucidate this aspect. More formal and more detailed comments can be found, for example, in the textbook by R.G. Newton.[4]

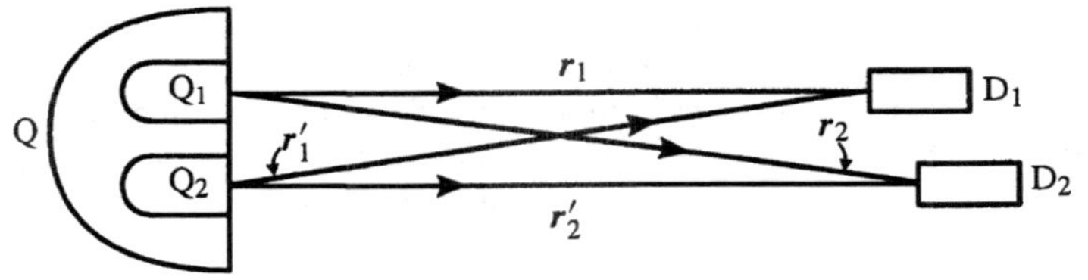

Fig. 8.4. Oversimplified setup of the Hanbury-Brown and Twiss experiment

For simplicity we assume that the source of the light beam consists of two closely adjacent emitters Q_1 and Q_2; see Fig. 8.4. Furthermore, Q_1 is required to emit coherent light with a wave-number vector $\boldsymbol{k}_1$ and an amplitude a_1; Q_2 emits coherent light with $\boldsymbol{k}_2$ and a_2. The light of the two emitters comes with the same polarization; the relative phases of the amplitudes vary randomly between 0 and π. From Q_1, light with the amplitude $a_1\,\mathrm{e}^{\mathrm{i}\boldsymbol{k}_1\cdot\boldsymbol{r}_1}$ reaches the detector D_1; here $\boldsymbol{r}_1$ represents the distance vector from Q_1 to D_1. In the same manner the light originating from Q_2 and falling into the detector D_1 positioned at $\boldsymbol{r}_1'$ has the amplitude $a_2\,\mathrm{e}^{\mathrm{i}\boldsymbol{k}_2\cdot\boldsymbol{r}_1'}$. According to the superposition principle the total field amplitude reaching D_1 is given by

$$A_1 = a_1\,\mathrm{e}^{\mathrm{i}\boldsymbol{k}_1\cdot\boldsymbol{r}_1} + a_2\,\mathrm{e}^{\mathrm{i}\boldsymbol{k}_2\cdot\boldsymbol{r}_1'}\,, \tag{8.33a}$$

where we did not write down the polarization vector. The intensity becomes

$$J_1 = |A_1|^2 = |a_1|^2 + |a_2|^2 + \Re\left(a_1^* a_2\,\mathrm{e}^{\mathrm{i}(\boldsymbol{k}_2\cdot\boldsymbol{r}_1' - \boldsymbol{k}_1\cdot\boldsymbol{r}_1)}\right)\,. \tag{8.33b}$$

Analogously, for the total amplitude and intensity at D_2 we find:

$$A_2 = a_1\,\mathrm{e}^{\mathrm{i}\boldsymbol{k}_1\cdot\boldsymbol{r}_2} + a_2\,\mathrm{e}^{\mathrm{i}\boldsymbol{k}_2\cdot\boldsymbol{r}_2'}\,, \tag{8.34a}$$

$$J_2 = |A_2|^2 = |a_1|^2 + |a_2|^2 + \Re\left(a_1^* a_2\,\mathrm{e}^{\mathrm{i}(\boldsymbol{k}_2\cdot\boldsymbol{r}_2' - \boldsymbol{k}_1\cdot\boldsymbol{r}_1')}\right)\,. \tag{8.34b}$$

$\Re$ denotes the real part of the corresponding quantity. As stated earlier the relative phases of a_1 and a_2 can vary randomly. A more suited measure for the corresponding total intensities is given by an average over the relative phases of the partial amplitudes; this average is equivalent to the time average performed in the Hanbury-Brown and Twiss experiment. Suppose $a_{1,2} = \alpha_{1,2}\,\mathrm{e}^{\mathrm{i}\beta_{1,2}}$, then the averaging implies

[4] R.G. Newton: *Scattering Theory of Waves and Particles* (Springer, Berlin, Heidelberg 1982).

$$\langle|a_{1,2}|^2\rangle = |\alpha_{1,2}|^2\frac{1}{2\pi}\int_0^{\pi} d\xi = |a_{1,2}|^2 ,$$
$$\langle a_1^* a_2\rangle = \alpha_1^*\alpha_2\frac{1}{2\pi}\int_0^{2\pi} d\xi\, e^{-i\xi} = 0 , \quad \xi = \beta_1 - \beta_2 . \tag{8.35}$$

As a consequence, terms containing products $a_1 a_2$ drop out after the average is taken. Hence, the averaged intensities become

$$\langle J_1\rangle = \langle J_2\rangle = |a_1|^2 + |a_2|^2 . \tag{8.36}$$

The product of the averaged intensities $\langle J_1\rangle\langle J_2\rangle$ is independent of the distance between the detectors. This does not hold for the average of the product of intensities; we obtain for the product $J_1 J_2$

$$J_1 J_2 = |A_1 A_2|^2 = \big|a_1^2\, e^{i\boldsymbol{k}_1\cdot(\boldsymbol{r}_1+\boldsymbol{r}_2)} + a_2^2\, e^{i\boldsymbol{k}_2\cdot(\boldsymbol{r}_1'+\boldsymbol{r}_2')} + a_1 a_2 (e^{\boldsymbol{k}_1\cdot\boldsymbol{r}_1}\, e^{i\boldsymbol{k}_2\cdot\boldsymbol{r}_2'} + e^{\boldsymbol{k}_2\cdot\boldsymbol{r}_1'}\, e^{i\boldsymbol{k}_1\cdot\boldsymbol{r}_2})\big|^2 . \tag{8.37}$$

Multiplying out and averaging over the relative phases eliminates terms of the form $a_1 a_2 |a_1|^2$, $a_1 a_2 |a_2|^2$, ..., and we obtain

$$\begin{aligned}\langle J_1 J_2\rangle &= |a_1|^4 + |a_2|^4 + |a_1|^2|a_2|^2 \left|e^{\boldsymbol{k}_1\cdot\boldsymbol{r}_1}\, e^{i\boldsymbol{k}_2\cdot\boldsymbol{r}_2'} + e^{\boldsymbol{k}_2\cdot\boldsymbol{r}_1'}\, e^{i\boldsymbol{k}_1\cdot\boldsymbol{r}_2}\right|^2 \\ &= \langle J_1\rangle\langle J_2\rangle + 2|a_1|^2|a_2|^2 \\ &\quad\times \cos\left[\boldsymbol{k}_2\cdot(\boldsymbol{r}_1' - \boldsymbol{r}_2') - \boldsymbol{k}_1\cdot(\boldsymbol{r}_1 - \boldsymbol{r}_2)\right] .\end{aligned} \tag{8.38}$$

If both emitters Q_1 and Q_2 are very close together, which should be the case for a well-collimated light beam, we can replace $\boldsymbol{r}_1 - \boldsymbol{r}_2$ with $\boldsymbol{r}_1' - \boldsymbol{r}_2'$ in the above calculation.

Hence, we deduce for the correlated intensity as a function of the relative distance between the detectors:

$$\langle J_1 J_2\rangle = \langle J_1\rangle\langle J_2\rangle + 2|a_1|^2|a_2|^2 \cos\left[\boldsymbol{p}\cdot(\boldsymbol{r}_1 - \boldsymbol{r}_2)\right] \tag{8.39}$$

with $\boldsymbol{p} = \boldsymbol{k}_2 - \boldsymbol{k}_1$. This function takes on a maximum for $\boldsymbol{r}_1 = \boldsymbol{r}_2$. In a realistic light beam, $\boldsymbol{p}$ has a distribution; a Gaussian, for example. Thus, an additional averaging over different $\boldsymbol{p}$ yields again a function of the form (13) of Example 8.1.

The quantum-mechanical interpretation of the separate terms in (8.37) would be as follows: The a_1^2 and the a_2^2 terms represent the probability amplitudes of observing a photon pair from Q_1 and Q_2, respectively. They lead to the position-independent terms of (8.39). The mixed term $a_1 a_2$ represents the probability amplitude of measuring a photon from Q_1 and another one from Q_2. The two probabilities of registering one photon from Q_1 at $\boldsymbol{r}_1$ and the other one from Q_2 at $\boldsymbol{r}_2$ or vice versa cannot be distinguished. The interference of the two amplitudes leads to the cosine term in (8.39). Therefore, the tendency of photons to clump together can be understood as a consequence of the superposition principle.

The HBT effect has been used in astrophysics for the analysis of the spacial extension of cosmic light sources (stars, galaxies, and so on). It has also become of importance for pions emitted from hot compressed nuclear matter regions as they occur in high-energy heavy-ion encounters where nuclear shock waves cause the heating and compression of the nuclear matter. It serves there as a tool to measure the spatial extend of the source.[5]

8.4 Cooper Pairs

A key ingredient for our understanding of superconductivity is the pairing of electrons during their motion in a crystal. At first, this is not to be expected because two equally charged particles repel each other, so pairing implies that they attract. Indeed, under certain conditions an attraction is accomplished by the interaction of the electrons with the lattice. With a simple but instructive model we want to find out the interaction mechanism leading to pairing.

For the moment we forget about all the structure effects in a crystal and consider a metal with volume V as a potential box (cube) filled with electrons. Furthermore, the interaction between the electrons is neglected for the time being, so that the normalized electron eigenfunctions with periodic boundary conditions are given by

$$\psi_n(\boldsymbol{r}) = \frac{1}{\sqrt{V}} \mathrm{e}^{\mathrm{i}\boldsymbol{k}_n \cdot \boldsymbol{r}} ; \tag{8.40}$$

see Fig. 1.1. The electrons possess the energy

$$\varepsilon_{\boldsymbol{k}_n} = \frac{\hbar k_n^2}{2m} . \tag{8.41a}$$

The possible $\boldsymbol{k}$ vectors inside a box of length L are [see (1.30)]

$$\boldsymbol{k}_n = \frac{2\pi}{L}\boldsymbol{n} = \frac{2\pi}{L}\{n_1, n_2, n_3\} , \qquad n_\nu \in \mathbb{Z} . \tag{8.41b}$$

n_ν are integers with $-\infty \le n_\nu \le +\infty$. Every single state (8.40) can be filled with two electrons (spin degeneracy); as a consequence the state with the lowest energy for the total system is characterized by

$$N = \sum_{\boldsymbol{n}\ \text{with}\ k_n < k_\mathrm{F}} 2 . \tag{8.42}$$

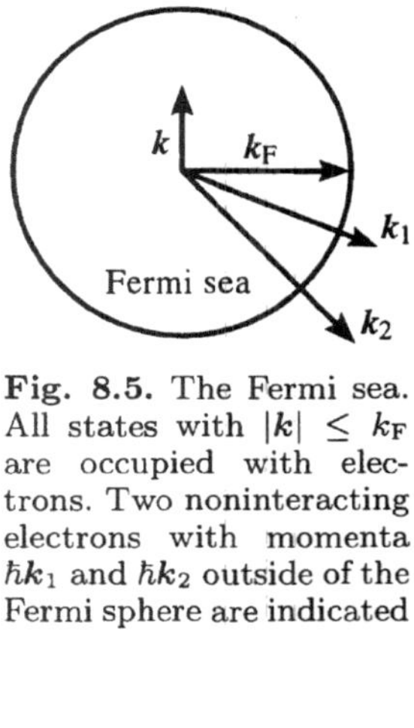

Fig. 8.5. The Fermi sea. All states with $|\boldsymbol{k}| \le k_\mathrm{F}$ are occupied with electrons. Two noninteracting electrons with momenta $\hbar\boldsymbol{k}_1$ and $\hbar\boldsymbol{k}_2$ outside of the Fermi sphere are indicated

N represents the total number of electrons within the box and $\boldsymbol{k}_\mathrm{F}$ is known as the *Fermi momentum*. From (8.41a) the Fermi momentum is related to the *Fermi energy*:

$$\varepsilon_\mathrm{F} = \frac{\hbar^2 k_\mathrm{F}^2}{2m} . \tag{8.43}$$

In $\boldsymbol{k}$ space all states within the sphere of radius $|\boldsymbol{k}_\mathrm{F}|$, i.e. $|\boldsymbol{k}| \le |\boldsymbol{k}_\mathrm{F}|$, are occupied; see Fig. 8.5.

These states constitute the *Fermi sea*. For the case that N and L become very large, the vectors $\boldsymbol{k}_n$ are very close together according to (8.41b) and the summation over $\boldsymbol{n}$ in (8.42) can be replaced:

[5] D. Ardouin: Int. J. Mod. Phys. E (1997).

$$\sum_{n_1} \rightarrow \frac{L}{2\pi}\int \mathrm{d}k_x \qquad \text{or}$$

$$\sum_{n} \rightarrow \frac{L^3}{(2\pi)^3}\int \mathrm{d}^3k = \frac{V}{(2\pi)^3}\int \mathrm{d}^3k\,. \tag{8.44}$$

Here, the normalization volume $V = L^3$ has been introduced. The quantity

$$\frac{V}{(2\pi)^3} \tag{8.45}$$

can be interpreted as the level density in k space. From (8.42) we get

$$N = 2\frac{V}{(2\pi)^3}\frac{4\pi}{3}k_\mathrm{F}^3\,, \tag{8.46}$$

from which

$$k_\mathrm{F} = \left(3\pi^2\frac{N}{V}\right)^{1/3} = (3\pi^2 n)^{1/3} \tag{8.47}$$

follows. $N/V = n$ represents the electron density.

Now we consider two electrons that are just outside the Fermi sphere (see again Fig. 8.5) and interact with each other via a weak attractive force. The interaction between the electrons inside the Fermi sphere and their interaction with the electron pair outside the Fermi sphere are neglected for the time being; however, for a realistic model of superconductivity these interactions have to be taken into account. The two electrons outside the Fermi sphere are assumed to have opposite spin: in accordance with the Pauli principle they can simultanously occupy the same spatial wavefunction. Without interaction the energy eigenfunction of the electron pair reads

$$\psi(\boldsymbol{r}_1,\boldsymbol{r}_2,t) = \frac{\mathrm{e}^{\mathrm{i}\boldsymbol{k}_1\cdot\boldsymbol{r}_1}}{\sqrt{V}}\frac{\mathrm{e}^{\mathrm{i}\boldsymbol{k}_2\cdot\boldsymbol{r}_2}}{\sqrt{V}}\mathrm{e}^{-(\mathrm{i}/\hbar)(\varepsilon_{\boldsymbol{k}_1}+\varepsilon_{\boldsymbol{k}_2})t}\,. \tag{8.48}$$

The interaction leads to the electrons scattering from each other; as a consequence (8.48) no longer represents an energy eigenstate. We have to take a superposition of unperturbed (free) states (8.48) of the form

$$\Psi(\boldsymbol{r}_1,\boldsymbol{r}_2,t) = \sum_{\boldsymbol{k}_1,\boldsymbol{k}_2} a_{\boldsymbol{k}_1\boldsymbol{k}_2}(t)\frac{\mathrm{e}^{\mathrm{i}\boldsymbol{k}_1\cdot\boldsymbol{r}_1}}{\sqrt{V}}\frac{\mathrm{e}^{\mathrm{i}\boldsymbol{k}_2\cdot\boldsymbol{r}_2}}{\sqrt{V}} \tag{8.49}$$

with

$$a_{\boldsymbol{k}_1\boldsymbol{k}_2}(t) = \mathrm{e}^{-\mathrm{i}Et/\hbar}a_{\boldsymbol{k}_1\boldsymbol{k}_2}\,. \tag{8.50}$$

Here E stands for the total energy of the electron pair. The correlated wavefunction (8.49) is a double Fourier series in $\boldsymbol{r}_1$ and $\boldsymbol{r}_2$, and $a_{\boldsymbol{k}_1\boldsymbol{k}_2}(t)$ represents the probability amplitude of finding particle 1 with momentum $\hbar\boldsymbol{k}_1$ and particle 2 with momentum $\hbar\boldsymbol{k}_2$. Since states inside the Fermi sphere are occupied it has to hold that

$$a_{\boldsymbol{k}_1\boldsymbol{k}_2} = 0 \qquad \text{for } k_1, k_2 \leq k_\mathrm{F} \tag{8.51}$$

because of Pauli's principle. We now study the behavior of the wavefunction in time. If no interaction between the two electrons is present, i.e. $E = \varepsilon_{\boldsymbol{k}_1} + \varepsilon_{\boldsymbol{k}_2}$, it follows that

$$\mathrm{i}\hbar\frac{\partial}{\partial t}a_{\boldsymbol{k}_1\boldsymbol{k}_2}(t) = (\varepsilon_{\boldsymbol{k}_1} + \varepsilon_{\boldsymbol{k}_2})a_{\boldsymbol{k}_1\boldsymbol{k}_2}(t) \tag{8.52}$$

directly from (8.49) and the Schrödinger equation, and thus

$$a_{\boldsymbol{k}_1\boldsymbol{k}_2}(t) = \exp\left[-\mathrm{i}\left(\frac{\varepsilon_{\boldsymbol{k}_1} + \varepsilon_{\boldsymbol{k}_2}}{\hbar}\right)t\right]a_{\boldsymbol{k}_1\boldsymbol{k}_2}(0)\,. \tag{8.53}$$

This is a pure phase factor, which is already known from the uncorrelated wavefunction (8.48). Once the particles interact they will constantly change their momentum because of the ever recuring scattering from each other: In one scattering event $\boldsymbol{k}_1, \boldsymbol{k}_2$ will become $\boldsymbol{k}_1', \boldsymbol{k}_2'$. The Hamiltonian of the two-electron pair with interaction $\mathcal{V}(\boldsymbol{r}_1, \boldsymbol{r}_2)$ reads

$$\hat{H} = -\frac{\hbar^2}{2m}\boldsymbol{\nabla}_1^2 - \frac{\hbar^2}{2m}\boldsymbol{\nabla}_2^2 + \mathcal{V}(\boldsymbol{r}_1, \boldsymbol{r}_2) \tag{8.54}$$

and we deduce from (8.49) and the time-dependent Schrödinger equation that

$$\begin{aligned}\mathrm{i}\hbar\frac{\partial}{\partial t}a_{\boldsymbol{k}_1\boldsymbol{k}_2}(t) &= (\varepsilon_{\boldsymbol{k}_1} + \varepsilon_{\boldsymbol{k}_2})a_{\boldsymbol{r}_1\boldsymbol{r}_2}(t)\\ &\quad + \sum_{\boldsymbol{k}_1',\boldsymbol{k}_2'}\langle\boldsymbol{k}_1\boldsymbol{k}_2|\mathcal{V}(\boldsymbol{r}_1, \boldsymbol{r}_2)|\boldsymbol{k}_1'\boldsymbol{k}_2'\rangle a_{\boldsymbol{k}_1'\boldsymbol{k}_2'}\,.\end{aligned} \tag{8.55}$$

Evidently, the matrix element $\langle\boldsymbol{k}_1\boldsymbol{k}_2|\mathcal{V}(\boldsymbol{r}_1, \boldsymbol{r}_2)|\boldsymbol{k}_1'\boldsymbol{k}_2'\rangle$ describes the elementary scattering process $\boldsymbol{k}_1\boldsymbol{k}_2 \to \boldsymbol{k}_1'\boldsymbol{k}_2'$ mentioned above. In detail, it becomes

$$\begin{aligned}\langle\boldsymbol{k}_1\boldsymbol{k}_2|\mathcal{V}(\boldsymbol{r}_1, \boldsymbol{r}_2)|\boldsymbol{k}_1'\boldsymbol{k}_2'\rangle &= \frac{1}{V^2}\int \mathrm{d}^3r_1\mathrm{d}^3r_2\,\mathrm{e}^{\mathrm{i}\boldsymbol{k}_1\cdot\boldsymbol{r}_1}\,\mathrm{e}^{\mathrm{i}\boldsymbol{k}_2\cdot\boldsymbol{r}_2}\\ &\quad \times \mathcal{V}(\boldsymbol{r}_1, \boldsymbol{r}_2)\,\mathrm{e}^{-\boldsymbol{k}_1'\cdot\boldsymbol{r}_1}\,\mathrm{e}^{-\mathrm{i}\boldsymbol{k}_2'\cdot\boldsymbol{r}_2}\,.\end{aligned} \tag{8.56}$$

In general, the Fourier transform of a matrix element $\langle\boldsymbol{k}_1\boldsymbol{k}_2|\mathcal{V}|\boldsymbol{k}_1'\boldsymbol{k}_2'\rangle$ is given by

$$\begin{aligned}\langle\boldsymbol{r}_1\boldsymbol{r}_2|\mathcal{V}|\boldsymbol{r}_1'\boldsymbol{r}_2'\rangle &= \frac{1}{V^2}\sum_{\boldsymbol{k}_1,\boldsymbol{k}_2,\boldsymbol{k}_1',\boldsymbol{k}_2'}\mathrm{e}^{\mathrm{i}\boldsymbol{k}_1\cdot\boldsymbol{r}_1}\,\mathrm{e}^{\mathrm{i}\boldsymbol{k}_2\cdot\boldsymbol{r}_2}\\ &\quad \times \langle\boldsymbol{k}_1\boldsymbol{k}_2|\mathcal{V}|\boldsymbol{k}_1'\boldsymbol{k}_2'\rangle\,\mathrm{e}^{-\boldsymbol{k}_1'\cdot\boldsymbol{r}_1}\,\mathrm{e}^{-\mathrm{i}\boldsymbol{k}_2'\cdot\boldsymbol{k}_2'}\,.\end{aligned} \tag{8.57}$$

This means that an arbitrary amplitude $\langle\boldsymbol{k}_1\boldsymbol{k}_2|\mathcal{V}|\boldsymbol{k}_1'\boldsymbol{k}_2'\rangle$ generally implies a *nonlocal potential* $\langle\boldsymbol{r}_1\boldsymbol{r}_2|\mathcal{V}|\boldsymbol{r}_1'\boldsymbol{r}_2'\rangle$. The corresponding *nonlocal Schrödinger equation* reads

$$\begin{aligned}\mathrm{i}\hbar\frac{\partial}{\partial t}\Psi(\boldsymbol{r}_1, \boldsymbol{r}_2, t) &= \left(-\frac{\hbar^2}{2m}\boldsymbol{\nabla}_1^2 - \frac{\hbar^2}{2m}\boldsymbol{\nabla}_1^2\right)\Psi(\boldsymbol{r}_1, \boldsymbol{r}_2, t)\\ &\quad + \int \mathrm{d}^3r_1'\,\mathrm{d}^3r_2'\,\langle\boldsymbol{r}_1\boldsymbol{r}_2|\mathcal{V}|\boldsymbol{r}_1'\boldsymbol{r}_2'\rangle\Psi(\boldsymbol{r}_1', \boldsymbol{r}_2', t)\,.\end{aligned} \tag{8.58}$$

Hence, a change in the wavefunction $\Psi(\boldsymbol{r}_1, \boldsymbol{r}_2, t)$ at positions $\boldsymbol{r}_1, \boldsymbol{r}_2$ also depends on the wavefunction at the distant positions $\boldsymbol{r}_1', \boldsymbol{r}_2'$. Only if

$$\langle\boldsymbol{r}_1\boldsymbol{r}_2|\mathcal{V}|\boldsymbol{r}_1'\boldsymbol{r}_2'\rangle = \mathcal{V}(\boldsymbol{r}_1, \boldsymbol{r}_2)\delta(\boldsymbol{r}_1 - \boldsymbol{r}_1')\delta(\boldsymbol{r}_2 - \boldsymbol{r}_2')\,, \tag{8.59}$$

does (8.58) turn again into a local Schrödinger equation. In this case

$$\begin{aligned}
&\langle \boldsymbol{k}_1\boldsymbol{k}_2|\mathcal{V}(\boldsymbol{r}_1,\boldsymbol{r}_2,\boldsymbol{r}_1',\boldsymbol{r}_2')|\boldsymbol{k}_1'\boldsymbol{k}_2'\rangle \\
&= \frac{1}{V^2}\int \mathrm{d}^3r_1\,\mathrm{d}^3r_2\,\mathrm{d}^3r_1'\,\mathrm{d}^3r_2'\,\mathrm{e}^{\mathrm{i}\boldsymbol{k}_1\cdot\boldsymbol{r}_1}\,\mathrm{e}^{\mathrm{i}\boldsymbol{k}_2\cdot\boldsymbol{r}_2} \\
&\quad\times \mathcal{V}(\boldsymbol{r}_1,\boldsymbol{r}_2)\delta(\boldsymbol{r}_1-\boldsymbol{r}_1')\delta(\boldsymbol{r}_2-\boldsymbol{r}_2')\,\mathrm{e}^{-\boldsymbol{k}_1'\cdot\boldsymbol{r}_1}\,\mathrm{e}^{-\mathrm{i}\boldsymbol{k}_2'\cdot\boldsymbol{k}_2'} \\
&= \int \mathrm{d}^3r_1\,\mathrm{d}^3r_2\,\mathrm{e}^{\mathrm{i}(\boldsymbol{k}_1-\boldsymbol{k}_1')\cdot\boldsymbol{r}_1}\,\mathrm{e}^{\mathrm{i}(\boldsymbol{k}_2-\boldsymbol{k}_2')\cdot\boldsymbol{r}_2}\mathcal{V}(\boldsymbol{r}_1,\boldsymbol{r}_2)\,, \qquad (8.60)
\end{aligned}$$

which can be simplified further if the potential $\mathcal{V}$ depends only on the relative distance (for two-body forces this is always the case), i.e.

$$\mathcal{V}(\boldsymbol{r}_1,\boldsymbol{r}_2) = \mathcal{V}(\boldsymbol{r}_1-\boldsymbol{r}_2)\,.$$

Then, it holds that

$$\begin{aligned}
&\langle \boldsymbol{k}_1\boldsymbol{k}_2|\mathcal{V}(\boldsymbol{r}_1,\boldsymbol{r}_2,\boldsymbol{r}_1',\boldsymbol{r}_2')|\boldsymbol{k}_1'\boldsymbol{k}_2'\rangle \\
&= \frac{1}{V^2}\int \exp\{\,\mathrm{i}\,[(\boldsymbol{k}_1-\boldsymbol{k}_1')\cdot(\boldsymbol{r}_1-\boldsymbol{r}_2)+(\boldsymbol{k}_2-\boldsymbol{k}_2'+\boldsymbol{k}_1-\boldsymbol{k}_1')\cdot\boldsymbol{r}_2]\} \\
&\quad\times \mathcal{V}(\boldsymbol{r}_1-\boldsymbol{r}_2)\,\mathrm{d}^3r_1\,\mathrm{d}^3r_2 \\
&= \int \exp(\mathrm{i}\boldsymbol{q}\cdot\boldsymbol{r})\,\mathcal{V}(\boldsymbol{r})\,\mathrm{d}^3r \int \exp\left[\mathrm{i}(\boldsymbol{k}_1+\boldsymbol{k}_2-\boldsymbol{k}_1'-\boldsymbol{k}_2')\cdot\boldsymbol{r}_2\right]\mathrm{d}^3r_2 \\
&= \frac{1}{V}\mathcal{V}(\boldsymbol{q})\delta(\boldsymbol{k}_1+\boldsymbol{k}_2-\boldsymbol{k}_1'-\boldsymbol{k}_2')\,, \qquad (8.61)
\end{aligned}$$

where the momentum transfer is

$$\boldsymbol{q} = \boldsymbol{k}_1-\boldsymbol{k}_1' = \frac{\boldsymbol{k}_1-\boldsymbol{k}_1'-(\boldsymbol{k}_2-\boldsymbol{k}_2')}{2} = \frac{\boldsymbol{k}_1-\boldsymbol{k}_2}{2}-\frac{\boldsymbol{k}_1'-\boldsymbol{k}_2'}{2} \equiv \boldsymbol{k}-\boldsymbol{k}'\,.$$

The total momentum $\boldsymbol{K}$ of the pair is given by $\boldsymbol{K}=\boldsymbol{k}_1+\boldsymbol{k}_2=\boldsymbol{k}_1'+\boldsymbol{k}_2'$. Indeed, the δ function in (8.61) expresses that the total momentum of the pair is conserved during the collision. Hence, we can also write down result (8.61) in the form

$$\langle \boldsymbol{k}_1\boldsymbol{k}_2|\mathcal{V}(\boldsymbol{r}_1,\boldsymbol{r}_2,\boldsymbol{r}_1',\boldsymbol{r}_2')|\boldsymbol{k}_1'\boldsymbol{k}_2'\rangle = \frac{1}{V}\mathcal{V}(\boldsymbol{k}-\boldsymbol{k}')\delta(\boldsymbol{K}-\boldsymbol{K}')\,. \qquad (8.62)$$

Since the total momentum is conserved during the scattering process, the matrix element of a nonlocal interaction in general will only connect states with identical total momentum $\boldsymbol{K}=\boldsymbol{k}_1+\boldsymbol{k}_2=\boldsymbol{k}_1'+\boldsymbol{k}_2'$. Thus,

$$\langle \boldsymbol{k}_1\boldsymbol{k}_2|\mathcal{V}(\boldsymbol{r}_1,\boldsymbol{r}_2,\boldsymbol{r}_1',\boldsymbol{r}_2')|\boldsymbol{k}_1'\boldsymbol{k}_2'\rangle = \frac{1}{V}V_{\boldsymbol{k}\boldsymbol{k}'}(\boldsymbol{K})\delta(\boldsymbol{K}-\boldsymbol{K}') \qquad (8.63)$$

and the nonlocal potential takes on the following form in coordinate space:

$$\begin{aligned}
\langle \boldsymbol{r}_1\boldsymbol{r}_2|\mathcal{V}|\boldsymbol{r}_1'\boldsymbol{r}_2'\rangle &= \int \mathrm{d}^3K\,\mathrm{d}^3k\,\mathrm{e}^{\mathrm{i}\boldsymbol{K}\cdot\boldsymbol{R}}\,\mathrm{e}^{\mathrm{i}\boldsymbol{k}\cdot\boldsymbol{r}} \\
&\quad\times V_{\boldsymbol{k}\boldsymbol{k}'}(\boldsymbol{K})\delta(\boldsymbol{k}-\boldsymbol{k}')\mathrm{e}^{-\mathrm{i}\boldsymbol{K}'\cdot\boldsymbol{R}'}\,\mathrm{e}^{-\mathrm{i}\boldsymbol{k}'\cdot\boldsymbol{r}'} \\
&= \langle \boldsymbol{r}|\mathcal{V}(\boldsymbol{R}-\boldsymbol{R}')|\boldsymbol{r}'\rangle\,, \qquad (8.64)
\end{aligned}$$

where

$$\begin{aligned} \boldsymbol{R} &= \frac{\boldsymbol{r}_1+\boldsymbol{r}_2}{2}, & \boldsymbol{R}' &= \frac{\boldsymbol{r}_1'+\boldsymbol{r}_2'}{2}, \\ \boldsymbol{r} &= \boldsymbol{r}_1-\boldsymbol{r}_2, & \boldsymbol{r}' &= \boldsymbol{r}_1'-\boldsymbol{r}_2' \end{aligned} \tag{8.65}$$

represent the center-of-mass and relative coordinates, respectively. Let us consider an eigenstate with energy E and sharp (conserved) total momentum. The amplitude (8.50) can be transformed into

$$a_{\boldsymbol{k}_1\boldsymbol{k}_2}(t) = a_{\boldsymbol{k}}(\boldsymbol{K})\,\mathrm{e}^{-\mathrm{i}Et/\hbar} \tag{8.66}$$

and (8.55) becomes

$$(E-\varepsilon_{\boldsymbol{k}_1}-\varepsilon_{\boldsymbol{k}_2})a_{\boldsymbol{k}}(\boldsymbol{K}) = \sum_{\boldsymbol{k}'} V_{\boldsymbol{k}\boldsymbol{k}'}(\boldsymbol{K})a_{\boldsymbol{k}'}(\boldsymbol{K})\,. \tag{8.67}$$

Although the total momentum $\hbar\boldsymbol{K}$ only appears as a parameter, these equations are very difficult to solve with a general force. Electrons in a metal experience two kinds of forces: electrons repel each other as a result of the Coulomb interaction and they are attracted by the ions in the crystal lattice. A moving electron pulls the ions a little bit in its direction and away from their corresponding crystal lattice sites. Since the ions are very heavy, the shifts remain very small. But the moving ions pull along other electrons which in this manner effectively follow the first electrons. In other words, two electrons that usually repel each other are able to attract each other in a crystal lattice. Of course, this depends on the structure of the crystal lattice and the kind of ions. The interaction resulting from these repulsions and attractions are in some metals attractive for electrons near the Fermi surface. This can be approximated as

$$V_{\boldsymbol{k}\boldsymbol{k}'}(\boldsymbol{K}) = \begin{cases} -\dfrac{v_0}{V} & \text{for } k_\mathrm{F} < k_1, k_2, k_1', k_2' < k_a\,, \\ 0 & \text{otherwise}\,, \end{cases} \tag{8.68}$$

where v_0 is positive and k_a is slightly larger than k_F (see Fig. 8.6).

Then, (8.67) becomes

$$(E-\varepsilon_{\boldsymbol{k}_1}-\varepsilon_{\boldsymbol{k}_2})a_{\boldsymbol{k}}(\boldsymbol{K}) = -\frac{v_0}{V}\sum_{\boldsymbol{k}'}{}' a_{\boldsymbol{k}'}(\boldsymbol{K})\,. \tag{8.69}$$

The prime (′) for the summation sign indicates that the $\boldsymbol{k}'$ summation only runs over those $\boldsymbol{k}'$ that lie inside the shaded spherical shell sketched in Fig. 8.6:

$$k_\mathrm{F} < \left|\frac{\boldsymbol{K}}{2} \pm \boldsymbol{k}'\right| < k_\mathrm{a}\,. \tag{8.70}$$

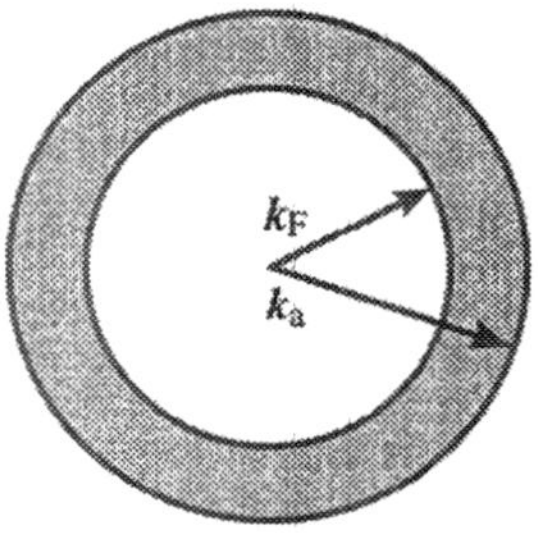

Fig. 8.6. Electrons inside a small spherical shell $k_\mathrm{F} < k < k_\mathrm{a}$ and at the same time outside the Fermi sphere k_F attract each other

From the relation between $\boldsymbol{k}'$ and $\boldsymbol{K}$ given after (8.61) it is clear that

$$\begin{aligned} \boldsymbol{k}_1 &= \frac{\boldsymbol{K}}{2}+\boldsymbol{k}\,, \\ \boldsymbol{k}_2 &= \frac{\boldsymbol{K}}{2}-\boldsymbol{k}\,. \end{aligned} \tag{8.71}$$

Condition (8.70) corresponds exactly to the spherical shell for the involved particle momenta noted in (8.68).

The sum $\sum'_{k'}$ in (8.69) cannot vanish. This can be seen as follows: The amplitude $a_{\boldsymbol{k}}(\boldsymbol{K})$ has to be different from zero for some $\boldsymbol{k}$; otherwise the correlated wavefunction (8.49) would vanish identically. If $\sum'_{k'}$ in (8.69) were equal to zero then it would follow for the corresponding $\boldsymbol{k}$ on the left side of (8.69) that

$$E = \varepsilon_{\boldsymbol{k}_1} + \varepsilon_{\boldsymbol{k}_2} = \frac{(\hbar \boldsymbol{k})^2}{m} + \frac{(\hbar \boldsymbol{K})^2}{4m} . \tag{8.72}$$

For a given $\boldsymbol{K}$ and E the last equation can only be fulfilled for one $\boldsymbol{k}$; consequently, only *one* $a_{\boldsymbol{k}}(\boldsymbol{K})$ can be different from zero. As a consequence the sum would not vanish, i.e. $\sum'_{k'} a_{\boldsymbol{k}'}(\boldsymbol{K}) \neq 0$, which contradicts the original assumption. Thus, we have to deduce $\sum'_{k'} a_{\boldsymbol{k}'}(\boldsymbol{K}) \neq 0$.

In order to solve (8.69), both sides are divided by $(E - \varepsilon_{\boldsymbol{k}_1} - \varepsilon_{\boldsymbol{k}_2})$ and are then summed over $\boldsymbol{k}$. This leads to

$$\sum_{\boldsymbol{k}}{}' a_{\boldsymbol{k}}(\boldsymbol{K}) = -\frac{v_0}{V} \sum_{\boldsymbol{k}}{}' \frac{1}{(E - \varepsilon_{\boldsymbol{k}_1} - \varepsilon_{\boldsymbol{k}_2})} \sum_{\boldsymbol{k}'}{}' a_{\boldsymbol{k}'}(\boldsymbol{K}) . \tag{8.73}$$

Since $\sum'_{k} a_{\boldsymbol{k}}(\boldsymbol{K}) = \sum'_{k'} a_{\boldsymbol{k}'}(\boldsymbol{K})$ we get

$$1 = -\frac{v_0}{V} \sum_{\boldsymbol{k}}{}' \frac{1}{(E - \varepsilon_{\boldsymbol{k}_1} - \varepsilon_{\boldsymbol{k}_2})} . \tag{8.74}$$

Graphically this equation can be solved easily: Let us depict the function

$$\chi(E) = \frac{1}{V} \sum_{\boldsymbol{k}}{}' \frac{1}{(E - \varepsilon_{\boldsymbol{k}_1} - \varepsilon_{\boldsymbol{k}_2})} \tag{8.75}$$

and look for intersections with the line $-1/v_0$, the latter being parallel to the E axis (see Fig. 8.7).

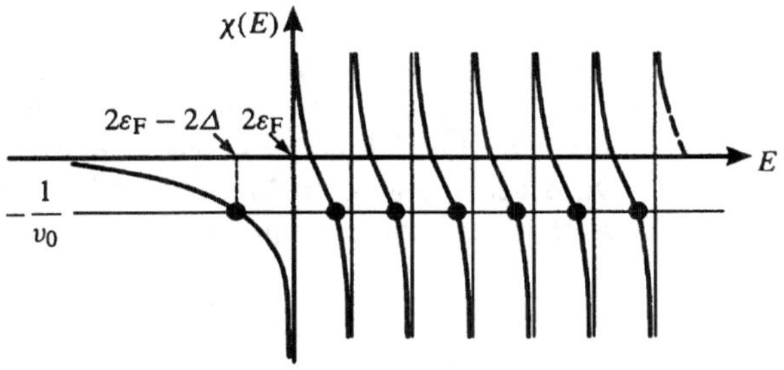

Fig. 8.7. Graphical solution of (8.74)

Evidently, the function $\chi(E)$ possesses poles at all possible energy values $E = \varepsilon_{\boldsymbol{k}_1} + \varepsilon_{\boldsymbol{k}_2}$ of a noninteracting electron pair outside the Fermi sphere. These electron pairs need to have a total momentum $\hbar \boldsymbol{K}$ and single momenta $\hbar \boldsymbol{k}_1, \hbar \boldsymbol{k}_2$ in the spherical shell between $\hbar k_{\mathrm{F}}$ and $\hbar k_{\mathrm{a}}$. Hence, the lowest pole position has to lie on the Fermi surface itself, i.e. at energy $E = 2\varepsilon_{\mathrm{F}} = 2(\hbar k_{\mathrm{F}})^2/2m$. The function $\chi(E)$ has many intersections with the straight line $-1/v_0$ above $2\varepsilon_{\mathrm{F}}$. Essentially these energies correspond to the energies of unperturbed (noninteracting) pairs; qualitatively, the corresponding states are similar to these uncorrelated pairs. Since v_0 is positive [see (8.68)] an additional intersection point below $2\varepsilon_F$ exists:

$$E_{bound} = 2\varepsilon_F - 2\Delta . \tag{8.76}$$

This implies that the interaction leads to a *bound state of the two electrons*, which is qualitatively different from the uncorrelated two-electron states outside the Fermi sphere.

In order to inspect the last result in more detail we consider the case $\boldsymbol{K} = 0$ of vanishing total momentum and look for its binding energy E_{bound}. Then,

$$\boldsymbol{k}_1 = \boldsymbol{k} \quad \text{and} \quad \boldsymbol{k}_2 = -\boldsymbol{k}$$

and the function $\chi(E)$ in (8.75) can be easily calculated with the volume element in k space:

$$\begin{aligned}\chi(E) &= \int_{k_F}^{k_a} \frac{d^3k}{(2\pi)^3} \frac{1}{E - 2\varepsilon_k} \\ &= \int_{k_F}^{k_a} \frac{d^3k}{(2\pi)^3} \frac{1}{E - (\hbar k)^2/m} \\ &= \frac{m}{2\pi^2\hbar^2} \int_{\varepsilon_F}^{\varepsilon_a} k\, d\varepsilon_k \frac{1}{E - 2\varepsilon_k} ,\end{aligned} \tag{8.77}$$

where $\varepsilon_a = (\hbar k_a)^2/2m$. Since k_a is close to k_F, the k in the integrand can be replaced by k_F. For $E < 2\varepsilon_F$ it follows that

$$\chi(E) = -\frac{N(0)}{2} \ln \left| \frac{2\varepsilon_a - E}{2\varepsilon_F - E} \right| , \tag{8.78}$$

where

$$N(0) = \frac{m(\hbar k_F)}{2\pi^2\hbar^3} = \int \frac{d^3k}{(2\pi)^3} \delta(\varepsilon_k - \varepsilon_F) \tag{8.79}$$

simply represents the density of states near the Fermi surface. The last identity can easily be verified by using the well-known relation $\delta(f(x)) = (1/|df/dx|)\delta(x - a)$ for δ functions, where a is the position x for which $f(x) = 0$. Once according to (8.76) $E = E_{bound} = 2\varepsilon_F - 2\Delta$ is inserted into (8.78), so that the group of equations

$$\begin{aligned}\chi(E) &= -\frac{1}{v_0} , \\ -\frac{N(0)}{2} \ln \left| \frac{2\varepsilon_a - 2\varepsilon_F + 2\Delta}{2\Delta} \right| &= -\frac{1}{v_0} , \\ \frac{2\Delta}{2(\varepsilon_a - \varepsilon_F) + 2\Delta} &= e^{-2/N(0)v_0}\end{aligned} \tag{8.80}$$

following from (8.74) is automatically solved. We get

$$\Delta = \frac{\varepsilon_a - \varepsilon_F}{e^{2/N(0)v_0} - 1} . \tag{8.81}$$

Now

$$\varepsilon_a - \varepsilon_F \approx \hbar\omega_D \tag{8.82}$$

is of the same order of magnitude as the *Debye energy* $\hbar\omega_D$. The Debye frequency ω_D characterizes the *maximum frequency* of lattice vibrations in metals; see Exercise 8.3.

EXERCISE

8.3 The Debye Frequency

Problem. Determine the maximum frequency of the lattice vibrations in a cubelike crystal of length L. The total number of atoms inside the cube is N, so the number of degrees of freedom is $3N$. Assume that the number of lattice-vibration modes is $3N$ at most.

Solution. The attractive interaction, which leads to the bonding of electrons into Cooper pairs, is a result of the exchange of phonons, i.e. the elementary vibrational excitations of the crystal lattice. The effect of these phonons on the electrons has been introduced approximately in (8.68) as effective interactions between the electrons with momenta $\boldsymbol{k}, \boldsymbol{k}'$:

$$V_{\boldsymbol{k}\boldsymbol{k}'} = \begin{cases} -\dfrac{v_0}{V} & \text{for } k_F < |\boldsymbol{k}|, |\boldsymbol{k}'| < k_a \\ 0 & \text{otherwise}. \end{cases} \tag{1}$$

k_F represents the momentum of the Fermi surface and k_a is a cutoff momentum for the interaction. We can understand this maximum momentum in such a way that for energy differences

$$\Delta\varepsilon \approx \varepsilon_a - \varepsilon_F = \frac{\hbar^2 k_a^2}{2m} - \frac{\hbar^2 k_F^2}{2m} \tag{2}$$

above the Fermi surface, excitation modes of the lattice that can propagate (transmit) the electron–electron interaction are no longer present. This maximum energy is called *Debye energy* ε_D,

$$\varepsilon_D = \hbar\omega_D ; \tag{3}$$

ω_D is called the *Debye frequency.* In the context of the *Debye model* we can quite easily give an approximate estimate for ω_D.

Suppose we have a solid cube of length L. The excitations (vibrations) of the lattice are given as standing waves, which can be constructed in a cube. Then the waves $\varphi(\boldsymbol{k})$ are

$$\varphi_{\boldsymbol{k}}(\boldsymbol{x}) \sim \sin(k_x x)\sin(k_y y)\sin(k_z z),$$

where boundary conditions are assumed in such a way that the excitations vanish at the edge of the solid system, i.e. we require

$$\sin(k_i x_i)|_{x_i=0} = \sin(k_i x_i)|_{x_i=L} = 0, \qquad i = 1,2,3. \tag{4}$$

This leads to a spectrum for the wavevectors

$$k_i = \frac{2\pi n_i}{L}, \tag{5}$$

Exercise 8.3.

where n_i represent integer numbers. This implies that the level density in k space for a volume L is given by

$$\rho_k = \frac{L^3}{(2\pi)^3} \,. \tag{6}$$

The maximum value for k is determined in such a way that we consider N particles in the solid-state system which have exactly $3N$ vibrational degrees of freedom. We divide the solid-state excitations into longitudinal and transverse oscillations. k_{max} is determined by requiring that the maximum number of longitudinal oscillation modes is equal to N and $2N$ for the transverse oscillation modes because of two polarization degrees of freedom. Hence, for longitudinal oscillations

$$\int_0^{|k_{\mathrm{max}}|} \mathrm{d}^3k \, \rho_k = N$$

or

$$\frac{4\pi}{3} \rho_k \, k_{\mathrm{max}}^3 = \frac{L^3}{6\pi^2} k_{\mathrm{max}}^3 = N \,. \tag{7}$$

The maximum wave number

$$k_{\mathrm{max}}^3 = 6\pi^2 \frac{N}{V} \tag{8}$$

is essentially given by the number of lattice atoms per volume. Making use of the dispersion relation

$$\omega = v_{\mathrm{l}} k \,, \tag{9}$$

with the propagation velocity v_{l} (speed of sound) of the longitudinal oscilations in the solid and a similar calculation for the transverse waves, we obtain for the Debye frequency

$$\omega_{\mathrm{D}} \equiv \omega_{\mathrm{max}} = v_{\mathrm{l}} v_{\mathrm{t}} \sqrt[3]{\frac{6\pi^2 N/V}{v_{\mathrm{l}}^3 + v_{\mathrm{t}}^3}} \,; \tag{10}$$

v_{t} represents the corresponding propagation velocity of the transverse waves. For metals, the Debye frequency is usually about 30 meV.

Typical values are $\hbar\omega_D/\varepsilon_{\mathrm{F}} \approx 1/100$ and $N(0)v_0 \approx 1/4$. Thus, with the help of (8.82) the relation (8.81) can be simplified to

$$\Delta \approx \hbar\omega_{\mathrm{D}} \, \mathrm{e}^{-2/N(0)v_0} \,. \tag{8.83}$$

This is the *binding energy per electron* in the bound (electron-pair) state. The extremely nonlinear dependence of this pair energy Δ on the force parameter v_0 is remarkable. This mechanism of pairing of electrons was discovered by L.N. Cooper,[6] hence the name *Cooper pairs.*

[6] L.N. Cooper: Phys. Rev. **104** (1956) 1189. This idea was most essential for understanding superconductivity. Together with J.R. Schrieffer and J. Bardeen, L.N. Cooper received in 1972 the Nobel Prize for physics.

Our analytical studies make use of the assumption that the total momentum of the Cooper pair is $\hbar\boldsymbol{K}=0$. If this is not the case, i.e. for $\hbar\boldsymbol{K}\neq 0$, the number of possible $\boldsymbol{k}$ values in (8.75) decreases rapidly. As already stated earlier, the poles of $\chi(E)$ characterize the energies of the noninteracting (or only weakly interacting) pairs. Since now fewer $\boldsymbol{k}$ values are available (allowed), the value of $\chi(E)$ for $E<2\varepsilon_{\mathrm{F}}$ (see Fig. 8.7) decreases. Hence $E_{\text{bound}}=2\varepsilon_{\mathrm{F}}-2\Delta$ approaches closer to $2\varepsilon_{\mathrm{F}}$ and the binding energy 2Δ of the Cooper pairs becomes smaller. This decrease is strong for an increasing total momentum $\hbar\boldsymbol{K}$. It is important to note that the smaller the number of states is, which are connected via the attractive interaction $\langle\boldsymbol{k}_1\boldsymbol{k}_2|\mathcal{V}|\boldsymbol{k}_1'\boldsymbol{k}_2'\rangle$, the smaller the binding energy of the Cooper pairs becomes. This also follows from (8.83) for $N(0)\to 0$. We obtain the largest binding energy for electrons that possess the total momentum $\hbar\boldsymbol{K}=0$; for them $\boldsymbol{k}_1=-\boldsymbol{k}_2$. Those single electron states lie outside the Fermi sphere in opposite directions (see Fig. 8.8).

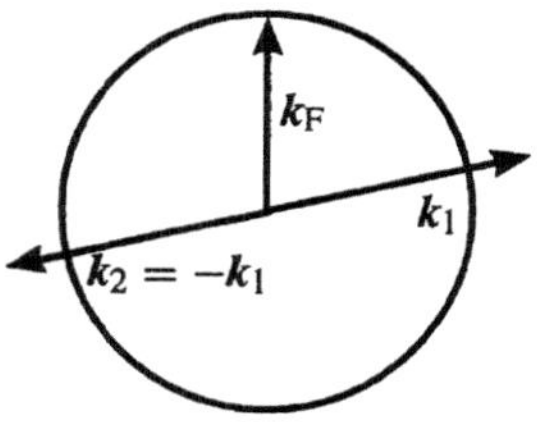

Fig. 8.8. Electrons that are in opposite directions outside the Fermi sphere, i.e. $(\boldsymbol{k}_1=-\boldsymbol{k}_2)$, have total momentum $\hbar\boldsymbol{K}=0$ and, hence, the largest level density. This implies that that their pair energy becomes largest

In order to determine the wavefunction of a Cooper pair we use (8.69), according to which

$$a_{\boldsymbol{K}}(\boldsymbol{K}) = \frac{1}{E-\varepsilon_{\boldsymbol{k}_1}-\varepsilon_{\boldsymbol{k}_2}}\text{const}\,. \tag{8.84}$$

This is so because the right-hand side of (8.69) is independent of k, i.e. it is constant with respect to k. Inserting (8.84) into (8.66) and this into (8.49) and using $\boldsymbol{k}_1=\boldsymbol{K}/2+\boldsymbol{k}$ and $\boldsymbol{k}_2=\boldsymbol{K}/2-\boldsymbol{k}$, results for a fixed total momentum $\hbar\boldsymbol{K}$, in

$$\psi(\boldsymbol{r}_1,\boldsymbol{r}_2) = N_1 \mathrm{e}^{\mathrm{i}\boldsymbol{K}\cdot(\boldsymbol{r}_1-\boldsymbol{r}_2)/2}\frac{1}{V}\sum_{\boldsymbol{k}}{}' \frac{\mathrm{e}^{\mathrm{i}\boldsymbol{k}\cdot(\boldsymbol{r}_1-\boldsymbol{r}_2)}}{E-\varepsilon_{\boldsymbol{k}_1}-\varepsilon_{\boldsymbol{k}_2}}\,. \tag{8.85}$$

Hence, with $\boldsymbol{r}=\boldsymbol{r}_1-\boldsymbol{r}_2$ the relative wavefunction of the Cooper pair reads

$$\varphi(\boldsymbol{r}) = N_2\int{}' \frac{\mathrm{d}^3k}{(2\pi)^3}\frac{\mathrm{e}^{\mathrm{i}\boldsymbol{k}\cdot\boldsymbol{r}}}{E-\varepsilon_{\boldsymbol{k}_1}-\varepsilon_{\boldsymbol{k}_2}}\,. \tag{8.86}$$

N_1 and N_2 are suitable normalization factors. For a pair with $\boldsymbol{K}=0$ this results in

$$\begin{aligned}
\varphi(\boldsymbol{r}) &= N_2\int{}' \frac{\mathrm{d}^3k}{(2\pi)^3}\frac{\mathrm{e}^{\mathrm{i}\boldsymbol{k}\cdot\boldsymbol{r}}}{2\varepsilon_{\mathrm{F}}-2\Delta-(\hbar\boldsymbol{k})^2/m}\\
&= N_2\int{}' \frac{\mathrm{d}^3k}{(2\pi)^3}\frac{\mathrm{e}^{\mathrm{i}\boldsymbol{k}\cdot\boldsymbol{r}}}{2\varepsilon_{\mathrm{F}}-2\Delta-2\varepsilon_{\mathrm{F}}}\\
&= N_2\int{}' \frac{\mathrm{d}^3k}{(2\pi)^3}\frac{\mathrm{e}^{\mathrm{i}\boldsymbol{k}\cdot\boldsymbol{r}}}{(\hbar^2/m)\,(k_{\mathrm{F}}^2-k^2-2m\Delta/\hbar^2)}\\
&= -N_2\frac{4\pi m}{\hbar^2}\left[\frac{\cos k_{\mathrm{F}}r}{\alpha r^2}+\frac{\sin k_{\mathrm{F}}r}{\alpha^2 r^3}\right]\,,\qquad \alpha = \frac{2m\Delta}{\hbar^2 k_{\mathrm{F}}}\,,
\end{aligned} \tag{8.87}$$

where (8.72) has been used; this will be calculated in detail in Exercise 8.4. This relative wavefunction has its first extremum at

$$\alpha r_{\max} \approx 1\,,$$

from which

$$d = r_{\max} = \frac{1}{\alpha} = \frac{\hbar^2 k_F}{2m\Delta} \approx 10^{-4}\,\text{cm} \tag{8.88}$$

follows. The correlation length d approximately characterizes the extension of a Cooper pair. It is remarkable that $\varphi(r)$ in (8.87) represents an s wavefunction; it does not depend on the angle and is thus proportional to $Y_{00}(\vartheta, \varphi)$. Hence, the angular momentum of a Cooper pair is $l = 0$.

Superconductivity is a result of the simultaneous interaction of *all* electrons. All electrons at the Fermi surface build up correlated pairs with the same total momentum. Their correlation can be such that a current is produced, which is then called a *super-current.* In order to stop it, all electron pairs have to be stopped at the same time. This is not the case for a conventional conductor: electrons contributing to the current are stopped one after the other. In a superconductor the correlated electron current has to be stopped in total and instantly. Of course, this is very difficult; this is the reason why the super-current flows for ever.[7]

EXERCISE

8.4 Correlation Length of a Cooper Pair

Problem. Determine the relative wavefunction $\varphi(\boldsymbol{r})$ of a Cooper pair for the total momentum $\boldsymbol{K} = 0$ [see (8.87)]. Estimate the extension of a Cooper pair.

Solution. According to (8.86), $\varphi(\boldsymbol{r})$ is given by

$$\varphi(\boldsymbol{r}) = N_2 \int \frac{\mathrm{d}^3 k}{(2\pi)^3} \frac{\mathrm{e}^{\mathrm{i}\boldsymbol{k}\cdot\boldsymbol{r}}}{E - \varepsilon_{\boldsymbol{k}_1} - \varepsilon_{\boldsymbol{k}_2}} . \tag{1}$$

From (8.72), for $\boldsymbol{K} = 0$ we have

$$\varepsilon_{\boldsymbol{k}_1} + \varepsilon_{\boldsymbol{k}_2} = \frac{(\hbar \boldsymbol{k})^2}{m} . \tag{2}$$

With (8.76) we obtain

$$\begin{aligned} \varphi(\boldsymbol{r}) &= N_2 \int \frac{\mathrm{d}^3 k}{(2\pi)^3} \frac{\mathrm{e}^{\mathrm{i}\boldsymbol{k}\cdot\boldsymbol{r}}}{2\varepsilon_F - 2\Delta - (\hbar \boldsymbol{k})^2/m} \\ &= N_2 \int \frac{\mathrm{d}^3 k}{(2\pi)^3} \frac{\mathrm{e}^{\mathrm{i}\boldsymbol{k}\cdot\boldsymbol{r}}}{(\hbar^2/m)\,(k_F^2 - k^2 - 2m\Delta/\hbar^2)} . \end{aligned} \tag{3}$$

We make use of a partial wave decomposition of $\mathrm{e}^{\mathrm{i}\boldsymbol{k}\cdot\boldsymbol{r}}$, i.e.

$$\mathrm{e}^{\mathrm{i}\boldsymbol{k}\cdot\boldsymbol{r}} = \sum_{n=0}^{\infty} i^n (\lambda_n + 1) j_n(kr) P_n(\cos\vartheta) . \tag{4}$$

The angle integration can then be performed. The $j_n(kr)$ are spherical Bessel functions. $P_n(\cos\vartheta)$ represent Legendre polynomials for which the orthogonality relation

[7] This argument goes back to J.R. Schrieffer: *Theory of Superconductivity* (Benjamin, New York 1964).

Exercise 8.4.

$$\int_0^{\pi} P_n(\cos\vartheta)P_m(\cos\vartheta)\sin\vartheta\,\mathrm{d}\vartheta = \frac{2}{2n+1}\delta_{nm} \tag{5}$$

holds. Inserting (4) into (3) yields

$$\varphi(\boldsymbol{r}) = N_2\sum_{n=0}^{\infty}\int\frac{k^2\,\mathrm{d}k}{(2\pi)^3}\frac{(2n+1)j_n(kr)}{(\hbar^2/2m)\,(k_{\mathrm{F}}^2-k^2-2m\Delta/\hbar^2)}$$
$$\times\int\mathrm{d}\varphi\int\mathrm{d}\vartheta\sin\vartheta P_n(\cos\vartheta)\,. \tag{6}$$

The ϑ integration can also be written as $[P_0(x)=1]$

$$\int\mathrm{d}\vartheta\sin\vartheta P_n(\cos\vartheta)P_0(\cos\vartheta) = 2\delta_{n0}\,, \tag{7}$$

where (5) has been used. With $j_0(x)=\sin x/x$ we obtain

$$\varphi(r) = N_2\frac{4\pi m}{\hbar^2}\int\frac{k^2\,\mathrm{d}k}{(2\pi)^3}\frac{\sin kr}{kr\,(k_{\mathrm{F}}^2-k^2-2m\Delta/\hbar^2)}\,. \tag{8}$$

Since the k integration should go only over a range $\Delta k = k_{\mathrm{a}}-k_{\mathrm{F}} \ll k_{\mathrm{F}}$ above the Fermi surface we can write

$$k_{\mathrm{F}}^2-k^2 \approx k_{\mathrm{F}}(k_{\mathrm{F}}-k) = -k_{\mathrm{F}}x\,. \tag{9}$$

Hence,

$$\sin kr = \sin[(k-k_{\mathrm{F}}+k_{\mathrm{F}})r] = \sin[(k_{\mathrm{F}}+x)r] \tag{10}$$

and insertion into (8) yields

$$\varphi(r) \approx -N_2\frac{4\pi m}{\hbar^2 r}\int_0^{\Delta k}\mathrm{d}x\,\frac{\sin[(k_{\mathrm{F}}+x)r]}{x+2m\Delta/\hbar^2k_{\mathrm{F}}}\,. \tag{11}$$

Mainly small x values contribute to the integral. Hence, we are allowed to shift the upper integration bound to approximately ∞; remember that $\alpha = 2m\Delta/\hbar^2k_{\mathrm{F}}$. This gives

$$\varphi(r) \approx -N_2\frac{4\pi m}{\hbar^2 r}\left(\sin k_{\mathrm{F}}r\int_0^{\infty}\mathrm{d}x\,\frac{\cos xr}{x+\alpha}+\cos k_{\mathrm{F}}r\int_0^{\infty}\mathrm{d}x\,\frac{\sin xr}{x+\alpha}\right) \tag{12}$$
$$\approx N_2\frac{4\pi m}{\hbar^2 r}\{\sin k_{\mathrm{F}}r[+\mathrm{ci}(\alpha r)\cos(\alpha r)+\mathrm{si}(\alpha r)\sin(\alpha r)]$$
$$+\cos k_{\mathrm{F}}r[-\mathrm{ci}(\alpha r)\sin(\alpha r)+\mathrm{si}(\alpha r)\cos(\alpha r)]\}\,. \tag{13}$$

The *integral sine* $\mathrm{si}(x)$ and the *integral cosine* $\mathrm{ci}(x)$ appear here; they are defined via the integrals

$$\mathrm{si}(x) = \int_{\infty}^{x}\frac{\sin t}{t}\,\mathrm{d}t\,, \tag{14}$$

$$\mathrm{ci}(x) = \int_{\infty}^{x}\frac{\cos t}{t}\,\mathrm{d}t\,. \tag{15}$$

The series expansion of these functions reveals their asymptotic behavior for large r:

Exercise 8.4.

$$\mathrm{si}(x) = -\cos x\left[\sum_{m=0}^{M-1}\frac{(-1)^m(2m)!}{x^{2m+1}} + O(|x|^{-2M-1})\right] + \sin x\left[\sum_{m=1}^{N-1}\frac{(-1)^m(2m-1)!}{x^{2m}} + O(|x|^{-2N})\right], \quad (16)$$

$$\mathrm{ci}(x) = \cos x\left[\sum_{m=1}^{N-1}\frac{(-1)^m(2m-1)!}{x^{2m}} + O(|x|^{-2N})\right] + \sin x\left[\sum_{m=0}^{M-1}\frac{(-1)^m(2m)!}{x^{2m+1}} + O(|x|^{-2M-1})\right] \quad (17)$$

with $M, N = 1, \dots$. For $x \to \infty$ we obtain

$$\mathrm{si}(x) \approx -\frac{\cos x}{x}, \qquad \mathrm{ci}(x) \approx \frac{\sin x}{x}. \quad (18)$$

Insertion into (13) yields, for large r,

$$\varphi(r) \approx -N_2\frac{4\pi m}{\hbar^2}\frac{\cos k_F r}{r^2} = -N_2\frac{2\pi k_F}{\Delta}\frac{\cos k_F r}{r^2}. \quad (19)$$

In order to get an impression of the extension of a Cooper pair we take into account the next terms from (17) for the expansion of the wavefunction (13) and obtain

$$\varphi(r) \approx -N_2\frac{4\pi m}{\hbar^2}\left(\frac{\cos k_F r}{\alpha r^2} + \frac{\sin k_F r}{\alpha^2 r^3}\right). \quad (20)$$

Determining the first extremum of this function $\varphi_{\max}(r_m)$, we arrive at the relation

$$\alpha r_{\max} \approx 1. \quad (21)$$

Thus, the extension of the Cooper pair is of the order of magnitude

$$d = r_{\max} \approx \frac{1}{\alpha} = \frac{\hbar^2 k_F}{2m\Delta} \quad (22)$$

with numerical values for r_m in the range $\approx 10^{-4}$ cm.

EXERCISE

8.5 Determination of the Coupling Strength of a Bound Cooper Pair

Problem. Determine the coupling strength v_0 needed to obtain a bound Cooper pair ($E < 0$) with total momentum $\boldsymbol{K} = 0$ for the case of vanishing Fermi momentum ($k_F = 0$).

Solution. We know from (8.74)–(8.80) that the following relation has to be fulfilled for a bound state ($\boldsymbol{K} = 0$):

Exercise 8.5.

$$-\frac{1}{v_0} = \int_{k_F}^{k_a} \frac{d^3k}{(2\pi)^3} \frac{1}{E - \hbar^2 k^2/m} . \tag{1}$$

With $E = 2\varepsilon_F - 2\Delta$ [see (8.76)] and $k_F = 0$ we obtain

$$\begin{aligned} \frac{1}{v_0} &= \int_0^{k_a} \frac{d^3k}{(2\pi)^3} \frac{1}{2\Delta + \hbar^2 k^2/m} \\ &= \frac{m}{2\pi^2\hbar^2} \int_0^{k_a} dk \frac{k^2}{2m\Delta/\hbar^2 + k^2} . \end{aligned} \tag{2}$$

The integral can be calculated in an elementary way:

$$\int dx \frac{x^2}{1+x^2} = x - \arctan x .$$

The result is

$$\frac{1}{v_0} = \frac{m}{2\pi^2\hbar^2} \left(k_a - \sqrt{\frac{2m\Delta}{\hbar^2}} \arctan \frac{k_a}{\sqrt{2m\Delta/\hbar^2}} \right) . \tag{3}$$

We want to find out the conditions under which a bound pair is able to exist; hence, we consider the limiting case $\Delta \to 0$ and obtain

$$v_0^{\min} = \frac{2\pi^2\hbar^2}{mk_a} \tag{4}$$

as the minimum interaction strength. Contrary to the case for $k_F \neq 0$ in (8.81), a bound pair is obtained only for interaction strenghts that are larger than a certain minimum value $v_0^{\min}$. This is because of suppression of the pairing effect as the result the small phase-space level density $d^3k/(2\pi)^3$ for $|\boldsymbol{k}| \to 0$. This becomes more clear once we look at the approximate relation (8.81), which in principle holds only for $k_F > 0$:

$$\Delta = \frac{\varepsilon_a - \varepsilon_F}{e^{2/N(0)v_0} - 1} . \tag{5}$$

Here, $N(0)$ represents the density of states near the Fermi surface. Of course, for $k_F \to 0$ this density becomes $N(0) \to 0$; thus, within this approximation the binding energy Δ vanishes for all couplings v_0.

9. Quasiparticles in Plasmas and Metals: Selected Topics

In the last chapters we saw that the concept of a quasiparticle in many-particle physics has many advantages: A system of interacting particles can be treated – in a suitable approximation – as a system of noninteracting quasiparticles.

Now we will make further use of this approach for an electrically neutral system consisting of electrons and positively charged ions, i.e. a *plasma*. Such plasmas appear manyfold in nature: the gas in a flame or in an electric discharge, the matter in plasma reactors or in a star. In all these cases we are dealing with some kind of a plasma. Here we will assume the plasma to be isotropic and homogeneous; then, the physical properties are translationally invariant.

It is convenient to simplify the problem by assuming a self-consistent field: the particles interact with an electrostatic potential $\varphi(\boldsymbol{x},t)$, which in return has to be calculated from the average charge density $\rho(\boldsymbol{x},t)$ of the plasma. With the techniques developed in Chap. 6, the Hamiltonian of the system becomes, in second quantization,

$$\hat{H} = \sum_i \int d^3x\, \hat{\Psi}_i^\dagger \left[\frac{\hbar^2}{2m_i}\nabla^2 + e_i\varphi\right] \Psi_i = \hat{H}_0 + \hat{H}_\mathrm{I}\,. \tag{9.1}$$

The index i distinguishes the particle species, i.e. electrons or ions. We expand $\hat{\Psi}^\dagger$ and $\hat{\Psi}$ into plane waves,

$$\hat{\Psi}_i = \sum_i \hat{b}_{i\boldsymbol{q}} \frac{e^{i\boldsymbol{q}\cdot\boldsymbol{x}}}{\sqrt{V}} \tag{9.2}$$

(box normalization), where $V = L^3$ represents the normalization volume. For the Hamiltonian we now get

$$\hat{H}_0 = \sum_i \sum_{\boldsymbol{q}} \frac{\hbar^2 \boldsymbol{q}^2}{2m_i} \hat{b}_{i\boldsymbol{q}}^\dagger \hat{b}_{i\boldsymbol{q}}\,. \tag{9.3}$$

With the help of the Fourier transform of the potential φ,

$$\tilde{\varphi}(\boldsymbol{q}_2 - \boldsymbol{q}_1) = \int \frac{d^3x}{V}\, e^{i(\boldsymbol{q}_2-\boldsymbol{q}_1)\cdot\boldsymbol{x}} \varphi(\boldsymbol{x}) \qquad V = L^3\,, \tag{9.4}$$

the interaction Hamiltonian can be written as [see (6.7) and the equations following it]

$$\hat{H}_{\mathrm{I}} = \sum_i \sum_{q_1} \sum_{q_2} e_i \hat{b}^\dagger_{iq_1} \hat{b}_{iq_2} \tilde{\varphi}(q_2 - q_1) \,. \tag{9.5}$$

Since the potential $\varphi(\boldsymbol{x}, t)$ is a one-body potential, only one creation and one annihilation operator appear; however, in nondiagonal form. Equations (9.3) and (9.5) express the Hamiltonian in second-quantized form, which enables us now to study such relevant physical quantities as, for example, the time dependence of the particle-number operator. Mathematically speaking, in its most general form we will study products of the form

$$\hat{N}_{iq'q} = \hat{b}^\dagger_{iq'} \hat{b}_{iq} \,.$$

The time dependence of an arbitrary operator $\hat{O}$ is governed by Heisenberg's equation of motion:

$$\frac{\hbar}{\mathrm{i}} \frac{\partial}{\partial t} \hat{O} = [\hat{H}, \hat{O}]_- \,,$$

which is in our case

$$\frac{\hbar}{\mathrm{i}} \frac{\partial}{\partial t} (\hat{b}^\dagger_{iq'} \hat{b}_{iq}) = \left[(\hat{H}_0 + \hat{H}_{\mathrm{I}}), (\hat{b}^\dagger_{iq'} \hat{b}_{iq}) \right]_- \,. \tag{9.6}$$

Here $[\hat{A}, \hat{B}]_- = \hat{A}\hat{B} - \hat{B}\hat{A}$ denotes the minus commutator. With the help of (9.3) and (9.5) we can calculate the commutator. The result does not depend on the choice of boson or fermion commutation relations for the operators $\hat{b}_{iq}$. We obtain

$$\begin{aligned} \frac{\partial}{\partial t} \hat{b}^\dagger_{iq'} \hat{b}_{iq} &= \frac{\mathrm{i}}{\hbar} (E_{iq'} - E_{iq}) \hat{b}^\dagger_{iq'} \hat{b}_{iq} \\ &\quad + \frac{\mathrm{i} e_i}{\hbar} \sum_{p} \left[\tilde{\varphi}(p - q) \hat{b}^\dagger_{iq'} \hat{b}_{ip} - \tilde{\varphi}(q' - p) \hat{b}^\dagger_{ip'} \hat{b}_{iq} \right] \,. \end{aligned} \tag{9.7}$$

The first part of (9.7) is a consequence of $\hat{H}_0$: For the case $\boldsymbol{q} = \boldsymbol{q}'$ the operator $\hat{b}^\dagger_{iq'} \hat{b}_{iq}$ represents the particle-number operator for particles of species i with momentum $\boldsymbol{q}$; for this case the first term in (9.7) vanishes, i.e. $\hat{H}_0$ does not effect any change in the particle number. Only the potential changes the particle number. The first term within the square brackets $\sim \hat{b}^\dagger_{sq'} \hat{b}_{sp}$ annhilates a particle with momentum $\boldsymbol{p}$ and creates another one with momentum $\boldsymbol{q}'$; it describes the increase in the particle number with momentum $\boldsymbol{q}'$ due to scattering from other momentum states. The second term within the brackets describes the decrease of the particle number with $\boldsymbol{q}'$ due to scattering to the momentum $\boldsymbol{p}$.

In order to find the relationship to classical quantities it is convenient to introduce the *distribution function of the particle species* i:

$$F_i(\boldsymbol{q}', \boldsymbol{q}, t) = \sum_\alpha P_\alpha \langle \alpha | \hat{b}^\dagger_{iq'} \hat{b}_{iq} | \alpha \rangle \,, \tag{9.8}$$

where $|\alpha\rangle$ represent the states of the system and P_α is the probability of finding the system in state $|\alpha\rangle$. Equation (9.7) becomes

$$\frac{\partial}{\partial t} F_i(\boldsymbol{q}', \boldsymbol{q}, t) = \frac{\mathrm{i}}{\hbar} (E_{iq'} - E_{iq}) F_i(\boldsymbol{q}', \boldsymbol{q}, t)$$

$$+\frac{\mathrm{i}e_i}{\hbar}\sum_{\boldsymbol{p}}[\tilde{\varphi}(\boldsymbol{p}-\boldsymbol{q})F_i(\boldsymbol{q}',\boldsymbol{p},t)-\tilde{\varphi}(\boldsymbol{q}'-\boldsymbol{p})F_i(\boldsymbol{p},\boldsymbol{q},t)]\ . \tag{9.9}$$

The physical significance of the distribution function $F_i(\boldsymbol{q}',\boldsymbol{q},t)$ becomes clear once we consider the particle-number operator $\hat{N}=\hat{\psi}^\dagger\hat{\psi}$. Its mean value with respect to the states $|\alpha\rangle$ is obtained by averaging with P_α over the expectation values $\langle\alpha|\hat{N}|\alpha\rangle$:

$$\langle\hat{N}_i(\boldsymbol{x},t)\rangle = \sum_\alpha P_\alpha\langle\alpha|\hat{\psi}_i^\dagger(\boldsymbol{x},t)\hat{\psi}_i(\boldsymbol{x},t)|\alpha\rangle\ . \tag{9.10}$$

We insert (9.2) and obtain

$$\begin{aligned}\langle\hat{N}_i(\boldsymbol{x},t)\rangle &= \sum_{\boldsymbol{q}}\sum_{\boldsymbol{p}}\sum_\alpha P_\alpha\langle\alpha|\hat{b}^\dagger_{i\boldsymbol{p}}\hat{b}_{i\boldsymbol{p}+\boldsymbol{q}}|\alpha\rangle\frac{\mathrm{e}^{\mathrm{i}\boldsymbol{q}\cdot\boldsymbol{x}}}{V}\\ &= \sum_{\boldsymbol{q}}\sum_{\boldsymbol{p}}\langle\hat{b}^\dagger_{i\boldsymbol{p}}\,\hat{b}_{i\boldsymbol{p}+\boldsymbol{q}}\rangle\frac{\mathrm{e}^{\mathrm{i}\boldsymbol{q}\cdot\boldsymbol{x}}}{V}\\ &= \sum_{\boldsymbol{q}}\sum_{\boldsymbol{p}}F_i(\boldsymbol{p},\boldsymbol{p}+\boldsymbol{q},t)\frac{\mathrm{e}^{\mathrm{i}\boldsymbol{q}\cdot\boldsymbol{x}}}{V}\ .\end{aligned} \tag{9.11}$$

This result suggests the introduction of the *distribution function $F_i(\boldsymbol{x},\boldsymbol{p},t)$ in phase space* (coordinate and momentum space) according to

$$F_i(\boldsymbol{x},\boldsymbol{p},t) = \sum_{\boldsymbol{q}}F_i(\boldsymbol{p},\boldsymbol{p}+\boldsymbol{q},t)\frac{\mathrm{e}^{\mathrm{i}\boldsymbol{q}\cdot\boldsymbol{x}}}{V}\ ; \tag{9.12}$$

thus,

$$\langle\hat{N}_i(\boldsymbol{x},t)\rangle = \sum_{\boldsymbol{p}}F_i(\boldsymbol{x},\boldsymbol{p},t)\ . \tag{9.13}$$

On the other hand we derive for the momentum distribution for the particle species i

$$\langle\hat{N}_i(\boldsymbol{p},t)\rangle = \int\mathrm{d}^3x\,F_i(\boldsymbol{x},\boldsymbol{p},t) = \langle\hat{b}^\dagger_{i\boldsymbol{p}}\hat{b}_{i\boldsymbol{p}}\rangle\ . \tag{9.14}$$

Hence, $F_i(\boldsymbol{x},\boldsymbol{p},t)$ is directly connected with the particle-number operator. The definition (9.8), however, means that F_i represents not a classical, but a *quantum-mechanical distribution function*;[1] frequently it is known as the *Wigner function*.

For a given potential φ [see (9.4) and (9.5)] the quantum-mechanical distribution function is determined from (9.9). On the other hand the potential φ depends on the Wigner function itself. In order to formulate this problem self-consistently, we simply insert $\langle\hat{N}(\boldsymbol{x},t)\rangle$ into the Poisson equation and obtain for φ

$$\nabla^2\varphi = -\sum_i\sum_{\boldsymbol{q}}\sum_{\boldsymbol{p}}\frac{4\pi e_i}{V}F_i(\boldsymbol{p},\boldsymbol{p}+\boldsymbol{q},t)\,\mathrm{e}^{\mathrm{i}\boldsymbol{q}\cdot\boldsymbol{x}}\ . \tag{9.15}$$

[1] If $F_i(\boldsymbol{x},\boldsymbol{p},t)$ were a classical distribution function, $F_i(\boldsymbol{x},\boldsymbol{p},t)$ would specify the probable number of particles in the volume d^3x and in the momentum range d^3p. The expressions (9.13) and (9.14) show that this is not the case for the quantum-mechanical distribution function; however, the Wigner function is analogous to the classical distribution function.

From these two equations, (9.9) and (9.15), the distribution function $F_i(\boldsymbol{p}, \boldsymbol{p}+\boldsymbol{q}, t)$ and the averaged potential φ have to be determined by successive iteration. Note, however, that (9.15) already represents a certain approximation because the true charge density has been replaced by the averaged one. By starting from an initial distribution function $F_i^{(0)}$, an averaged potential $\varphi^{(0)}$ can be calculated. With this potential $\varphi^{(0)}$ another new approximation for F_i can be determined, and so forth. In atomic physics this approximation is known as the *Hartree approximation.* Equation (9.15) and (9.9) are the *quantum-mechanical analogy to the so-called Vlasov equations in classical plasma physics.*

To gain more insight into this system of equations let us consider the equilibrium case for which the charge density and the potential φ vanish. In equilibrium ($\varphi = \varphi^{(0)} = 0 \rightarrow \tilde{\varphi}^{(0)} = 0$), (9.9) takes on the form

$$\frac{\partial F_i^{(0)}(\boldsymbol{q}', \boldsymbol{q}, t)}{\partial t} = \frac{\mathrm{i}}{\hbar}(E_{i\boldsymbol{q}'} - E_{i\boldsymbol{q}})F_i^{(0)}(\boldsymbol{q}', \boldsymbol{q}, t) \, ,$$

and it follows that

$$F_i^{(0)}(\boldsymbol{q}', \boldsymbol{q}, t) = F_i^{(0)}(\boldsymbol{q}', \boldsymbol{q})\, \mathrm{e}^{(\mathrm{i}/\hbar)(E_{i\boldsymbol{q}'} - E_{i\boldsymbol{q}})t} \, .$$

For $\varphi = 0$, the states $|\alpha\rangle$ in (9.8) have a fixed particle number; as a consequence $F_i^{(0)}(\boldsymbol{q}', \boldsymbol{q}, t) \sim \delta_{\boldsymbol{q}'\boldsymbol{q}}$, which leads to

$$F_i^{(0)}(\boldsymbol{q}', \boldsymbol{q}, t) = F_i^{(0)}(\boldsymbol{q})\, \delta_{\boldsymbol{q}'\boldsymbol{q}} \, .$$

We now consider the case in which F_i deviates only slightly from equilibrium and oscillates around it. Hence, we set

$$F_i(\boldsymbol{q}', \boldsymbol{q}, t) = F_i^{(0)}(\boldsymbol{q})\delta_{\boldsymbol{q}\boldsymbol{q}'} + F_i^{(1)}(\boldsymbol{q}', \boldsymbol{q})\, \mathrm{e}^{-\mathrm{i}\omega t} \, , \qquad (9.16\mathrm{a})$$

where $F_i^{(1)}$ should stay small compared to $F_i^{(0)}$. Also for the potential φ we set

$$\varphi = \varphi^{(0)} + \varphi^{(1)}\, \mathrm{e}^{-\mathrm{i}\omega t} \, . \qquad (9.16\mathrm{b})$$

We insert this into the equation of motion (9.9) and deduce

$$\begin{aligned} \mathrm{i}\omega F_i^{(1)}(\boldsymbol{q}', \boldsymbol{q})\, \mathrm{e}^{-\mathrm{i}\omega t} &= i\nu(\boldsymbol{q}', \boldsymbol{q})F_i^{(1)}(\boldsymbol{q}', \boldsymbol{q})\, \mathrm{e}^{-\mathrm{i}\omega t} \\ &+ \frac{\mathrm{i}e_i}{\hbar}\sum_{\boldsymbol{p}} \Big[\tilde{\varphi}^{(1)}(\boldsymbol{p}-\boldsymbol{q})\, \mathrm{e}^{-\mathrm{i}\omega t}\left(F_i^{(0)}(\boldsymbol{q}')\delta_{\boldsymbol{q}'\boldsymbol{p}} + F_i^{(1)}(\boldsymbol{q}', \boldsymbol{p})\, \mathrm{e}^{-\mathrm{i}\omega t}\right) \\ &- \tilde{\varphi}^{(1)}(\boldsymbol{q}'-\boldsymbol{p})\, \mathrm{e}^{-\mathrm{i}\omega t}\left(F_i^{(0)}(\boldsymbol{q})\delta_{\boldsymbol{p}\boldsymbol{q}} + F_i^{(1)}(\boldsymbol{p}, \boldsymbol{q})\, \mathrm{e}^{-\mathrm{i}\omega t}\right)\Big] \, , \end{aligned}$$

where $\nu_i(\boldsymbol{q}', \boldsymbol{q}) = (1/\hbar)(E_{i\boldsymbol{q}'} - E_{i\boldsymbol{q}})$. This equation can be rearranged as

$$\begin{aligned} &\mathrm{i}[\omega - \nu_i(\boldsymbol{q}', \boldsymbol{q})]F_i^{(1)}(\boldsymbol{q}', \boldsymbol{q}) \\ &\quad \approx \frac{\mathrm{i}e_i}{\hbar}\left(\tilde{\varphi}^{(1)}(\boldsymbol{q}'-\boldsymbol{q})F_i^{(0)}(\boldsymbol{q}') - \tilde{\varphi}^{(1)}(\boldsymbol{q}'-\boldsymbol{q})F_i^{(0)}(\boldsymbol{q})\right) \, , \end{aligned}$$

once the terms on the right-hand side proportional to $\tilde{\varphi}^{(1)}F^{(1)}$ are neglected. $\tilde{\varphi}^{(1)}$ and $F^{(1)}$ are treated as small quantities, so quadratically small terms, such as the term $\tilde{\varphi}^{(1)}F^{(1)}$ just mentioned, can be neglected. This yields

$$F_i^{(1)}(\boldsymbol{q}',\boldsymbol{q}) = \frac{e_i}{\hbar}\tilde{\varphi}^{(1)}(\boldsymbol{q}',\boldsymbol{q})\frac{F_i^{(0)}(\boldsymbol{q}') - F_i^{(0)}(\boldsymbol{q})}{\omega - \nu_i(\boldsymbol{q}',\boldsymbol{q})} \tag{9.17}$$

with the frequency

$$\nu_i(\boldsymbol{q}',\boldsymbol{q}) = \frac{1}{\hbar}\left(E_{i\boldsymbol{q}'} - E_{i\boldsymbol{q}}\right) . \tag{9.18}$$

Again we insert the solution (9.17) into the Poisson equation (9.15) and arrive at

$$\nabla^2\varphi = \sum_i\sum_{\boldsymbol{q}}\sum_{\boldsymbol{p}}\frac{4\pi e_i^2}{\hbar V}\tilde{\varphi}^{(1)}(-\boldsymbol{q})\frac{F_i^{(0)}(\boldsymbol{p}) - F_i^{(0)}(\boldsymbol{p}+\boldsymbol{q})}{\omega - \nu_i(\boldsymbol{p},\boldsymbol{p}+\boldsymbol{q})}\,\mathrm{e}^{\mathrm{i}\boldsymbol{q}\cdot\boldsymbol{x}} . \tag{9.19}$$

Owing to (9.4) it holds that

$$\varphi(\boldsymbol{x}) = \sum_{\boldsymbol{q}}\tilde{\varphi}^{(1)}(-\boldsymbol{q})\,\mathrm{e}^{\mathrm{i}\boldsymbol{q}\cdot\boldsymbol{x}} \tag{9.20}$$

and we have

$$\nabla^2\varphi(\boldsymbol{x}) = -\sum_{\boldsymbol{q}} q^2\tilde{\varphi}^{(1)}(-\boldsymbol{q})\,\mathrm{e}^{\mathrm{i}\boldsymbol{q}\cdot\boldsymbol{x}} . \tag{9.21}$$

Then, (9.19) can be written as

$$\sum_{\boldsymbol{q}}\epsilon(\boldsymbol{q},\omega)q^2\tilde{\varphi}^{(1)}(-\boldsymbol{q}) = 0 , \tag{9.22}$$

where

$$\epsilon(\boldsymbol{q},\omega) = 1 + \sum_i\sum_{\boldsymbol{p}}\frac{4\pi e_i^2}{q^2\hbar V}\frac{F_i^{(0)}(\boldsymbol{p}) - F_i^{(0)}(\boldsymbol{p}-\boldsymbol{q})}{\omega - \nu_i(\boldsymbol{p},\boldsymbol{p}-\boldsymbol{q})} . \tag{9.23}$$

$\epsilon(\boldsymbol{q},\omega)$ is called the *dielectric function* of the plasma. In addition to the trivial case $\tilde{\varphi}^{(1)}(-\boldsymbol{q}) = 0$, (9.22) has the solution

$$\epsilon(\boldsymbol{q},\omega) = 0 . \tag{9.24}$$

The solutions $\omega(\boldsymbol{q})$ of this equation yield the frequencies of the waves propagating with momentum $\boldsymbol{q}$ in the plasma. For the solution of (9.24) it is convenient to express the momentum distribution $F_i^{(0)}(\boldsymbol{p})$ in terms of the velocity distribution $f_i^{(0)}(\boldsymbol{v})$ with $\boldsymbol{v} = \hbar\boldsymbol{p}/m$, to replace the summation over $\boldsymbol{p}$ by an integration, and to let the volume V go to infinity:

$$\sum_{\boldsymbol{p}} \to V\int \mathrm{d}^3v ;$$

we get

$$\epsilon(\boldsymbol{q},\omega) = 1 + \sum_i\frac{4\pi e_i^2}{q^2\hbar}\int\mathrm{d}^3v\,\frac{f_i^{(0)}(\boldsymbol{v}) - f_i^{(0)}(\boldsymbol{v} - \hbar\boldsymbol{q}/m_i)}{\omega - \boldsymbol{q}\cdot\boldsymbol{v} + \hbar q^2/2m_i} . \tag{9.25}$$

In order to find the classical limit of (9.25), we expand the nominator of the integrand into a Taylor series and let $\hbar \to 0$:

$$\epsilon_{\text{class}}(\boldsymbol{q},\omega) = 1+\sum_i \frac{4\pi e_i^2}{m_i q^2}\int \mathrm{d}^3v \frac{\boldsymbol{q}\cdot\partial f_i^{(0)}/\partial \boldsymbol{v}}{\omega-\boldsymbol{q}\cdot\boldsymbol{v}} . \tag{9.26}$$

An interpretation of $\epsilon(\boldsymbol{q},\omega)$ in the present form is not possible because the denominator of the integrand in (9.26) may have zeros; the integrand is singular. However, it is possible to regularize the integral: ω becomes complex valued and the conventional methods of complex variables can be applied. With a suitable choice of the integration path around the singularity on the real axis the integral becomes regular. With an analytical continuation the desired result is finally obtained. Hence, we replace ω with $\omega+i\eta$ and make use of Plemelj's well-known formula (see Exercise 2.13)

$$\lim_{\eta\to 0^+}\frac{1}{x+\mathrm{i}\eta} = P\left(\frac{1}{x}\right)-\mathrm{i}\pi\delta(x) , \tag{9.27}$$

where P denotes the principle value of the integral. We obtain

$$\epsilon(\boldsymbol{q},\omega) = \epsilon_1(\boldsymbol{q},\omega)+\mathrm{i}\epsilon_2(\boldsymbol{q},\omega) \tag{9.28}$$

with the real part

$$\epsilon_1(\boldsymbol{q},\omega) = 1+\sum_i \frac{4\pi e_i^2}{m_i q^2} P\int \mathrm{d}^3v \frac{f_i^{(0)}(\boldsymbol{v})-f_i^{(0)}(\boldsymbol{v}-\hbar\boldsymbol{q}/m_i)}{\omega-\boldsymbol{q}\cdot\boldsymbol{v}+\hbar^2q^2/2m_i} \tag{9.29a}$$

and the imaginary part

$$\begin{aligned}\epsilon_2(\boldsymbol{q},\omega) &= -\sum_i \frac{4\pi^2 e_i^2}{m_i q^2}\int \mathrm{d}^3v\left[f_i^{(0)}(\boldsymbol{v})-f_i^{(0)}\left(\boldsymbol{v}-\frac{\hbar\boldsymbol{q}}{m_i}\right)\right] \\ &\quad\times\delta\left(\omega-\boldsymbol{q}\cdot\boldsymbol{v}+\frac{\hbar^2q^2}{2m_i}\right) .\end{aligned} \tag{9.29b}$$

This prescription regularizes the integral at the zeros of the denominator. Hence, in general the solutions of the equation $\epsilon(\boldsymbol{q},\omega)=0$ are complex valued. As a consequence the plasma waves either increase exponentially or decrease exponentially. With more advanced methods it can be shown that the solutions ω have a negative imaginary part and are thus damped exponentially if $f_i^{(0)}(\boldsymbol{v})$ is a monotonous decreasing function in $v=|\boldsymbol{v}|$. This condition is always realized in thermal equilibrium because then $f(\boldsymbol{v})\sim \mathrm{e}^{-v^2}$. If the plasma is removed far from equilibrium, solutions with a positive imaginary part may appear and the perturbations increase exponentially. In this way one obtains *plasma instabilities*, which have also been observed experimentally.

9.1 Plasmons and Phonons

After these introductory remarks we will now study in more detail the importance of the quasiparticle concept for a plasma. We neglect quantum corrections to the frequency and employ the classical dielectric function (9.26).

Furthermore, for the equilibrium distribution functions of the electrons and ions we assume Fermi functions:

$$f_i^{(0)}(v) = \left\{ \begin{array}{ll} \dfrac{3}{4\pi v_{F_i}^3} & \text{for} \quad v < v_{F_i} \\ 0 & \text{for} \quad v > v_{F_i} \end{array} \right\} = \frac{3}{4\pi \nu_{F_i}^3} \Theta(v - v_{F_i}) \,, \tag{9.30}$$

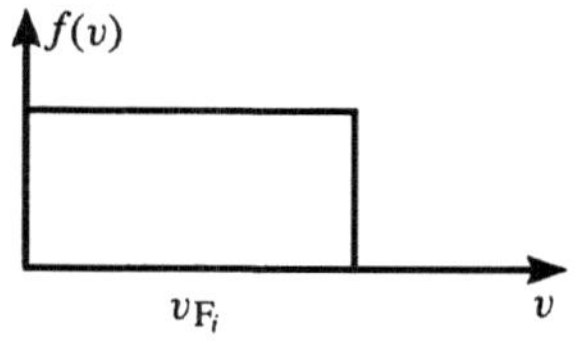

Fig. 9.1. Fermi distribution function of the electrons or the ions

where v_{F_i} represents the Fermi velocity of the particle species i. It is determined from

$$\rho = \frac{4\pi}{3} k_{F_i}^3 = \frac{4\pi}{3} \frac{p_{F_i}^3}{\hbar^3} = \frac{4\pi}{3} \frac{m_i^3 v_{F_i}^3}{\hbar^3} \,,$$

$$v_{F_i} = \frac{\hbar}{m_i} \left(\frac{3\rho}{4\pi} \right)^{1/3} \,, \tag{9.31}$$

where ρ represents the particle density, which is identical for electrons and ions because of homogeneity. Since we can rewrite $F_i^{(0)}(v)$ according to (9.30), $f_i^{(0)} = (3\rho/4\pi v_{F_i}^3)\Theta(v - v_{F_i})$, where Θ denotes the step function. The derivative of $f_i^{(0)}$ becomes

$$\boldsymbol{q} \cdot \frac{\partial f_i^{(0)}}{\partial \boldsymbol{v}} = -\frac{\boldsymbol{q} \cdot \boldsymbol{v}}{v} \left(\frac{3\rho}{4\pi v_{F_i}^3} \right) \delta(v - v_{F_i}) \,. \tag{9.32}$$

For convenience let us introduce the abbreviations

$$z_i = \frac{\omega}{q v_{F_i}} \,, \qquad q = |\boldsymbol{q}| \,,$$
$$\omega_{\mathrm{p}i} = \left(\frac{4\pi \rho e_i^2}{m_i} \right)^{1/2} \,, \tag{9.33}$$
$$u(z) = \left\{ \begin{array}{ll} 1 & \text{for} \quad z < 1 \\ 0 & \text{for} \quad z > 1 \end{array} \right\} = \Theta(z - 1) \,.$$

Here $\omega_{\mathrm{p}i}$ (i = electron or ion) are characteristic frequencies, which, as we shall see, characterize the plasma oscillations to zero order. This enables us to perform the integration (9.26) and we obtain

$$\epsilon(\boldsymbol{q}, \omega) = 1 + \sum_i \frac{4\pi e_i^2}{m_i q^2} \int \mathrm{d}^3 v \, \frac{\boldsymbol{q} \cdot \partial f_i^{(0)} / \partial \boldsymbol{v}}{\omega - \boldsymbol{q} \cdot \boldsymbol{v} + \mathrm{i}\eta}$$
$$= 1 - \sum_i \frac{4\pi e_i^2}{m_i q^2} \left(\frac{3\rho}{4\pi v_{F_i}^3} \right) \int \mathrm{d}^3 v \, \frac{\boldsymbol{q} \cdot \boldsymbol{v}}{v} \frac{\delta(v - v_{F_i})}{\omega - \boldsymbol{q} \cdot \boldsymbol{v} + \mathrm{i}\eta} \,. \tag{9.34}$$

We use (9.27) and arrive at (compare with Exercise 9.3)

$$\epsilon(\boldsymbol{q}, \omega) = 1 + \sum_i \frac{3}{2} \frac{\omega_{\mathrm{p}i}^2}{q^2 v_{F_i}^2} \left(2 - z_i \ln \left| \frac{1 + z_i}{1 - z_i} \right| + \mathrm{i}\pi z_i u(z_i) \right) \,. \tag{9.35}$$

In order to search for the zeros we assume $z_{\mathrm{el}} \gg 1$, i.e. $\omega \gg q v_{F_{\mathrm{el}}}$ for the electrons; for the ions this holds all the more because $v_{F_{\mathrm{ion}}} = v_{F_{\mathrm{el}}}(m_{\mathrm{el}}/m_{\mathrm{ion}})$, i.e. $v_{F_{\mathrm{ion}}} \ll v_{F_{\mathrm{el}}}$.

Furthermore,

$$2 - z \ln \left| \frac{1 + z}{1 - z} \right| \approx -\frac{2}{3} \frac{1}{z^2} - \frac{2}{5} \frac{1}{z^4} \qquad \text{for } z \gg 1 \,. \tag{9.36}$$

We insert this into (9.35) and obtain

$$\epsilon(\boldsymbol{q},\omega) \approx 1-\frac{\omega_{\mathrm{P_{el}}}^2}{\omega^2}\left(1-\frac{m_{\mathrm{el}}}{m_{\mathrm{ion}}}\right)-\frac{3}{5}\frac{\omega_{\mathrm{P_{el}}}^2 v_{\mathrm{F_{el}}}^2(1+m_{\mathrm{el}}^3/m_{\mathrm{ion}}^3)q^2}{\omega^4} . \tag{9.37}$$

In order to find the eigenmodes of the plasma we have – according to (9.24) – to set $\epsilon(\boldsymbol{q},\omega)=0$ and solve approximately for ω^2:

$$\omega^2 \approx \omega_{\mathrm{P_{el}}}^2\left(1-\frac{m_{\mathrm{el}}}{m_{\mathrm{ion}}}\right)+\frac{3}{5}v_{\mathrm{F_{el}}}^2 q^2 . \tag{9.38}$$

These oscillations are known as *plasma oscillations.* The corresponding quanta are called *plasmons.* Their frequency is approximately constant and equal to the plasma frequency $\omega_{\mathrm{P_{el}}}$. The recoil correction is negligible ($m_{\mathrm{el}}/m_{\mathrm{ion}} \ll 1$). The correction term proportional to q^2 the frequency to be momentum dependent and causes *dispersion.* The plasma oscillations (9.38) are not damped because the imaginary part of (9.35) proportional to $u(z)$ vanishes because $z \gg 1$.

Yet another solution for $\epsilon(\boldsymbol{q},\omega)=0$ exists with $z_{\mathrm{ion}} \gg 1$ but $z_{\mathrm{el}} \ll 1$; this is possible because $z_{\mathrm{el}}/z_{\mathrm{ion}} = m_{\mathrm{el}}/m_{\mathrm{ion}}$. However, the solution has a completely different physical interpretation. At first we use the approximations

$$\begin{aligned} 2-z_{\mathrm{ion}}\ln\left|\frac{1+z_{\mathrm{ion}}}{1-z_{\mathrm{ion}}}\right| &\approx -\frac{2}{3z_{\mathrm{ion}}^2} , \\ 2-z_{\mathrm{el}}\ln\left|\frac{1+z_{\mathrm{el}}}{1-z_{\mathrm{el}}}\right| &\approx 2 \end{aligned} \tag{9.39}$$

and deduce from (9.35) that

$$\epsilon(\boldsymbol{q},\omega) = 1-\frac{\omega_{\mathrm{P_{ion}}}^2}{\omega^2}+3\frac{\omega_{\mathrm{P_{el}}}^2}{q^2 v_{\mathrm{F_{el}}}^2}+\mathrm{i}\frac{3\pi}{2}\frac{\omega_{\mathrm{P_{el}}}^2\omega}{q^2 v_{\mathrm{F_{el}}}^3} . \tag{9.40}$$

We set $\epsilon(\boldsymbol{q},\omega)=0$, solve it, taking into account the outlined approximations, and get

$$\omega = \overline{\omega}\left(1-\frac{\mathrm{i}\overline{\omega}}{2qv_{\mathrm{F_{el}}}}\frac{(\omega_{\mathrm{P_{el}}}/qv_{\mathrm{F_{el}}})^2}{1+3\omega_{\mathrm{P_{el}}}^2/q^2v_{\mathrm{F_{el}}}^2}\right) , \tag{9.41}$$

where

$$\overline{\omega} = \frac{\omega_{\mathrm{P_{ion}}}}{\sqrt{1+3\omega_{\mathrm{P_{el}}}^2/q^2v_{\mathrm{F_{el}}}^2}} . \tag{9.42}$$

For long waves (small q) we have

$$\overline{\omega} = \frac{v_{\mathrm{F_{el}}}}{\sqrt{3}}\left(\frac{\omega_{\mathrm{P_{ion}}}}{\omega_{\mathrm{P_{el}}}}\right)q = \frac{v_{\mathrm{F_{el}}}}{\sqrt{3}}\left(\frac{m_{\mathrm{el}}}{m_{\mathrm{ion}}}\right)^{1/2} q$$

and hence

$$\omega \approx \overline{\omega} = \frac{v_{\mathrm{F_{el}}}}{\sqrt{3}}\left(\frac{m_{\mathrm{el}}}{m_{\mathrm{ion}}}\right)^{1/2} q . \tag{9.43}$$

This is the typical relation for sound waves (proportionality between frequency and momentum), i.e.

$$\omega = c_s q \,,$$

where the *speed of sound* c_s is given by

$$c_s = \frac{v_{F_{el}}}{\sqrt{3}} \left(\frac{m_{el}}{m_{ion}} \right)^{1/2} . \qquad (9.44)$$

The experimental results for longitudinal sound waves in metals are in good agreement with (9.44); this holds particularly for alkali metals where a 20% agreement is found. The oscillations associated with these quanta are denoted as phonons. For large wavelenghts we obtain from (9.41)

$$\omega \approx \overline{\omega} \left(1 - \frac{\mathrm{i}}{3\sqrt{3}} \left(\frac{m_{el}}{m_{ion}} \right)^{1/2} \right) , \qquad (9.45)$$

i.e. a small negative imaginary part. Here the phonons are weakly damped. In general the weak damping of waves in plasmas is called *Landau damping.*[2]

EXERCISE

9.1 Electrostatic Potential of a Charge in a Plasma

Problem. Use (9.35) to determine the electrostatic potential of a charge Q in a plasma.

Solution. Without a plasma the potential would be Q/r and the corresponding Fourier transform would then be

$$\varphi(\boldsymbol{q}) = \frac{4\pi Q}{V q^2} \qquad (\mathrm{V} = L^3 = \text{volume}) .$$

In the presence of a dielectric medium the potential becomes

$$\varphi(\boldsymbol{q}) \rightarrow \varphi(\boldsymbol{q}, \omega) = \frac{\varphi(\boldsymbol{q})}{\epsilon(\boldsymbol{q}, \omega)} .$$

Since the charge Q is stationary we have $\omega = 0$, i.e.

$$\varphi(q) = \frac{4\pi Q}{V q^2 \epsilon(q, 0)} ,$$

where according to (9.35)

$$\epsilon(q, 0) = 1 + \frac{1}{q^2 \lambda^2}$$

with

$$\frac{1}{\lambda^2} = \sum_i 3 \frac{\omega_{p_i}^2}{v_{F_i}} .$$

[2] See e.g. A.L. Fetter, J.D. Walecka: *Quantum Theory of Many Particle Systems* (McGraw Hill, New York 1971); U.C. Kittel: *Quantum Theory of Solids* (Wiley, New York 1966).

Exercise 9.1.

We invert the Fourier transformation and get

$$\begin{aligned}\varphi(\boldsymbol{r}) &= \frac{V}{(2\pi)^3}\int d^3q\,\varphi(q)e^{iq\cdot x}\\ &= \frac{V}{(2\pi)^3}\int_0^\infty q^2\,dq\int_0^\pi \sin\vartheta_q\,d\vartheta_q\int_0^{2\pi} d\varphi_q\varphi(q)\,e^{iqr\cos\vartheta_q}\,.\end{aligned}$$

Since the integrand is independent of φ_q the integration over φ_q can be performed:

$$\varphi(r) = \frac{V}{(2\pi)^3}\int_0^\infty q^2\,dq\int_0^\pi \sin\vartheta_q\,d\vartheta_q\,\varphi(q)\,e^{iqr\cos\vartheta_q}\,.$$

Substitution of $v = qr\cos\vartheta_q$ and $dv = qv(-\sin\vartheta_q)\,d\vartheta_q$ yields

$$\begin{aligned}\varphi(r) &= \frac{V}{4\pi^2}\int q^2\,dq\,\varphi(q)\int(-dv)\frac{e^{iv}}{qr}\\ &= \frac{V}{4\pi^2}\int q^2\,dq\,\varphi(q)\frac{1}{iqr}(e^{-iqr}-e^{iqr})\\ &= \frac{V}{2\pi^2}\int \frac{q^2}{qr}\,dq\,\varphi(q)\sin qr\,.\end{aligned}$$

Insertion of $\varphi(q)$ gives

$$\varphi(r) = \frac{2Q}{\pi}\int_0^\infty \frac{q}{r}\frac{dq\sin qr}{q^2+(1/\lambda)^2}\,.$$

This integral can be solved analytically:

$$\int_0^\infty \frac{x\sin bx}{a^2+x^2}\,dx = \frac{\pi}{2}e^{-(ab)}\,,$$

to give

$$\varphi(r) = \frac{Q}{r}e^{-r/\lambda}\,.$$

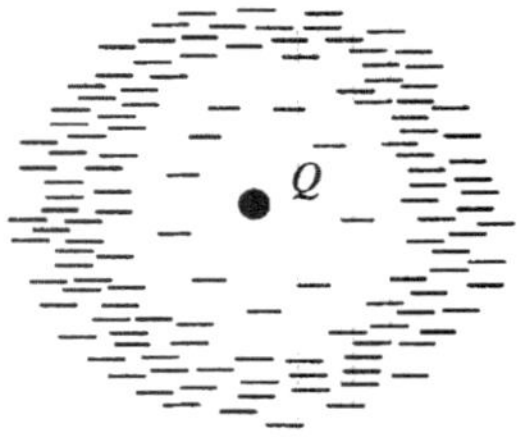

Fig. 9.2. Illustration of Debye screening of a charge Q in a plasma

Within the distance λ the charge is screened. If instead of the Fermi distribution a Maxwell–Boltzmann distribution is used for the derivation of (9.35) we arrive at

$$\lambda = \left(\frac{k_B T}{4\pi\rho e^2}\right)\,.$$

This quantity is denoted as the *Debye screening length*. It characterizes the screening of a charge by a plasma charge cloud which is built up around the charge Q (see Fig. 9.2 and Exercise 9.2).

EXERCISE

9.2 Classical Dielectric Function

Problem. Use (9.26) to determine the classical dielectric function for particles obeying a Maxwell distribution.

Exercise 9.2.

Solution. The normalized Maxwell distribution is given by

$$f^{(0)}(v) = \frac{N}{\pi^{3/2}\alpha^3}\,\mathrm{e}^{-v^2/\alpha^2} .$$

$\alpha = (2k_\mathrm{B}T/m)^{1/2}$ represents a thermal velocity. Furthermore, according to (9.26)

$$\epsilon_\text{class} = 1 + \sum_i \frac{4\pi e_i^2}{m_i q^2}\int \mathrm{d}^3v \frac{\boldsymbol{q}\cdot(\partial f_i^{(0)}/\partial \boldsymbol{v})}{\omega - \boldsymbol{q}\cdot\boldsymbol{v}} .$$

We consider only one-particle species and choose $\boldsymbol{q}$ along the z direction: $\boldsymbol{q} = q\boldsymbol{e}_z$. Then the x–y integration can be easily performed.

Insertion of the Maxwell distribution yields, for $s = 0$,

$$\epsilon(\boldsymbol{q},\omega) = 1 - \frac{8\pi e^2\rho}{mq^2\pi^{3/2}\alpha^5}\int \mathrm{d}^3v \frac{\boldsymbol{q}\cdot\boldsymbol{v}\,\mathrm{e}^{-v^2/\alpha^2}}{\omega - \boldsymbol{q}\cdot\boldsymbol{v}} .$$

The choice of cylindrical coordinates leads us to

$$\begin{aligned}
\epsilon(\boldsymbol{q},\omega) &= 1 - \frac{8\pi e^2\rho}{mq^2\pi^{3/2}\alpha^5}\int_0^{2\pi}\mathrm{d}v_\varphi \int_0^\infty v_\varrho\,\mathrm{d}v_\varrho \\
&\quad\times \int_{-\infty}^{\infty}\mathrm{d}v_z \frac{qv_z\,\mathrm{e}^{-v_z^2/\alpha^2}\,\mathrm{e}^{-v_\varrho^2/\alpha^2}}{\omega - q\cdot v_z} \\
&= 1 - \frac{16\pi e^2\rho}{mq^2\pi^{1/2}\alpha^5}\int_0^\infty v_\varrho\,\mathrm{d}v_\varrho\,\mathrm{e}^{-v_\varrho^2/\alpha^2}\int_{-\infty}^{\infty}\mathrm{d}v_z \frac{qv_z\,\mathrm{e}^{-v_z^2/\alpha^2}}{\omega - q\cdot v_z} .
\end{aligned}$$

The elementary integral over v_ϱ is of the type

$$\int_0^\infty x^{2n+1}\,\mathrm{e}^{-px^2} = \frac{n!}{2p^{n+1}} \qquad (p > 0) ,$$

and therefore

$$\epsilon(\boldsymbol{q},\omega) = 1 - \frac{8\pi e^2\rho}{mq^2\pi^{1/2}\alpha^3}\int_{-\infty}^{\infty}\mathrm{d}v_z \frac{qv_z\,\mathrm{e}^{-v_z^2/\alpha^2}}{\omega - qv_z} .$$

Adding a zero in the form of adding and subtracting an identical term yields

$$\begin{aligned}
\epsilon(\boldsymbol{q},\omega) &= 1 - \frac{8\pi e^2\rho}{mq^2\pi^{1/2}\alpha^3}\int_{-\infty}^{\infty}\mathrm{d}v_z \frac{(qv_z - \omega + \omega)}{\omega - qv_z}\,\mathrm{e}^{-v_z^2/\alpha^2} \\
&= 1 - \frac{8\pi e^2\rho}{mq^2\pi^{1/2}\alpha^3}\int_{-\infty}^{\infty}\mathrm{d}v_z\left(-\,\mathrm{e}^{-v_z^2/\alpha^2} + \frac{\omega\,\mathrm{e}^{-v_z^2/\alpha^2}}{\omega - qv_z}\right) \\
&= 1 - \frac{8\pi e^2\rho}{mq^2\pi^{1/2}\alpha^3}\left(\int_{-\infty}^{\infty}\frac{\omega\,\mathrm{e}^{-v_z^2/\alpha^2}\,\mathrm{d}v_z}{\omega - qv_z} - \alpha\sqrt{\pi}\right) \\
&= 1 + \frac{1}{\lambda^2q^2} - \frac{1}{\lambda^2q^2}\left(\frac{\omega}{q\alpha}\right)Z\left(\frac{\omega}{q\alpha}\right) ,
\end{aligned}$$

where

$$\lambda = \left(\frac{m\alpha^2}{8\pi\rho e^2}\right)^{1/2} = \left(\frac{k_\mathrm{B}T}{4\pi\rho e^2}\right)^{1/2}$$

Exercise 9.2.

and

$$Z(z) = \int_{-\infty}^{\infty} \mathrm{d}x \frac{\mathrm{e}^{-x^2}}{z-x}$$

represents the so called *Fried–Conte function*,[3] which can be tabulated numerically.

EXERCISE

9.3 Details of Calculating the Dielectric Function $\varepsilon(\boldsymbol{q},\omega)$

Problem. Use expression (9.34) to verify the result (9.35) for $\epsilon(\boldsymbol{q},\omega)$.

Solution. According to (9.34) we have

$$\epsilon(\boldsymbol{q},\omega) = 1 - \sum_i \frac{4\pi e_i^2}{m_i q^2}\left(\frac{3\rho}{4\pi v_{\mathrm{F}_i}^3}\right) \underbrace{\int \mathrm{d}^3 v \frac{\boldsymbol{q}\cdot\boldsymbol{v}}{v} \frac{\delta(v - v_{\mathrm{F}_i})}{\omega - \boldsymbol{q}\cdot\boldsymbol{v} + \mathrm{i}\eta}}_{=I} .$$

We now determine I:

$$\begin{aligned} I &= \int_0^{2\pi} d\varphi \int_0^{\pi} \sin\vartheta\, \mathrm{d}\vartheta \int_0^{\infty} v^2\, \mathrm{d}v \frac{qv\cos\vartheta}{v} \frac{\delta(v - v_{\mathrm{F}_i})}{\omega - qv\cos\vartheta + \mathrm{i}\eta} \\ &= 2\pi \int_0^{\pi} \sin\vartheta\, \mathrm{d}\vartheta\, q v_{\mathrm{F}_i}^2 \frac{\cos\vartheta}{\omega - q v_{\mathrm{F}_i}\cos\vartheta + \mathrm{i}\eta} . \end{aligned}$$

The substitution $u = \cos\vartheta$, $\mathrm{d}u = -\sin\vartheta\, \mathrm{d}\vartheta$ yields

$$I = 2\pi q v_{\mathrm{F}_i}^2 \int_{-1}^{+1} \frac{u\, \mathrm{d}u}{\omega - q v_{\mathrm{F}_i} u + \mathrm{i}\eta} .$$

Making use of (9.27) for the solution of the integral leads to

$$\begin{aligned} I &= 2\pi q v_{\mathrm{F}_i}^2 \left(P \int_{-1}^{+1} \frac{u\, \mathrm{d}u}{\omega - q v_{\mathrm{F}_i} u} - \mathrm{i}\pi \int u\, \mathrm{d}u\, \delta(\omega - q v_{\mathrm{F}_i} u) \right) \\ &= 2\pi q v_{\mathrm{F}_i}^2 \left\{ P \int_{-1}^{+1} \frac{u\, \mathrm{d}u}{\omega - q v_{\mathrm{F}_i} u} - \mathrm{i}\pi \int u\, \mathrm{d}u\, \delta\left[\left(u - \frac{\omega}{q v_{\mathrm{F}_i}}\right) q v_{\mathrm{F}_i}\right] \right\} \\ &= 2\pi q v_{\mathrm{F}_i}^2 \left(P \int_{-1}^{+1} \frac{u\, \mathrm{d}u}{\omega - q v_{\mathrm{F}_i} u} - \mathrm{i}\pi \frac{\omega}{(q v_{\mathrm{F}_i})^2} \right) . \end{aligned}$$

With the integral

$$\int \frac{x\, \mathrm{d}x}{a + bx} = \frac{x}{b} - \frac{a}{b^2} \ln x$$

it follows that

[3] B.D. Fried, C. Conte: *The Plasma Dispersion Function* (Academic Press, New York 1961).

Exercise 9.3.

$$\begin{aligned} I &= 2\pi q v_{F_i}^2 \left\{ \left[\frac{u}{(-q v_{F_i})} - \frac{\omega}{(q v_{F_i})^2} \ln |\omega - q v_{F_i} u| \right]_{-1}^{+1} - i\pi \frac{\omega}{(q v_{F_i})^2} \right\} \\ &= 2\pi q v_{F_i}^2 \left(-\frac{2}{q v_{F_i}} - \frac{\omega}{q v_{F_i}} \ln \left| \frac{\omega - q v_{F_i}}{\omega + q v_{F_i}} \right| - i\pi \frac{\omega}{(q v_{F_i})^2} \right) . \end{aligned}$$

Using the substitutions

$$\begin{aligned} \frac{\omega}{v_{F_i}} &= Z_i \, , \\ \frac{4\pi \rho e_i^2}{m_i} &= \omega_{p_i}^2 \, , \\ u(z_i) &= \begin{Bmatrix} 1 & \text{for} & z_i < 1 \\ 0 & \text{for} & z_i > 1 \end{Bmatrix} = \Theta(z-1) \end{aligned}$$

we obtain, for (9.34), the desired result:

$$\begin{aligned} \epsilon(\boldsymbol{q}, \omega) &= 1 - \sum_i \frac{3}{2} \frac{\omega_{p_i}^2}{q^2 v_{F_i}^2} \left(-2 - z_i \ln \left| \frac{\omega - \omega/z_i}{\omega + \omega/z_i} \right| + i\pi z_i u(z_i) \right) \\ &= 1 + \sum_i \frac{3}{2} \frac{\omega_{p_i}^2}{q^2 v_{F_i}^2} \left(2 - z_i \ln \left| \frac{z_i + 1}{z_i - 1} \right| + i\pi z_i u(z_i) \right) . \end{aligned}$$

10. Basics of Quantum Statistics

In this chapter we present some basics on the quantum-theoretical description of macroscopic many-particle systems. A detailed disclosure of thermodynamics and quantum statistics is out of reach for this book.[1] Here we will first introduce the concept of a *quantum-theoretical ensemble* and the so-called *statistical operator*, or *density operator* in short, of a many-particle system, and discuss its properties. We will see that with this approach not only can we calculate expectation values of observables, but we are also led to new notions such as entropy. Looking back on the chapter on quantum gases we will discuss in detail the density operator of a canonical ensemble.

10.1 Concept of Quantum Statistics and the Notion of Entropy

In classical mechanics as well as in quantum theory the dynamics of a physical many-particle system is described by differential equations for the time evolution of those quantities assigned to observables. Hamilton's equations for position and momentum coordinates are only one example in classical mechanics. In quantum theory the time evolution of observables is governed by Heisenberg's equations of motion for the assigned operators, or, equivalently, by the Schrödinger equation of the state vector. Knowledge of all initial conditions is crucial for the fixing of the dynamics of the system. In classical mechanics, for example, $6N$ space and momentum coordinates have to be known at an initial time for an N-particle system. The quantum-theoretical state vector of a system is fixed only by a measurement of a complete set of commutable observables at an initial time. Once all possible initial conditions are given in principle, so that the information about a system is completely known, the state vector of the system is unmistakably fixed. We say that the system is in a *pure state*. However, in practice a pure state cannot be realized experimentally, neither in the classical nor in the quantum-mechanical case. The reading of initial positions and momenta of classical systems and the measuring of all observables, which characterize the state vector of a quantum system, always come with uncertainties. Errors that occur in the measurement for the initial conditions cause a "wrong" state vector to be fixed. Thus, in

[1] We refer to W. Greiner, L. Neise, H. Stöcker: *Thermodynamics and Statistical Mechanics*, 2nd ed. (Springer, Berlin, Heidelberg 1994).

W. Greiner, *Quantum Mechanics*

principle the information about the system is incomplete. As a consequence of these uncertainties with respect to initial conditions we always have to deal with a whole group of pure states; all of them represent possible candidates for the description of the "true" state of the system. Anyhow, for a macroscopic system with $N \approx 10^{23}$ degrees of freedom it is impossible to fix a pure state; in addition, all this information would be extremely useless. Therefore we will attempt to describe the state of a many-particle system with a few global observables such as the total energy, pressure, and so on.

If one pure state cannot be given for a system anyhow, so that only a group of possible pure states are realized, we speak of a *mixture*. The multitude of pure states of the system realized under defined conditions is denoted as an *ensemble*. A quantum-mechanical description of a many-particle system is given in the form of *mixed states* that give information about the probabilities of single pure states realized in the ensemble. It is the task of quantum statistics to record the properties of a quantum-theoretical mixture.

10.2 Density Operator of a Many-Particle State

We have to develop a mathematical formalism that takes into account the incomplete information about the state vector of a quantum-mechanical many-particle system.

We denote the energy eigenstates of a many-particle system as $|\Psi^n_{E(n)}\rangle$, which are the solution of the stationary Schrödinger equation

$$\hat{H}|\Psi^n_{E(n)}\rangle = E(n)|\Psi^n_{E(n)}\rangle \,. \tag{10.1}$$

These many-body state vectors are enumerated by the index n, which abbreviates all required quantum numbers necessary to identify a special state, the corresponding energy of which is $E(n)$. In some cases n will also contain continuous quantum numbers like the total momentum of the system etc., but for simplicity we will take n here as one discrete variable. Complications due to the continuous character of some quantum numbers or due to the continuous part of the energy spectrum can then be easily taken care of later on. The state vectors $|\Psi^n_{E(n)}\rangle$ are assumed to form a complete orthonormal basis of Hilbert space, characterized by the completeness and orthonormality relations

$$\langle\Psi^n_{E(n)}|\Psi^m_{E(m)}\rangle = \delta_{n,m} \quad \text{and} \quad \sum_n |\Psi^n_{E(n)}\rangle\langle\Psi^n_{E(n)}| = \hat{\mathbf{1}} \,, \tag{10.2}$$

respectively. In the last relation the operator

$$\hat{\rho}_{|\Psi^n\rangle} = |\Psi^n_{E(n)}\rangle\langle\Psi^n_{E(n)}| \,, \tag{10.3}$$

denotes the projector on the particular state number n, which denotes the property

$$\hat{\rho}_{|\Psi^n\rangle}\hat{\rho}_{|\Psi^n\rangle} = \hat{\rho}_{|\Psi^n\rangle} \,, \tag{10.4}$$

since $|\Psi^n\rangle\langle\Psi^n|\Psi^n\rangle\langle\Psi^n| = |\Psi^n\rangle\langle\Psi^n|$ according to (10.2). If we now assume that the system is exactly in the state number n, we say the system is in a pure state, and the projector $\hat{\rho}_{|\Psi^n\rangle}$ is called the density matrix of this pure state. However, as stated above, one usually cannot determine exactly in which state n the system is. For example, if we allow an uncertainty of ΔE for the measurement of the energy E of the system there will be very many state vectors $|\Psi^n_{E(n)}\rangle$ which fulfill $E < E(n) < E + \Delta E$. The reason is that in a macroscopic system with many particles ($N \to \infty$, $V \to \infty$, $N/V = \text{const}$) the energy levels are very close together. We therefore do not have the information about exactly which state vector the system is in. This lack of information about the exact state vector can be accounted for if we allow each state vector to be assumed with a certain probability P^n. We write

$$\langle \hat{O} \rangle = \langle \Psi^n_{E(n)} | \hat{O} | \Psi^n_{E(n)} \rangle P^n . \tag{10.5}$$

The probabilities P^n shall be normalized according to

$$\sum_n P^n = 1 . \tag{10.6}$$

This average can be rewritten, by inserting the completeness relation from (10.2), as

$$\begin{aligned} \langle \hat{O} \rangle &= \sum_{nm} \langle \Psi^n_{E(n)} | \Psi^m_{E(m)} \rangle \langle \Psi^m_{E(m)} | \hat{O} | \Psi^n_{E(n)} \rangle P^n \\ &= \sum_m \langle \Psi^m_{E(m)} | \hat{O} \left(\sum_n | \Psi^n_{E(n)} \rangle P^n \langle \Psi^n_{E(n)} | \right) | \Psi^m_{E(m)} \rangle . \end{aligned}$$

In the last step we have rearranged the bra and ket vectors such that we can furthermore write

$$\langle \hat{O} \rangle_n = \mathrm{Tr}(\hat{O}\hat{\rho}) , \tag{10.7}$$

where the density matrix $\hat{\rho}$ is defined as

$$\hat{\rho} = \sum_n | \Psi^n_{E(n)} \rangle P^n \langle \Psi^n_{E(n)} | , \tag{10.8}$$

i.e. as a sum of all possible projection operators weighted by the probability P^n with which the actual physical system can be in state number n. We then say the system is in a mixed state. For example, if we know that the energy of the system is in the range E, $E + \Delta E$, according to its preparation or due to a measurement, then we know

$$\begin{aligned} P^n &= 0 \quad \text{for} \quad E(n) < E \quad \text{or} \quad E(n) > E + \Delta E \\ P^n &\neq 0 \quad \text{for} \quad E < E(n) < E + \Delta E . \end{aligned}$$

However, the actual values of the probabilities P^n with $E < E(n) < E + \Delta E$ are still not known to us. Here we make an important assumption that cannot be justified from first principles. We assume that all state vectors that are compatible with our knowledge about the system (here, energy in the range E, $E + \Delta E$) are equally probable. This corresponds exactly to the assumption of equal probabilities for the six possible outcomes in an experiment of tossing a dice. This assumption can of course be wrong, if the dice is not ideal, but it is the most probable assumption as long as we do not have any further information. Then we find in our example

$$P^n_{\text{mc}} = \begin{cases} \dfrac{1}{\Omega(E, \Delta E)} & \text{if } E < E(n) < E + \Delta E\,, \\ 0 & \text{otherwise}\,, \end{cases} \tag{10.9}$$

where $\Omega(E, \Delta E)$ is the number of energy eigenstates in the interval E, $E+\Delta E$, corresponding to the six possible outcomes in a dice-throwing experiment. In the case of a continuous energy spectrum one needs to know the so-called level density or density of states $g(E)$, i.e. the number of states per energy interval, and then one can calculate $\Omega(E, \Delta E) \approx g(E)\Delta E$, as long as $\Delta E \ll E$.

One calls the exact state vector of a physical system the microstate. On the other hand our knowledge about the systems defines the so-called macrostate. In our example the macrostate is given by the energy of the system and in most cases by the particle number and volume. The number $\Omega(E, N, V)$, which turns out to depend only weakly on ΔE as long as $\Delta E \ll E$, is then the number of microstates of the system, each of which is represented by a single pure state $|\Psi^n_{E(n)}\rangle$, which are compatible with the given macrostate. Our fundamental assumption is that all microstates within the small energy interval are equally probable. And this leads to (10.9), which is called the microcanonical probability distribution. We denote this by the subscript "mc".

If we prepare a large number of systems, say $\mathcal{N}$, with the same Hamiltonian, energy interval, particle number, and volume, then an exact measurement of all microstates would yield a certain number $\mathcal{N}_n$ of systems that are exactly on microstate (or pure state vector) number n. One generally calls such a large number of identical systems with the same macrostate an ensemble. Up to now we have dealt with a macrostate defined by the energy, particle number, and volume of the system and the corresponding ensemble is called the microcanonical ensemble. If the number of systems is very large, $\mathcal{N} \to \infty$, we expect that

$$\frac{\mathcal{N}_n}{\mathcal{N}} \to P^n \quad \text{for} \quad \mathcal{N} \to \infty\,, \tag{10.10}$$

i.e. that the relative frequency of microstate n in the ensemble tends to the probabilities P^n. Therefore the averages (10.5) and (10.7) are also called the ensemble averages of the observable $\hat{O}$. Now mathematical statistics shows that there exists a unique function of the probabilities which is a measure for the predictability of a probability experiment: the so-called uncertainty function. It is defined as

$$\mathcal{H}(p_1, p_2, \ldots) = -\sum_i p_i \ln p_i\,, \tag{10.11}$$

where p_i is the probability of the possible outcome i and the sum runs over all of them. This function is exactly zero if and only if there is no uncertainty about the outcome of the experiment, i.e. if one of the p_i is 1 and all others are zero. This means the outcome of the experiment is then only the special event i. Note that $1 \ln 1 = 0$ and $\lim_{x\to\infty} x \ln x = 0$. On the other hand $\mathcal{H}$ is constructed such that the case of equal probabilities has the maximum uncertainty. For an event with Ω possible outcomes we obtain $p_i = 1/\Omega$ and

$$\mathcal{H} = -\sum_{i=1}^{\Omega} \frac{1}{\Omega} \ln \Omega = -\Omega \frac{1}{\Omega} \ln \Omega = \ln \Omega . \tag{10.12}$$

The maximum uncertainty is thus given by the logarithm of the number of possible outcomes of the experiment.

We show now that the constant probability distribution (all p_i equal) is indeed the distribution with the highest unpredictability. We therefore calculate the variation of $\mathcal{H}$ with a variation of the p_i, and for $\mathcal{H}$ to have a maximum this must vanish:

$$\delta\mathcal{H} = -\sum_{i=1} \delta p_i \ln p_i - \sum_{i=1} \delta p_i = 0 . \tag{10.13}$$

However, the p_i are not independent of each other and from the normalization we have a constraint for the allowed variations δp_i:

$$\sum_{i=1} p_i = 1 \longrightarrow \sum_{i=1} \delta p_i = 0 . \tag{10.14}$$

If we multiply (10.14) with a Lagrange multiplier, α, and add the resulting equation to (10.13) we find

$$\sum_{i=1} \delta p_i (\ln p_i - 1 + \alpha) = 0 .$$

Now we can argue that the variations δp_i may be assumed to be independent of each other, such that each term of the sum must vanish itself if we account later on for the normalization condition by a suitable choice of α. Then we find

$$\ln p_i = 1 - \alpha \quad \text{or} \quad p_i = \mathrm{e}^{1-\alpha} . \tag{10.15}$$

The important fact is that the right-hand side of (10.15) is independent of i so that all p_i are equal and the normalization then yields $p_i = 1/\Omega$ for Ω possible outcomes. We have shown in this way that our assumption of equal probabilities in the microcanonical ensemble in a given small-energy shell corresponds to the assumption of maximum uncertainty.

Now, if the macrostate of the system is not specified by the energy but rather by the temperature, T, of the surroundings of the system (e.g. a heat bath) the system can assume all possible energy values and not only a small interval around E (we have assumed $\Delta E \ll E$!). The reason is that the heat bath is assumed to be very large compared to the system of interest ($E_{\text{bath}} \gg E$) and the exchange of energy between the heat bath and system therefore leads to fluctuations of the energy of the system. Fluctuations are generally deviations of an observable from its mean value. However, the mean value of the energy, i.e. of the Hamiltonian, is still fixed:

$$\begin{aligned} U = \langle \hat{H} \rangle &= \sum_n \langle \Psi^n_{E(n)} | \hat{H} | \Psi^n_{E(n)} \rangle P^n \\ &= \sum_n P^n E(n) . \end{aligned} \tag{10.16}$$

An ensemble of systems in which the macrostate is specified by a fixed mean value of the energy and fixed particle number and volume is called a canonical ensemble. The corresponding canonical probability distribution can now be calculated by requiring maximum uncertainty (10.13),

$$\delta\mathcal{H} = -\sum_n \delta P^n(\ln P^n + 1)\,, \tag{10.17}$$

where the normalization (10.14) restricts the variations of the P^n:

$$0 = -\sum_n \delta P^n\,. \tag{10.18}$$

Now we have the additional constraint (10.16), which leads to the variational condition

$$0 = -\sum_n \delta P^n E(n)\,. \tag{10.19}$$

If (10.18) is multiplied by a Lagrange multiplier α and (10.19) by $-\beta$, and if these two equations are added to (10.17), we obtain (the minus sign is arbitrary, but makes β positive)

$$-\sum_n \delta P^n + 1 - \alpha + \beta E(n)\,.$$

Here we can again assume all variations δP^n to be independent of each other, so that

$$P^n = c\,\mathrm{e}^{-\beta E(n)}\,,$$

where $c = \exp(\alpha - 1)$ can be determined from the normalization condition (10.14) which yields the canonical probability distribution

$$P^n_{\mathrm{c}} = \frac{\mathrm{e}^{-\beta E(n)}}{\sum_k \mathrm{e}^{-\beta E(k)}}\,. \tag{10.20}$$

Here the probabilities of the states drop exponentially with the energy. In Thermodynamics and Statistical Mechanics[2] it is shown that from (10.16) and the fact that the entropy of the system is given by

$$S = k_{\mathrm{B}}\mathcal{H} = -k_{\mathrm{B}}\sum_n P^n \ln P^n\,, \tag{10.21}$$

where k_{B} is Boltzmann's constant, one can identify the up to now unknown Lagrange parameter β to be

$$\beta = \frac{1}{k_{\mathrm{B}}T}\,, \tag{10.22}$$

where T is the temperature of the heat bath that fixes the mean energy of the system.

It is very interesting to rewrite the probabilities P^n as the matrix elements of the operator $\hat{\rho}$ defined in (10.8). From that follows

$$P^n_{\mathrm{c}} = \langle \Psi^n_{E(n)}|\hat{\rho}_{\mathrm{c}}|\Psi^n_{E(n)}\rangle\,. \tag{10.23}$$

[2] W. Greiner, L. Neise, H. Stöcker: *Thermodynamics and Statistical Mechanics*, 2nd ed. (Springer, New York 1997).

Now we see immediately that the P^n of (10.20) can be obtained as the expectation value of an abstract operator

$$\hat{\rho}_c = \frac{e^{-\beta \hat{H}}}{\mathrm{Tr}(e^{-\beta \hat{H}})}, \tag{10.24}$$

the canonical density operator. Here the trace can be rewritten in the basis of the $|\Psi^n_{E(n)}\rangle$ as

$$\mathrm{Tr}\left(e^{-\beta \hat{H}}\right) = \sum_n \langle \Psi^n_{E(n)}| e^{-\beta \hat{H}} |\Psi^n_{E(n)}\rangle = \sum_n e^{-\beta E(n)}, \tag{10.25}$$

and in the last step we used the fact that the abstract operator $e^{-\beta \hat{H}}$, which is to be understood as

$$\exp(-\beta \hat{H}) = \sum_{k=0}^{\infty} \frac{(-\beta)^k}{k!} \hat{H}^k,$$

can directly be replaced by an eigenvalue, since the $|\Psi^n_{E(n)}\rangle$ are eigenstates to $\hat{H}^k$ with the eigenvalues $[E(n)]^k$. In the canonical ensemble we have assumed that the macrostate of the system is defined by the temperature, volume, and particle number. However, in many systems one does not even know exactly the particle number. Consider, for example, vapor in equilibrium with its corresponding liquid phase. The actual particle number in the vapor will then no longer be a fixed quantity, rather the actual number of particles in the vapor will fluctuate around a mean value $\langle N \rangle$. The liquid phase acts here as a particle reservoir for the vapor in the same way as the heat bath in our canonical ensemble acts as an energy reservoir. If we are interested only in the properties of the vapor, disregarding its direct interaction with the liquid, we must take into account that we do not have knowledge about its actual particle number. We can only say that there will be an average particle number which must be calculated from the probabilities $P^{n,N}$ that define the probability of finding the system with exactly N particles in the N-particle state number n. We require the normalization

$$\sum_{N=0}^{\infty} \sum_n P^{n,N} = 1. \tag{10.26}$$

The mean particle number is then given as the average of the actual particle number, N,

$$\sum_{N=0}^{\infty} \sum_n N P^{n,N} = \langle N \rangle. \tag{10.27}$$

Furthermore, in most cases a particle reservoir acts also as a heat bath that fixes a certain temperature, so our macroscopic knowledge about the system is only the corresponding mean value, exactly as in the canonical ensemble:

$$\sum_{N=0}^{\infty} \sum_n E(n, N) P^{n,N} = \langle \hat{H} \rangle. \tag{10.28}$$

Here the energy eigenvalues must be numbered by two indices: N enumerates the different possibilities for no particle at all in the system ($N = 0$), one particle, two particles, and so on, whereas n still enumerates all energy eigenstates at given particle number. These eigenstates then also carry these two indices: $|\Psi^{n,N}_{E(n,N)}\rangle$. If we now again ask for that probability distribution, $P^{n,N}$, that maximizes our uncertainty about the system, we have to vary the uncertainty function (10.11), which now reads

$$-\sum_{N=0}^{\infty}\sum_{n} P^{n,N} \ln P^{n,N} = \mathcal{H}\,, \tag{10.29}$$

under the subsidiary conditions (10.26), (10.27), and (10.28), with respect to the up to now unknown probabilities. Completely analogously to the method for the canonical ensemble, this can be done by varying (10.26), (10.27), and (10.28) and multiplying the resulting variations by three unknown Lagrange multipliers, say α, $-\beta$, and γ, and adding these equations to the variation of (10.29). One obtains

$$-\sum_{N=0}^{\infty}\sum_{n} \delta P^{n,N} \left(\ln P^{n,N} + 1 - \alpha + \beta E(n,N) - \gamma N\right) = 0\,.$$

Again we can argue that each of the brackets must vanish separately if we fulfill the subsidiary conditions (10.26)–(10.28) later on by appropriate choice of the Lagrange multipliers. We find

$$P^{n,N}_{gc} = \frac{\exp\left[-\beta(E(n,N) - \mu N)\right]}{\sum_{N'=0}^{\infty}\sum_{n'} \exp\left[-\beta(E(n',N') - \mu N')\right]}\,, \tag{10.30}$$

where we have determined α from (10.26) and introduced the chemical potential μ by the relation $\gamma = \beta\mu$. The physical meaning of $\mu = \gamma/\beta$ as the chemical potential must be derived from condition (10.27), which determines this parameter and we refer the reader again to the lecture on thermodynamics and statistical mechanics for this calculation.[3] The meaning of β is the same as in the canonical ensemble and is given by (10.22). A system whose macrostate is determined by a mean energy or temperature, a certain volume, and a mean particle number or chemical potential is called a grand canonical system and (10.30) is the corresponding probability distribution of the grand canonical ensemble. This means that in a set of $\mathcal{N}$ systems ($\mathcal{N} \to \infty$) with common temperature, volume, and chemical potential (T, V, μ) the probability of finding one of these $\mathcal{N}$ systems exactly with N particles and in the special microstate number n is given by (10.30). Exactly as for (10.24) for the canonical density operator, we can also find the representation of the free density operator of the grand canonical ensemble:

$$\hat{\rho}_{gc} = \frac{\exp[-\beta(\hat{H} - \mu\hat{N})]}{\mathrm{Tr}\left\{\exp[-\beta(\hat{H} - \mu\hat{N})]\right\}}\,, \tag{10.31}$$

where the trace now contains not only a sum over all expectation values with a complete set of basis vectors in the N-particle Hilbert space, but also a sum

[3] W. Greiner, L. Neise, H. Stöcker: *Thermodynamics and Statistical Mechanics*, 2nd ed. (Springer, New York 1997).

over all particle numbers [see (10.30)]:

$$\mathrm{Tr}\, e^{-\beta(\hat{H}-\mu\hat{N})} = \sum_{N=0}^{\infty}\sum_{n}\langle\Psi^{n,N}_{E(n,N)}|\,\mathrm{e}^{-\beta(\hat{H}-\mu\hat{N})}|\Psi^{n,N}_{E(n,N)}\rangle\,, \qquad (10.32)$$

where we used here the complete set of energy eigenfunctions but, as we know, any other complete set can also be used to evaluate the trace even though the actual calculation will be more complicated.

It is clear that for other definitions of the macrostate other probability distributions and other ensembles will be needed, but for most situations the canonical or grand canonical ensemble is sufficient. Before we proceed it is useful to state some general properties of the density matrix, which follow directly from the general definition (10.8) and hold in all ensembles:

$$\mathrm{Tr}\,\hat{\rho} = 1\,, \qquad (10.33)$$

since the trace of the exponential operators in (10.24) or (10.31) exactly cancels the c number in the denominator, which is given explicitly by (10.25) or (10.32), respectively.

$$\hat{\rho}^{\dagger} = \hat{\rho}\,, \qquad (10.34)$$

i.e. the density matrix is an Hermitean operator, which follows from the fact that $\hat{\rho}$ is a sum of Hermitean projection operators, (10.8), where the weights, P^n, are real probabilities.

$$\hat{\rho}^2 = \hat{\rho} \quad\Leftrightarrow\quad \hat{\rho} = |\Psi^{n,N}_{E(n,N)}\rangle\langle\Psi^{n,N}_{E(n,N)}|\,, \qquad (10.35)$$

i.e. if $\hat{\rho}^2 = \hat{\rho}$ then $\hat{\rho}$ describes a pure state of a system in an exactly known microstate (here state number n, and the system has exactly N particles). In this case $\hat{\rho}$ is a projector onto this microstate, otherwise $\hat{\rho}^2 \neq \hat{\rho}$. Finally we remember that averages of observables have to be calculated as

$$\langle\hat{O}\rangle = \mathrm{Tr}(\hat{\rho}\hat{O}) = \mathrm{Tr}(\hat{O}\hat{\rho})\,, \qquad (10.36)$$

where the cyclic invariance under the trace may be used. Note that this average contains a purely quantum-mechanical average such as $\langle\Psi^{n,N}_{E(n,N)}|\hat{O}|\Psi^{n,N}_{E(n,N)}\rangle$ and additionally a statistical average over all these quantities weighted with the probabilities, $P^{n,N}$, according to our starting point (10.5).

We can now write the entropy according to (10.21) as the average of $-k\ln\hat{\rho}$, i.e.[4]

$$\begin{aligned} S &= \langle -k\ln\hat{\rho}\rangle = -k\,\mathrm{Tr}(\hat{\rho}\ln\hat{\rho}) \\ &= -k\sum_{n}\langle\Psi^{n}_{E(n)}|\hat{\rho}\ln\hat{\rho}|\Psi^{n}_{E(n)}\rangle\,, \end{aligned} \qquad (10.37)$$

where the logarithm operator is to be understood as a series expansion. However, we can use

[4] The expression originates from ideas of L.E. Boltzmann and M. Planck; the general interpretation in terms of information theory has been formulated by C. Shannon and E.T. Jaynes (see Example 6.9).

$$\begin{aligned}
\hat{\rho}|\Psi^n_{E(n)}\rangle &= \sum_m |\Psi^m_{E(m)}\rangle P^m \langle \Psi^m_{E(m)}|\Psi^n_{E(n)}\rangle \\
&= \sum_m |\Psi^m_{E(m)}\rangle P^m \delta_{m,n} \\
&= P^n |\Psi^n_{E(n)}\rangle ,
\end{aligned} \tag{10.38}$$

i.e. the states $|\Psi^n_{E(n)}\rangle$ are eigenstates of the density operator and the probabilities P^n are its eigenvalues. Therefore the matrix element in (10.37) can be evaluated to be the number $P^n \ln P^n$ such that (10.37) is only a more complicated formulation of (10.21). This formulation has the advantage that it is representation free and holds for all kinds of ensembles. Special forms of the density operator and the density matrix will be exemplified in the following problems. However, here we note some of the results here. Expressed by field operators $\hat{\Psi}^\dagger(\boldsymbol{x})$ and $\hat{\Psi}(\boldsymbol{x})$ the *density operator* takes on the form

$$\hat{\rho}(\boldsymbol{x}) = \hat{\Psi}^\dagger(\boldsymbol{x})\hat{\Psi}(\boldsymbol{x}) . \tag{10.39}$$

Analogously, the operator for the *density matrix* reads

$$\hat{\rho}(\boldsymbol{x}, \boldsymbol{x}') = \hat{\Psi}^\dagger(\boldsymbol{x}')\hat{\Psi}(\boldsymbol{x}) . \tag{10.40}$$

EXAMPLE

10.1 Density Operators in Second Quantization

The starting point for our consideration is an element $|\Psi\rangle$ of the N-particle Hilbert space. In order to represent this rather abstract state vector we need to have a complete basis of the Fock space. For this, arbitrary one-particle states $|\beta\rangle$ can be used; we assume that they are eigenstates of an operator $\hat{B}$. β stands for a corresponding set of quantum numbers. Out of states $|\beta\rangle$ an arbitrary n-particle state can be constructed by taking direct products: symmetric for bosons and antisymmetric for fermions. It is convenient to know about the "unity operator" $\hat{\mathbf{1}}$ in Fock space:

$$\hat{\mathbf{1}} = |0\rangle\langle 0| + \sum_{\beta_1} |\beta_1\rangle\langle\beta_1| + \sum_{\beta_1\beta_2} |\beta_1\beta_2\rangle\langle\beta_1\beta_2| + \cdots . \tag{1}$$

The various terms express the contributions from the ground state, the one-particle, two-particle, ... states to the complete $\hat{\mathbf{1}}$. They represent symmetric or antisymmetric many-particle states for bosons or fermions, respectively. An explicit distinction becomes necessary only once commutation relations are derived; see Exercise 10.3.

We project the state $|\Psi\rangle$ onto the Fock space $\hat{\mathbf{1}}$:

$$\begin{aligned}
|\Psi\rangle &= \hat{\mathbf{1}}|\Psi\rangle \\
&= |0\rangle\langle 0|\Psi\rangle + \sum_{\beta_1} |\beta_1\rangle\langle\beta_1|\Psi\rangle + \sum_{\beta_1\beta_2} |\beta_1\beta_2\rangle\langle\beta_1\beta_2|\Psi\rangle \\
&\quad + \cdots + \sum_{\beta_1\cdots\beta_N} |\beta_1\cdots\beta_N\rangle\langle\beta_1\cdots\beta_N|\Psi\rangle .
\end{aligned} \tag{2}$$

Example 10.1.

The quantum numbers β stand, for example, for momentum or position quantum numbers. The expansion (2) reduces to exactly one term for a pure n-particle state: $|\beta_1 \cdots \beta_n\rangle = |\Psi^n\rangle$. The quantities $|\langle\beta_1 \cdots \beta_n|\Psi\rangle|^2$ represent the probabilities for the realization of a n-particle state with quantum numbers $\beta_1 \cdots \beta_n$. Creation and annihilation operators $\hat{b}^\dagger_\beta$ and $\hat{b}_\beta$ are defined via their action on a many-particle state, i.e.

$$\hat{b}^\dagger_\beta|0\rangle = |\beta\rangle\,, \qquad \hat{b}_\beta|0\rangle = 0\,, \tag{3}$$

$$b^\dagger_{\beta'}|\beta\rangle = |\beta'\beta\rangle\,, \qquad \hat{b}_{\beta'}|\beta'\rangle = |0\rangle\,, \tag{4}$$

and so forth. The operator $\hat{b}^\dagger_\beta$ transforms an (n-1)-particle state into an n-particle state; the opposite holds for the adjoint operator $\hat{b}_\beta$. This leads to the interpretation in terms of creation and annihilation operators. Applying $\hat{b}^\dagger_\beta$ onto the Fock space $\hat{\mathbf{1}}$ leads with the help of (3), (4), and (2) to the decomposition

$$\begin{aligned}\hat{b}^\dagger_\beta &= \hat{b}^\dagger_\beta\hat{\mathbf{1}} \\ &= |\beta\rangle\langle 0| + \sum_{\beta_1}|\beta\beta_1\rangle\langle\beta_1| + \sum_{\beta_1\beta_2}|\beta\beta_1\beta_2\rangle\langle\beta_1\beta_2| + \cdots\,;\end{aligned} \tag{5}$$

for the adjunct operator $\hat{b}_\beta$ we get

$$\begin{aligned}\hat{b}_\beta &= \hat{b}_\beta\hat{\mathbf{1}} \\ &= |0\rangle\langle\beta| + \sum_{\beta_1}|\beta_1\rangle\langle\beta_1\beta| + \sum_{\beta_1\beta_2}|\beta_1\beta_2\rangle\langle\beta_1\beta_2\beta| + \cdots\,.\end{aligned} \tag{6}$$

A special class of creation and annihilation operators is represented by the field operators $\hat{\Psi}^\dagger(\boldsymbol{x})$ and $\hat{\Psi}(\boldsymbol{x})$. They are already known from (3.43a,b). In analogy to (5) and (6) they can be defined via an expansion with respect to the complete Fock-space basis of position states:

$$\begin{aligned}\hat{\Psi}^\dagger(\boldsymbol{x}) &= |\boldsymbol{x}\rangle\langle 0| + \int \mathrm{d}^3x_1|\boldsymbol{x}\boldsymbol{x}_1\rangle\langle\boldsymbol{x}_1| \\ &\quad + \int \mathrm{d}^3x_1\,\mathrm{d}^3x_2|\boldsymbol{x}\boldsymbol{x}_1\boldsymbol{x}_2\rangle\langle\boldsymbol{x}_1\boldsymbol{x}_2|\cdots\end{aligned} \tag{7}$$

$$\begin{aligned}\hat{\Psi}(\boldsymbol{x}) &= |0\rangle\langle\boldsymbol{x}| + \int \mathrm{d}^3x_1|\boldsymbol{x}_1\rangle\langle\boldsymbol{x}_1\boldsymbol{x}| \\ &\quad + \int \mathrm{d}^3x_1\,\mathrm{d}^3x_2|\boldsymbol{x}_1\boldsymbol{x}_2\rangle\langle\boldsymbol{x}_1\boldsymbol{x}_2\boldsymbol{x}|\cdots\,.\end{aligned} \tag{8}$$

These operators create and annihilate a quantum at position $\boldsymbol{x}$. They obey the commutation relation (3.44), i.e. $[\hat{\Psi}(\boldsymbol{x}), \hat{\Psi}^\dagger(\boldsymbol{x}')]_\pm = \delta(\boldsymbol{x} - \boldsymbol{x}')$. The operators $\hat{b}^\dagger_\beta, \hat{b}_\beta$ and $\hat{\Psi}^\dagger(\boldsymbol{x}), \hat{\Psi}(\boldsymbol{x})$ are related via a unitary transformation of the form

$$\hat{\Psi}^\dagger(\boldsymbol{x}) = \sum_\beta \hat{b}^\dagger_\beta\langle\beta|\boldsymbol{x}\rangle \equiv \sum_\beta \hat{b}^\dagger_\beta\Psi^\dagger_\beta(\boldsymbol{x})\,, \tag{9}$$

$$\hat{\Psi}(\boldsymbol{x}) = \sum_\beta \hat{b}_\beta\langle\boldsymbol{x}|\beta\rangle \equiv \sum_\beta \hat{b}_\beta\Psi_\beta(\boldsymbol{x})\,; \tag{10}$$

see (3.43a,b) and Exercise 10.2.

Example 10.1.

It is possible to define the density operator with the help of the field operators (9) and (10). With (7) and (8) we get

$$\begin{aligned}\hat{\rho}(\boldsymbol{x}) &= \hat{\Psi}^{\dagger}(\boldsymbol{x})\hat{\Psi}(\boldsymbol{x}) \\ &= |\boldsymbol{x}\rangle\langle\boldsymbol{x}| + \int \mathrm{d}^3x_1 |\boldsymbol{x}\boldsymbol{x}_1\rangle\langle\boldsymbol{x}_1\boldsymbol{x}| \\ &\quad + \int \mathrm{d}^3x_1\,\mathrm{d}^3x_2 |\boldsymbol{x}\boldsymbol{x}_1\boldsymbol{x}_2\rangle\langle\boldsymbol{x}_2\boldsymbol{x}_1\boldsymbol{x}| + \cdots \,. \end{aligned} \tag{11}$$

Again, the various contributions result from the one-particle, two-particle, ..., n-particle subspace of the Fock space. The *one-particle density* describes the probability of finding a particle in an *n-particle state* $|\Psi^n\rangle$ at position $\boldsymbol{x}$ and is represented by the matrix element

$$\begin{aligned}\rho_{|\Psi^n\rangle}(\boldsymbol{x}) &= \langle\Psi^n|\hat{\rho}(\boldsymbol{x})|\Psi^n\rangle \\ &= \int \mathrm{d}^3x_2\ldots\mathrm{d}^3x_n \langle\Psi^n|\boldsymbol{x}\boldsymbol{x}_2\ldots\boldsymbol{x}_n\rangle\langle\boldsymbol{x}\boldsymbol{x}_2\ldots\boldsymbol{x}_n|\Psi^n\rangle \\ &= n\int \mathrm{d}^3x_2\ldots\mathrm{d}^3x_n \Psi^{\dagger}(\boldsymbol{x},\boldsymbol{x}_2,\ldots,\boldsymbol{x}_n)\Psi(\boldsymbol{x},\boldsymbol{x}_2,\ldots,\boldsymbol{x}_n)\,. \end{aligned} \tag{12}$$

Here, the density has been normalized to the particle number n.

Equation (12) can be written in another, equivalent form, namely

$$\begin{aligned}\rho_{|\Psi^n\rangle}(\boldsymbol{x}) &= \int \mathrm{d}^3x_1\ldots\mathrm{d}^3x_n \Psi^{\dagger}(\boldsymbol{x}_1\boldsymbol{x}_2\cdots\boldsymbol{x}_n)\left(\sum_{i=1}^{n}\delta(\boldsymbol{x}-\boldsymbol{x}_i)\right) \\ &\quad \times \Psi(\boldsymbol{x}_1,\boldsymbol{x}_2\ldots\boldsymbol{x}_n)\,. \end{aligned} \tag{13}$$

In this form (12) and (13) are valid for fermions and bosons. From (13) we read off the *local density operator in coordinate representation*:

$$\hat{\rho}_{|\Psi^n\rangle}(\boldsymbol{x}) = \sum_{i=1}^{n}\delta(\boldsymbol{x}-\boldsymbol{x}_i)\,. \tag{14}$$

In view of (12)–(14) it becomes clear that the operator $\hat{\rho}(\boldsymbol{x})$ in (11) acts on n-particle Hilbert spaces. We also stress that the form of the density operator (11) does not imply a space representation of $\hat{\rho}$. The special appearance of the density operator comes from its expression in terms of field operators. The arguments $\boldsymbol{x}$ have to be viewed merely as parameters (position quantum numbers) which just characterize this class of creation and annihilation operators.

In order to make this point even clearer we discuss another form of the density operator. According to the unitary transformation (9) and (10) this leads to

$$\hat{\rho}(\boldsymbol{x}) = \sum_{\alpha,\beta}\hat{b}^{\dagger}_{\alpha}\hat{b}_{\beta}\langle\alpha|\boldsymbol{x}\rangle\langle\boldsymbol{x}|\beta\rangle\,. \tag{15}$$

Let's take the expectation value in analogy to (12):

$$\rho_{|\Psi^n\rangle}(\boldsymbol{x}) = \sum_{\alpha\beta}\langle\Psi^n|\hat{b}^{\dagger}_{\alpha}\hat{b}_{\beta}|\Psi^n\rangle\Psi^{\dagger}_{\alpha}(\boldsymbol{x})\Psi_{\beta}(\boldsymbol{x}) \equiv \sum_{\alpha\beta}n_{\alpha\beta}\Psi^{\dagger}_{\alpha}(\boldsymbol{x})\Psi_{\beta}(\boldsymbol{x})\,. \tag{16}$$

Due to the hermiticity of the matrix $n_{\alpha\beta}$ a unitary transformation $\hat{U}$ can be found which diagonalizes $(n_{\alpha\beta})$: *Example 10.1.*

$$n_\alpha \delta_{\alpha\beta} = \sum_{\mu\nu} (\hat{U}^\dagger)_{\alpha\mu} n_{\mu\nu} \hat{U}_{\nu\beta} .$$

The diagonal elements result to be

$$n_\alpha = \sum_{\mu\nu} U^*_{\mu\alpha} U_{\nu\alpha} n_{\mu\nu} = \langle \Psi^n | \left(\sum_\mu U^*_{\mu\alpha} \hat{b}^\dagger_\mu \right) \left(\sum_\nu U_{\nu\alpha} \hat{b}_\nu \right) | \Psi^n \rangle .$$

This suggests the introduction of "rotated" creation and annihilation operators, i.e.

$$\hat{b}^\dagger_\alpha = \sum_\mu U^*_{\mu\alpha} \hat{b}^\dagger_\mu ,$$

$$\hat{b}_\beta = \sum_\nu U_{\nu\alpha} \hat{b}_\nu ,$$

and (16) takes on diagonal form:

$$\begin{aligned} \rho_{|\Psi^n\rangle}(\boldsymbol{x}) &= \sum_\alpha n_\alpha \Psi^\dagger_\alpha(\boldsymbol{x}) \Psi_\alpha(\boldsymbol{x}) \\ &= \sum_\alpha n_\alpha \langle \alpha | \boldsymbol{x} \rangle \langle \boldsymbol{x} | \alpha \rangle \\ &= \langle \boldsymbol{x} | \sum_\alpha n_\alpha | \alpha \rangle \langle \alpha | \boldsymbol{x} \rangle . \end{aligned}$$

Then, the density operator becomes

$$\hat{n}_{|\Psi^n\rangle} = \sum_\alpha n_\alpha |\alpha\rangle\langle\alpha| . \tag{17}$$

A comparison with (12) yields the identity

$$\langle \Psi^n | \hat{\rho}(\boldsymbol{x}) | \Psi^n \rangle = \langle \boldsymbol{x} | \hat{n}_{\Psi^n} | \boldsymbol{x} \rangle . \tag{18}$$

The numbers n_α have a very simple meaning: they are the occupation numbers, or occupation probabilities, for the one-particle state $|\alpha\rangle$ in the many-particle state $|\Psi^n\rangle$. As a distinction to the form (11) the density operator (17) acts in the one-particle Hilbert space.

Now we focus on the *density matrix.* One of its forms is defined as

$$\hat{\rho}(\boldsymbol{x}, \boldsymbol{x}') = \hat{\Psi}^\dagger(\boldsymbol{x}') \hat{\Psi}(\boldsymbol{x}) . \tag{19}$$

Forming the matrix element with respect to an n-particle state $|\Psi^n\rangle$ we are led to the *density matrix normalized to n*:

$$\begin{aligned} \rho_{|\Psi^n\rangle}(\boldsymbol{x}, \boldsymbol{x}') &= \langle \Psi^n | \hat{\rho}(\boldsymbol{x}, \boldsymbol{x}') | \Psi_n \rangle \qquad (20) \\ &= \int \mathrm{d}^3x_2 \dots \mathrm{d}^3x_n \langle \Psi^n | \boldsymbol{x}' \boldsymbol{x}_2 \dots \boldsymbol{x}_n \rangle \langle \boldsymbol{x} \boldsymbol{x}_2 \dots \boldsymbol{x}_n | \Psi^n \rangle \\ &= n \int \mathrm{d}^3x_2 \dots \mathrm{d}^3x_n \Psi^\dagger(\boldsymbol{x}', \boldsymbol{x}_2, \dots, \boldsymbol{x}_n) \Psi(\boldsymbol{x}, \boldsymbol{x}_2, \dots, \boldsymbol{x}_n) . \end{aligned}$$

Example 10.1.

The one-particle density $\rho_{|\Psi^n\rangle}(\boldsymbol{x})$ in the n-particle state from (12) results as a special case (diagonal element): $\rho_{|\Psi^n\rangle}(\boldsymbol{x}) = \rho_{|\Psi^n\rangle}(\boldsymbol{x}, \boldsymbol{x})$. From (20) we deduce the spatial representation for the nonlocal operator of the density matrix [compare with (13)]:

$$\hat{\rho}_{\Psi^n}(\boldsymbol{x}, \boldsymbol{x}') = \sum_{i=1}^{n} \delta(\boldsymbol{x}' - \boldsymbol{x}_i)\delta(\boldsymbol{x} - \boldsymbol{x}_i) . \tag{21}$$

As in the case of the density operator the form

$$\hat{\rho}(\boldsymbol{x}, \boldsymbol{x}') = \sum_{\alpha\beta} \hat{b}^\dagger_\alpha \hat{b}_\beta \langle\alpha|\boldsymbol{x}'\rangle\langle\boldsymbol{x}|\beta\rangle \tag{22}$$

can be taken as a basis. It follows after insertion of (9) and (10) into (19). The relation

$$\begin{aligned} \rho_{|\Psi^n\rangle}(\boldsymbol{x}, \boldsymbol{x}') &= \sum_{\alpha\beta} \langle\Psi^n|\hat{b}^\dagger_\alpha \hat{b}_\beta|\Psi^n\rangle\langle\alpha|\boldsymbol{x}'\rangle\langle\boldsymbol{x}|\beta\rangle \\ &= \sum_{\alpha\beta} \langle\boldsymbol{x}|\beta\rangle n_{\alpha\beta}\langle\alpha|\boldsymbol{x}'\rangle \end{aligned} \tag{23}$$

leads to

$$\hat{n}_{\Psi^n} = \sum_{\alpha\beta} |\beta\rangle n_{\alpha\beta}\langle\alpha| . \tag{24}$$

Obviously, the same one-particle operator is assigned to both the density and the density matrix [compare (24) with (16)]. According to (23) for the density matrix the nondiagonal matrix elements $n_{\alpha\beta}$ of $\hat{n}_{|\Psi^n\rangle}$ are essential:

$$\rho_{|\Psi^n\rangle}(\boldsymbol{x}, \boldsymbol{x}') = \langle\boldsymbol{x}'|\hat{n}_{|\Psi^n\rangle}|\boldsymbol{x}\rangle = \sum_{\alpha\beta} n_{\alpha\beta}\Psi^\dagger_\alpha(\boldsymbol{x}')\Psi_\beta(\boldsymbol{x}) . \tag{25}$$

EXERCISE

10.2 Transformation Equations for Field Operators

Problem. Derive the equations for transformation between the operators $\hat{b}^\dagger_\beta, \hat{b}_\beta$ and the field operators $\hat{\Psi}^\dagger(\boldsymbol{x}), \hat{\Psi}(\boldsymbol{x})$.

Solution. We begin with the transformation equations for the operators $\hat{b}^\dagger_\beta$ and $\hat{\Psi}^\dagger(\boldsymbol{x})$. Making use of the Fock space $\hat{\mathbf{1}}$,

$$\hat{\mathbf{1}} = |0\rangle\langle 0| + \sum_\beta |\beta\rangle\langle\beta| + \sum_{\beta\beta_1} |\beta\beta_1\rangle\langle\beta_1\beta| + \cdots ,$$

and the expansion (7) for the field operator $\hat{\Psi}^\dagger(\boldsymbol{x})$ (see Example 10.1) we find

$$\begin{aligned} \hat{\Psi}^\dagger(\boldsymbol{x}) &= \hat{\mathbf{1}}\hat{\Psi}^\dagger(\boldsymbol{x}) \\ &= \left(|0\rangle\langle 0| + \sum_\beta |\beta\rangle\langle\beta| + \sum_{\beta\beta_1} |\beta\beta_1\rangle\langle\beta_1\beta| + \cdots\right) \end{aligned}$$

Exercise 10.2.

$$\times\left(|\boldsymbol{x}\rangle\langle 0| + \int d^3x_1 |\boldsymbol{x}\boldsymbol{x}_1\rangle\langle\boldsymbol{x}_1| + \int d^3x_1\, d^3x_2 |\boldsymbol{x}\boldsymbol{x}_1\boldsymbol{x}_2\rangle\langle\boldsymbol{x}_2\boldsymbol{x}_1| + \cdots\right).$$

This yields

$$\begin{aligned}\hat{\Psi}^\dagger(\boldsymbol{x}) &= \sum_\beta |\beta\rangle\langle\beta|\boldsymbol{x}\rangle\langle 0| \\ &+ \sum_{\beta\beta_1}\int d^3x_1 |\beta\beta_1\rangle\langle\beta|\boldsymbol{x}\rangle\langle\beta_1|\boldsymbol{x}_1\rangle\langle\boldsymbol{x}_1| \\ &+ \sum_{\beta\beta_1\beta_2}\int d^3x_1\, d^3x_2 |\beta\beta_1\beta_2\rangle\langle\beta|\boldsymbol{x}\rangle \\ &\times \langle\beta_1\beta_2|\boldsymbol{x}_1\boldsymbol{x}_2\rangle\langle\boldsymbol{x}_1\boldsymbol{x}_2| + \cdots ;\end{aligned}$$

all the other matrix elements like $\langle\beta|0\rangle$ and $\langle\beta\beta_1|x\rangle$ vanish. Only the matrix elements between states having the same number of particles are different from zero. Keeping this in mind, we easily understand that the equation above can also be written as

$$\begin{aligned}\hat{\Psi}^\dagger(\boldsymbol{x}) &= \sum_\beta \langle\beta|\boldsymbol{x}\rangle\Big(|\beta\rangle\langle 0| \\ &+ \sum_{\beta_1}|\beta\beta_1\rangle\langle\beta_1| + \sum_{\beta_1\beta_2}|\beta\beta_1\beta_2\rangle\langle\beta_2\beta_1| + \cdots\Big) \\ &\times\Big(|0\rangle\langle 0| + \int d^3x_1 |\boldsymbol{x}_1\rangle\langle\boldsymbol{x}_1| \\ &+ \int d^3x_1\, d^3x_2 |\boldsymbol{x}_1\boldsymbol{x}_2\rangle\langle\boldsymbol{x}_2\boldsymbol{x}_1| + \cdots\Big).\end{aligned}$$

Remembering (5) of Example 10.1, we identify the expression within the first bracket as the decomposition of the operator $\hat{b}^\dagger_\beta$; the second term yields just the $\hat{\mathbf{1}}$ [see (1) in Example 10.1]. Thus, we obtain the wanted transformation from $\hat{b}^\dagger_\beta$ to $\hat{\Psi}^\dagger(\boldsymbol{x})$:

$$\hat{\Psi}^\dagger(\boldsymbol{x}) = \sum_\beta \hat{b}^\dagger_\beta\langle\beta|\boldsymbol{x}\rangle . \tag{1}$$

For the inversion of (1) it is necessary to project with $\langle\boldsymbol{x}|\beta\rangle$:

$$\begin{aligned}\int d^3x\, \hat{\Psi}^\dagger(\boldsymbol{x})\langle\boldsymbol{x}|\beta\rangle &= \int d^3x \sum_{\beta_1}\hat{b}^\dagger_{\beta_1}\langle\beta_1|\boldsymbol{x}\rangle\langle\boldsymbol{x}|\beta\rangle \\ &= \sum_{\beta_1}\hat{b}^\dagger_{\beta_1}\langle\beta_1|\beta\rangle = \hat{b}^\dagger_\beta ;\end{aligned}$$

this leads to

$$\hat{b}^\dagger_\beta = \int d^3x\, \hat{\Psi}^\dagger(\boldsymbol{x})\langle\boldsymbol{x}|\beta\rangle . \tag{2}$$

Exercise 10.2.

The corresponding transformation equations for the annihilation operators are obtained by taking the adjoint of (1) and (2):

$$\hat{\Psi}(\boldsymbol{x}) = \sum_{\beta} \hat{b}_{\beta} \langle \boldsymbol{x}|\beta\rangle \,, \tag{3}$$

$$\hat{b}_{\beta} = \int \mathrm{d}^3x \, \hat{\Psi}(\boldsymbol{x}) \langle \beta|\boldsymbol{x}\rangle \,. \tag{4}$$

EXERCISE

10.3 Commutation Relations for Fermion Field Operators

Problem. Derive the commutation relations between fermion creation and annihilation operators.

Solution. For the case of fermions we have to take the Fock space of the antisymmetric many-particle states as a basis. The action of the *antisymmetrization operator* $\hat{A}$ on a one-particle product state $|\beta_1\beta_2\dots\beta_N\rangle$ creates a completely antisymmetric many-particle state normalized to the particle number N:

$$\begin{aligned} |\beta_1\beta_2\cdots\beta_N\rangle_{\mathrm{a}} &= \sqrt{N!}\hat{A}|\beta_1\rangle|\beta_2\rangle\cdots|\beta_N\rangle \\ &= \sqrt{N!}\hat{A}|\beta_1\beta_2\dots\beta_N\rangle \,, \end{aligned} \tag{1}$$

where the index 'a' symbolizes antisymmetry. We find for the matrix element between two arbitrary many-particle states:

$${}_{\mathrm{a}}\langle \beta_1'\beta_2'\dots\beta_{N'}'|\beta_1\beta_2\dots\beta_N\rangle_{\mathrm{a}} = 0 \qquad \text{for} \quad N' \neq N \tag{2}$$

and

$${}_{\mathrm{a}}\langle \beta_1'\beta_2'\dots\beta_N'|\beta_1\beta_2\dots\beta_N\rangle_{\mathrm{a}} = \begin{vmatrix} \delta_{\beta_1'\beta_1} & \delta_{\beta_1'\beta_2} & \cdots & \delta_{\beta_1'\beta_N} \\ \delta_{\beta_2'\beta_1} & \delta_{\beta_2'\beta_2} & \cdots & \delta_{\beta_2'\beta_N} \\ \vdots & & & \vdots \\ \delta_{\beta_N'\beta_1} & \delta_{\beta_N'\beta_2} & \cdots & \delta_{\beta_N'\beta_N} \end{vmatrix} \qquad \text{for} \quad N = N' \,. \tag{3}$$

The unity operator in the Fock space of the antisymmetric states is given by

$$\hat{\mathbf{1}} = |0\rangle\langle 0| + \sum_{\beta_1} |\beta_1\rangle\langle\beta_1| + \sum_{\beta_1\beta_2} \frac{1}{2!}|\beta_1\beta_2\rangle_{\mathrm{a}}\,{}_{\mathrm{a}}\langle\beta_1\beta_2| + \cdots \tag{4}$$

from which the expressions for the fermion creation and annihilation operators follow. We get for the creation operator

$$\begin{aligned} \hat{b}_{\beta}^{\dagger} = \hat{b}_{\beta}^{\dagger}\hat{\mathbf{1}} &= |\beta\rangle\langle 0| + \sum_{\beta_1'} |\beta\beta_1'\rangle_{\mathrm{a}}\,\langle\beta_1'| \\ &\quad + \sum_{\beta_1'\beta_2'} \frac{1}{2!}\,|\beta\beta_1'\beta_2'\rangle_{\mathrm{a}}\,{}_{\mathrm{a}}\langle\beta_1'\beta_2'| + \cdots \end{aligned} \tag{5}$$

Exercise 10.3.

and for the annihilation operator

$$\hat{b}_\beta = |0\rangle\langle\beta| + \sum_{\tilde{\beta}_1} |\tilde{\beta}_1\rangle\,{}_{\mathrm{a}}\langle\beta\tilde{\beta}_1| + \sum_{\tilde{\beta}_1\tilde{\beta}_2} \frac{1}{2!}\,|\tilde{\beta}_1\tilde{\beta}_2\rangle_{\mathrm{a}}\,{}_{\mathrm{a}}\langle\beta\tilde{\beta}_1\tilde{\beta}_2| + \cdots \,. \tag{6}$$

Here we have used the idea that the action of the operator $\hat{b}^\dagger_\beta$ on an antisymmetric N-particle state leads to an antisymmetric $(N+1)$-particle state, i.e.

$$\hat{b}^\dagger_\beta|\beta_1\beta_2\ldots\beta_N\rangle_{\mathrm{a}} = |\beta\beta_1\beta_2\ldots\beta_N\rangle_{\mathrm{a}}\,. \tag{7}$$

In order to derive commutation relations we determine the operator products $\hat{b}_{\beta'}\hat{b}^\dagger_\beta$ and $\hat{b}^\dagger_\beta\hat{b}_{\beta'}$ according to (5) and (6). Let us have an explicit look on the first terms:

$$\hat{b}_{\beta'}\hat{b}^\dagger_\beta = |0\rangle\langle\beta'|\beta\rangle\langle 0| + \sum_{\beta_1\tilde{\beta}_1} |\tilde{\beta}_1\rangle_{\mathrm{a}}\,\langle\beta'\tilde{\beta}_1|\beta\beta'_1\rangle_{\mathrm{a}}\,\langle\beta'_1| \tag{8}$$

$$+ \sum_{\beta'_1\beta'_2\tilde{\beta}_1\tilde{\beta}_2} \frac{1}{(2!)^2}|\tilde{\beta}_1\tilde{\beta}_2\rangle_{\mathrm{a}}\,{}_{\mathrm{a}}\langle\beta'\tilde{\beta}_1\tilde{\beta}_2|\beta\beta'_1\beta'_2\rangle_{\mathrm{a}}\,{}_{\mathrm{a}}\langle\beta'_1\beta'_2| + \cdots \,. \tag{9}$$

The appearing matrix elements have to be analyzed according to (3) and yield

$${}_{\mathrm{a}}\langle\beta'\tilde{\beta}_1|\beta\beta'_1\rangle_{\mathrm{a}} = \delta_{\beta'\beta}\delta_{\tilde{\beta}_1\beta'_1} - \delta_{\beta'\beta'_1}\delta_{\tilde{\beta}_1\beta}$$

and

$$\begin{aligned}{}_{\mathrm{a}}\langle\beta'\tilde{\beta}_1\tilde{\beta}_2|\beta\beta'_1\beta'_2\rangle_{\mathrm{a}} &= \delta_{\beta'\beta}(\delta_{\tilde{\beta}_1\beta'_1}\delta_{\tilde{\beta}_2\beta'_2} - \delta_{\tilde{\beta}_1\beta'_2}\delta_{\tilde{\beta}_2\beta'_1})\\ &\quad - \delta_{\beta'\beta'_1}(\delta_{\tilde{\beta}_1\beta}\delta_{\tilde{\beta}_2\beta'_2} - \delta_{\tilde{\beta}_1\beta'_2}\delta_{\tilde{\beta}_2\beta})\\ &\quad + \delta_{\beta'\beta'_2}(\delta_{\tilde{\beta}_1\beta}\delta_{\tilde{\beta}_2\beta'_1} - \delta_{\tilde{\beta}_1\beta'_1}\delta_{\tilde{\beta}_2\beta})\,.\end{aligned}$$

Inserting this into (8) and performing the summation leads us to

$$\begin{aligned}\hat{b}_{\beta'}\hat{b}^\dagger_\beta &= \delta_{\beta'\beta}|0\rangle\langle 0| + \delta_{\beta'\beta}\sum_{\beta'_1}|\beta'_1\rangle\langle\beta'_1| - |\beta\rangle\langle\beta'|\\ &\quad + \delta_{\beta'\beta}\sum_{\beta'_1\beta'_2}\frac{1}{(2!)^2}\left(|\beta'_1\beta'_2\rangle_{\mathrm{a}}\,{}_{\mathrm{a}}\langle\beta'_1\beta'_2| - |\beta'_2\beta'_1\rangle_{\mathrm{a}}\,{}_{\mathrm{a}}\langle\beta'_1\beta'_2|\right)\\ &\quad - \sum_{\beta'_2}\frac{1}{(2!)^2}\left(|\beta\beta'_2\rangle_{\mathrm{a}}\,{}_{\mathrm{a}}\langle\beta'\beta'_2| - |\beta'_2\beta\rangle_{\mathrm{a}}\,{}_{\mathrm{a}}\langle\beta'\beta'_2|\right)\\ &\quad + \sum_{\beta'_1}\frac{1}{(2!)^2}\left(|\beta\beta'_1\rangle_{\mathrm{a}}\,{}_{\mathrm{a}}\langle\beta'_1\beta'| - |\beta'_1\beta\rangle_{\mathrm{a}}\,{}_{\mathrm{a}}\langle\beta'_1\beta'|\right) + \cdots\\ &= \delta'_{\beta'\beta}\left(|0\rangle\langle 0| + \sum_{\beta'_1}|\beta'_1\rangle\langle\beta'_1| + \sum_{\beta'_1\beta'_2}\frac{1}{2!}|\beta'_1\beta'_2\rangle_{\mathrm{a}}\,{}_{\mathrm{a}}\langle\beta'_1\beta'_2| + \cdots\right)\\ &\quad - |\beta\rangle\langle\beta'| - \sum_{\beta'_1}|\beta\beta'_1\rangle_{\mathrm{a}}\,{}_{\mathrm{a}}\langle\beta'\beta'_1| - \cdots\,.\end{aligned}$$

Exercise 10.3.

Calculating the next higher-order terms in (8) we convince ourselves that those terms proportional to the factor $\delta_{\beta'\beta}$ add up to the Fock space $\hat{\mathbf{1}}$ [see (4)].

We obtain

$$\hat{b}_{\beta'}\hat{b}^\dagger_\beta = \delta_{\beta\beta'}\hat{\mathbf{1}} - |\beta\rangle\langle\beta'| - \sum_{\beta'_1} |\beta\beta'_1\rangle_{\mathrm{a}\,\mathrm{a}}\langle\beta'\beta'_1| - (\text{more remaining terms}) , \tag{10}$$

where the remaining terms in the operator product $\hat{b}^\dagger_\beta \hat{b}_{\beta'}$ originate with opposite sign. Furthermore,

$$\begin{aligned}\hat{b}^\dagger_\beta \hat{b}_{\beta'} &= |\beta\rangle\langle 0|0\rangle\langle\beta'| + \sum_{\beta'_1\tilde{\beta}_1} |\beta\beta'_1\rangle_{\mathrm{a}}\,\langle\beta'_1|\tilde{\beta}_1\rangle_{\mathrm{a}}\,\langle\beta'\tilde{\beta}_1| + \cdots \\ &= |\beta\rangle\langle\beta'| + \sum_{\beta'_1} |\beta\beta'_1\rangle_{\mathrm{a}\,\mathrm{a}}\langle\beta'\beta'_1| + (\text{more remaining terms}) .\end{aligned} \tag{11}$$

Thus, with (10) and (11) we are led to the well-known commutation relations for fermions:

$$[\hat{b}^\dagger_\beta, \hat{b}_{\beta'}]_+ = \delta_{\beta\beta'}\hat{\mathbf{1}} . \tag{12}$$

10.3 Dynamics of a Quantum-Statistical Ensemble

The dynamics, i.e. the time evolution, of a quantum mixture is determined by the Hamiltonian of the system. If no measurements are performed on the system over an arbitrary time interval the information about the system, which is reflected in the weights P^n, does not change. This implies that the P^n remain constant over the corresponding time interval, i.e.

$$\frac{\mathrm{d}}{\mathrm{d}t}P^n = \frac{\partial}{\partial t}P^n = 0 . \tag{10.41}$$

The dynamics of each (pure) state $|\Psi^n\rangle$ of the ensemble is described by the Schrödinger equation (Schrödinger picture)

$$\mathrm{i}\partial_t|\Psi^n\rangle = \hat{H}|\Psi^n\rangle . \tag{10.42}$$

Here, we have set $\hbar = 1$, which is the same as replacing $\hat{H}$ by $\hat{H}/\hbar$: i.e. $\hat{H} \to \hat{H}/\hbar$. As we ask for the *evolution of the whole ensemble* we have to find the evolution equation for the density operator. Directly from the Schrödinger equation (10.42) and its adjoint equation it follows that the density operator of the pure state (10.3) is

$$\begin{aligned}(\mathrm{i}\partial_t|\Psi^n\rangle)\langle\Psi^n| &= (\hat{H}|\Psi^n\rangle)\langle\Psi^n| , \\ |\Psi^n\rangle(-\mathrm{i}\partial_t\langle\Psi^n|) &= |\Psi^n\rangle(\langle\Psi^n|\hat{H}) ,\end{aligned}$$

so that

$$\begin{aligned}
\mathrm{i}\partial_t\hat{\rho}_{|\Psi^n\rangle} &= \mathrm{i}\partial_t(|\Psi^n\rangle\langle\Psi^n|) \\
&= \mathrm{i}[(\partial_t|\Psi^n\rangle)\langle\Psi^n| + |\Psi^n\rangle(\partial_t\langle\Psi^n|)] \\
&= (\hat{H}|\Psi^n\rangle)\langle\Psi^n| - |\Psi^n\rangle(\hat{H}\langle\Psi^n|) \\
&= \hat{H}|\Psi^n\rangle\langle\Psi^n| - |\Psi^n\rangle\langle\Psi^n|\hat{H} \\
&= [\hat{H}|\Psi^n\rangle\langle\Psi^n|]_-
\end{aligned}$$

and therefore

$$\mathrm{i}\partial_t\hat{\rho}_{|\Psi^n\rangle} = [\hat{H},\hat{\rho}_{|\Psi^n\rangle}]_- \,. \tag{10.43}$$

Since the weights P^n are constants, the ensemble averaging (10.8) over (10.43) can be performed directly:

$$\sum_{n\in\mathrm{E}} P^n\,\mathrm{i}\partial_t\hat{\rho}_{|\Psi^n\rangle} = \sum_{n\in\mathrm{E}} P^n[\hat{H},\hat{\rho}_{|\Psi^n\rangle}] \,.$$

Finally we arrive with (10.8) at

$$\mathrm{i}\partial_t\hat{\rho}_{\mathrm{E}} = [\hat{H},\hat{\rho}_{\mathrm{E}}] \,. \tag{10.44}$$

Hence, we have derived an *evolution equation* for the density operator of the quantum-theoretical mixture. It can be understood as the *quantum-mechanical analogue to Liouville's equation* of classical many-particle mechanics.

With the help of the evolution equation for the density operator we succeed in deriving the equation for the time evolution of an arbitrary observable. $\hat{O}$ denotes the Schrödinger operator, which is assigned to the observable $\langle\hat{O}\rangle_{\mathrm{E}}$ and does not explicitly depend on time, i.e. $\mathrm{d}\hat{O}/\mathrm{d}t = \partial\hat{O}/\partial t = 0$. We consider now the change with time of the ensemble average (10.7)

$$\langle\hat{O}\rangle_{\mathrm{E}} = \mathrm{Tr}(\hat{\rho}_{\mathrm{E}}\hat{O}) \,,$$

and indicate some of the intermediate steps. First,

$$\begin{aligned}
\partial_t\langle\hat{O}\rangle_{\mathrm{E}} = \sum_{m\in\mathrm{E}} &\big[(\partial_t\langle\Psi^m|)\hat{\rho}_{\mathrm{E}}\hat{O}|\Psi^m\rangle + \langle\Psi^m|(\partial_t\hat{\rho}_{\mathrm{E}})\hat{O}|\Psi^m\rangle \\
&+\langle\Psi^m|\hat{\rho}_{\mathrm{E}}\hat{O}(\partial_t|\Psi^m\rangle)\big] \,.
\end{aligned}$$

Making use of the Schrödinger equation (10.42) and the evolution equation (10.44) we deduce

$$\begin{aligned}
\partial_t\langle\hat{O}\rangle_{\mathrm{E}} &= \sum_{m\in\mathrm{E}} -\mathrm{i}\big[-\langle\Psi^m|[\hat{H},\hat{\rho}_{\mathrm{E}}\hat{O}]|\Psi^m\rangle + \langle\Psi^m|[\hat{H},\hat{\rho}_{\mathrm{E}}]\hat{O}|\Psi^m\rangle\big] \\
&= \sum_{m\in\mathrm{E}} -\mathrm{i}\langle\Psi^m|\hat{\rho}_{\mathrm{E}}[\hat{H},\hat{O}]|\Psi^m\rangle \,.
\end{aligned}$$

This leads us to the *evolution equation for the averages*:

$$\partial_t\langle\hat{O}\rangle_{\mathrm{E}} = -\,\mathrm{i}\,\mathrm{Tr}(\hat{\rho}_{\mathrm{E}}[\hat{O},\hat{H}]) \,. \tag{10.45}$$

EXERCISE

10.4 Density Operator of a Mixture

Problem. Determine the form of the density operator of a mixture for the case in which information is given about its pure states.

Solution. We assume that the ensemble consists of a finite number N of pure states. If we have no information about the pure state of the system we have to proceed in such a way that all possible state vectors $|\Psi^n\rangle$ are realized with the same probability. This implies that all weights have to be equal, i.e. $P^n = p$. This leads us to the following form of $\hat{\rho}$:

$$\hat{\rho} = p\sum_{n=1}^{N} |\Psi^n\rangle\langle\Psi^n| \; ; \tag{1}$$

it is the sum over projectors onto the pure states of the ensemble.

We can also provide another ansatz for the density operator, which is equivalent to (1). For this we only have to take into account that on the one hand the sum occuring in (1) represents a part of the complete unity operator in Fock space and that on the other hand all expectation values always imply ensemble averaging, i.e. taking the trace with the density operator. We write (1) as

$$\hat{\rho} = p\hat{\mathbf{1}} - p\hat{Q} \, , \tag{2}$$

where the operator $\hat{Q}$ simply represents the remaining sum over all projectors onto pure states, which do not belong to the ensemble. Thus, it holds that

$$\hat{Q}|\Psi^n\rangle = 0 \qquad \text{for} \quad n \in \mathrm{E} \, . \tag{3}$$

Once the action of $\hat{Q}$ onto one $|\Psi^n\rangle$ from the ensemble vanishes, this part in the density operator does not contribute in all possible trace expressions that contain $\hat{\rho}$. In this respect, the form of the density operator equivalent to (1) is simply given by

$$\hat{\rho}' = p\hat{\mathbf{1}} \, . \tag{4}$$

We still have to determine the weight p. It follows in a straightforward manner from the general property $\mathrm{Tr}(\hat{\rho}) = 1$ of density operators. We arrive at the result

$$\hat{\rho}' = \frac{\hat{\mathbf{1}}}{\mathrm{Tr}(\hat{\mathbf{1}})} = \frac{\hat{\mathbf{1}}}{N} \, , \tag{5}$$

where N reflects the number of possible pure states of the ensemble.

With absolutely no information about the pure state of the system the density operator can be set proportional to the unity operator.

EXERCISE

10.5 Construction of the Density Operator for a System of Unpolarized Electrons

Problem. Construct the density operator for a system of unpolarized electrons by using the spin eigenstates of $\hat{\sigma}_z$.

Solution. Only the spin part of a possible pure many-electron state is of importance. The pure spin states of the system are the two possible eigenspinors of the operator $\hat{\sigma}_z$. The two spinors

$$\chi_{\frac{1}{2}\frac{1}{2}} = \begin{pmatrix} 1 \\ 0 \end{pmatrix} \mathrel{\hat{=}} |\Psi^1\rangle\,, \qquad \chi_{\frac{1}{2}-\frac{1}{2}} = \begin{pmatrix} 0 \\ 1 \end{pmatrix} \mathrel{\hat{=}} |\Psi^2\rangle \tag{1}$$

form the ensemble. For the construction of the density operator we need a representation of the projectors $|\Psi^n\rangle\langle\Psi^n|$ onto the pure spinstates $\chi_{\frac{1}{2}\pm\frac{1}{2}}$ in first place. They result as dyadic products:

$$\hat{P}_\pm = \chi^\dagger_{\frac{1}{2}\pm\frac{1}{2}} \otimes \chi_{\frac{1}{2}\pm\frac{1}{2}}\,, \tag{2}$$

or more explicitly:

$$\begin{aligned} \hat{P}_+ &= (1,\ 0)\begin{pmatrix} 1 \\ 0 \end{pmatrix} = \begin{pmatrix} 1 & 0 \\ 0 & 0 \end{pmatrix}, \\ \hat{P}_- &= (0,\ 1)\begin{pmatrix} 0 \\ 1 \end{pmatrix} = \begin{pmatrix} 0 & 0 \\ 0 & 1 \end{pmatrix}. \end{aligned} \tag{3}$$

Hence, we obtain the form

$$\hat{\rho} = p_+\hat{P}_+ + p_-\hat{P}_- \tag{4}$$

for the density operator. Now we have to determine the weights p_+ and p_-, i.e. the probabilities for the realization of a pure state with a z projection of the spins of $+\frac{1}{2}$ and $-\frac{1}{2}$, respectively. In general it has to hold that

$$\mathrm{Tr}(\hat{\rho}) = \sum_{m_s} \chi^\dagger_{\frac{1}{2}m_s}\hat{\rho}\chi_{\frac{1}{2}m_s} = 1\,. \tag{5}$$

In more detail this becomes

$$\begin{aligned} \mathrm{Tr}(\hat{\rho}) &= (1,\ 0)\left[p_+\begin{pmatrix} 1 & 0 \\ 0 & 0 \end{pmatrix} + p_-\begin{pmatrix} 0 & 0 \\ 0 & 1 \end{pmatrix}\right]\begin{pmatrix} 1 \\ 0 \end{pmatrix} \\ &\quad + (0,\ 1)\left[p_+\begin{pmatrix} 1 & 0 \\ 0 & 0 \end{pmatrix} + p_-\begin{pmatrix} 0 & 0 \\ 0 & 1 \end{pmatrix}\right]\begin{pmatrix} 0 \\ 1 \end{pmatrix}, \end{aligned}$$

which leads to the condition

$$p_+ + p_- = 1\,. \tag{6}$$

However, in an unpolarized many-electron system no spin projection is favored, so the probabilities p_+ and p_- are equal. As a result of (6), (4) takes on the form

Exercise 10.5.

$$\hat{\rho} = \frac{1}{2}\hat{P}_+ + \frac{1}{2}\hat{P}_- = \frac{1}{2}\begin{pmatrix} 1 & 0 \\ 0 & 1 \end{pmatrix} = \frac{\hat{\mathbf{1}}}{\mathrm{Tr}(\hat{\mathbf{1}})} ; \tag{7}$$

note that in our case $\mathrm{Tr}(\hat{\mathbf{1}}) = N = 2$.

This result demonstrates that an unpolarized electron system represents the physical realization of an ensemble for which we have no knowledge about its pure state. $p_+ = p_-$ reflects the absence of information.

10.4 Ordered and Disordered Systems: The Density Operator and Entropy

In the preceding subsections we have learned how the existing lack of knowledge about the state vector $|\Psi\rangle$ is reflected in the density operator for the description of a state of a physical many-particle system. We have also realized that the uncertainty determined by $\hat{\rho}$ is restrained from below and above by two, so-to-say, extreme density operators. The lower bound is fixed by the density operator of the pure state:

$$\hat{\rho}_{|\Psi\rangle} = |\Psi\rangle\langle\Psi| . \tag{10.46}$$

This implies that the state of the system is entirely known. On the other hand we have experienced the density operator

$$\hat{\rho}_{\mathrm{U}} = \frac{\hat{\mathbf{1}}}{\mathrm{Tr}(\hat{\mathbf{1}})} = \frac{\hat{\mathbf{1}}}{N} , \tag{10.47}$$

which has to be chosen when there is full ignorance about the state of the system (see Exercise 10.4 and 10.5); the index 'U' stands for 'unawareness'. Often it is important to possess a meaningful measure for the 'deviation' of the density operator $\hat{\rho}$ from its pure state $\hat{\rho}_{|\Psi\rangle}$. This measure can also be understood as a measure for the missing information about the system. In accordance with information theory this measure, which we will simply denote as *deviation* in the following and which is assumed to be a positive number σ, is defined as a functional $\sigma[\hat{\rho}]$ of the density operator in the form

$$\sigma[\hat{\rho}] = -\,\mathrm{Tr}[\hat{\rho}\ln(\hat{\rho})] . \tag{10.48}$$

The *deviation* $\sigma[\hat{\rho}]$ can also be understood as a measure for the disorder of the system. Here we will not justify (10.48) from the information theory point of view and refer to Example 6.9 instead, but we will straightforwardly prove the most important properties of $\sigma[\hat{\rho}]$. We proceed with the eigen representation of the density operator:

$$\hat{\rho}|\Psi^n\rangle = P^n|\Psi^n\rangle \quad \text{with} \quad \sum_n P^n = 1 \quad \text{and} \quad 0 \le P^n \le 1 .$$

Within this representation (10.48) becomes:

$$\sigma[\hat{\rho}] = -\sum_n P^n \ln(P^n) ; \tag{10.49}$$

compare with (12) of Example 6.9. Since $0 \leq P^n \leq 1$ holds for all weights, we deduce at first positive definiteness:

$$\sigma[\hat{\rho}] \geq 0 . \tag{10.50}$$

Next we show that the measures $\sigma[\hat{\rho}_{\mathrm{U}}]$ and $\sigma[\hat{\rho}_{|\Psi\rangle}]$ corresponding to the density operators $\hat{\rho}_{\mathrm{U}}$ and $\hat{\rho}_{|\Psi\rangle}$ are indeed upper and lower bounds for every other $\sigma[\hat{\rho}]$, so that

$$\sigma[\hat{\rho}_{|\Psi\rangle}] \leq \sigma[\hat{\rho}] \leq \sigma[\hat{\rho}_{\mathrm{U}}] . \tag{10.51}$$

Once all weights P^n except one ($p' = p = 1$) are set equal to zero we obtain from the eigen representation (10.49) the density operator $\hat{\rho}_{|\Psi\rangle}$ of a pure state. The limit $\lim_{x\to 0} x\ln(x) = 0$ leads to

$$\sigma[\hat{\rho}_{|\Psi\rangle}] = 0 . \tag{10.52}$$

As it should do, the deviation σ vanishes exactly for the density operator of the pure state since it implies a complete knowledge about the state vector of the system. In order to determine the upper bound for σ we have to calculate $\sigma[\hat{\rho}_{\mathrm{U}}]$. The density operator $\hat{\rho}_{\mathrm{U}}$ originates from (10.49) in such a way that we set all weights equal to $P^n = p = 1/N$. This yields

$$\begin{aligned}
\sigma[\hat{\rho}_{\mathrm{U}}] &= -\sum_{n=1}^{N} \frac{1}{N} \ln\left(\frac{1}{N}\right) = -\frac{1}{N} \ln \frac{1}{N} \sum_{n=1}^{N} 1 \\
&= -\frac{1}{N}(\ln 1 - \ln N)\, N = \ln(N) .
\end{aligned} \tag{10.53}$$

Hence, the claim (10.51) explicitly reads

$$0 \leq \sigma[\hat{\rho}] \leq \ln(N) . \tag{10.54}$$

This relation can be proven by rewriting the expression (10.49) for $\sigma[\hat{\rho}]$ in the following form:

$$\begin{aligned}
\sigma[\hat{\rho}] &= \sum_n P^n \ln\left(\frac{1}{P^n}\right) = \sum_n P^n \ln\left(\frac{N}{P^n N}\right) \\
&= \ln(N) \sum_n P^n + \sum_n P^n \ln\left(\frac{1}{P^n N}\right) \\
&\leq \ln(N) + \sum_n P^n \left(\frac{1}{P^n N} - 1\right) = \ln(N) ,
\end{aligned}$$

where we have used the relation $\ln(x) \leq x - 1$. This is the proof of statement (10.54).

In Sect. 10.3 we have learned that the time evolution of an isolated system is only governed by its corresponding Hamiltonian $\hat{H}$ and that for this case the weights P^n are constant [see (10.41)]. Hence, also the deviation σ [see (10.49)] remains constant:

$$\frac{\mathrm{d}}{\mathrm{d}t}\sigma[\hat{\rho}] = 0 . \tag{10.55}$$

We should mention another aspect here. The deviation σ also represents a measure for the disorder within a system. This becomes plausible once we remember, for instance, the example of the unpolarized electron system (Exercise 10.5). For this system the density operator was $\hat{\rho}_{\mathrm{U}} = \hat{1}/\,\mathrm{Tr}(\hat{1})$, which led to the largest deviation $\sigma = \ln(2)$. Unpolarization means arbitrary orientation of the single spins. With respect to spin orientation the system is completely disordered. However, once a certain spin orientation is favored (magnetization) this reflects a state of less disorder. Hence, the deviation σ also becomes smaller:

$$\sigma = -[(1-p)\ln(1-p) + p\ln(p)] \leq \ln(2)\,, \qquad p \leq 1/2\,,$$

where p represents the weight of the less favored spin orientation. Once the system is in a pure state, i.e. all spins are oriented in the same direction (complete magnetization), σ vanishes. Then the system is in a state of largest order.

Another conclusion is at hand: (10.55) implies that for the case of an isolated system, which is left to its own, the degree of disorder does not change. Detailed studies show that the deviation σ increases because of the influence of external stochastic perturbations. The measure σ behaves like an order parameter. In fact, σ is related to the *thermodynamic entropy*. The quantum-mechanical definition of entropy is given by

$$S = k_{\mathrm{B}}\sigma\,, \tag{10.56}$$

where k_{B} denotes Boltzmann's constant (see Examples 6.6 and 6.9). From experience we can say that a system always tends to a state of maximum entropy. In other words, the density operator $\hat{\rho}$ of a quantum-mechanical mixture always maximizes the deviation $\sigma[\hat{\rho}]$. In the next subsection we will consider the extremal properties of σ to pin down the density operators for stationary ensembles. For further information on these subjects we refer the reader to the lectures on thermodynamics and statistical mechanics.[5]

10.5 Stationary Ensembles

According to (10.44) the vanishing of the commutator between the density operator and the Hamiltonian,

$$[\hat{\rho}, \hat{H}]_- = 0\,, \tag{10.57}$$

is a necessary and sufficient condition for the presence of a *stationary ensemble*. For example, the density operator $\hat{\rho}_{\mathrm{U}}$ (10.47) of a completely disordered system is stationary because the unity operator commutes with every arbitrary Hamiltonian. A sufficient condition for stationarity is given once $\hat{\rho}$ becomes a function of the Hamiltonian or a function of the operators $\hat{O}$ that commute with $\hat{H}$:

$$\hat{\rho} = \hat{\rho}(\hat{H}, \hat{O})\,, \qquad [\hat{O}, \hat{H}]_- = 0\,. \tag{10.58}$$

[5] W. Greiner, L. Neise, H. Stöcker: *Thermodynamics and Statistical Mechanics*, 2nd ed. (Springer, New York 1997).

Of particular importance are stationary ensembles that maximize the deviation σ; examples are the operator $\hat{\rho}_U$ and the generalized grand canonical ensembles. They are determined from the variational principle

$$\delta\sigma[\hat{\rho}] = \delta\,\mathrm{Tr}[\hat{\rho}\ln(\hat{\rho})] = 0\,, \tag{10.59a}$$

with the single restriction

$$\mathrm{Tr}(\hat{\rho}) = 1\,, \tag{10.59b}$$

i.e.

$$\delta\left\{\mathrm{Tr}[\hat{\rho}\ln(\hat{\rho})] + A\,\mathrm{Tr}(\hat{\rho})\right\} = 0\,. \tag{10.59c}$$

A represents a Lagrange multiplier. The consideration of fixed expectation values for certain observables $\hat{O}_i$ (additional information about the system), such as

$$\mathrm{Tr}(\hat{\rho}\hat{O}_i) = \langle\hat{O}_i\rangle_{\mathrm{E}} \equiv \bar{O}_i\,, \tag{10.60}$$

leads to further constraints for the variation (10.59a–c); with Lagrange multipliers λ_i the variational equation then becomes

$$\delta\left[\mathrm{Tr}[\hat{\rho}\ln(\hat{\rho})] + A\,\mathrm{Tr}(\hat{\rho}) + \sum_i \lambda_i\,\mathrm{Tr}(\hat{\rho}\hat{O}_i)\right] = 0 \tag{10.61}$$

and we arrive at various grand canonical ensembles. If we assume an eigen representation of $\hat{\rho}$, i.e.

$$\langle\Psi^n|\hat{\rho}|\Psi^m\rangle \equiv \rho^{nm} = \sum_i p^i\delta^{ni}\delta^{mi} = p^n\delta^{nm}\,, \tag{10.62}$$

the general variational equation (10.61) reads

$$\sum_n \delta\rho^{nn}\left(\ln(\rho^{nn}) + 1 + A + \sum_i \lambda_i O_i^{nn}\right) = 0\,, \tag{10.63}$$

where $O_i^{nn} = \langle\Psi^n|\hat{O}_i|\Psi^n\rangle$. Since the variations $\delta\rho^{nn}$ fulfill only the condition

$$\sum_n \delta\rho^{nn} = 0\,,$$

and are otherwise arbitrary, the relation (10.63) can be fulfilled only once every square bracket vanishes for its own. This immediately leads to

$$\rho^{nn} = \exp\left[-\left(A + \sum_i \lambda_i O_i^{nn}\right)\right] = \frac{1}{\mathcal{Z}}\exp\left(-\sum_i \lambda_i O_i^{nn}\right) \tag{10.64}$$

and to the density operator

$$\hat{\rho} = \frac{1}{\mathcal{Z}}\exp\left(-\sum_i\right)\lambda_i\hat{O}_i\,. \tag{10.65a}$$

Here we have formally set $1/\mathcal{Z} = \mathrm{e}^{-A}$. Clearly A and therefore also $\mathcal{Z}$ are normalization factors. The normalization factor Z is known as the *partition function* and follows from $\mathrm{Tr}(\hat{\rho}) = 1$:

$$\mathcal{Z} = \mathrm{Tr}\left[\exp\left(-\sum_i \lambda_i \hat{O}_i\right)\right] . \tag{10.65b}$$

The Lagrange parameters λ_i are denoted as *fugacities*; their physical interpretation is coupled to the corresponding observables $\hat{O}_i$.

Finally, we consider two important special cases of (10.63). The *canonical ensemble* maximizes σ with the additional constraint that the average energy $\langle \hat{H} \rangle_{\mathrm{E}} = \bar{E}$ of the ensemble possesses a given fixed value. It is determined by

$$\delta\sigma \;=\; \delta\,\mathrm{Tr}[\hat{\rho}\ln(\hat{\rho})] \;=\; 0 \tag{10.66a}$$

and the conditions

$$\mathrm{Tr}(\hat{\rho}) \;=\; 1\,, \qquad \mathrm{Tr}(\hat{\rho}\hat{H}) \;=\; \bar{E}\,. \tag{10.66b}$$

According to (10.65a–b) the *canonical density operator* becomes

$$\hat{\rho}_k \;=\; \frac{1}{\mathcal{Z}_k}\,\mathrm{e}^{-\beta\hat{H}}\,, \tag{10.67a}$$

with

$$\mathcal{Z}_k \;=\; \mathrm{Tr}\left(\mathrm{e}^{-\beta\hat{H}}\right), \qquad \beta \;=\; \frac{1}{kT}\,. \tag{10.67b}$$

In addition to the mean energy the mean particle number $\langle \hat{N} \rangle_{\mathrm{E}} = \bar{N}$ is also kept fixed for the so-called *grand canonical ensemble.* The corresponding variational equations

$$\delta\sigma[\hat{\rho}] \;=\; \delta\,\mathrm{Tr}[\hat{\rho}\ln(\hat{\rho})] \;=\; 0 \tag{10.68a}$$

together with the conditions

$$\mathrm{Tr}(\hat{\rho}) \;=\; 1\,, \qquad \mathrm{Tr}(\hat{\rho}\hat{H}) \;=\; \bar{E}\,, \qquad \mathrm{Tr}(\hat{\rho}\hat{N}) \;=\; \bar{N} \tag{10.68b}$$

lead to the *grand canonical density operator*

$$\hat{\rho}_{\mathrm{G}k} \;=\; \frac{1}{\mathcal{Z}_{\mathrm{G}k}}\exp\left[-\beta(\hat{H}-\mu\hat{N})\right] \tag{10.69a}$$

with the partition function

$$\mathcal{Z}_{\mathrm{G}k} = \mathrm{Tr}\left\{\exp\left[-\beta(\hat{H}-\mu\hat{N})\right]\right\}. \tag{10.69b}$$

The parameter μ is called the *chemical potential.* This should do for some of the most important ensembles. In Chap. 6, about quantum gases, we have already discussed one application of the grand canonical density operator. For further applications of these methods we refer to *Thermodynamics and Statistical Mechanics.*[6]

[6] W. Greiner, L. Neise, H. Stöcker: *Thermodynamics and Statistical Mechanics*, 2nd ed. (Springer, New York 1997).

EXAMPLE

10.6 Systems of Noninteracting Fermions and Bosons

We consider a canonical ensemble of free noninteracting fermions and bosons. For each case we will determine the partition function and the mean occupation numbers. The latter will lead us to the Fermi–Dirac and Bose–Einstein statistics.

We use the occupation-number representation to express the Hamiltonian $\hat{H}$ and the particle-number operator $\hat{N}$:

$$\hat{H} = \sum_\alpha \epsilon_\alpha \hat{n}_\alpha \tag{1}$$

and

$$\hat{N} = \sum_\alpha \hat{n}_\alpha , \tag{2}$$

respectively. $\hat{n}_\alpha$ represents the number operator for particles in the quantum state α, where α stands for a characteristic set of quantum numbers such as momentum $\boldsymbol{k}$ and spin projection σ. Furthermore, ϵ_α denotes the corresponding one-particle energy. Pauli's principle means that each one-particle state for fermions can be occupied at most only once, i.e.

$$n_\alpha = 0\,,1 \qquad \text{(fermions)}\,, \tag{3}$$

whereas for bosons each occupation number can be a natural number, i.e.

$$n_\alpha = 0, 1, 2, \dots \qquad \text{(bosons)}\,. \tag{4}$$

The grand canonical density operator takes the form

$$\rho_{\mathrm{G}k} = \frac{1}{\mathcal{Z}_{\mathrm{G}k}} \exp\left[-\beta(\hat{H} - \mu\hat{N})\right] = \frac{1}{\mathcal{Z}_{\mathrm{G}k}} \exp\left(-\beta \sum_\alpha (\epsilon_\alpha - \mu)\hat{n}_\alpha\right) . \tag{5}$$

At first we calculate the grand canonical partition function:

$$\mathcal{Z}_{\mathrm{G}k} = \mathrm{Tr}\left\{ \exp\left[-\beta(\hat{H} - \mu\hat{N})\right]\right\} . \tag{6}$$

$|n_{\alpha_1} n_{\alpha_2} \cdots n_{\alpha_i} \cdots\rangle$ represents an arbitrary many-particle state; then, the trace-operation implies:

$$\begin{aligned} \mathcal{Z}_{\mathrm{G}k} &= \sum_{\{\cdots n_{\alpha_i} \cdots\}} \langle n_{\alpha_1} n_{\alpha_2} \cdots n_{\alpha_i} \cdots | \exp\left[-\beta(\hat{H} - \mu\hat{N})\right] | n_{\alpha_1} n_{\alpha_2} \cdots n_{\alpha_i} \cdots\rangle \\ &= \sum_{\{\cdots n_{\alpha_i} \cdots\}} \prod_{\alpha_i} \exp\left[-\beta(\epsilon_{\alpha_i} - \mu)n_{\alpha_i}\right] , \end{aligned} \tag{7}$$

where the sum runs over all possible configurations of occupation numbers n_{α_i}. The ith one-particle state may have quantum numbers α_i. For a general many-particle state every one-particle state contained in it appears with occupation numbers according to (3) or (4). Hence, we are allowed to interchange the summation with the product series: first we sum over possible individual

Example 10.6.

occupation numbers n and then we take the product over all allowed sets of quantum numbers α_i. We get:

$$\mathcal{Z}_{\mathrm{G}k} = \prod_{\alpha}\sum_{n} \exp\left[-\beta(\epsilon_\alpha - \mu)n\right] . \tag{8}$$

For fermions we arrive at

$$\mathcal{Z}^{\mathrm{f}}_{\mathrm{G}k} = \prod_{\alpha} \{1 + \exp\left[-\beta(\epsilon_\alpha - \mu)\right]\} \tag{9}$$

and for bosons we get

$$\mathcal{Z}^{\mathrm{b}}_{\mathrm{G}k} = \prod_{\alpha} \{1 - \exp\left[-\beta(\epsilon_\alpha - \mu)\right]\}^{-1} . \tag{10}$$

Let us now calculate the mean particle number

$$\bar{n}_\nu = \mathrm{Tr}(\hat{n}_\nu \hat{\rho}_{\mathrm{G}k}) \tag{11}$$

for a system of independent fermions and bosons in thermodynamic equilibrium in the state with energy ϵ_ν. The corresponding grand canonical ensemble comes with a mean energy

$$\bar{E} = \mathrm{Tr}(\hat{H}\hat{\rho}_{\mathrm{G}k})$$

and a mean particle number

$$\bar{N} = \mathrm{Tr}(\hat{N}\hat{\rho}_{\mathrm{G}k}) .$$

The trace operation (11) implies that

$$\begin{aligned}\bar{n}_\nu &= \frac{1}{\mathcal{Z}_{\mathrm{G}k}} \sum_{\{\cdots n_\nu \cdots n_{\alpha_i} \cdots\}} n_\nu \exp\left[-\beta(\epsilon_\nu - \mu)n_\nu\right] \\ &\quad\times \prod_{\alpha_i \neq \nu} \exp\left[-\beta(\epsilon_{\alpha_i} - \mu)n_{\alpha_i}\right] .\end{aligned} \tag{12}$$

As with the determination of the partition function we interchange the product series with the summation:

$$\bar{n}_\nu = \frac{1}{\mathcal{Z}_{\mathrm{G}k}} \left[\sum_n n \exp\left[-\beta(\epsilon_\nu - \mu)n\right]\right] \prod_{\alpha_i \neq \nu} \sum_n \exp\left[-\beta(\epsilon_{\alpha_i} - \mu)n\right] . \tag{13}$$

We insert the partition function according to (8); then all factors in this equation containing an $\alpha_i \neq \nu$ drop out and

$$\bar{n}_\nu = \frac{\sum_n n \exp\left[-\beta(\epsilon_\nu - \mu)n\right]}{\sum_n \exp\left[-\beta(\epsilon_\nu - \mu)n\right]} . \tag{14}$$

Here we have to distinguish between fermions and bosons. For fermionic systems, (14) takes on the form

$$\bar{n}^{\mathrm{f}}_\nu = \frac{\exp\left[-\beta(\epsilon_\nu - \mu)\right]}{1 + \exp\left[-\beta(\epsilon_\nu - \mu)\right]} = \frac{1}{\exp\left[(\beta_\nu - \mu)\right] + 1} . \tag{15}$$

This is the so-called *Fermi distribution.*

Example 10.6.

For the case of bosons we obtain from (14)

$$\bar{n}_\nu^{\mathrm{b}} = \frac{\sum_{n=0}^{\infty} n \exp\left[-\beta(\epsilon_\nu - \mu)n\right]}{\sum_{n=0}^{\infty} \exp\left[-\beta(\epsilon_\nu - \mu)n\right]}$$

$$= \frac{-\frac{\partial}{\partial x}\sum_{n=0}^{\infty} \mathrm{e}^{-xn}}{\sum_{n=0}^{\infty} \mathrm{e}^{-xn}}\,, \qquad x = \beta(\epsilon_\nu - \mu)\,.$$

Summation of the geometric series yields the Bose–Einstein distribution:

$$\bar{n}_\nu^{\mathrm{b}} = \frac{1}{\exp\left[\beta(\epsilon_\nu - \mu)\right] - 1}\,. \tag{16}$$

The temperature and chemical potential appearing in the two distributions (15) and (16) are determined by the fixing of the average total energy

$$\bar{E} = \sum_\nu \bar{n}_\nu \epsilon_\nu$$

and the average particle number

$$\bar{N} = \sum_\nu \bar{n}_\nu$$

of the ensemble considered.

11. Structure of Atoms

In this chapter we discuss several approaches to the calculation of properties of atoms. Exact calculations are extremely difficult. This is because of the repulsive Coulomb interaction between electrons and the spin–orbit interaction of the electrons, which results from the interaction of the electron spin with the electric field of the atom.

First we discuss atoms with two electrons, which represent the simplest system next to the hydrogen atom. Then we will consider systems with many electrons and introduce different approaches (Hartree, Hartree–Fock, and Thomas–Fermi).

11.1 Atoms with Two Electrons

For the case of light atoms, the spin–orbit coupling can be neglected to a good degree. This leads to a considerable simplification. Typical atoms with two elecrons are for example He, H^-, Li^+, and Be^{++}. Nowadays it is possible to completely strip heavy ions, e.g. uranium, of all electrons in heavy-ion accelerators. Once uranium is shot through a thin foil with an energy larger than about 500 MeV per nucleon, i.e. with a total energy of more than $235 \times 500\,\mathrm{MeV} = 117.5\,\mathrm{GeV}$, it loses most of its electrons. In this way it is possible to produce a heavy ion with no, one, two, or more electrons with a nonnegligible probability. This offers the opportunity to study systems with few electrons in the strong central fields of heavy ions experimentally and, in particular, to check the effects of quantum electrodynamics in these systems (self-energy, vacuum polarization). The Schrödinger-type Hamiltonian for a light atom with two electrons is given by

$$\hat{H} = \frac{\hat{\boldsymbol{p}}_1^2}{2m} + \frac{\hat{\boldsymbol{p}}_2^2}{2m} - \frac{Ze^2}{r_1} - \frac{Ze^2}{r_2} + \frac{e^2}{|\boldsymbol{r}_1 - \boldsymbol{r}_2|} \,. \tag{11.1}$$

The first two terms in (11.1) describe the kinetic energy of the two electrons with mass m. The next two terms represent the attractive interaction between the electrons and the nucleus. Z is the nuclear charge and $\boldsymbol{r}_1$ and $\boldsymbol{r}_2$ represent the position vectors of the electrons from the nucleus (see Fig. 11.1). The last term appears because of the repulsive Coulomb interaction between the two electrons. The kinetic energy of the nucleus is neglected in (11.1), because the nucleus is assumed to stay motionless in space as a result of its large mass. The Hamiltonian $\hat{H}$ is supplemented by $\hat{H}_\sigma$, indicating the spin of the

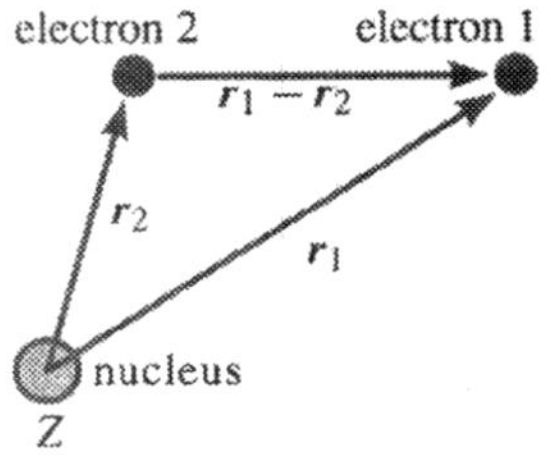

Fig. 11.1. An atom with two electrons. For simplicity the nucleus is assumed to be large and at rest

electrons. Because neither spin–orbit or spin–spin interactions are important for light atoms, they will be neglected here. Nevertheless we always have to remember that the total wavefunction for the two electron system also contains a spin wavefunction. It is important for the exchange symmetry of the total wavefunction, as we shall see soon.

Because of the electron–electron interaction in (11.1) it is impossible to solve the problem of the atom with two electrons exactly. Therefore we have to apply an approximation scheme, which will be discussed in the following. In the first place the electron–electron interaction is neglected. This approximation allows the problem to be solved. Afterwards the influence of the electron–electron interaction [the last term in (11.1)] is treated in the framework of perturbation theory. In zeroth order the electrons feel only the usual Coulomb potential $-Ze^2/r$. Then the lowest-energy state the two electrons occupy is the 1s state. The wavefunction then reads

$$\psi = \phi_{1s}(r_1)\,\phi_{1s}(r_2)\,\chi_0^0 \,. \tag{11.2}$$

ϕ_{1s} describes the 1s state and χ_m^j is the spin wavefunction of the two electrons, which are coupled to the spin j with projection m. The spatial wavefunction $\phi_{1s}(r)$ is given by

$$\phi_{1s}(r) = \frac{1}{\sqrt{\pi}}\left(\frac{Z}{a_0}\right)^{3/2} \mathrm{e}^{-Zr/a_0} \,. \tag{11.3}$$

$a_0 = \hbar^2/me^2$ is Bohr's atomic radius. Because the spatial part of the wavefunction is symmetric under exchange of the electron indices, the spin part has to be antisymmetric. This can be fulfilled only by the *singlet state*

$$\chi_0^0 = \frac{1}{\sqrt{2}}\left[\chi_\uparrow(1)\,\chi_\downarrow(2) - \chi_\downarrow(1)\,\chi_\uparrow(2)\right] \,. \tag{11.4}$$

The antisymmetry of the total wavefunction against electron exchange ensures the Pauli principle is fulfilled. The energy of an electron in a Coulomb potential is given by

$$-\frac{(Ze)^2}{2a_0} = -\frac{mZ^2e^4}{2} \,.$$

For two electrons it is

$$E_0^0 = -\frac{Z^2e^2}{a_0} \,. \tag{11.5}$$

The superscript 0 refers to the zeroth order of perturbation theory. The subscript 0 reflects the ground state.

Now we will calculate the perturbative contribution of the electron–electron interaction. In first order the change in the energy is

$$\Delta E = \langle\psi|\frac{e^2}{|r_1 - r_2|}|\psi\rangle \,, \tag{11.6}$$

where ψ is given by (11.2). Inserting (11.3) into (11.6) leads to

$$\begin{aligned}\Delta E &= \int d^3r_1\, d^3r_2 \left[\frac{1}{\sqrt{\pi}}\left(\frac{Z}{a_0}\right)^{\frac{3}{2}}\right]^2 e^{-Z(r_1+r_2)/a_0} \frac{e^2}{|\boldsymbol{r}_1-\boldsymbol{r}_2|} \\ &\quad\times \left[\frac{1}{\sqrt{\pi}}\left(\frac{Z}{a_0}\right)^{\frac{3}{2}}\right]^2 e^{-Z(r_1+r_2)/a_0} \\ &= \int d^3r_1\, d^3r_2 \frac{1}{\pi^2}\left(\frac{Z}{a_0}\right)^6 \frac{e^2}{|\boldsymbol{r}_1-\boldsymbol{r}_2|} e^{-2Z(r_1+r_2)/a_0} . \end{aligned} \tag{11.7}$$

To solve the integrals, we use the Fourier representation of $1/|\boldsymbol{r}_1-\boldsymbol{r}_2|$ (see Exercise 11.1):

$$\frac{1}{|\boldsymbol{r}_1-\boldsymbol{r}_2|} = \int \frac{d^3k}{(2\pi)^3} e^{i\boldsymbol{k}\cdot(\boldsymbol{r}_1-\boldsymbol{r}_2)} \frac{4\pi}{k^2} . \tag{11.8}$$

For the energy shift we then obtain

$$\Delta E = \int d^3k \frac{e^2}{2\pi^4 k^2}\left(\frac{Z}{a_0}\right)^6 \left|\int d^3r\, e^{i\boldsymbol{k}\cdot\boldsymbol{r}-2Zr/a_0}\right|^2 . \tag{11.9}$$

Here the integrals over $\boldsymbol{r}_1$ and $\boldsymbol{r}_2$ have been exchanged with the integration over $\boldsymbol{k}$. The integral over $\boldsymbol{r}_1$ is complex conjugated to the integral over $\boldsymbol{r}_2$, which explains the last factor in (11.9). Now the following relation holds (see Exercise 11.1):

$$\int d^3r\, e^{-2Zr/a_0} e^{i\boldsymbol{k}\cdot\boldsymbol{r}} = \frac{16\pi Z}{a_0\left[k^2+(2Z/a_0)^2\right]^2} . \tag{11.10}$$

Inserting this into (11.9) yields

$$\Delta E = \int d^3k \frac{128e^2}{\pi^2 k^2}\left(\frac{Z}{a_0}\right)^8 \frac{1}{\left[k^2+(2Z/a_0)^2\right]^4} . \tag{11.11}$$

We integrate over the angular part of the volume element $d^3k = k^2\, dk \sin\vartheta\, d\vartheta d\varphi$. Because the integrand depends only on k and not on the angle, the integration over the angle is easily performed. With the substitution of $x = a_0/2Zk$ we get (see Exercise 11.1)

$$\Delta E = \frac{4Ze^2}{\pi a_0}\int_0^\infty \frac{dx}{[x^2+1]^4} = \frac{5}{8}\frac{Ze^2}{a_0} . \tag{11.12}$$

Thus the energy of the ground state in first-order perturbation theory is given by

$$E_0 = -\frac{Ze^2}{a_0}\left(Z-\frac{5}{8}\right) . \tag{11.13}$$

The calculated energy of the ground state represents an upper value for the real energy. This can be seen as follows. The wavefunction in first-order perturbation theory can be regarded as a test wavefunction. If the expectation value of the corresponding Hamiltonian is generated, an upper limit is always obtained. This statement is nothing more than the variational principle from Ritz.[1]

EXERCISE

11.1 Calculation of Some Frequently used Integrals

Problem. Calculate the following integrals.

$$\text{(a)} \quad \int \frac{d^3k}{(2\pi)^3} e^{i\boldsymbol{k}\cdot\boldsymbol{r}} \frac{4\pi}{k^2} = \frac{1}{|\boldsymbol{r}|}\,.$$

$$\text{(b)} \quad \int d^3r\, e^{-\alpha r} e^{i\boldsymbol{k}\cdot\boldsymbol{r}} = \frac{4\pi\alpha}{(k^2+\alpha^2)^2}\,.$$

$$\text{(c)} \quad \int_0^\infty \frac{dx}{(1+x^2)^4} = \frac{5\pi}{32}\,.$$

Solution.

(a) We calculate the integral

$$I_1 = \int \frac{d^3k}{(2\pi)^3} e^{i\boldsymbol{k}\cdot\boldsymbol{r}} \frac{4\pi}{k^2}\,. \tag{1}$$

It is more comfortable to introduce spherical coordinates

$$\begin{aligned} k_x &= k\cos\theta\cos\varphi\,, & x &= r\cos\theta'\cos\varphi'\,,\\ k_y &= k\cos\theta\sin\varphi\,, & y &= r\cos\theta'\sin\varphi'\,,\\ k_z &= k\sin\theta\,, & z &= r\sin\theta'\,. \end{aligned} \tag{2}$$

It follows that

$$\begin{aligned} I_1 &= \frac{4\pi}{(2\pi)^3}\int_0^\infty k^2\,dk \int_{-\frac{\pi}{2}}^{+\frac{\pi}{2}} \cos\theta\,d\theta \int_0^{2\pi} d\varphi\,\frac{1}{k^2}\\ &\quad\times \exp\left\{ikr\left[\cos\theta\cos\theta'\left(\cos\varphi\cos\varphi' + \sin\varphi\sin\varphi'\right) + \sin\theta\sin\theta'\right]\right\}. \end{aligned} \tag{3}$$

If we adjust the k-coordinate system in a way such that $\boldsymbol{k}$ falls onto the z axis, (3) can be simplified to

$$I_1 = \frac{4\pi}{(2\pi)^3}\int_0^\infty k^2\,dk \int_{-\frac{\pi}{2}}^{+\frac{\pi}{2}} \cos\theta\,d\theta \int_0^{2\pi} d\varphi\,\frac{1}{k^2}\,e^{ikr\sin\theta}\,. \tag{4}$$

After the intergration over φ is performed, the substitution $x = \sin\theta$ gives

$$\begin{aligned} I_1 &= \frac{4\pi}{(2\pi)^3}\,2\pi\int_0^\infty dk \int_{-1}^{+1} dx\,e^{ikrx}\\ &= \frac{4\pi}{(2\pi)^3}\,2\pi\int_0^\infty dk\,\frac{1}{ikr}\,e^{ikrx}\bigg|_{x=-1}^{x=+1}\,. \end{aligned} \tag{5}$$

[1] Consult, for example, W. Greiner: *Quantum Mechanics – An Introduction*, 2nd ed. (Springer, Berlin, Heidelberg 1994).

Exercise 11.1.

Because $e^{i\varphi} = \cos\varphi + i\sin\varphi$ it results in

$$\begin{aligned} I_1 &= \frac{4\pi}{(2\pi)^3}\, 2\pi\, 2 \int_0^\infty dk\, \frac{\sin kr}{kr} \\ &= \frac{4\pi}{(2\pi)^3}\, 2\pi\, 2\, \frac{1}{r} \int_0^\infty d\tau\, \frac{\sin\tau}{\tau} \\ &= \frac{4\pi}{(2\pi)^3}\, 2\pi\, 2\, \frac{1}{r}\, \frac{\pi}{2} = \frac{1}{r}\,. \end{aligned} \tag{6}$$

(b) For the calculation of the integral

$$I_2 = \int d^3r\, e^{-\alpha|\boldsymbol{r}|}\, e^{i\boldsymbol{k}\cdot\boldsymbol{r}} \tag{7}$$

we use an analogous strategy to that for (a). We apply spherical coordinates and perform the integration over the angle:

$$\begin{aligned} I_2 &= \int r^2 dr\, \cos\theta\, d\theta\, d\varphi\, e^{-\alpha r}\, e^{ikr\sin\theta} \\ &= 4\pi \int_0^\infty r^2 dr\, e^{-\alpha r}\, \frac{\sin kr}{kr}\, \frac{4\pi}{k} \int_0^\infty r\, dr\, e^{-\alpha r} \sin kr\,. \end{aligned} \tag{8}$$

This results in

$$\begin{aligned} I_2 &= -\frac{4\pi}{k}\frac{\partial}{\partial\alpha} \int_0^\infty dr\, e^{-\alpha r} \sin kr \\ &= -\frac{4\pi}{k}\frac{\partial}{\partial\alpha} \left[\frac{e^{-\alpha r}}{\alpha^2+k^2}\left(-\alpha\sin kr - k\cos kr\right)\right]_0^\infty \\ &= -\frac{4\pi}{k}\frac{\partial}{\partial\alpha}\frac{k}{\alpha^2+k^2} = -4\pi\frac{\partial}{\partial\alpha}\left(\alpha^2+k^2\right)^{-1} \\ &= -4\pi(-1)\left(\alpha^2+k^2\right)^{-2} 2\alpha = \frac{8\pi\alpha}{(\alpha^2+k^2)^2}\,. \end{aligned}$$

(c) We employ the analytic continuation of the integrand to the complex plane and apply the residue theorem in the following way:

$$\lim_{R\to\infty} \int_{-R}^{+R} f(x)\, dx = 2\pi i \sum_{k=1}^{R} \mathrm{Res}\left(f(z), z = z_k\right),$$

where $\Im(z_k) > 0$ is valid for all z_k, i.e. we close the integration contour in the upper half-plane.

If z_k is a pole of order n, the residue is determined by

$$\mathrm{Res}\left(f(z), z = z_k\right) = \frac{1}{(n-1)!} g^{(n-1)}(z_k)$$

with

$$g(z) = (z - z_k)^n f(z)\,,$$

which can be easily derived by taking the Laurent series representation of $f(z)$. We have

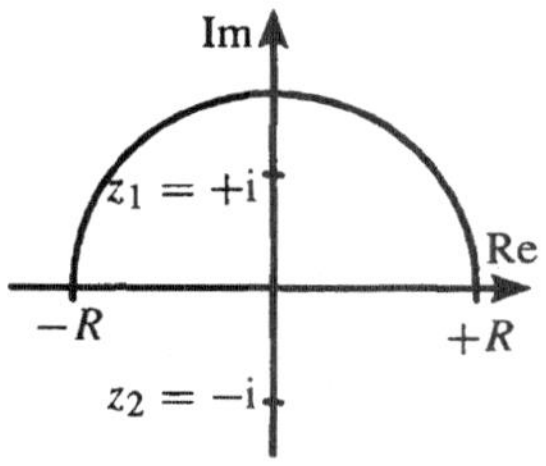

Fig. 11.2. Poles of the function $f(z) = 1/(1+z^2)^4$ and integration contour

Exercise 11.1.

$$\int_0^{\infty} \frac{dx}{(1+x^2)^4} = \frac{1}{2}\int_{-\infty}^{+\infty} \frac{dx}{(1+x^2)^4},$$

because the integrand is an even function. Once we investigate the complex function $f(z) = 1/(1+z^2)^4$ with respect to singularities, we find two poles of order $n = 4$ at $z_1 = +\mathrm{i}$ and $z_2 = -\mathrm{i}$. For the residue in the upper half-plane we get with

$$g(z) = (z-\mathrm{i})^4 \frac{1}{(z-\mathrm{i})^4(z+\mathrm{i})^4} = (z+\mathrm{i})^{-4}$$

and

$$g(z) = (-4)(-5)(-6)(z+\mathrm{i})^{-7}$$

the result

$$\mathrm{Res}\,\big(f(z); z = z_1\big) = \frac{1}{3!}(-4)(-5)(-6)(2\mathrm{i})^{-7} = \frac{5}{2^5\mathrm{i}}.$$

The integral thus has the value

$$\int_0^{\infty} \frac{\mathrm{d}x}{(1+x^2)^4} = \frac{1}{2}2\pi\mathrm{i}\frac{5}{2^5\mathrm{i}} = \frac{5}{32}\pi.$$

A very interesting exercise is the determination of the ionization energy. This is the minimum energy needed to remove an electron from an atom. A hydrogen-like atom remains thereafter. Hence, the ionization energy is the difference between the energies of the hydrogen-like atom and the atom with two electrons:

$$E_{\mathrm{ion}} = E_0^{1\mathrm{e}} - E_0^{2\mathrm{e}} \tag{11.14}$$

with $E_0^{1\mathrm{e}} = -Z^2e^2/2a_0$ and $E_0^{2\mathrm{e}}$ as in (11.13). As an example we consider He. The measured ionization energy amounts to 1.807 Rydberg (Ry). One Rydberg is the ionization energy of a hydrogen atom, or $1\,\mathrm{Ry} = e^2/2a_0 = me^4/2\hbar^2 = 13.606\,\mathrm{eV}$. For $E_0^{1\mathrm{e}}$ we know the exact result

$$E_0^{1\mathrm{e}} = -\frac{Z^2e^2}{2a_0} = -Z^2\frac{me^4}{2\hbar^2} = -Z^2\,\mathrm{Ry} \tag{11.15}$$

with $a_0 = \hbar^2/me^2$. Thus for the energy $E_0^{2\mathrm{e}}$ of the ground state of the atom with two electrons we have

$$E_0^{2\mathrm{e}} = E_0^{1\mathrm{e}} - E_{\mathrm{ion}}^{\mathrm{measured}} = -4 - 1.807 = -5.807\,\mathrm{Ry} \tag{11.16a}$$

for $Z = 2$. Using the approximate result (11.13) we obtain

$$E_0^{2\mathrm{e}} = -5.5\,\mathrm{Ry}. \tag{11.16b}$$

As expected this lies above the exact result. The value presented in (11.16b) in not very accurate with a deviation of about 5%. The reason for this is that the electron does not feel the complete nuclear charge, since the other electron shields it somewhat. In order to improve the result an effective nuclear charge

Z_{eff} has to be introduced. In other words: our trial wavefunction must be altered. One electron in a 1s state now possesses the wavefunction

$$\phi_{1s}(r) = \frac{1}{\sqrt{\pi}}\left(\frac{Z_{\text{eff}}}{a_0}\right)^{3/2} e^{-Z_{\text{eff}}r/a_0} \tag{11.17}$$

instead of (11.3). Z_{eff} is varied until the minimum energy value is reached.

The expectation value of $\hat{H}$ for the new trial wavefunction can be estimated in the following way. The electron–electron interaction provides the energy shift $5Z_{\text{eff}}e^2/8a_0$, i.e. the same value as in (11.12), but Z has to be replaced by Z_{eff}. The other two kinetic contributions each yield the energy $+Z_{\text{eff}}^2e^2/2a_0$, i.e. the kinetic energy one electron has in a Coulomb potential $Z_{\text{eff}}e^2/r$. Care has to be taken when dealing with for the expectation value of the Coulomb potential in the Hamiltonian, since there the true Z appears. It is

$$-\frac{Ze^2}{r} = \frac{Z}{Z_{\text{eff}}}\left(-\frac{Z_{\text{eff}}e^2}{r}\right). \tag{11.18}$$

Now the effective charge appears in the expectation value. The energy contribution of the potential, i.e. the expectation value, is

$$\left\langle\left|-\frac{Z_{\text{eff}}e^2}{r}\right|\right\rangle = -\frac{Z_{\text{eff}}^2e^2}{a_0}.$$

We summarize all this and arrive at

$$E_0^{2e}(Z, Z_{\text{eff}}) = \frac{5}{8}\frac{Z_{\text{eff}}e^2}{a_0} + 2\left(\frac{Z_{\text{eff}}^2e^2}{2a_0} - \frac{ZZ_{\text{eff}}e^2}{a_0}\right). \tag{11.19}$$

The number 2 in front of the second term states that the energy contribution is the same for the two electrons. The result (11.19) can be rewritten as

$$\begin{aligned} E_0^{2e}(Z, Z_{\text{eff}}) &= \left(\frac{5}{8}Z_{\text{eff}} - 2ZZ_{\text{eff}} + Z_{\text{eff}}^2\right)\frac{e^2}{a_0} \\ &= \left[Z_{\text{eff}} - \left(Z - \frac{5}{16}\right)\right]^2\frac{e^2}{a_0} - \left(Z - \frac{5}{16}\right)^2\frac{e^2}{a_0}. \end{aligned} \tag{11.20}$$

Obviously $E_0(Z_{\text{eff}})$ has a minimum at

$$Z_{\text{eff}} = Z - \frac{5}{16}. \tag{11.21}$$

We take this value and the energy of the ground state becomes

$$E_0^{2e}(Z) = -\left(Z - \frac{5}{16}\right)^2\frac{e^2}{a_0} = -2\left(Z - \frac{5}{16}\right)^2 \text{Ry}. \tag{11.22}$$

For He we obtain

$$E_0^{2e}(Z = 2) = 5.695\,\text{Ry}. \tag{11.23}$$

Compared to the preceding calculation (11.16b) the additional energy shift is enormous. However, the accuracy is not sufficient, because the same method leads to an incorrect result for the H^- atom, for which we obtain

$$E_0(H^-) = -0.945\,\text{Ry}, \tag{11.24a}$$

whereas the H atom has a ground-state energy of -1 Ry. According to (11.14) the ionization energy becomes

$$E_{\rm ion}^{\rm H^-} = -0.055\,{\rm Ry}\,. \tag{11.24b}$$

This energy is negative and implies instability for H^-; in other words the energy of the ground state of the H^- atom is higher than that of a H atom plus an electron. The calculated instability of the H^- atom is in contrast to the experimental findings: the H^- atom is stable. This means that the calculational method has to be further improved. In practice a many-parameter dependent trial wavefunction is constructed and minimized towards these parameters. Since this process is rather involved, we do not further discuss this approach.[2]

11.2 The Hartree Method

The Hartree method is constructed in such a way that it works for large atoms. It is a "mean-field" method. This means that the electrons create a central potential as a mean field, in which they are moving. The mean potential is a function of the electron wavefunctions. First we construct an ansatz for the mean potential and solve the Schrödinger equation with it. Then a new mean potential is reconstructed with the help of the wavefunctions obtained. With the new mean potential the Schrödinger equation is solved again, to obtain improved wavefunctions. These iteration steps are repeated several times until convergence is obtained, i.e. until we reach consistency. We are now going to explain this procedure in more detail.

To begin with we construct an ansatz for the wavefunction of an N-electron system, which fulfills our physical expectations as well as possible. The simplest ansatz is a product ansatz, which constructs the entire wavefunction from N one-particle wavefunctions:

$$\psi\,(1,\ldots,N) = \varphi_1\,(1)\,\varphi_2\,(2)\ldots\varphi_N\,(N)\ . \tag{11.25}$$

$\varphi_i(k)$ represents the single-particle wavefunction of the kth electron. The wavefunction (11.25) is not antisymmetric, so one has to be aware of unphysical states. This can be taken into consideration: for example, each electron is placed into another single-particle state. In this way the Pauli principle is fulfilled.

Initially, the mean potential is constructed as follows. The mean charge density of the electron in state φ_j at the point $\boldsymbol{r}'$ is given by $-e|\varphi_j(\boldsymbol{r}')|^2$. Considering the Coulomb interaction between the electrons, we find that the potential energy of another electron at position $\boldsymbol{r}$ is

$$\int \mathrm{d}^3\boldsymbol{r}'\frac{e^2}{|\boldsymbol{r}-\boldsymbol{r}'|}\,|\varphi_j(\boldsymbol{r})|^2\ . \tag{11.26}$$

[2] See, e.g., H.A. Bethe, R. Jackiw: *Intermediate Quantum Mechanics*, 2nd ed. (Benjamin, London 1968).

Summing over all electrons yields

$$\begin{aligned} V_i'(\boldsymbol{r}) &= \int d^3r' \frac{e^2}{|\boldsymbol{r}-\boldsymbol{r}'|} \sum_{j\neq i} |\varphi_j(\boldsymbol{r}')|^2 - \frac{Ze^2}{r} \\ &\equiv V_i(\boldsymbol{r}) - \frac{Ze^2}{r} \end{aligned} \tag{11.27}$$

for the potential for the electron in state φ_i. The last term describes the contribution of the interaction with the nucleus. The wavefunction of the ith electon is determined by the Schrödinger equation

$$\left[-\frac{\hbar^2}{2m}\nabla^2 - \frac{Ze^2}{r} + V_i(\boldsymbol{r})\right]\varphi_i(\boldsymbol{r}) = \epsilon_i \varphi_i(\boldsymbol{r}) \ . \tag{11.28}$$

Equations (11.27) and (11.28) represent a system of coupled equations, which unfortunately can only be solved numerically. The solutions $\varphi_i(\boldsymbol{r})$ are in general not orthogonal, so they have to be orthogonalized. The calculation can be further simplified once $V_i(\boldsymbol{r})$ is replaced by the potential averaged over all angles

$$V_i'(r) = \frac{1}{4\pi}\int d\varphi \sin\vartheta d\vartheta \, V_i(\boldsymbol{r}) \ . \tag{11.29}$$

This method was introduced the first time by **Hartree**. It is therefore called the *Hartree method.*

11.3 Thomas–Fermi Method

To determine the mean behavior of the potentials $V_i(\boldsymbol{r})$, Fermi and Thomas developed a semi-classical approach, which can be applied to many-electron atoms. First the mean potential (11.27) for the ith electron is approximately given by

$$V(\boldsymbol{r}) = \int d^3r' \frac{e^2}{|\boldsymbol{r}-\boldsymbol{r}'|} \sum_{j} |\varphi_j(\boldsymbol{r}')|^2 - \frac{Ze^2}{r} \ . \tag{11.30}$$

Contrary to (11.27) the sum also runs over $j = i$: it is supposed that for a many-electron atom the influence of one electron can be neglected. The electron density is given by

$$\sum_j |\varphi_j(\boldsymbol{r}')|^2 = \rho(\boldsymbol{r}') \ . \tag{11.31}$$

How does this density depend on the potential $V(\boldsymbol{r})$? This is a self-consistent problem, since $V(\boldsymbol{r})$ again is given by $\rho(\boldsymbol{r}')$. According to the Thomas–Fermi method, $\varphi_i(\boldsymbol{r})$ is replaced by a plane wave

$$\varphi_i(\boldsymbol{r}) \sim e^{(i/\hbar)\boldsymbol{p}_i(\boldsymbol{r})\cdot\boldsymbol{r}} \ . \tag{11.32}$$

The justification for this is that the kinetic energy of the electrons in heavy atoms is very large; the potential varies only slowly compared to the spatial change (oscillation) of the electrons. The total value of the linear momentum $p_i(\boldsymbol{r})$ depends on the energy ϵ_i of the ith electron via

$$\epsilon_i = \frac{p_i^2(\boldsymbol{r})}{2m} + V(r) . \tag{11.33}$$

The above approximation is valid only if

$$\epsilon_i \gg V(\boldsymbol{r}) , \tag{11.34}$$

so the kinetic energy dominates. Consequently, the potential can be neglected and we are allowed to consider free electrons as in (11.32). Within this semi-classical approach an electron in state φ_i is more or less uniformly distributed over the classically allowed space and has a small probability in the classically forbidden region. This is illustrated in Fig. 11.3. Electrons occupy all states up to a maximum energy ε_F, which is the Fermi energy. The distance from the potential curve $V(r)$ (we assume a spherical potential) to the energy (ϵ_i or ϵ_F) of the electron provides the kinetic energy of the electron.

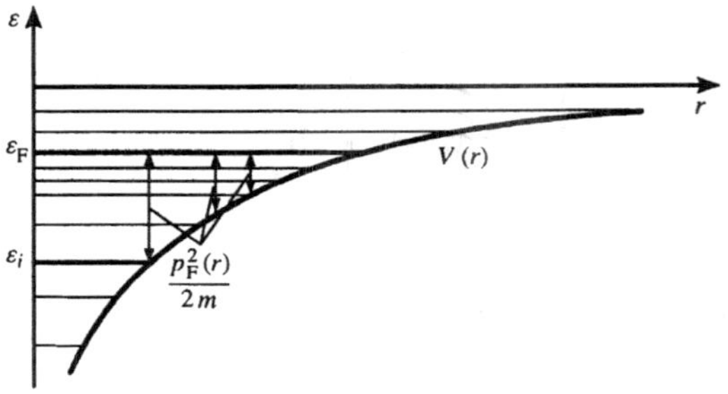

Fig. 11.3. Electron distribution in the ground state of the Thomas–Fermi model. The kinetic energy of an electron at the Fermi edge ϵ_F is marked for various distances r from the center

For a given $\boldsymbol{r}$, electron states exist for which the kinetic energies are between $0 \leq E_{\text{kin}}(r) \leq \epsilon_F - V(r)$. The maximum momentum is given by the kinetic energy of the electron with energy ϵ_F, i.e.

$$p_F(\boldsymbol{r}) = \{2m[\epsilon_F - V(r)]\}^{1/2} . \tag{11.35}$$

Only electrons with energy $\epsilon_i \geq V(r)$ are considered. Energetically lower lying electrons are hardly contributing to the density at position $\boldsymbol{r}$. The electron density at $\boldsymbol{r}$ is then given by the sum over all electrons allowed to stay at $\boldsymbol{r}$, i.e.

$$\begin{aligned}\rho(\boldsymbol{r}) &= 2\frac{1}{(2\pi)^3}\int \mathrm{d}^3k = 2\frac{1}{(2\pi\hbar)^3}\int \mathrm{d}^3p = 2\frac{1}{(2\pi\hbar)^3}4\pi\int_0^{p_F} p^2\,\mathrm{d}p \\ &= \frac{p_F^3}{3\pi^2\hbar^3} .\end{aligned} \tag{11.36}$$

In correspondence with (11.32) we have used $\int \mathrm{d}^3n = \int \mathrm{d}^3kL^3/(2\pi)^3$ for the number of plane waves in volume L^3; consequently the density becomes $\int \mathrm{d}^3n/L^3 = (2/(2\pi)^3)\int \mathrm{d}^3k$. The factor 2 takes into account that two electrons (with spin projection $+1/2$ and $-1/2$) can occupy a plane wave state. With (11.35) and (11.36) we obtain

$$\rho(r) = \frac{\{2m[\epsilon_F - V(r)]\}^{3/2}}{3\pi^2\hbar^3} \tag{11.37}$$

for the electron density. Insertion of (11.37) and (11.31) into (11.30) determines the mean potential $V(r)$ of a many-electron system. The following *integral equation* results:

$$\begin{aligned} V(\boldsymbol{r}) &= \int d^3r' \frac{e^2}{|\boldsymbol{r}-\boldsymbol{r}'|}\rho(\boldsymbol{r}') \\ &= \int d^3r' \frac{e^2\{2m[\epsilon_F - V(\boldsymbol{r}')]\}^{3/2}}{|\boldsymbol{r}-\boldsymbol{r}'|\,3\pi^2\hbar^3} . \end{aligned} \tag{11.38}$$

There is also the possiblity to determine the potential $V(r)$ with a differential equation. To do this we transform (11.30) into a Poisson equation:

$$\Delta V(r) = -4\pi e^2 \rho(r) \qquad \text{for} \quad r > 0 , \tag{11.39a}$$

where $\Delta = \boldsymbol{\nabla}\cdot\boldsymbol{\nabla}$ represents the Laplace operator. Here we have used the relation $\Delta|\boldsymbol{r}-\boldsymbol{r}'|^{-1} = -4\pi\delta|\boldsymbol{r}-\boldsymbol{r}'|$. If we additionally require a rotationally symmetric potential $V(r)$, we obtain

$$\frac{1}{r^2}\frac{\partial}{\partial r}r^2\frac{\partial}{\partial r}[-V(r)] = \frac{4e^2}{3\pi\hbar^3}\{2m[\epsilon_F - V(r)]\}^{3/2} \tag{11.39b}$$

by using (11.37) and (11.39a). Notice that the angular part of the Laplace operator disappears once applied to $V(r)$. Equation (11.39b) represents the determining equation for $V(r)$. It is a nonlinear differential equation. To solve the problem we use the asymptotic behavior of $V(r)$ for $r \to 0$ and $r \to \infty$. In the first case we expect $V(r) \approx -Ze^2/r$, since for $r \to 0$ the screening of the Coulomb potential of the central nucleus by the electrons decreases. If r is equal to the atomic radius R, we require $\epsilon_F - V(R) = 0$ for neutral atoms, i.e. from outside no charge is observed. In this case the electrons totally screen the central charge. If the atom is k times ionized and possesses only $(Z-k)$ electrons, a Coulomb potential of the form $(Z-k)e^2/r$ has to show up outside the atom. For simplicity we restrict ourselves to neutral atoms. Since for neutral atoms we set $V(R) = 0$, $\epsilon_F = 0$ is fixed, too. This will be of use later. Let us make the ansatz

$$V(r) = -\frac{Ze^2}{r}\phi(x) \tag{11.40a}$$

and introduce the new combination for the variable r and of the parameter a_0

$$\begin{aligned} x &= \frac{r}{b}Z^{1/3} \\ b &= \frac{1}{2}\left(\frac{3\pi}{4}\right)^{2/3} a_0 , \qquad a_0 = \frac{\hbar^2}{me^2} . \end{aligned} \tag{11.40b}$$

With this we obtain for (11.38) and (11.39b) the differential equation

$$x^{1/2}\frac{d^2\phi(x)}{dx^2} = \phi^{3/2}(x) . \tag{11.41}$$

One boundary condition for $\phi(x)$ is $\phi(0) = 1$, because for $r \to 0$ the potential $V(r)$ has to agree with the Coulomb potential. Unfortunately (11.41) can be solved only numerically. The solution has the following form:

$$\phi(x) = \begin{cases} 1 - 1.59x & \text{for} \quad x \to 0 \\ 144/x^3 & \text{for} \quad x \to \infty \end{cases} . \tag{11.42}$$

For $x \to \infty$ this solution behaves unphysically. $\phi(x)$ vanishes only for $x \to \infty$ and not for $x \gtrsim (R/b)Z^{1/3}$, i.e. for x larger than the radius R of the electron distribution. For x (or r) $\to 0$ the potential has the form

$$V(r) \approx -\frac{Ze^2}{r} + 1.80\, Z^{4/3} \frac{e^2}{a_0} . \tag{11.43}$$

The reason for the appearance of the last term in (11.43) originates from the finite contribution of the electrons near the origin. The potential at the origin increases because of the outer electron shell. There is another interesting property predicted by the Thomas–Fermi model: Since $\phi(x)$ in (11.40a) does not depend on atomic parameters (radius, charge, etc.), the form of the atom is always the same. Only the length scale is changed. Because the scale parameter r is proportional to $Z^{-1/3}$ in correspondence with (11.40b), the extension of the atom decreases with Z. According to (11.43) the second term dominates for large Z and $r \approx a_0$, i.e. $V(r)$ can be approximately set equal to $V(r) = 1.80(e^2/a_0)Z^{4/3}$ near $r \approx a_0$. With this we obtain for the electron density $\rho(\boldsymbol{r})$ [see (11.37)] of neutral atoms ($\epsilon_F \equiv 0$) the behavior $\rho(r) \sim Z^2$. Hence, according to (11.36) it follows a behavior $\sim Z^{2/3}$ for the mean electron momentum. A typical atomic density distribution, calculated with the Thomas–Fermi model, is sketched in Fig. 11.4.

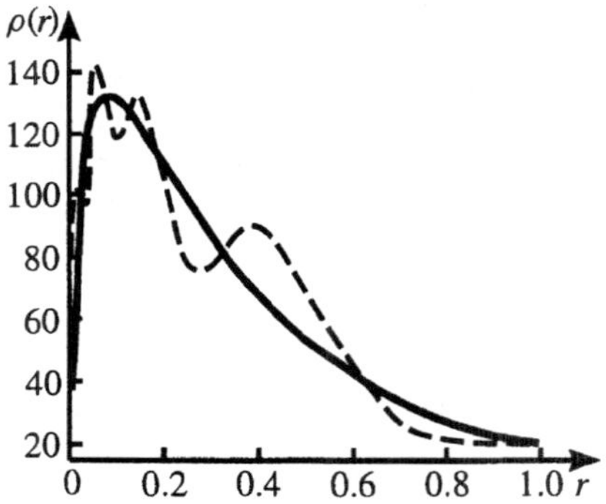

Fig. 11.4. A typical density distribution of a large atom (mercury) within the Thomas–Fermi model. r is given in atomic length units

So far, we have not considered *angular momentum*. If one takes angular momentum states into account, $V(r)$ has to be modified. The simplest way to do this is to add the angular momentum barrier (centrifugal potential) to the Coulomb potential $V(r)$. Such an effective potential then reads

$$V_{\text{eff}}(r) = V(r) + \frac{l(l+1)\hbar^2}{mr^2} . \tag{11.44}$$

A bound state exists only if $V_{\text{eff}}(r)$ has only one minimum for $l = 0$ states for $Z \leq 4$; for $4 < Z \leq 19$ it has a minimum for $l = 0$ and $l = 1$, and so forth. This means that for $Z \leq 4$ only bound s states exist, but no p states, etc. All this is in good agreement with experiments, e.g. the first bound p state appears at $Z = 5$.

Finally it should be mentioned that for small and large distances from the atomic nucleus the Thomas–Fermi approximation breaks down. For small distances the solution of the potential varies too strongly and for large distances the behavior of the potential is unphysical (see remark given above). A drawback of the Thomas–Fermi model is the neglection of the exchange interaction of the electrons.[3]

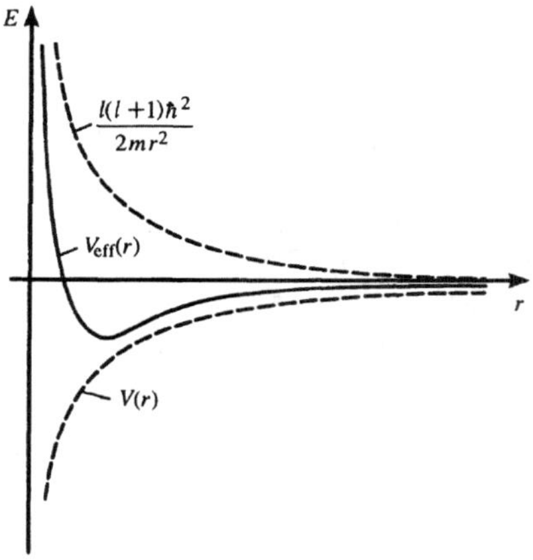

Fig. 11.5. The sum of the Coulomb potential $V(r)$ and centrifugal potential $[l(l+1)\hbar^2]/2mr^2$ yields the effective potential $V_{\text{eff}}(r)$. If a minimum arises at $E < 0$, then stable atoms with angular momentum exist. For very large angular momenta this is not the case; the atoms are unstable

11.4 The Hartree–Fock Method

We have not yet considered the *exchange effect* of the electrons; the Hartree–Fock method will take care of that. For that purpose a test wavefunction is constructed, which is already entirely antisymmetric. One uses the *Slater determinant* of one-particle states:

$$\psi(1,2,\ldots,N) = \frac{1}{\sqrt{N!}}\begin{vmatrix} \varphi_1(1) & \ldots & \varphi_1(N) \\ \vdots & & \vdots \\ \varphi_N(1) & \ldots & \varphi_N(N) \end{vmatrix} . \tag{11.45}$$

All one-particle wavefunctions are already orthonormal, i.e. $\langle\varphi_i|\varphi_j\rangle = \delta_{ij}$. The expectation value of the Hamiltonian is calculated in the following way. The kinetic energy is given by

[3] Modern versions of the Thomas–Fermi model can be found in the book by R.M. Dreizler, E.K.U. Gross: *Density Functional Theory – An Approach to the Quantum Many-Body Problem* (Springer, Berlin, Heidelberg 1995).

$$-\sum_{i=1}^{N}\int \mathrm{d}^3r\,\varphi_i^*(\boldsymbol{r})\,\frac{\hbar^2}{2m}\Delta\varphi_i(\boldsymbol{r}) \;=\; -\sum_{i=1}^{N}\int \mathrm{d}^3r\,\frac{\hbar^2\,|\boldsymbol{\nabla}\varphi_i(\boldsymbol{r})|^2}{2m}\,. \tag{11.46}$$

A partial integration has been used and Gauss' law was applied. The interaction energy of the electrons with the nucleus is given by

$$-\sum_{i=1}^{N}\int \mathrm{d}^3r\,|\varphi_i(\boldsymbol{r})|^2\,\frac{Ze^2}{r}\,. \tag{11.47}$$

In order to calculate the contribution of the electron–electron interaction it is of advantage to employ second quantization, which we encountered in the third chapter. There the electron–electron interaction takes on the following form [see Chaps. 4 and 6, especially (6.1) and Exercise 6.1]:

$$\frac{1}{2}\sum_{ss'}\int \mathrm{d}^3r\,\mathrm{d}^3r'\,\frac{e^2}{|\boldsymbol{r}-\boldsymbol{r}'|}\langle\hat{\psi}_s^\dagger(\boldsymbol{r})\,\hat{\psi}_{s'}^\dagger(\boldsymbol{r}')\,\hat{\psi}_{s'}(\boldsymbol{r}')\,\hat{\psi}_s(\boldsymbol{r})\,\rangle\,. \tag{11.48}$$

The subscripts s and s' attached to the field operators $\hat{\psi}_s^\dagger(\boldsymbol{r}), \hat{\psi}_s(\boldsymbol{r})$ refer to the spins. The expression inside the bracket "$\langle\cdots\rangle$" describes the annihilation and subsequent creation of two electrons at points $\boldsymbol{r}$ and $\boldsymbol{r}'$ with spin s and s', respectively. This interaction is weighted with the factor $e^2/|\boldsymbol{r}-\boldsymbol{r}'|$. The bracket itself gives the expectation value. This expectation value provides the so-called *pair-correlation function*:

$$\begin{aligned}&\langle\hat{\psi}_s^\dagger(\boldsymbol{r})\,\hat{\psi}_{s'}^\dagger(\boldsymbol{r}')\,\hat{\psi}_{s'}(\boldsymbol{r}')\,\hat{\psi}_s(\boldsymbol{r})\,\rangle\\ &\quad= \langle\hat{\psi}_s^\dagger(\boldsymbol{r})\,\hat{\psi}_s(\boldsymbol{r})\,\rangle\langle\hat{\psi}_{s'}^\dagger(\boldsymbol{r}')\,\hat{\psi}_{s'}(\boldsymbol{r}')\,\rangle\\ &\qquad-\delta_{ss'}\langle\hat{\psi}_s^\dagger(\boldsymbol{r})\,\hat{\psi}_s(\boldsymbol{r}')\,\rangle\langle\hat{\psi}_s^\dagger(\boldsymbol{r}')\,\hat{\psi}_s(\boldsymbol{r})\,\rangle\,.\end{aligned} \tag{11.49}$$

This will be proven in Exercise 11.2. The second term allows for the possible exchange of electrons at positions $\boldsymbol{r}$ and $\boldsymbol{r}'$. The density of the electrons at point $\boldsymbol{r}$ is

$$\rho(\boldsymbol{r}) \;=\; \sum_s\langle\hat{\psi}_s^\dagger(\boldsymbol{r})\,\hat{\psi}_s(\boldsymbol{r})\,\rangle \;=\; \sum_i|\varphi_i(\boldsymbol{r})|^2 \tag{11.50}$$

and the "*correlation density*"

$$\rho_s(\boldsymbol{r},\boldsymbol{r}') \;=\; \langle\hat{\psi}_s^\dagger(\boldsymbol{r})\,\hat{\psi}_s(\boldsymbol{r}')\,\rangle \;=\; \sum_{\substack{i\\ \text{with spin } s}}\varphi_i^*(\boldsymbol{r})\,\varphi_i(\boldsymbol{r}')\,. \tag{11.51}$$

EXERCISE

11.2 Proof of (11.49)

Problem. Prove relation (11.49), i.e.

$$\begin{aligned}&\langle\hat{\psi}_s^\dagger(\boldsymbol{r})\hat{\psi}_{s'}^\dagger(\boldsymbol{r}')\hat{\psi}_{s'}(\boldsymbol{r}')\hat{\psi}_s(\boldsymbol{r})\rangle\\ &\quad= \langle\hat{\psi}_s^\dagger(\boldsymbol{r})\hat{\psi}_s(\boldsymbol{r})\rangle\langle\hat{\psi}_{s'}^\dagger(\boldsymbol{r}')\hat{\psi}_{s'}(\boldsymbol{r}')\rangle\\ &\qquad-\delta_{ss'}\langle\hat{\psi}_s^\dagger(\boldsymbol{r})\hat{\psi}_s(\boldsymbol{r}')\rangle\langle\hat{\psi}_s^\dagger(\boldsymbol{r}')\hat{\psi}_s(\boldsymbol{r})\rangle\,.\end{aligned} \tag{1}$$

The expectation value refers to a Slater determinant.

Exercise 11.2.

Solution. If $\varphi_i(\boldsymbol{r})$ are the orbital wavefunctions and i includes all quantum numbers needed, then the field operators $\hat{\psi}_s^\dagger(\boldsymbol{r})$ and $\hat{\psi}_s(\boldsymbol{r})$ are given by

$$\hat{\psi}_s^\dagger(\boldsymbol{r}) = \sum_i \varphi_i^*(\boldsymbol{r})\hat{b}_{is}^\dagger \tag{2}$$

and

$$\hat{\psi}_s(\boldsymbol{r}) = \sum_j \varphi_j(\boldsymbol{r})\hat{b}_{js} \tag{3}$$

[compare with (3.43)]. The left-hand side of (1) can then be written as

$$\begin{aligned}&\langle \hat{\psi}_s^\dagger(\boldsymbol{r})\hat{\psi}_{s'}^\dagger(\boldsymbol{r})\hat{\psi}_{s'}(\boldsymbol{r}')\hat{\psi}_s(\boldsymbol{r})\rangle \\ &\quad = \sum_{ijkl} \varphi_i^*(\boldsymbol{r})\varphi_j^*(\boldsymbol{r}')\varphi_k(\boldsymbol{r}')\varphi_l(\boldsymbol{r})\langle \hat{b}_{is}^\dagger \hat{b}_{js'}^\dagger \hat{b}_{ks'}\hat{b}_{ls}\rangle .\end{aligned} \tag{4}$$

In the following, we limit our consideration to the calculation of $\langle \hat{b}_{is}^\dagger \hat{b}_{js'}^\dagger \hat{b}_{ks'}\hat{b}_{ls}\rangle$. For this we have to consider the expectation value of the Slater determinant. The corresponding state is given by

$$\prod_{ps} \hat{b}_{ps}^\dagger|0\rangle \equiv |\psi\rangle . \tag{5}$$

All states, from the lowest up to the highest, are occupied. By anticommutation on both sides (*bra* and *ket*) we find that $|\psi\rangle$ can be written as

$$|\psi\rangle = \hat{b}_{ks'}^\dagger \hat{b}_{ls}^\dagger|\psi'\rangle , \tag{6}$$

which are the corresponding creation operators belonging to the two annihilation operators in (4). Since we apply the anticommutation on *bra* and *ket*, no sign is changed. The combination (ks') and (ls) does not appear in $|\psi'\rangle$, since otherwise two electrons would be in the same orbit. Applying $\hat{b}_{ks'}\hat{b}_{ls}$ on $|\psi\rangle$ gives

$$\begin{aligned}\hat{b}_{ks'}\hat{b}_{ls}|\psi\rangle &= \hat{b}_{ks'}\hat{b}_{ls}\hat{b}_{ks'}^\dagger \hat{b}_{ls}^\dagger|\psi'\rangle \\ &= \hat{b}_{ks'}\delta_{lk}\delta_{ss'}\hat{b}_{ls}^\dagger|\psi'\rangle - \hat{b}_{ks'}\hat{b}_{ks'}^\dagger|\psi'\rangle \\ &= (\delta_{lk}\delta_{ss'} - 1)|\psi'\rangle .\end{aligned} \tag{7}$$

The same result is obtained once $\hat{b}_{is}^\dagger \hat{b}_{js'}^\dagger$ is applied to the left-hand side. There is another simplification: $\hat{b}_{ks'}\hat{b}_{ls}$ annihilates states (ks') and (ls). In order to have the same state in the expectation value on the left-hand side (otherwise the expectation value is 0), it has to hold that $(is) = (ls)$ and $(js') = (ks')$, or $(is) = (ks')$ and $(js') = (ls)$. The latter gives an additional minus sign, because one has to transform the expectation value into the form

$$\langle \hat{b}_{ls}^\dagger \hat{b}_{ks'}^\dagger \hat{b}_{ks'}\hat{b}_{ls}\rangle . \tag{8}$$

This leads to

$$(\delta_{il}\delta_{jk} - \delta_{jl}\delta_{ik}\delta_{ss'})(\delta_{lk}\delta_{ss'} - 1)(\delta_{lk}\delta_{ss'} - 1) = \delta_{il}\delta_{jk} - \delta_{jl}\delta_{ik}\delta_{ss'} . \tag{9}$$

Inserting (9) into (4) gives

Exercise 11.2.

$$\begin{aligned}&\langle\hat{\psi}_s^\dagger(\boldsymbol{r})\hat{\psi}_{s'}^\dagger(\boldsymbol{r}')\hat{\psi}_{s'}(\boldsymbol{r}')\hat{\psi}_s(\boldsymbol{r})\rangle\\ &\quad=\sum_{ij}\varphi_i^*(\boldsymbol{r})\varphi_j^*(\boldsymbol{r}')\varphi_j(\boldsymbol{r}')\varphi_i(\boldsymbol{r})-\delta_{ss'}\sum_{ij}\varphi_i^*(\boldsymbol{r})\varphi_j^*(\boldsymbol{r}')\varphi_i(\boldsymbol{r}')\varphi_j(\boldsymbol{r})\\ &\quad=\sum_i|\varphi_i(\boldsymbol{r})|^2\sum_j|\varphi_j(\boldsymbol{r}')|^2-\delta_{ss'}\sum_i\varphi_i^*(\boldsymbol{r})\varphi_i(\boldsymbol{r}')\sum_j\varphi_j^*(\boldsymbol{r}')\varphi_j(\boldsymbol{r})\,.\end{aligned}\tag{10}$$

Furthermore,

$$\langle\hat{\psi}_s^\dagger(\boldsymbol{r})\hat{\psi}_s(\boldsymbol{r}')\rangle=\sum_{ij}\varphi_i^*(\boldsymbol{r})\varphi_j(\boldsymbol{r}')\langle\hat{b}_{is}^\dagger\hat{b}_{js}\rangle=\sum_i\varphi_i^*(\boldsymbol{r})\varphi_j(\boldsymbol{r}')\,.\tag{11}$$

Here we have used the fact that the expectation value is only different from zero if the same state is annihilated in *bra* and *ket*. For $\boldsymbol{r}'=\boldsymbol{r}$ in (11) the squared modulus appears. Inserting this formula into (10) yields

$$\begin{aligned}&\langle\hat{\psi}_s^\dagger(\boldsymbol{r})\hat{\psi}_{s'}^\dagger(\boldsymbol{r}')\hat{\psi}_{s'}(\boldsymbol{r}')\hat{\psi}_s(\boldsymbol{r})\rangle\\ &\quad=\langle\hat{\psi}_s^\dagger(\boldsymbol{r})\hat{\psi}_s(\boldsymbol{r})\rangle\langle\hat{\psi}_{s'}^\dagger(\boldsymbol{r}')\hat{\psi}_{s'}(\boldsymbol{r}')\rangle\\ &\qquad-\delta_{ss'}\langle\hat{\psi}_s^\dagger(\boldsymbol{r})\hat{\psi}_s(\boldsymbol{r}')\rangle\langle\hat{\psi}_s^\dagger(\boldsymbol{r}')\hat{\psi}_s(\boldsymbol{r})\rangle\,.\end{aligned}\tag{12}$$

Note that the subscripts i and j are connected with s and s', respectively.

If we summarize (11.46)–(11.51), we obtain for the expectation value of the Hamiltonian

$$\begin{aligned}\langle\hat{H}\rangle=&-\sum_i\int\mathrm{d}^3r\left(\frac{\hbar^2|\boldsymbol{\nabla}\varphi_i(\boldsymbol{r})|^2}{2m}+\frac{Ze^2}{r}|\varphi_i(\boldsymbol{r})|^2\right)\\ &+\frac{1}{2}\sum_{ij}\int\mathrm{d}^3r\,\mathrm{d}^3r'\frac{e^2}{|\boldsymbol{r}-\boldsymbol{r}'|}|\varphi_i(\boldsymbol{r})|^2|\varphi_j(\boldsymbol{r}')|^2\\ &-\frac{1}{2}\sum_{ij}\delta_{s_is_j}\int\mathrm{d}^3r\,\mathrm{d}^3r'\frac{e^2}{|\boldsymbol{r}-\boldsymbol{r}'|}\varphi_i^*(\boldsymbol{r})\varphi_i(\boldsymbol{r}')\varphi_j^*(\boldsymbol{r}')\varphi_j(\boldsymbol{r})\,.\end{aligned}\tag{11.52}$$

Here $\delta_{s_is_j}$ means that the summation is only over such states i and j, that have the same spin quantum number. In order to obtain the equation of motion for $\varphi_i(\boldsymbol{r})$, (11.52) has to be minimized. It needs to be guaranteed, that $\{\varphi_i(\boldsymbol{r})\}$ remains an orthonormal set. The variation with respect to $\varphi_i^*(\boldsymbol{r})$ immediately provides the following nonlinear equation:

$$\begin{aligned}&\left(-\frac{\hbar^2\Delta}{2m}-\frac{Ze^2}{r}\right)\varphi_i(\boldsymbol{r})+\int\mathrm{d}^3r'\frac{e^2}{|\boldsymbol{r}-\boldsymbol{r}'|}\sum_j\varphi_j^*(\boldsymbol{r}')\\ &\quad\times\left[\varphi_j(\boldsymbol{r}')\varphi_i(\boldsymbol{r})-\varphi_j(\boldsymbol{r})\varphi_i(\boldsymbol{r}')\delta_{s_is_j}\right]=\epsilon_i\varphi_i(\boldsymbol{r})\,,\quad i=1,\dots,N\,.\end{aligned}\tag{11.53}$$

These are the *Hartree–Fock equations*.[4] They differ from the Hartree equations (11.28) by the additional $-\varphi_j(\boldsymbol{r})\varphi_i(\boldsymbol{r}')$ part in (11.53). The antisymmetric part

[4] D.R. Hartree: Proc. Cambridge Phil. Soc. **24** (1928) 89, III; V. Fock: Z. Physik **61** (1930) 126; J.C. Slater: Phys. Rev. **35** (1930) 250.

$$[\varphi_j(\boldsymbol{r}')\varphi_i(\boldsymbol{r}) - \varphi_j(\boldsymbol{r})\varphi_i(\boldsymbol{r}')]$$

indicates the amplitude with which the electrons in states i and j can be found at positions $\boldsymbol{r}$ and $\boldsymbol{r}'$, respectively. This term automatically fulfills the Pauli priciple in (11.53).

EXERCISE

11.3 The Hartree–Fock Equation as a Nonlocal Schrödinger Equation

Problem. Transform the Hartree–Fock equation (11.53) into the form of a nonlocal Schrödinger equation.

Solution. The Hartree–Fock equation (11.53) is

$$\begin{aligned}\epsilon_i\varphi_i(\boldsymbol{r}) &= \left(-\frac{\hbar^2}{2m}\Delta - \frac{Ze^2}{r}\right)\varphi_i(\boldsymbol{r}) + \int \mathrm{d}^3r' \frac{e^2}{|\boldsymbol{r}-\boldsymbol{r}'|}\sum_j \varphi_j^*(\boldsymbol{r}') \\ &\quad \times [\varphi_j(\boldsymbol{r}')\varphi_i(\boldsymbol{r}) - \varphi_j(\boldsymbol{r})\varphi_i(\boldsymbol{r}')\delta_{s_is_j}] \,. \end{aligned} \tag{1}$$

Obviously

$$V(\boldsymbol{r}) = -\frac{Ze^2}{r} + \int \mathrm{d}^3r' \frac{e^2}{|\boldsymbol{r}-\boldsymbol{r}'|}\left(\sum_{j=1}^{Z}\varphi_j^*(\boldsymbol{r}')\varphi_j(\boldsymbol{r}')\right) \tag{2}$$

represents the *local potential* energy consisting of the central potential energy $-Ze^2/r$ and of the mean energy of the electrons in the field of all electrons (*Hartree energy*)

$$\int \mathrm{d}^3r' \frac{e^2}{|\boldsymbol{r}-\boldsymbol{r}'|}\left(\sum_{j=1}^{Z}\varphi_j^*(\boldsymbol{r}')\varphi_j(\boldsymbol{r}')\right) \,. \tag{3}$$

The (last) *exchange term* in (1) can be written as

$$\int U(\boldsymbol{r},\boldsymbol{r}')\varphi_i(\boldsymbol{r}')\,\mathrm{d}^3r' \,, \tag{4}$$

where

$$U(\boldsymbol{r},\boldsymbol{r}') = -\frac{e^2}{|\boldsymbol{r}-\boldsymbol{r}'|}\sum_{j=1}^{Z}\varphi_j^*(\boldsymbol{r}')\varphi_j(\boldsymbol{r})\delta_{s_is_j} \,. \tag{5}$$

It characterizes the nonlocal interaction energy. The exchange energy is nonlocal and the Hartree–Fock equation (1) can be rewritten in the form

$$\epsilon_i\varphi_i(\boldsymbol{r}) = \left(-\frac{\hbar^2}{2m}\Delta + V(\boldsymbol{r})\right)\varphi_i(\boldsymbol{r}) + \int U(\boldsymbol{r},\boldsymbol{r}')\varphi_i(\boldsymbol{r}')\,\mathrm{d}^3r' \,. \tag{6}$$

Equation (5) guarantees hermitecity,

$$U(\boldsymbol{r},\boldsymbol{r}') = U^*(\boldsymbol{r}',\boldsymbol{r}) \,, \tag{7}$$

Exercise 11.3.

which enforces the reality of the eigenvalues ϵ_i. In order to convince ourselves we rewrite (6) for $\varphi_k^*(\boldsymbol{r})$:

$$\epsilon_k^* \varphi_k^*(\boldsymbol{r}) = \left(-\frac{\hbar^2}{2m}\Delta + V(\boldsymbol{r})\right) + \int U(\boldsymbol{r},\boldsymbol{r}')^* \varphi_k^*(\boldsymbol{r}')\,\mathrm{d}^3r' \,. \tag{8}$$

After multiplication of (6) with $\varphi_k^*(\boldsymbol{r})$ and (8) with $\varphi_i(\boldsymbol{r})$, subsequent integration over $\boldsymbol{r}$ and subtraction of both equations yields

$$\begin{aligned}(\epsilon_i - \epsilon_k^*) \int \varphi_k^*(\boldsymbol{r})\varphi_i(\boldsymbol{r})\,\mathrm{d}^3r \\ = \int [\varphi_k^*(\boldsymbol{r})U(\boldsymbol{r},\boldsymbol{r}')\varphi_i(\boldsymbol{r}') - \varphi_i(\boldsymbol{r})U^*(\boldsymbol{r},\boldsymbol{r}')\varphi_k^*(\boldsymbol{r}')]\,\mathrm{d}^3r\,\mathrm{d}^3r' \,.\end{aligned} \tag{9}$$

Because of (7) the right-hand side vanishes and it follows that

$$\epsilon_i = \epsilon_k^* \tag{10}$$

for $i = k$ because $\int |\varphi_i(\boldsymbol{r})|^2\,\mathrm{d}^3r = 1$.

In (5) the nonlocal density (also called the density matrix) appears:

$$\rho(\boldsymbol{r},\boldsymbol{r}') = \sum_{j=1}^{Z} \varphi_j^*(\boldsymbol{r})\varphi_j(\boldsymbol{r}') \,. \tag{11}$$

If the sum runs over all states (i.e. $Z \to \infty$) and not only over the occupied states, it would result in

$$\rho(\boldsymbol{r},\boldsymbol{r}') = \sum_{j=1}^{\infty} \varphi_j^*(\boldsymbol{r})\varphi_j(\boldsymbol{r}') = \delta(\boldsymbol{r}-\boldsymbol{r}') \tag{12}$$

and

$$\int \rho(\boldsymbol{r},\boldsymbol{r}')\,\mathrm{d}^3r' = \int \delta(\boldsymbol{r}-\boldsymbol{r}')\,\mathrm{d}^3r' = 1 \,. \tag{13}$$

The nonlocal interaction in (5) or (6) would then be local. Equation (13) remains approximately valid for the nonlocal density (11). Often it is practical to consider the mean potential

$$\overline{U}(\boldsymbol{r}) = \int U(\boldsymbol{r},\boldsymbol{r}')\,\mathrm{d}^3r' \tag{14}$$

and the *effective* potential

$$U_{\mathrm{eff}}(\boldsymbol{r})\varphi_i(\boldsymbol{r}) = \int U(\boldsymbol{r},\boldsymbol{r}')\varphi_i(\boldsymbol{r}')\,\mathrm{d}^3r' \,. \tag{15}$$

These relations define $\overline{U}(\boldsymbol{r})$ and $U_{\mathrm{eff}}(\boldsymbol{r})$. In Example 11.4 the exchange potential will be considered in more detail and a handy expression will be derived within a reasonable approximation.

The energy eigenvalue ϵ_i of (11.53) is about the same as the negative value of the energy needed to transport an electron from the ith state out of the atom, i.e. equal to the ionization energy of the ith electron. This can be seen from the following. Multiplying (11.53) from the left side with $\varphi_i^*(\boldsymbol{r})$ and integrating over $\boldsymbol{r}$ yields

$$\begin{aligned}\epsilon_i &= -\int \mathrm{d}^3r \left(\frac{|\nabla\varphi_i(\boldsymbol{r})|^2}{2m} + \frac{Ze^2}{r}|\varphi_i(\boldsymbol{r})|^2\right) \\ &\quad + \sum_j \int \mathrm{d}^3r\, \mathrm{d}^3r' \frac{e^2}{|\boldsymbol{r}-\boldsymbol{r}'|}\varphi_i^*(\boldsymbol{r})\varphi_j^*(\boldsymbol{r}') \\ &\quad \times [\varphi_j(\boldsymbol{r}')\varphi_i(\boldsymbol{r}) - \varphi_j(\boldsymbol{r})\varphi_i(\boldsymbol{r}')]\,\delta_{s_i s_j}\,. \end{aligned} \quad (11.54)$$

A comparison with (11.52) shows that this is exactly the amount by which $\langle\hat{H}\rangle$ is reduced once all terms refering to the fixed "i" are canceled. $-\epsilon_i$ is the energy needed to remove an electron from the ith state. Notice that in (11.52) the sum is over all indices i, j. If the term refering to the ith state is removed, then the fixed i appears twice in the double sum. Here we assume that the wavefunctions of the other particles are not changed. This is of course only approximately true; for this reason ϵ_i can only be approximately identified with the ionization energy. These facts are known as "*Koopmans' theorem*".[5]

The Hartree–Fock state (11.45) represents an improved test wavefunction in comparison to the Hartree state. Therefore the Hartree–Fock state is lower in energy than the analogous state calculated with the Hartree method. However, it turns out that the calculated energy is still about 1 eV per electron pair too large. Corrections have to be introduced.

Instead of looking deeper into details, it is worth studying the form of the general solution: As a first, good approximation to start with we assume the mean potential experienced by the electron to be spherically symmetric. The wavefunction of the electron can then be described by quantum numbers n, l, m_l, m_s. Here n represents the radial quantum number and l the orbital angular momentum with projection m_l. Furthermore, m_s is the magnetic projection of the spin ($\pm 1/2$) and $n - l - 1$ gives the number of nodes in the radial wavefunction. The wavefunction then takes on the form

$$\varphi_i = R_{nl}(r) Y_{lm_l}(\theta, \varphi) \chi_{m_s}\,. \quad (11.55)$$

In the hydrogen atom all orbital angular momenta l belonging to the same n are degenerate; this is a special property of the $1/r$ potential. With several electrons present, this is no longer valid. Identical orbital angular momenta belonging to the same main quantum number are *no longer degenerate.*

We consider a fixed l. For a given l there are $(2s+1)(2l+1) = 2(2l+1) = 4l+2$ states. These $4l+2$ states constitute a shell. The Hartree–Fock trial wavefunction is then constructed in such a way that all one-particle states are filled from below; we have to indicate how many electrons sit in an orbit. In general the last shell is not entirely filled. As a result of the Pauli principle only 2 electrons fit into an s shell ($l = 0$), only 6 into an p shell ($l = 1$), 10 electrons into a d shell ($l = 2$), and so on. For example, the nitrogen atom N

[5] T.H. Koopmans: Physica **1** (1933) 104.

has two electrons in the 1s orbit, two in the 2s shell, and three in the 2p shell. The electron configuration is then given by $(1s)^2(2s)^2(2p)^3$. The number in front of s or p indicates the main quantum number n.

EXAMPLE

11.4 An Approximation for the Hartree–Fock Exchange Term

We intend to determine an approximation for the exchange term of the expectation value (11.52) of $\hat{H}$ in the Hartree–Fock approach. For that purpose we have to rewrite this term:

$$\begin{aligned} X &= -\frac{1}{2}\sum_{i,j}\delta_{s_i s_j}\int \mathrm{d}^3r\,\mathrm{d}^3r'\frac{e^2}{|\boldsymbol{r}-\boldsymbol{r}'|}\varphi_i^*(\boldsymbol{r})\varphi_i(\boldsymbol{r}')\varphi_j^*(\boldsymbol{r}')\varphi_j(\boldsymbol{r}) \\ &= -e^2\int \mathrm{d}^3r_1\,\mathrm{d}^3r_2\,\frac{1}{r_{12}}\,\rho_s^2(\boldsymbol{r}_1,\boldsymbol{r}_2)\,, \end{aligned} \tag{1}$$

where $r_{12} = |\boldsymbol{r}_1 - \boldsymbol{r}_2|$ and $\rho_s(\boldsymbol{r}_1,\boldsymbol{r}_2)$ is given by (11.51). A factor 2 results, because the combination $s_i = s_j$ appears twice in the sum over i and j. We now introduce relative ($\boldsymbol{r} = \boldsymbol{r}_1 - \boldsymbol{r}_2$) and center-of-mass coordinates $[\boldsymbol{R} = (\boldsymbol{r}_1 + \boldsymbol{r}_2)/2]$:

$$X = -e^2\int \mathrm{d}^3r\,\mathrm{d}^3R\,\frac{1}{r}\,\bar{\rho}_s^2(\boldsymbol{R},\boldsymbol{r})\,. \tag{2}$$

For $\bar{\rho}_s(\boldsymbol{R},\boldsymbol{r})$ we want to find a suitable substitute within the Thomas–Fermi approach. For that purpose we replace in the sum over the product $\varphi_i^*(\boldsymbol{r}_1)\varphi_i(\boldsymbol{r}_j)$ of occupied states (11.51) by a sum over plane waves with their wave numbers fulfilling $|\boldsymbol{k}| < |\boldsymbol{k}_\mathrm{F}|$:

$$\begin{aligned} \rho_s(\boldsymbol{r}_1,\boldsymbol{r}_2) &\approx \frac{1}{V}\sum_j \mathrm{e}^{\mathrm{i}\boldsymbol{k}_j\cdot(\boldsymbol{r}_1-\boldsymbol{r}_2)} \\ &\approx \frac{1}{(2\pi)^3}\int \mathrm{e}^{\mathrm{i}\boldsymbol{k}\cdot\boldsymbol{r}}\,\mathrm{d}^3k \\ &= \frac{4\pi}{(2\pi)^3}\int_0^{k_\mathrm{F}(R)}\frac{\sin kr}{kr}k^2\,\mathrm{d}k \\ &= \frac{1}{2\pi^2}\frac{1}{r^3}(\sin k_\mathrm{F}(R)r - k_\mathrm{F}(R)r\cos k_\mathrm{F}(R)r) \\ &\approx \bar{\rho}_s(R,r)\,. \end{aligned} \tag{3}$$

Inserting (3) into (2) and integrating over $\boldsymbol{r}$ leads to

$$\begin{aligned} X &\approx -e^2\frac{1}{4\pi^4}4\pi\int r^2\,\mathrm{d}r\,\mathrm{d}^3R\,\frac{1}{r}\left[\frac{1}{r^3}(\sin k_\mathrm{F}r - k_\mathrm{F}r\cos k_\mathrm{F}r)\right]^2 \\ &= -e^2\frac{1}{4\pi^3}\int \mathrm{d}^3R\,k_\mathrm{F}^4(R)\,. \end{aligned} \tag{4}$$

Using (11.36) with $k_\mathrm{F} = \hbar p_\mathrm{F}$ we get

$$X \approx -e^2 \frac{3}{4}\left(\frac{3}{\pi}\right)^{1/3} \int d^3R\, \rho^{4/3}(R) . \tag{5}$$

Example 11.4.

If we now take into account that

$$\rho(\boldsymbol{r}) = \sum_i \varphi_i^*(\boldsymbol{r})\varphi_i(\boldsymbol{r}) ,$$

insert (5) into (11.52) and vary with respect to $\varphi_i(\boldsymbol{r})$, we obtain

$$-\int d^3r' \frac{e^2}{|\boldsymbol{r}-\boldsymbol{r}'|} \sum_j \varphi_j^*(\boldsymbol{r}')\varphi_j(\boldsymbol{r})\varphi_i(\boldsymbol{r}')\delta_{s_i,s_j} \longrightarrow -e^2\left(\frac{3}{\pi}\rho(\boldsymbol{r})\right)^{1/3} \tag{6}$$

instead of the nonlocal term in (11.53). The idea of this approximation goes back to Slater,[6] the correct implementation leading to (6) to Kohn and Sham.[7]

11.5 On the Periodic System of the Elements

The periodic system of the elements is depicted in Fig. 11.8 on page 331. Each element is assigned to a box. The charge number is given in the upper right corner of each element symbol. Below the element symbol the last occupied shells are indicated. In the lower part of each box the configuration of the ground state is described. Here one should notice that the lower angular momentum in a many-particle atom is energetically lower than the others. The explanation for this is that the probability of finding an electron with lower angular momentum is larger in the vicinity of the atomic nucleus. For this reason the electrons feel more the Coulomb part and less the screening due to the other electrons.

With these arguments we get a better understanding of the periodic system. In a hydrogen atom there is one electron in a 1s shell, in a He atom the 1s shell is filled, in Li and Be the 2s shell fills up and since all electrons are paired the ground state has zero angular momentum. In the next row, from Na to Ar, first the 3s orbit and then the 3p orbit is filled. From the fourth row on problems arise. First of all one expects, that the 3d orbit is filled after the 3p orbit. It turns out that the 4s orbit collapses in energy in such a way that 3d and 4s are almost energetically degenerate. This leads to a complicated situation, in which for one time the 4s orbit and for the next time the 3d orbit predominates. The fourth row ends with the 4p orbit entirely filled.

The 3d electrons do not reach out of the atom as far as the 4s or 4p electrons do. This is due to the radial wavefunction $R_{nl}(r)$. Therefore the 3d orbit hardly plays a role for chemical properties. However, a completely filled 3d orbit is responsible for magnetic properties, e.g. in Fe or Ni. These considerations can be repeated for the remaining rows.

The outer electrons determine the chemical properties of an atom. If there is a similiar configuration in the outer shell of two atoms, then both atoms

[6] J.C. Slater: Phys. Rev. **81** (1951) 386.

[7] W. Kohn, L.V. Sham: Phys. Rev. **140** (1965) A1133.

have similiar chemical properties. Hence, these atoms are arranged in the same column of the periodic system. The last column on the right-hand side indicates rare gases. For rare gases the outer shell is entirely filled, i.e. all electrons are paired. For this reason the chemical activity is low, or rather is nonexistant. Alkali elements are presented in the first column of the periodic system. They are characterized by a weakly bound s electron in the outer shell. The alkali metals are for that reason chemically very active.

We draw particular attention to the recent discovery of elements with $Z = 104$ and $Z = 112$.[8] They are precursers of the predicted superheavy elements,[9], expected to be centered around $Z = 114$. Their chemical properties had also been predicted by utilizing Dirac–Hartree–Fock calculations. The word "Dirac" in this context means that Hartree–Fock calculations as described here are based on the relativistic Dirac–Hamiltonian. It will be interesting to watch the discovery of further superheavy elements in the near future.[10]

11.6 Splitting of Orbital Multiplets

In order to obtain the state of an atom it is not sufficient to know the electron configuration. For example, for Be the first excited configuration is 2.5 eV above the ground state $(1s)^2(2s)^2$. The excited configuration is given by $(1s)^2(2s)(2p)$. Altogether there are 12 possible states, that is $2(2l+1) = 4 \times 1 + 2 = 6$ for the 2p orbit multiplied with $2 = 4 \times 0 + 2$ for the 2s orbit. The energies of these 12 states differ only slightly. This is because for different states of the same configuration the electron–electron interaction is only slightly different, but it is different. In addition the spin–orbit interaction has to be taken into account; it is small for light atoms and can be treated pertubatively once the electron–electron interaction is calculated. In order to see how the states of a configuration split, we first think about the quantum numbers serving for the classification of the states. The total angular momentum operator commutes with the Hamiltonian. As a result the total angular momentum J is a good quantum number. If we were to neglect the spin–orbit interaction, the total orbital angular momentum $\boldsymbol{L} = \sum_i \hat{\boldsymbol{l}}_i$ and the total spin $\boldsymbol{S} = \sum_i \hat{\boldsymbol{s}}_i$ would commute with the Hamiltonian, i.e. L and S would be nearly good quantum numbers. Taking the projection of the angular momenta onto the z axis into account, we can classify a state belonging to a fixed J by

$$|JM, LM_L, SM_S\rangle \, . \tag{11.56}$$

Here, $M_L = \sum_i m_L(i)$ and $M_S = \sum_i m_S(i)$. The spin projections $m_S(i)$ of the various single particles are not good quantum numbers, because the

[8] See e.g. S. Hofmann, G. Münzenberg, et al.: Z. Phys. A **350** (1955) 277 and 288, and further references therein.

[9] W. Greiner: Int. J. Mod. Phys. E **5** (1996) 1.

[10] The nuclear aspects of superheavy nuclei are discussed in W. Greiner, J.A. Maruhn: *Nuclear Models* (Springer, Berlin, Heidelberg 1996); J. Eisenberg, W. Greiner: *Nuclear Theory*, Vols. 1–3 (North-Holland, Amsterdam 1985).

entire spin wavefunction is a linear combination of different projections; for two electrons we have, for example,

$$\chi_{SM_S} = \sum_{m_S(1),m_s(2)} \left(\frac{1}{2}\,\frac{1}{2}\,S\,|m_S(1)\,m_S(2)\,M_S\right) \chi^{\frac{1}{2}}_{m_S(1)}(1)\chi^{\frac{1}{2}}_{m_S(2)}(2)\,. \tag{11.57}$$

Here $(\frac{1}{2}\frac{1}{2}S|m_s(1)m_s(2)M_s)$ represents a Clebsch–Gordan coefficient for the SO(3) of the spin group. Applying $\hat{S} = \hat{S}_{1z} + \hat{S}_{2z}$ on (11.57), where $\hat{S}_{1z}$ acts only on the spin function of the first and $\hat{S}_{2z}$ on the spin function of the second electron, results in the eigenvalue $M_s = m_s(1) + m_s(2)$. The entire spin wavefunction χ_{SM_s} in (11.57) has to be antisymmetric. For a two-electron state we therefore have only one singlet:

$$\chi_{00} = \chi^{\frac{1}{2}}_{\frac{1}{2}}(1)\chi^{\frac{1}{2}}_{-\frac{1}{2}}(2) - \chi^{\frac{1}{2}}_{-\frac{1}{2}}(1)\chi^{\frac{1}{2}}_{\frac{1}{2}}(2)\,. \tag{11.58}$$

The triplet spin function χ_{1M_s} is symmetric under exchange of both particles. This situation is depicted in Fig. 11.6. If the electron–electron interaction is neglected, then all states of the same main quantum number would be degenerate. The electron–electron interaction causes states belonging to the same configuration but with different L or S to differ somewhat in energy. If various configurations differ by several eV, then the difference between the states belonging to one configuration is given by several 10^{-1} eV. States with the same L and S are degenerate. Once the spin–orbit interaction is added, this degeneration vanishes. The splitting only amounts to $10^{-2}-10^{-3}$ eV for light atoms. For heavy atoms the spin–orbit coupling becomes important and cannot be neglected or treated perturbatively.

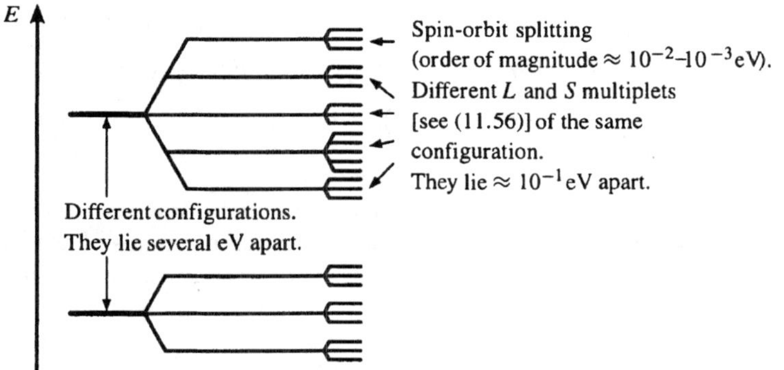
E
Spin-orbit splitting
(order of magnitude ≈ 10^{-2}–10^{-3}eV).
Different L and S multiplets
[see (11.56)] of the same
configuration.
They lie ≈ 10^{-1}eV apart.
Different configurations.
They lie several eV apart.

Fig. 11.6. Configurations and their splitting into $L-S$ multiplets. The latter split further as a result of the spin–orbit coupling

In spectroscopy, states are described by ${}^{2S+1}L_J$. J is the total angular momentum, L the total orbital momentum, and S the total spin. This means, that, for example, ${}^3\mathrm{P}_2$ corresponds to a state with $L = 1$, $S = 1$, and $J = 2$. It is called "triplet P two". At the beginning of the periodic table one can still clearly assign an L, S, and J. Thus a hydrogen atom (H) is described by ${}^2\mathrm{S}_{1/2}$, because a spin-1/2 electron is found in a 1s orbit. The total angular momentum is $J = 1/2$. For the case of a He atom a second electron is added to the 1s orbit. The Pauli principle means it has to be antisymmetric and therefore antiparallel with respect to the first one [see (11.57)]. Therefore the total spin is $S = 0$, the total angular momentum is $J = 0$, and the state is described by $^1S_0$ ("singlet S zero"), i.e. a singlet state. In general, filled shells are described by $^1S_0$ states.

For the next heavier elements the assignment of states becomes more complicated. For example, boron (B) has the configuration $(1s)^2(2s)^2(2p)$, i.e. $L = 1$, $S = 1/2$. The total angular momentum possesses two possible values $J = 1/2$ or $3/2$. The ground state is given by either $^2P_{1/2}$ or $^2P_{3/2}$. The degeneracy for $J = 1/2$ is twofold $(2J + 1 = 1 + 1 = 2)$ and for $J = 3/2$ fourfold $(2J + 1 = 2 \times (3/2) + 1 = 4)$. For carbon (C) with the ground-state configuration $(1s)^2(2s)^2(2p)^2$ it is worse. One finds six possible states for the first electron in the 2p orbit (notice the factor 2 for the spin) and only five for the second electron as a result of the Pauli principle. Altogether there are $6 \times 5/2 = 15$ different states. The factor 1/2 comes from the nondistinction of the electrons, i.e. we cannot say which one is the first and which one is the second electron. What are the possible LSJ combinations? Let us begin with the simplest one: Two electrons in a p orbit can be at most coupled to the total orbital angular momentum $L = 2$. The orbital wavefunctions then read

$$\begin{aligned} &R_{n=2\,L=1}(r_1)R_{21}(r_2)(112|m_1m_2M_L)Y_{1m_1}(1)Y_{1m_2}(2) \\ &\quad = R_{21}(r_1)R_{21}(r_2) \\ &\quad\times \begin{cases} Y_{11}(\Omega_1)Y_{11}(\Omega_2) & \text{for} \quad M_2 = 2 \\ \sqrt{\frac{1}{2}}\,[Y_{10}(\Omega_1)Y_{11}(\Omega_2) \ + Y_{11}(\Omega_1)Y_{10}(\Omega_2)] & \text{for} \quad M_L = 1 \\ \sqrt{\frac{1}{6}}\,[Y_{1-1}(\Omega_1Y_{11}(\Omega_2) \ + 2Y_{00}(\Omega_1)Y_{00}(\Omega_2) & \\ \qquad\qquad + Y_{11}(\Omega_1)Y_{1-1}(\Omega_2)] & \text{for} \quad M_2 = 0 \\ \sqrt{\frac{1}{2}}\,[Y_{10}(\Omega_1)Y_{1-1}(\Omega_2) + Y_{1-1}(\Omega_1)Y_{10}(\Omega_2)] & \text{for} \quad M_2 = -1 \\ Y_{1-1}(\Omega_1)Y_{1-1}(\Omega_2) & \text{for} \quad M_2 = -2\,. \end{cases} \end{aligned} \tag{11.59}$$

Here $\Omega_{1,2}$ denote the polar angles of the electron 1 or 2, respectively. Obviously the wavefuntions (11.59) are symmetric against exchange of the two electrons. For the entire wavefunction to be antisymmetric the electrons have to be coupled to spin 0:

$$\chi_{S=0\,M_S=0} = \frac{1}{\sqrt{2}}\left(\chi_{\frac{1}{2}\frac{1}{2}}(1)\chi_{\frac{1}{2}-\frac{1}{2}}(2) - \chi_{\frac{1}{2}-\frac{1}{2}}(1)\chi_{\frac{1}{2}\frac{1}{2}}(2)\right)\,, \tag{11.60}$$

since this spin wavefunction is antisymmetric. For this case we have $J = L = 2$, i.e. the state is labeled by $^1D_2$. It includes five states $(2J + 1 = 5)$. From 15 possible states ten remain. Next we couple the two electrons to $L = 1$ in the spatial part of their wavefunction, i.e. in their orbital angular momenta. Owing to the symmetry properties of the Clebsch–Gordan coefficients for $L = l_1 = l_2 = 1$, the resulting orbital state is antisymmetric. Thus, the entire spin wavefunction has to possess $S = 1$, i.e.

$$\chi_{S=1\,M_S} = \sum_{m,\nu} \left(\frac{1}{2}\frac{1}{2}1|m\nu M_s\right) \chi_m(1)\chi_\nu(2)$$

$$= \begin{cases} \chi_{\frac{1}{2}}(1)\chi_{\frac{1}{2}}(2) & \text{for} \quad M_S = 1 \\ \frac{1}{\sqrt{2}}\left(\chi_{\frac{1}{2}}(1)\chi_{-\frac{1}{2}}(2) + \chi_{-\frac{1}{2}}(1)\chi_{\frac{1}{2}}(2)\right) & \text{for} \quad M_S = 0 \\ \chi_{-\frac{1}{2}}(1)\chi_{-\frac{1}{2}}(2) & \text{for} \quad M_S = -1 \,. \end{cases} \tag{11.61}$$

These spin wavefunctions are symmetric. With $L = S = 1$ it is possible to generate three different total angular momenta because of the couplings $[L = 1 \otimes S = 1]^{[J]}$ $(J = 0, 1, 2)$. The corresponding labels are

$$3\mathrm{P}_0,\ 3\mathrm{P}_1,\ \text{and}\ 3\mathrm{P}_2\,,$$

i.e. altogether there are nine $[(2 \times 0 + 1) + (2 \times 1 + 1) + (2 \times 2 + 1) = 9)]$. With them we have constructed $5 + 9 = 14$ out of 15 possible states so far. The missing state has $L = 0$. The spatial wavefunction is symmetric, therefore the spin part has to be antisymmecric ($S = 0$). With this $^1\mathrm{S}_0$ state we finally cover all 15 states.

Which energies do these states have? Which state is the lowest? The answers to these questions are given by *empirical rules*, the so-called *Hund's rules*:

1) The LS multiplet with largest spin S is lowest in energy.
2) If there are several multiplets with different orbital angular momenta L but with identical spin S, then the state with largest L has the lowest energy.
3) The total angular momentum J is then fixed by the spin–orbit coupling as follows. Once the last shell is filled to less than half, the total angular momentum J of the ground state is given by $J = |L - S|$. Once the shell is more than half full, it is $J = L + S$.

There is no simple proof of these rules, except by explicit detailed calculations or by "experience" (i.e. experiments). But one can give some arguments contributing to a better understanding. In order to understand the first rule, one has to note that the larger S becomes, the more symmetric the spin wavefunction becomes. The maximum value of S is $S = (1/2)n_e$, where n_e is equal to the number of the valence electrons. The corresponding spin wavefunction reads

$$\chi^{\frac{1}{2}}_{\frac{1}{2}}(1)\chi^{\frac{1}{2}}_{\frac{1}{2}}(2)\ldots\chi^{\frac{1}{2}}_{\frac{1}{2}}(n_e)\,. \tag{11.62}$$

The spin wavefunction is entirely symmetric. In order to guarantee antisymmetry of the total wavefunction, the orbital part has to become "more antisymmetric" as the spin part becomes "more symmetric". This implies that the spatial distance between the electrons becomes increasingly larger. The correlations due to the Pauli principle (here: antisymetric orbital wavefunctions) drive electrons apart. This leads to an effective decrease in the electron–electron repulsion. Thus it is understandable that the state with largest S is lowest in energy. In order to understand the second rule, we look at a classical

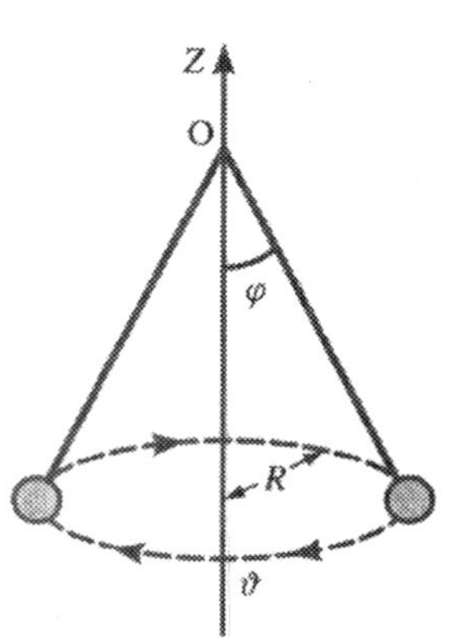

Fig. 11.7. Classical mechanical model for illustration of Hund's rules: The faster the balls rotate (i.e. the larger the angular momentum becomes) the further they fly apart: φ and, consequently, R increase

analogy, illustrated in Fig. 11.7. Two balls are connected with two strings at point O. The two balls are kicked and obtain an orbital angular momentum. It depends only on the velocity v, because the distance R between the balls and the z axis is determined by the velocity v (interplay between graviational and centrifugal force). If the orbital angular momentum increases, the distance between the two balls becomes larger. The same situation holds in an atom. The larger the orbital angular momentum is, the larger the distance between the electrons becomes. The role of the gravitational force is adopted by the central Coulomb force of the nucleus. The electrons behave similarly to the balls of our classical example. An increased distance leads to a lower repulsion between the electrons.

Before we now consider the spin–orbit interaction more closely, a technical point has to be mentioned. For the construction of the Hartree–Fock state, the wavefunction is built from a determinant involving the orbital wavefunctions. However, often many angular momentum states exist for the same configuration. The entire wavefunction cannot be described by a single determinant, but rather it is described by a linear combination of several determinants. Generally the Hartree–Fock state therefore is a linear combination of determinants. As an example we consider the $L = S = 0$ state of two p electrons [consult the preceding discussion, (11.59) and (11.60)]. Except for the normalization, the wavefunction is given by

$$\begin{aligned}&[Y_{11}(1)Y_{1-1}(2) + Y_{1-1}(1)Y_{11}(2) - Y_{10}(1)Y_{10}(2)]\\&\quad\times\left[\chi^{\frac{1}{2}}_{\frac{1}{2}}(1)\chi^{\frac{1}{2}}_{-\frac{1}{2}}(2) - \chi^{\frac{1}{2}}_{-\frac{1}{2}}(1)\chi^{\frac{1}{2}}_{\frac{1}{2}}(2)\right]\,.\end{aligned} \tag{11.63}$$

The factors attached to each term are given by the Clebsch–Gordan coefficients [see (11.59) and (11.60)]. The wavefunction (11.63) can be expressed by the following linear combination of determinants:

$$\begin{aligned}&\begin{vmatrix} Y_{11}(1)\chi^{\frac{1}{2}}_{\frac{1}{2}}(1) & Y_{1-1}(1)\chi^{\frac{1}{2}}_{-\frac{1}{2}}(1)\\ Y_{11}(2)\chi^{\frac{1}{2}}_{\frac{1}{2}}(2) & Y_{1-1}(2)\chi^{\frac{1}{2}}_{-\frac{1}{2}}(2)\end{vmatrix}\\&\quad-\begin{vmatrix} Y_{11}(1)\chi^{\frac{1}{2}}_{-\frac{1}{2}}(1) & Y_{1-1}(1)\chi^{\frac{1}{2}}_{\frac{1}{2}}(1)\\ Y_{11}(2)\chi^{\frac{1}{2}}_{-\frac{1}{2}}(2) & Y_{1-1}(2)\chi^{\frac{1}{2}}_{\frac{1}{2}}(2)\end{vmatrix}\\&\quad-\begin{vmatrix} Y_{10}(1)\chi^{\frac{1}{2}}_{\frac{1}{2}}(1) & Y_{10}(1)\chi^{\frac{1}{2}}_{-\frac{1}{2}}(1)\\ Y_{10}(2)\chi^{\frac{1}{2}}_{\frac{1}{2}}(2) & Y_{10}(2)\chi^{\frac{1}{2}}_{-\frac{1}{2}}(2)\end{vmatrix}\,.\end{aligned} \tag{11.64}$$

In practice, one first constructs a set of Hartree–Fock one-particle states and forms a linear combination of determinants from them. Then the expectation value of the Hamiltonian is calculated.

EXERCISE

11.5 Application of Hund's Rules

Problem. Apply Hund's rules to the atoms B, C, N, O, F, and Ne.

Solution.
Boron (B): As can be deduced from the periodic table there is a p electron. This can be coupled to $J = 1/2$ or $3/2$ ($L = 1$, $S = 1/2$). Since L and S are fixed, only the third of Hund's rules is important. It states that the minimum value of J has to be taken because the shell is filled to less than half. As a consequence $^2\mathrm{P}_{\frac{1}{2}}$ represents the ground state.
Carbon (C): The following orbits are allowed:

$$^1\mathrm{D}_2,\ ^3\mathrm{P}_0,\ ^3\mathrm{P}_1,\ ^3\mathrm{P}_2,\ \text{and}\ ^1\mathrm{S}_0\ ,$$

which altogether amounts to 15 states. According to the first of Hund's rules the state with the largest S is lying lowest. This is the case for the states $^3\mathrm{P}_0$, $^3\mathrm{P}_1$, and $^3\mathrm{P}_2$. The second of Hund's rules makes no difference since it is $L = 1$ for all states. The third rule states that $^3\mathrm{P}_0$ is the energetically lowest state, because the 2p shell (six possible states) is filled only to one third (two electrons in the p shell).
Nitrogen (N): Nitrogen contains three electrons in the p shell. Altogether we have $(6 \times 5 \times 4)/(1 \times 2 \times 3) = 20$ possible states. This number can be deduced as follows. In a p orbit (two spin multiplied by three orbital degrees of freedom) there are six possible states. The first electron then has six states to choose from. The second can find a place in only five states (Pauli's principle) and the third in only four. The denominator takes into consideration that the electrons are undistinguishable (3!). Three p electrons can be coupled to $L = 0, 1, 2$. $L = 3$ is not possible, because the wavefunction is totally symmetric with respect to exchange of coordinates and we are not able to construct an antisymmetric state with three electrons (this can be done for two electrons at most). For the total wavefunction to be antisymmetric, $L = 1$ and 2 can belong only to $S = 1/2$ and $L = 0$ only to $S = 3/2$. This can be seen if the states are explicitly coupled to $L = 0, 1$ (or $S = 1/2$, $3/2$ in spin) and a permutation is applied to the functions. $L = 2$, $S = 1/2$ splits into $^2\mathrm{D}_{3/2}$ and $^2\mathrm{D}_{5/2}$ as a result of the spin–orbit interaction. $L = 1$ and $S = 1/2$ splits into $^2\mathrm{P}_{1/2}$ and $^2\mathrm{P}_{3/2}$; $L = 0$, $S = 3/2$ leads to $^4\mathrm{S}_{3/2}$. In total we have 20 states. The $^4\mathrm{S}_{3/2}$ state possesses the largest value and therefore it is lowest in energy.
Oxygen (O): Oxygen has four electrons in a p orbit, which therefore is more than half full. Always two electrons have to be paired. The property of the ground state is then determined by the two remaining electrons. The problem therefore is equivalent to the C atom once we consider the holes instead of the electrons. There are exactly two. By applying the first of Hund's rules, we find that the $^3\mathrm{P}_0$, $^3\mathrm{P}_1$, and $^3\mathrm{P}_2$ states are lowest. But now the shell is more than half full, so according to the third of Hund's rules the state $^3\mathrm{P}_2$ is lowest.

Exercise 11.5.

Fluorine (F): F is equivalent to the B atom. There are five electrons in the p orbit, i.e. only 1 hole exists. It is $L = 1$, $S = 1/2$, i.e. the possible states are $^2P_{1/2}$ and $^2P_{3/2}$. Since the shell is more than the half full, the $^2P_{3/2}$ state lies energetically lowest.
Neon (Ne): In Ne the p shell is full, i.e. all electrons are paired. There is only one state: $^1S_0$.

11.7 Spin–Orbit Interaction

The spin–orbit interaction is due to the magnetic moment of the electron with the magnetic field caused by the electric current of the central nucleus. In the electron's rest frame the nucleus is orbiting and thus causing a magnetic field at the position of the electron. One may also understand it by remembering that a particle (electron) moving with the velocity $\boldsymbol{v}_e$ through a electric field $\boldsymbol{E}$ (of the central nucleus) sees a magnetic field proportional to $\boldsymbol{v}_e \times \boldsymbol{E}$. We shall derive the spin–orbit interaction from the relativistic Dirac equation in Example 11.7. The result is

$$\hat{H}_{SO} = \sum_i \hat{\boldsymbol{G}}(\boldsymbol{r}_i) \cdot \hat{\boldsymbol{s}}_i \,. \tag{11.65}$$

Here "i" represents the particle index and $\boldsymbol{s}_i$ the spin of the ith electron. $\hat{\boldsymbol{G}}(\boldsymbol{r}_i)$ has the form

$$\hat{\boldsymbol{G}}(\boldsymbol{r}_i) = -\frac{e}{2m_0c^2}\hat{\boldsymbol{v}}_i \times \boldsymbol{E}_i = \frac{1}{2m_0^2c^2}\left(\frac{1}{r_i}\frac{\mathrm{d}V(r_i)}{\mathrm{d}r_i}\right)\hat{\boldsymbol{L}}\,. \tag{11.66}$$

Here $\hat{\boldsymbol{v}}_i$ is the velocity operator of the ith electron, $\boldsymbol{E}_i$ is the electric field acting on the ith electron and $V(r)$ is the corresponding potential with $e\boldsymbol{E}(r) = -(\mathrm{d}V/\mathrm{d}r)(\boldsymbol{r}/r)$ for the central symmetric case. $\boldsymbol{L} = \boldsymbol{r}\times\boldsymbol{p}$ stands for the orbital angular momentum and $\boldsymbol{S} = (\hbar/2)\boldsymbol{\sigma}$ for the spin. The form (11.65) of the spin–orbit interaction is discussed in detail in Example 11.7. At the moment we take the form of this force for granted and show the resulting consequences. Exercise 11.8 presents the details of the transformations necessary for (11.66). For light atoms the spin–orbit coupling (11.65) can be treated perturbatively and we need only the matrix elements of

$$\langle QLSM_LM_S|\hat{H}_{SO}|QLSM_L'M_S'\rangle\,. \tag{11.67}$$

The orbital angular momentum L and the spin S are approximately good quantum numbers. Q stands for the remaining quantum numbers further classifying the state. For given Q, L, and S several values M_L, M_S exist. The corresponding states are degenerate.

Without spin–orbit coupling the Hamiltonian does not depend on the electron spin. The exact wavefunction is then given by a linear combination of products between the spatial wavefunctions $\Psi^{(\alpha)}_{QLM_L}(r_1,\dots,r_k)$ of k electrons and the spin wavefunction $\chi^{(\alpha)}_{SM_S}(S_1,\dots,S_k)$. The wavefunction then takes on the form

$$\langle 1,\dots,k|QLSM_LM_S\rangle = \sum_\alpha a_\alpha \Psi^\alpha_{QLM_L}\chi^\alpha_{SM_S} \tag{11.68}$$

with certain coefficients a_α. The matrix element $\langle QLSM_LM_S|\hat{G}_i \cdot \hat{s}_i|QLSM'_LM'_S\rangle$ then becomes

$$\sum_{\alpha\alpha' i} a^*_\alpha a_{\alpha'}\langle \alpha NLM_L|\hat{G}_i|\alpha' NLM'_L\rangle \cdot \langle \alpha SM_S|\hat{s}_i|\alpha' SM'_S\rangle\,. \tag{11.69}$$

Here we have separated the spatial part from the spin part. The first matrix element in (11.69) is the element of $\hat{G}_i$ between states $\Psi^{(\alpha)}_{QLM_L}$ and $\Psi^{(\alpha')}_{QLM'_L}$. The second matrix element is related to the spin part. Here it is useful to apply the Wigner–Eckart theorem. It states, that

$$\langle \alpha SM_s|\hat{s}_i|\alpha SM'_S\rangle \sim \langle SM_S|\hat{s}|SM'_S\rangle\,, \tag{11.70}$$

i.e. the matrix element of the vector operator $\hat{S}_i$ is proportional to the matrix element of the total spin $\hat{S}$ with respect to the same states. This is explained in more detail in Example 11.6. The proportionality factor does not depend on M_S and $M_{S'}$. A similar relation holds for the spatial part:

$$\langle \alpha QLM_L|\hat{G}_i|\alpha' QLM'_L\rangle \sim \langle LM_L|\hat{L}|LM'_L\rangle\,. \tag{11.71}$$

With this, the matrix elements for the spin–orbit interaction of the ith electron read

$$\begin{aligned}&\langle NLSM_LM_S|\hat{G}_i\cdot\hat{S}_i|NLSM'_LLM'_S\rangle\\ &\quad\sim \langle LM_L|\hat{L}|LM'_L\rangle\cdot\langle SM_S|\hat{S}|SM'_S\rangle\,.\end{aligned} \tag{11.72}$$

Summing over all electrons we obtain

$$\begin{aligned}&\langle QLSM_LM_S|\hat{H}_{\mathrm{SO}}|QLSM'_SM'_L\rangle\\ &\quad= \xi(QLS)\langle LM_L|\hat{L}|LM'_L\rangle\cdot\langle SM_S|\hat{S}|SM'_S\rangle\\ &\quad= \xi(QLS)\langle QLSM_LM_S|\boldsymbol{L}\cdot\boldsymbol{S}|QLSM'_LM'_S\rangle\,.\end{aligned} \tag{11.73}$$

The proportionality factor $\xi(QLS)$ depends on the form of the operator $\hat{G}_i$, but it is the same for all states of the LS multiplet. We realize the usefulness of the Wigner–Eckart theorem: From the relative complicated sum over one-particle spin–orbit couplings (11.65), the structure $\hat{L}\cdot\hat{S}$ with total spin $\hat{S}$ and total orbital angular momentum $\hat{L}$ is peeled off in (11.73); we would not have expected such a remarkable relationship from the beginning. The $\hat{L}\cdot\hat{S}$ operator is easy to diagonalize once the relationship to the total angular momentum $\hat{J}^2 = (\hat{L}+\hat{S})^2$ is noted. We obtain

$$\hat{L}\cdot\hat{S} = \frac{(\hat{J}^2-\hat{L}^2-\hat{S}^2)}{2}\,. \tag{11.74}$$

The matrix element vanishes, if $J' \neq J$ and/or $M' \neq M$. This is simply a consequence of the fact that $\hat{H}_{\mathrm{SO}}$ is rotationally invariant, i.e. a scalar of SU(2) [or O(3)], while J and M classify the states of the various multiplets and those within these multiplets, respectively. The value of the matrix element (11.73) is

$$\langle QLSJM|\hat{H}_{\mathrm{SO}}|QLSJM\rangle = \frac{\xi(QLS)}{2}[J(J+1) - L(L+1) - S(S+1)]\,. \tag{11.75}$$

This equation describes the splitting of the energy levels resulting from the spin–orbit interaction. The splitting of two neighboring levels J and $J-1$ within the same LS multiplet (i.e. levels with identical L and S) is now easy to calculate:

$$\begin{aligned} E_{QLS}(J) - E_{QLS}(J-1) &= \frac{\xi(QLS)}{2}[J(J+1) - J(J-1)] \\ &= \xi(QLS)\,J\,. \end{aligned} \tag{11.76}$$

This relation is called *Landé's interval rule.* For $\xi > 0$ the multiplet is *regular*. For $\xi < 0$ the multiplet is *inverted.* A connection to the third of Hund's rules exists: if the shell is less than half full with electrons, $\xi(QLS)$ has to be positive, so the level with $J-1$ lies lower than the one with J that the minimum value (energetically lowest state) appears for the total angular momentum $J = |L-S|$. If the shell is more than half full, $\xi(QLS)$ has to be negative and the energetically lowest state has the total angular momentum $J = L + S$. The empirical rule therefore is traced back to the sign and the precise structure of $\xi(QLS)$.

EXAMPLE

11.6 The Wigner–Eckart Theorem

The states

$$|JM\rangle\,, \qquad M = -J, -J+1, \ldots, J \tag{1}$$

represent the $2J+1$ eigenvectors of the squared angular momentum operator $\hat{\boldsymbol{J}}^2$ and $\hat{J}_z$:

$$\hat{\boldsymbol{J}}^2|JM\rangle = \hbar^2 J(J+1)|JM\rangle\,, \tag{2}$$

$$\hat{J}_z|JM\rangle = \hbar M|JM\rangle\,. \tag{3}$$

They form a multiplet of SU(2) [or O(3)] and J can be considered as the Casimir operator classifying multiplets. If the representation of $|JM\rangle$ is given in a fixed coordinate system with the z axis as the quantization axis, then in a system rotated by the Euler angles α, β, γ these vectors read[11]

$$|JM\rangle' = \sum_{M'} D^J_{MM'}(\alpha,\beta,\gamma)|JM'\rangle\,. \tag{4}$$

Here $D^J_{MM'}(\alpha,\beta,\gamma)$ is the expectation value of the rotation operator $\hat{D}(\alpha,\beta,\gamma)$ with respect to $|JM\rangle$ and $|JM'\rangle$ in the unrotated system:

$$D^J_{MM'}(\alpha,\beta,\gamma) = \langle JM|\hat{D}(\alpha,\beta,\gamma)|JM'\rangle \tag{5}$$

[11] J.M. Eisenberg, W. Greiner: *Nuclear Theory Vol. 1: Nuclear Models*, 3rd ed. (North-Holland, Amsterdam 1987).

Example 11.6.

with

$$\hat{D}(\alpha,\beta,\gamma) \;=\; \mathrm{e}^{-\mathrm{i}\alpha\hat{J}_z}\,\mathrm{e}^{-\mathrm{i}\beta\hat{J}_y}\,\mathrm{e}^{-\mathrm{i}\gamma\hat{J}_z}\,. \tag{6}$$

For a fixed J the vectors (1) span an invariant subspace[12] $\epsilon^J \subset \epsilon$. This means that the repeated application of shift operators [generators of SU(2)]

$$\hat{J}_\pm \;=\; \hat{J}_x \pm \mathrm{i}\hat{J}_y \tag{7}$$

onto the vectors $|JM\rangle$ of ϵ^J leads either to the zero vector or to the vectors belonging to ϵ^J.[13] Equation (4) determines the transformation properties of these vectors.

Sets of operators are now called *irreducible tensor operators* once they transform under rotation according to

$$\hat{D}\hat{T}_M^{(J)}\hat{D}^{-1} \;=\; \sum_{M'=-J}^{J} \hat{D}^J_{MM'}\hat{T}_{M'}^{(J)}\,, \tag{8}$$

where $\hat{T}_M^{(J)}$ with $M = -J,\dots,J$ represents the $(2J+1)$ components of the irreducible tensor operator of order J.

With

$$|\tau JM\rangle \tag{9}$$

we denote quantum-mechanical states that fulfill the eigenvalue equations (2) and (3). The quantity τ includes all remaining quantum numbers and the following is valid:

$$\langle\tau'J'M'|\tau JM\rangle \;=\; \delta_{\tau\tau'}\delta_{JJ'}\delta_{MM'}\,. \tag{10}$$

The *Wigner–Eckart theorem* now states the following about matrix elements of vectors (9) with irreducible tensor operators:

$$\langle\tau JM|\hat{T}_q^{(k)}|\tau'J'M'\rangle \;=\; \frac{1}{\sqrt{2J+1}}\langle\tau J\|\hat{T}^{(k)}\|\tau'J'\rangle(J'kJ|M'qM)$$
$$q = -k, -k+1, \dots, k\,. \tag{11}$$

The quantity

$$\langle\tau J\|\hat{T}^{(k)}\|\tau'J'\rangle \tag{12}$$

is the *reduced matrix element.* It depends only on τ, τ', J, J', and k and nolonger on q. Its brackets are indicated by a double bar, like $\langle\|\ \|\rangle$. $(J'kJ|M'qM)$ denotes a *Clebsch–Gordan coefficient.* Obviously the Wigner–Eckart theorem states that dependence of a matrix element on the quantum numbers MqM is solely contained in a Clebsch–Gordan coefficient. For its proof see the literature.[14]

Equivalent to the transformation property (8) of the irreducible tensor operators is the validity of the following commutation relations:

---

[12] ϵ is the entire space, which is built up from vectors $|JM\rangle$ for all J and $M \in \{-J, -J+1, \dots, J\}$.

[13] See also W. Greiner, B. Müller: *Quantum Mechanics – Symmetries*, 2nd ed. (Springer, Berlin, Heidelberg 1994).

[14] M.E. Rose: *Theory of Angular Momentum* (Wiley, New York 1957).

Example 11.6.

$$\left[\hat{J}_{\pm}, \hat{T}_q^{(k)}\right] = \hbar\left[k(k+1) - q(q\pm 1)\right]^{\frac{1}{2}} \hat{T}_{q\pm 1}^{(k)} ,$$
$$\left[\hat{J}_0, \hat{T}_q^{(k)}\right] = \hbar q \hat{T}_q^{(k)} , \tag{13}$$

where $\hat{J}_0 = \hat{J}_z$ is the z component of the total angular momentum operator and $\hat{J}_{\pm}$ are the shift operators as given in definition (7). For these operators the following relations also hold:

$$\begin{aligned} \hat{J}_{\pm}|\tau JM\rangle &= \langle JM\pm 1|J_{\pm}|JM\rangle|\tau JM\pm 1\rangle , \\ \langle\tau JM|\hat{J}_{\pm} &= \langle JM|J_{\pm}|JM\mp 1\rangle\langle\tau JM\mp 1| , \\ \hat{J}_0|JM\rangle &= \hbar M|JM\rangle . \end{aligned} \tag{14}$$

We now go back to (11.70). We easily check that the spherical components of the spin operator of the ith particle

$$\begin{aligned} (\hat{S}_{\pm 1})_i &\equiv \hat{T}_{\pm 1} = \mp\frac{1}{\sqrt{2}}(\hat{S}_x \pm \mathrm{i}\hat{S}_y)_i , \\ (\hat{S}_0)_i &\equiv \hat{T}_0 = (\hat{S}_z)_i \end{aligned} \tag{15}$$

fulfill the commutation relations given in (13) with $k = 1$. Hence, $(\hat{S}_q)_i$, $q = -1, 0, +1$, represents the components of the irreducible *vector operators* $\hat{\boldsymbol{S}}_i$ and $\hat{\boldsymbol{S}} = \sum_{i=1}^{n}\hat{\boldsymbol{S}}_i$ (n is the particle number) is the total spin operator.

Using the commutation relations (13) and the relations (14) we obtain

$$\langle\alpha SM_S|(\hat{S}_q)_i|\alpha' SM'_S\rangle = \frac{1}{\hbar}\langle\alpha S\|\hat{S}_i\|\alpha' S\rangle\langle SM_S|\hat{S}_q|SM_S\rangle \tag{16}$$

with $q = -1, 0, +1$. The first matrix element on the right-hand side of (16) is the reduced matrix element, which is independent of M_S, M'_S, and q. We will now show the validity of (16). First we derive the following relations from (13):

$$\begin{aligned} \left[\hat{J}_0, \hat{T}_{\pm 1}\right] &= \pm\hbar\hat{T}_{\pm 1} , \\ \left[\hat{J}_{\pm}, \hat{T}_{\pm 1}\right] &= 0 , \\ \left[\hat{J}_{\pm}, \hat{T}_{\mp 1}\right] &= \hbar\sqrt{2}\hat{T}_0 . \end{aligned} \tag{17}$$

In (17) we have used the abbreviations introduced in (15).

From (17) and (14) it follows that

$$\begin{aligned} \langle\tau JM|\hat{T}_{\pm}|\tau' JM'\rangle &= \pm(M - M')\langle\tau JM|\hat{T}_{\pm 1}|\tau' JM'\rangle \\ &= \delta_{M,M'\pm 1}\langle\tau JM|\hat{T}_{\pm 1}|\tau' JM'\rangle . \end{aligned}$$

Therefore this matrix element has to be proportional to $\langle\tau JM|\hat{J}_{\pm}|\tau' JM'\rangle$. Furthermore, by using (14) again it follows from (17) that

$$\begin{aligned} \langle\tau JM|[\hat{J}_{\pm}, \hat{T}_{\pm 1}]|\tau' JM'\rangle &= 0 \\ = \delta_{M,M'\pm 1}(\langle\tau JM'\pm 1|\hat{T}_{\pm 1}|\tau' JM'\rangle&\langle JM'\pm 2|\hat{J}_{\pm}|JM'\pm 1\rangle \\ - \langle\tau JM'\pm 2|\hat{T}_{\pm 1}|\tau' JM'\pm 1\rangle&\langle JM'\pm 1|\hat{J}_{\pm}|JM'\rangle) . \end{aligned} \tag{18}$$

This equation can be rewritten in another way:

Example 11.6.

$$\frac{\langle \tau JM' \pm 1|\hat{T}_{\pm 1}|\tau' JM'\rangle}{\langle JM' \pm 1|\hat{J}_{\pm}|JM'\rangle} = \frac{\langle \tau JM' \pm 2|\hat{T}_{\pm 1}|\tau' JM' \pm 1\rangle}{\langle JM' \pm 2|\hat{J}_{\pm}|JM' \pm 1\rangle} . \tag{19}$$

It immediately follows that

$$\langle \tau JM|\hat{T}_{\pm 1}|\tau' JM'\rangle = \mp\frac{1}{\sqrt{2}}C(\hat{T}_{\pm 1}, J, \tau, \tau')\langle JM'|\hat{J}_{\pm}|JM\rangle . \tag{20}$$

With (17), (14), and (20) we get

$$\begin{aligned}
&\langle \tau JM|\hat{T}_0|\tau' JM'\rangle \\
&= \frac{1}{\hbar\sqrt{2}}(\langle JM|\hat{J}_{\mp}|JM \pm 1\rangle\langle \tau JM \pm 1|\hat{T}_{\pm 1}|\tau' JM'\rangle \\
&\quad - \langle \tau JM|\hat{T}_{\pm 1}|\tau' JM \mp 1\rangle\langle JM|\hat{J}_{\mp}|JM' \mp 1\rangle) \\
&= \frac{\mp 1}{\hbar\sqrt{2}}\, C(\hat{T}_{\pm 1}, J, \tau, \tau')\delta_{MM'} \\
&\quad \times(\langle JM'|\hat{J}_{\mp}|JM' \pm 1\rangle\langle JM' \pm 1|\hat{J}_{\pm}|JM'\rangle \\
&\quad - \langle JM'|\hat{J}_{\pm}|JM' \mp 1\rangle\langle JM'|\hat{J}_{\mp}|JM' \mp 1\rangle) \\
&= C(\hat{T}_{\pm 1}, J, \tau, \tau')\delta_{MM'}\langle JM|\hat{J}_0|JM'\rangle .
\end{aligned} \tag{21}$$

We realize that the coefficient $C(\hat{T}_{\pm 1}, J, \tau, \tau')$ has to be independent of the component $\hat{T}_z$. Therefore one can write

$$\begin{aligned}
\langle \tau JM|\hat{T}_q|\tau' JM'\rangle &= C(\hat{T}_{\pm 1}, J, \tau, \tau')\langle JM|\hat{J}'_q|JM\rangle \\
&= \langle \tau J\|\hat{T}\|\tau' J\rangle\langle JM|\hat{J}'_q|JM'\rangle ,
\end{aligned}$$

where the standard components of $\hat{J}_q$ have to be taken:

$$\hat{J}'_{\pm} = \mp\frac{1}{\sqrt{2}}\hat{J}_{\pm} , \qquad \hat{J}'_0 = \hat{J}_0 .$$

This proves the validity of (16).

EXAMPLE

11.7 Derivation of the Spin–Orbit Interaction

Spin-1/2 particles moving in an external vector field experience a spin–orbit interaction. We will derive this interaction from the Dirac equation, which reads

$$\mathrm{i}\hbar\frac{\partial \Psi}{\partial t} = \hat{H}\Psi \tag{1}$$

with

$$\hat{H} = \hat{H}_\mathrm{D} - e[\hat{\boldsymbol{\alpha}} \cdot \boldsymbol{A}(\boldsymbol{r}) - A_0(\boldsymbol{r})] , \tag{2}$$

$$\hat{H}_\mathrm{D} = e\hat{\boldsymbol{\alpha}} \cdot \hat{\boldsymbol{p}} + m_0c^2\hat{\beta} . \tag{3}$$

$\hat{H}_\mathrm{D}$ is the free Dirac Hamiltonian and $\hat{\boldsymbol{\alpha}}$ and $\hat{\boldsymbol{\beta}}$ represent the well-known Dirac matrices

Example 11.7.

$$\hat{\boldsymbol{\alpha}} = \begin{pmatrix} 0 & \hat{\boldsymbol{\sigma}} \\ \hat{\boldsymbol{\sigma}} & 0 \end{pmatrix}, \quad \hat{\beta} = \begin{pmatrix} \hat{1} & 0 \\ 0 & -\hat{1} \end{pmatrix}, \tag{4}$$

where the 2×2 Pauli matrices are given by

$$\hat{\boldsymbol{\sigma}} = \left\{ \begin{pmatrix} 0 & 1 \\ 1 & 0 \end{pmatrix}, \begin{pmatrix} 0 & i \\ -i & 0 \end{pmatrix}, \begin{pmatrix} 1 & 0 \\ 0 & -1 \end{pmatrix} \right\}.$$

$\hat{H}$ in (2) includes the interaction with the external electromagnetic field, where $\boldsymbol{A}$ represents the vector potential and $A_0(\boldsymbol{r})$ the Coulomb potential. The charge of the Dirac particle is designated as e. Finally, $\hat{\boldsymbol{p}} = -\mathrm{i}\hbar\boldsymbol{\nabla}$ corresponds to the momentum operator.

The Dirac equation (1) includes the interaction with the external fields $\boldsymbol{A}(\boldsymbol{r})$ and $A_0(\boldsymbol{r})$, and in particular also the spin–orbit interaction. In order to extract the latter we investigate the Dirac equation (1) in the nonrelativistic limit. The spin–orbit coupling will then appear automatically.

Equation (1) includes only the coupling to the electromagnetic field. But elementary particles experience forces that have no electromagnetic origin (e.g. strong interaction, weak interaction, gravity). An immediate example is given by nucleons, which interact with each other through π-meson and ρ-meson fields and also show pronounced spin–orbit coupling.[15] In order to take into account such additional interactions of the Dirac particle we rewrite the Dirac equation (1) in the form

$$\left(\mathrm{i} \sum_{\mu} \hat{\gamma}^{\mu} \hat{p}_{\mu} + m_0 c \right) \psi = \frac{1}{c} \sum_{a} \hat{Q}_a \psi \tag{5}$$

with $\hat{\gamma} = -\mathrm{i}\hat{\beta}\hat{\boldsymbol{\alpha}}$, $\gamma_4 = \beta$.

The $\hat{Q}_a$ operators on the right-hand side of (5) characterize the different kinds of interactions. The following possibilities exist.

1) The interaction with a scalar field $V(\boldsymbol{r})$. The interaction operator then is simply given by

$$\hat{Q}_s = V(\boldsymbol{r}). \tag{6}$$

This kind of interaction acts like a space-dependent mass, because $V(\boldsymbol{r})c/c^2$ can be directly added to the mass m in (5) in order to obtain this coupling.

2) The interaction with a vector field B_{μ}. The interaction operator reads

$$\hat{Q}_v = -\mathrm{i} \sum_{\mu} \hat{\gamma}^{\mu} B_{\mu}, \qquad B_{\mu} = \{B_0(\boldsymbol{r}), -\boldsymbol{B}(\boldsymbol{r})\}. \tag{7}$$

A special vector field of this kind is given by the electromagnetic field; it is described by the four-potential

$$A_{\mu} = \{A_0(\boldsymbol{r}), -\boldsymbol{A}(\boldsymbol{r})\}. \tag{8}$$

A glimpse back at (2) and (5) reveals that in this case we have

$$B_{\mu} = e A_{\mu}. \tag{9}$$

---

15 (See footnote 12 on page 315).

Example 11.7.

3) The interaction with a tensor field of the form

$$\hat{Q}_{\mathrm{T}} = \sum_{\mu,\nu} \gamma^{\mu}\gamma^{\nu}C_{\mu\nu} . \tag{10}$$

An example is given by the interaction of a non-Dirac-like magnetic dipole with the electromagnetic fields

$$F_{\mu\nu} = \frac{\partial A_{\mu}}{\partial x^{\nu}} - \frac{\partial A_{\nu}}{\partial x^{\mu}} . \tag{11}$$

In this case it is

$$C_{\mu\nu} = \tfrac{1}{2} g\, \mu_0\, F_{\mu\nu} , \tag{12}$$

where $\mu_0 = e\hbar/2m_0c$, m_0 is the rest mass and g the coupling constant. For electrons $g = 0$, but for nucleons $g \neq 0$.

We now consider the case of a particle that simultaneously interacts with a scalar and a vector field. Inserting (6) and (7) into (5) and multiplying from the left with $\hat{\beta}$ yields, after rearrangement,

$$\mathrm{i}\hbar\frac{\partial\psi}{\partial t} = \left\{c\hat{\boldsymbol{\alpha}}\cdot\left[\hat{\boldsymbol{p}} - \frac{1}{c}\boldsymbol{B}(\boldsymbol{r})\right] + \hat{\beta}\left[m_0c^2 - V(\boldsymbol{r})\right] + B_0(\boldsymbol{r})\right\}\psi . \tag{13}$$

Next we consider the nonrelativistic limit of this Dirac equation. It can be done either very elegantly with the help of the Foldy–Wouthuysen transformation,[16] or in a more elementary fashion, as will be described next. We make the ansatz

$$\psi(\boldsymbol{r},t) = \psi(\boldsymbol{r})\exp\left(-\mathrm{i}\frac{E}{\hbar}t\right) \tag{14}$$

and write the four-component spinor as

$$\psi(r) = \begin{pmatrix}\varphi(r)\\ \chi(r)\end{pmatrix} . \tag{15}$$

$\varphi(r)$ and $\chi(r)$ describe two-component (Pauli) spinors, i.e. column vectors with two components. For simplicity we choose

$$B(r) = 0 , \tag{16}$$

so that in (13) only the two terms $V(r)$ and $B_0(r)$ remain. We insert (14) and (15) into (13), and arrive at a system of coupled differential equations:

$$\left\{E - B_0(r) - \left[m_0c^2 - V(r)\right]\right\}\varphi(r) = c(\hat{\boldsymbol{\sigma}}\cdot\hat{\boldsymbol{p}})\chi(r) , \tag{17}$$

$$\left\{E - B_0(r) + \left[m_0c^2 - V(r)\right]\right\}\chi(r) = c(\hat{\boldsymbol{\sigma}}\cdot\hat{\boldsymbol{p}})\varphi(r) . \tag{18}$$

We split the rest mass from the total energy, i.e. $E = E' + mc^2$, and get

$$\left[E' - B_0(r) + V(r)\right]\varphi(r) = c(\hat{\boldsymbol{\sigma}}\cdot\hat{\boldsymbol{p}})\chi(r) , \tag{19}$$

$$\left[2m_0c^2 + E' - B_0(r) - V(r)\right]\chi(r) = c(\hat{\boldsymbol{\sigma}}\cdot\hat{\boldsymbol{p}})\varphi(r) . \tag{20}$$

[16] See, for example, W. Greiner: *Relativistic Quantum Mechanics – Wave Equations*, 2nd ed. (Springer, Berlin, Heidelberg 1994).

Example 11.7.

These equations are still exact. Now we intend to investigate the nonrelativistic motion of a spin-1/2 particle influenced by the external fields $B_0(r)$ and $V(r)$. For this we take into consideration only terms up to order v^2/c^2. An expansion up to the first order in $[E' - B_0(r) - V(r)]/2m_0c^2$ yields for $\chi(r)$ from (20)

$$\chi(r) = \left(1 - \frac{E' - B_0(r) - V(r)}{2m_0c^2}\right) \frac{\hat{\boldsymbol{\sigma}} \cdot \hat{\boldsymbol{p}}}{2m_0c} \varphi(r) . \tag{21}$$

Inserting this expression into (19), we obtain the differential equation

$$[E' - B_0(r) + V(r)]\, \varphi(r) = \frac{\hat{\boldsymbol{\sigma}} \cdot \hat{\boldsymbol{p}}}{2m_0} \left(1 - \frac{E' - B_0(r) - V(r)}{2m_0c^2}\right) \hat{\boldsymbol{\sigma}} \cdot \hat{\boldsymbol{p}}\, \varphi(r) \tag{22}$$

for the two-component spinor function $\varphi(r)$. Using the relations

$$(\hat{\boldsymbol{\sigma}} \cdot \boldsymbol{A})(\hat{\boldsymbol{\sigma}} \cdot \boldsymbol{B}) = \boldsymbol{A} \cdot \boldsymbol{B} + \mathrm{i}\,(\hat{\boldsymbol{\sigma}} \cdot [\boldsymbol{A} \times \boldsymbol{B}]) \tag{23}$$

and

$$\begin{aligned}(\hat{\boldsymbol{\sigma}} \cdot \hat{\boldsymbol{p}}) f(r) (\hat{\boldsymbol{\sigma}} \cdot \hat{\boldsymbol{p}}) &= f(r)(\hat{\boldsymbol{\sigma}} \cdot \hat{\boldsymbol{p}})(\hat{\boldsymbol{\sigma}} \cdot \hat{\boldsymbol{p}}) - \mathrm{i}\hbar(\hat{\boldsymbol{\sigma}} \cdot \boldsymbol{\nabla} f)(\hat{\boldsymbol{\sigma}} \cdot \hat{\boldsymbol{p}}) \\ &= f(r)\hat{\boldsymbol{p}}^2 - \mathrm{i}\hbar\,\{(\boldsymbol{\nabla} f \cdot \hat{\boldsymbol{p}}) + \mathrm{i}\,[\hat{\boldsymbol{\sigma}} \cdot (\boldsymbol{\nabla} f \times \hat{\boldsymbol{p}})]\}\,,\end{aligned} \tag{24}$$

we can transform (22) into

$$\hat{H}'\varphi(r) = E'\varphi(r)\,, \tag{25}$$

where

$$\begin{aligned}\hat{H}' =& \left(1 - \frac{E' - B_0(r) - V(r)}{2m_0c^2}\right) \frac{\hat{\boldsymbol{p}}^2}{2m_0} + B_0(r) - V(r) \\ &+ \frac{\hbar\,\{\hat{\boldsymbol{\sigma}} \cdot [\boldsymbol{\nabla} B_0(r) \times \hat{\boldsymbol{p}}]\}}{4m_0^2c^2} + \frac{\hbar\,\{\hat{\boldsymbol{\sigma}} \cdot [\boldsymbol{\nabla} V(r) \times \hat{\boldsymbol{p}}]\}}{4m_0^2c^2} \\ &- \frac{\mathrm{i}\hbar}{4m_0^2c^2}\,[\boldsymbol{\nabla} B_0(r) \cdot \hat{\boldsymbol{p}}] - \frac{\mathrm{i}\hbar}{4m_0^2c^2}\,[\boldsymbol{\nabla} V(r) \cdot \hat{\boldsymbol{p}}]\,.\end{aligned} \tag{26}$$

For the theory to be consistent up to the order v^2/c^2 we have to pay attention to the normalization of $\varphi(r)$. In Dirac theory the charge density

$$\rho = e(\varphi^\dagger\varphi + \chi^\dagger\chi) \tag{27}$$

is normalized to

$$\int \rho\, \mathrm{d}^3r = 1\,. \tag{28}$$

In lowest order it follows from (21) that

$$\chi \approx \frac{\hat{\boldsymbol{\sigma}} \cdot \hat{\boldsymbol{p}}}{2m_0c}\varphi\,. \tag{29}$$

With the help of (23) we then obtain for the charge density

$$\rho \approx e\varphi^\dagger \left(1 + \frac{\hat{\boldsymbol{p}}^2}{4m_0^2c^2}\right) \varphi \tag{30}$$

and for the normalization integral

Example 11.7.

$$\int \rho \, \mathrm{d}^3 r \;=\; \int \varphi^\dagger \left(1 + \frac{\hat{\boldsymbol{p}}^2}{4m_0^2c^2}\right) \varphi \, \mathrm{d}^3 r \;=\; 1\,, \tag{31}$$

which represents the proper normalization condition for $\varphi(r)$. On the other hand, the normalization for a Schrödinger wavefunction ψ is given by

$$\int \psi^\dagger \psi \, \mathrm{d}^3 r \;=\; 1\,. \tag{32}$$

Hence, we have to introduce a proper rescaling of the wavefunction φ in the nonrelativistic range:

$$\psi \;=\; \hat{g}\varphi \tag{33}$$

with the property

$$\int \psi^\dagger \psi \, \mathrm{d}^3 r \;=\; \int \varphi^\dagger \hat{g}^\dagger \hat{g} \varphi \, \mathrm{d}^3 r \;=\; 1\,. \tag{34}$$

$\hat{g}$ is in general an operator. Comparing (34) with (31) we obtain

$$\hat{g} \;=\; \left(1 + \frac{\hat{\boldsymbol{p}}^2}{4m_0^2c^2}\right)^{1/2} \;\approx\; 1 + \frac{\hat{\boldsymbol{p}}^2}{8m_0^2c^2}\,. \tag{35}$$

The Hamiltonian belonging to ψ is obtained by multiplying (25) with $\hat{g}$ from the left:

$$\hat{g}\hat{H}'\hat{g}^{-1}\hat{g}\varphi \;=\; E'\hat{g}\varphi\,. \tag{36}$$

Thus,

$$\hat{H}\psi \;=\; E\psi\,, \tag{37}$$

where $\hat{H}$ is given by

$$\begin{aligned}
\hat{H} \;&=\; \left(1 + \frac{\hat{\boldsymbol{p}}^2}{8m_0^2c^2}\right) \hat{H}' \left(1 - \frac{\hat{\boldsymbol{p}}^2}{8m_0^2c^2}\right) \\
&=\; \frac{\hat{\boldsymbol{p}}^2}{2m_0} + B_0(r) - V(r) - \frac{[E' - B_0(r)]^2 - V^2(r)}{2m_0^2c^2} \\
&\quad + \frac{\hbar\,\{\hat{\boldsymbol{\sigma}} \cdot [\boldsymbol{\nabla} B_0(r) \times \hat{\boldsymbol{p}}]\}}{4m_0^2c^2} + \frac{\hbar\,\{\hat{\boldsymbol{\sigma}} \cdot [\boldsymbol{\nabla} V(r) \times \hat{\boldsymbol{p}}]\}}{4m_0^2c^2} \\
&\quad - \frac{\hbar^2}{8m_0^2c^2}\nabla^2 B_0(r) + \frac{\hbar^2}{8m_0^2c^2}\nabla^2 V(r)
\end{aligned} \tag{38}$$

up to order v^2/c^2. We have used the relations

$$\hat{\boldsymbol{p}}^2 V(r) - V(r)\hat{\boldsymbol{p}}^2 \;=\; -\hbar^2\nabla^2 V(r) - 2i\hbar\,(\boldsymbol{\nabla} V(r) \cdot \hat{\boldsymbol{p}})$$

and

$$\left(1 - \frac{E' - B_0(r) + V(r)}{2m_0c^2}\right)\hat{\boldsymbol{p}}^2 \;\approx\; \hat{\boldsymbol{p}}^2 - \frac{[E' - B_0(r)]^2 - V^2(r)}{c^2}\,. \tag{39}$$

The first three terms in (38) correspond to the known nonrelativistic Hamiltonian. The additional terms are relativistic corrections of order v^2/c^2. We distinguish them as

$$U \;=\; U_1 + U_2 + U_3\,. \tag{40}$$

Example 11.7.

The first term

$$U_1 = -\frac{\hbar^2}{8m_0^2c^2}\nabla^2 B_0(r) + \frac{\hbar^2}{8m_0^2c^2}\nabla^2 V(r) \tag{41}$$

was introduced for the first time by *Darwin*. For the case of the Coulomb field we have $B_0 = Ze^2/r$ and with $\nabla^2(1/r) = -4\pi\,\delta(r)$ it follows for U_1 that

$$(U_1)_{\text{Coul}} = \frac{\pi\hbar^2 e^2 Z}{2m_0^2c^2}\delta(r)\,. \tag{42}$$

The term

$$U_2 = -\frac{[E' - B_0(r)]^2 - V^2(r)}{2m_0^2c^2} \tag{43}$$

represents a correction to the kinetic energy as a result of the velocity dependence of the particle mass. The last term

$$U_3 = \frac{\hbar\{\hat{\boldsymbol{\sigma}}\cdot[\boldsymbol{\nabla} B_0(r)\times\hat{\boldsymbol{p}}]\}}{4m_0^2c^2} + \frac{\hbar\{\hat{\boldsymbol{\sigma}}\cdot[\boldsymbol{\nabla} V(r)\times\hat{\boldsymbol{p}}]\}}{4m_0^2c^2} \tag{44}$$

describes the spin–orbit interaction. In a spherical symmetric field we find

$$\begin{aligned}\boldsymbol{\nabla} B_0(r) &= \frac{\mathrm{d}B_0(r)}{\mathrm{d}r}\frac{\boldsymbol{r}}{r}\,,\\ \boldsymbol{\nabla} V(r) &= \frac{\mathrm{d}V(r)}{\mathrm{d}r}\frac{\boldsymbol{r}}{r}\,.\end{aligned} \tag{45}$$

With the help of the known expressions for the angular momentum operator, $\hat{\boldsymbol{L}} = \hbar\hat{\boldsymbol{l}} = [\hat{\boldsymbol{r}}\times\hat{\boldsymbol{p}}]$, and the spin operator, $\hat{\boldsymbol{s}} = (\hbar/2)\hat{\boldsymbol{\sigma}}$, we obtain, with an insertion of (45) into (44),

$$U_3 = \frac{\mathrm{d}B_0(r)}{\mathrm{d}r}\frac{\hat{\boldsymbol{s}}\cdot\hat{\boldsymbol{l}}}{2m_0^2c^2r} + \frac{\mathrm{d}V(r)}{\mathrm{d}r}\frac{(\hat{\boldsymbol{s}}\cdot\hat{\boldsymbol{l}})}{2m_0^2c^2r}\,. \tag{46}$$

The first term describes the spin–orbit interaction with the electromagnetic field; the second one represents the interaction with the nuclear field. Since the nuclear interaction is stronger for smaller distances than the electromagnetic one, the second term provides the main contribution. This is the reason for the relativistic contribution to the nuclear spin–orbit coupling. We estimate its order of magnitude with the help of the oscillator model. The potential in this case is given by $V(r) = (1/2)M\omega^2r^2$, so $\mathrm{d}V/\mathrm{d}r = M\omega^2 r$. When all factors for the second term in (46) are taken into account it follows that

$$\left|\frac{\mathrm{d}V(r)}{\mathrm{d}r}\frac{(\hat{\boldsymbol{s}}\cdot\hat{\boldsymbol{l}})}{2m_0^2c^2r}\right| \approx \frac{(\hbar\omega)^2}{2m_0c^2} \approx \frac{(6\,\mathrm{MeV})^2}{2000\,\mathrm{MeV}} \approx 0.02\,\mathrm{MeV}\,. \tag{47}$$

From this consideration alone we would deduce that the spin–orbit strength, which is caused by relativistic effects of the central potential (which simulates the mesonic exchange), provides a contribution that is too small to explain the strong spin–orbit strenght in the shell model. The coupling has to be caused by the mesonic field directly. At the moment the mesonic field theory is not far enough developed to confirm or to exclude such additional terms with

Example 11.7.

certainty. The spin–orbit coupling in the nucleon–nucleon interaction leads to a corresponding term in the potential; its influence increases with the number of nucleons. This agrees with the empirically secured observation that the spin–orbit coupling for nearly completely full shells is stronger than at the beginning of a new shell. For a continued discussion of the nuclear interaction see the book by Eisenberg and Greiner.[17]

EXERCISE

11.8 Transformation of the Spin–Orbit Interaction

Problem. Show that the spin–orbit coupling

$$H_{\mathrm{SO}} = \sum_i \frac{e}{2mc^2}(\boldsymbol{\nabla}_i \times \boldsymbol{E}_i)\cdot \hat{\boldsymbol{s}}_i$$

can be rewritten in the form

$$\hat{H}_{\mathrm{SO}} = \sum_i \frac{1}{2m^2c^2 r_i}\frac{\mathrm{d}V}{\mathrm{d}r_i}\hat{\boldsymbol{L}}_i \cdot \hat{\boldsymbol{s}}_i\,,$$

where $\boldsymbol{L}_i$ and $\boldsymbol{s}_i$ are the orbital angular momenta and the spins of the electrons.

Solution. We assume that the potential, which generates the mean electric field in the atom, is a central potential $V(r)$. Then the electric field experienced by the ith electron becomes

$$e\boldsymbol{E}_i = -\boldsymbol{\nabla}_i V(r) = -\boldsymbol{r}_i \frac{1}{r_i}\frac{\mathrm{d}V}{\mathrm{d}r_i}\,. \tag{1}$$

For H_{SO} we obtain

$$\hat{H}_{\mathrm{SO}} = \sum_i \frac{1}{2mc^2}\hat{\boldsymbol{s}}_i\cdot(\boldsymbol{r}_i\times\hat{\boldsymbol{v}}_i)\frac{1}{r_i}\frac{\mathrm{d}V}{\mathrm{d}r_i}\,. \tag{2}$$

Here the order of $\boldsymbol{r}_i$ and $\boldsymbol{v}_i$ have been exchanged. The orbital angular momentum $\boldsymbol{L}_i$ of the ith electron is $\hat{\boldsymbol{L}}_i = m(\boldsymbol{r}_i \times \hat{\boldsymbol{v}}_i)$. Therefore we obtain for H_{SO}

$$\hat{H}_{\mathrm{SO}} = \frac{1}{2m^2c^2}\sum_i \frac{1}{r_i}\frac{\mathrm{d}V}{\mathrm{d}r_i}(\hat{\boldsymbol{s}}_i\cdot\hat{\boldsymbol{L}}_i)\,. \tag{3}$$

This is the desired expression.

[17] M. Eisenberg, W. Greiner: *Nuclear Theory Vol. I: Nuclear Models*, 3rd ed. (North-Holland, Amsterdam 1987).

11.8 Treatment of the Spin–Orbit Splitting in the Hartree–Fock Approach

We want to throw some light on the observation that $\xi < 0$ for a less than half full shell and $\xi > 0$ otherwise, and calculate an explicit expression for $\hat{\boldsymbol{G}}_i$ in (11.66) for a central potential $V(r_i)$. Then $\boldsymbol{E}_i = -\boldsymbol{\nabla}_i V(r_i) = -(\mathrm{d}V\mathrm{d}r_i)(\boldsymbol{r}_i/r_i)$ and with (11.66) we have

$$\begin{aligned}\hat{\boldsymbol{G}}(r_i) &= -\frac{e}{2mc^2}\frac{1}{r_i}\frac{\mathrm{d}V}{\mathrm{d}r_i}\hat{\boldsymbol{v}}_i \times \boldsymbol{r}_i \\ &= \frac{|e|}{2m^2c^2}\frac{1}{r_i}\frac{\mathrm{d}V(r_i)}{\mathrm{d}r_i}\boldsymbol{r}_i \times \hat{\boldsymbol{p}}_i \\ &= \frac{|e|}{2m^2c^2}\frac{1}{r_i}\frac{\mathrm{d}V(r_i)}{\mathrm{d}r_i}\hat{\boldsymbol{L}}_i\,. \end{aligned} \tag{11.77}$$

$\hat{\boldsymbol{L}}_i$ represents the orbital angular momentum of the ith electron. The spin–orbit coupling part of the Hamiltonian is then written as

$$\hat{H}_{\mathrm{SO}} = \sum_i \frac{1}{2m^2c^2r_i}\frac{\mathrm{d}V}{\mathrm{d}r_i}\hat{\boldsymbol{L}}_i \cdot \hat{\boldsymbol{s}}_i\,. \tag{11.78}$$

In order to determine $\xi(QLS)$ [see (11.73) and (11.76)] it is sufficient to take special M_S and M_L values, since $\xi(QLS)$ is independent of them. We choose $M_L' = M_L$ and $M_S' = M_S$, and with (11.73) obtain

$$\begin{aligned}&\langle QLSM_LM_S|\hat{H}_{SB}|QLSM_LM_S\rangle \\ &\quad = \xi(QLS)\langle QLSM_LM_S|\hat{\boldsymbol{L}}\cdot\hat{\boldsymbol{S}}|QLSM_LM_S\rangle \\ &\quad = \xi(QLS)\langle QLSM_LM_S|\hat{L}_z\hat{S}_z|QLSM_LM_S\rangle \\ &\quad = \xi(QLS)M_LM_S\,. \end{aligned} \tag{11.79}$$

Only the diagonal matrix elements of both spin and angular momentum operators contribute; only $\hat{S}_Z$ and $\hat{L}_Z$ are diagonal. In general the state $|QLSM_LM_S\rangle$ is a linear combination of Slater determinants, which we express in abbreviated form as $\sum_\beta a_\beta|Q\beta\rangle$ with $\sum_\beta |a_\beta|^2 = 1$. The symbol β runs over all Slater determinants denoted by $|Q\beta\rangle$. The many-particle functions $|Q\beta\rangle$ are functions of the single-particle wavefunctions $\varphi_{nlm_l}\chi_{ms}$, where φ_{nlm_l} represents the angular momentum and radial part and χ_{ms} the spin part. By construction $|Q\beta\rangle$ is an eigenstate of $\hat{L}_Z$ and $\hat{S}_Z$ with eigenvalues M_L and M_S. We get:

$$\begin{aligned}&\langle QLSM_LM_S|\hat{H}_{\mathrm{SO}}|QLSM_LM_S\rangle \\ &\quad = \sum_{\beta\beta'} a_\beta^* a_{\beta'}\langle Q\beta|\sum_i \frac{1}{2m^2c^2r_i}\frac{\mathrm{d}V}{\mathrm{d}r_i}\boldsymbol{l}_i\cdot\boldsymbol{s}_i|Q\beta'\rangle\,. \end{aligned} \tag{11.80}$$

Only for $\beta = \beta'$ do we obtain nonzero matrix elements. This can be seen as follows. The matrix element goes with a sum of one-particle operators. If we pick one term, for example, the ith term, then it acts on the ith electron and does not modify the states of the other electrons. This means that in $|Q\beta\rangle$ and $|Q\beta'\rangle$ all electrons must have the same one-particle wavefunctions except for the ith electron. However, the ith electron also has to possess the same wavefunction, since $|Q\beta\rangle$ and $|Q\beta'\rangle$ are eigenstates of $\hat{L}_Z$. If the ith electron were not in the same state, M_L would be different for $|Q\beta\rangle$ and for $|Q\beta'\rangle$, so that the matrix element would vanish.

In order to calculate $\langle Q\beta|\hat{H}_{\text{SO}}|Q\beta\rangle$ we perform the sum over all occupied states:

$$\langle Q\beta|\hat{H}_{\text{SO}}|Q\beta\rangle = \sum_{nlm_lm_s} \int \mathrm{d}^3r \varphi_{nlm_l}(\boldsymbol{r}) \frac{1}{2m^2c^2r} \frac{\mathrm{d}V}{\mathrm{d}r} \hat{\boldsymbol{L}}\varphi_{nlm_l}(\boldsymbol{r}) \cdot \langle m_s|\hat{\boldsymbol{s}}|m_s\rangle \,. \tag{11.81}$$

$\hat{\boldsymbol{s}}$ has diagonal matrix elements $\langle m_s|\hat{s}_z|m_s\rangle = \hbar m_s$ only for $\hat{s}_z$. Therefore only $\hat{L}_z$ appears in (11.81) and we have

$$\langle Q\beta|\hat{H}_{\text{SO}}|Q\beta\rangle = \frac{\hbar^2}{2m^2c^2} \sum_{nlm_lm_s} \eta_{nl} m_l m_s \tag{11.82a}$$

with

$$\eta_{nl} = \int \mathrm{d}^3r |\varphi_{nlm_l}(\boldsymbol{r})|^2 \frac{1}{r}\frac{\mathrm{d}V}{\mathrm{d}r} \,. \tag{11.82b}$$

Since V is assumed to be spherical symmetric, η_{nl} is independent of m_l. From this we immediately conclude that for closed shells the spin–orbit coupling disappears. Indeed, for closed shells we have

$$\sum_{m_lm_s} m_l m_s = \sum_{m_l} m_l \sum_{m_s} m_s = 0 \,, \tag{11.83}$$

because the sum is performed over all m_l and m_s within a shell. In fact, for a completely occupied shell for any given occupied state with m_l, another occupied state with $-m_l$ exists. Consequently (11.82a) vanishes. Therefore we only need to focus our attention on the last open shell. We denote it as n_l and we get

$$\langle Q\beta|\hat{H}_{\text{SO}}|Q\beta\rangle = \frac{\hbar^2}{2m^2c^2} \eta_{nl} M_L M_S \,, \tag{11.84}$$

where M_L and M_S represent total projections of the orbital angular momentum and the spin, respectively. As a result of Hund's rule the lowest lying state is described by the maximum spin. Therefore we set $M_S = S$, where all electrons are alligned parallel. First we consider a shell less than half full with electrons. From (11.80) or (11.84) we then obtain

$$\langle QLSM_LM_S|\hat{H}_{\text{SO}}|QLSM_LM_S\rangle = \frac{\hbar^2\eta_{nl}}{2m^2c^2} M_L S \,. \tag{11.85}$$

Here we have used $\sum_\beta |a_\beta|^2 = 1$. Comparing this with (11.79) we deduce

$$\xi(QLS) = \frac{\hbar^2\eta_{nl}}{2m^2c^2} \,. \tag{11.86}$$

According to (11.82b) η_{nl} includes $(1/r)(\mathrm{d}V\mathrm{d}r)$. Since in general $V(r)$ increases with r, η_{nl} and, consequently, $\xi(QLS)$ are positive. This is in accordance with the first part of the third of Hund's rules. For a shell that is more than half full, not all spins can be parallel. It holds that

$$\sum_{\substack{\text{occupied}\\\text{states}}} m_l m_s + \sum_{\substack{\text{nonoccupied}\\\text{states}}} m_l m_s = \sum_{\substack{\text{all states}\\\text{of a shell}}} m_l m_s = 0\,. \tag{11.87}$$

In other words, M_L and M_S of the nonoccupied states are the same as the negative values of the corresponding values for the occupied states. Hence, we switch from a particle to a hole picture, for which all holes are coupled parallel in spin. Consequently, we obtain

$$\sum_{\text{occ.states}} m_l m_s = -M_L M_S\,. \tag{11.88}$$

In comparison with (11.86) we obtain

$$\xi(QLS) = -\frac{\hbar^2 \eta_{nl}}{2m^2c^2}\,. \tag{11.89}$$

Now the third of Hund's rules is completely explained. A special situation is given for a half-full shell. All electrons are coupled parallel to spin S. Hence, in every state there is exactly one electron with $m_s = +1/2$ *without* a partner with $m_s = -1/2$. This means that all orbital states have to be occupied. As a consequence every occupied one-particle wavefunction with m_l has as its counterpart an occupied one-particle wavefunction with $-m_l$. Therefore the component M_L vanishes, i.e. $M_L = 0$. This is the only possible value and, consequently, there is only one $L = 0$. The spin–orbit coupling disappears for half-open shells.

In addition, the strength of the spin–orbit coupling can be estimated. The factor η_{nl} behaves as

$$\eta_{nl} \sim \int \mathrm{d}^3r |\varphi_{nl}|^2 \frac{e^2}{r^3} \sim \frac{e^2}{(Z^{-\frac{1}{3}}a_0)^3}\,. \tag{11.90}$$

Here the Bohr radius $a_0 = \hbar^2/(me^2)$ has been inserted (see Sect. 11.1). For ξ we then get

$$\xi \sim \frac{Z}{a_0^2 m^2 c^2}\left(\frac{e^2}{2a_0}\right) = Z\alpha^2 Ry \tag{11.91}$$

with the fine-structure constant $\alpha = e^2/\hbar c \approx 1/137$. The splitting strength is proportional to $\alpha^2 \approx 5 \times 10^{-5}$ and increases linearly with Z. For small Z the perturbative treatment of the spin–orbit coupling is justified. This approach is known as LS or *Russell–Saunders* coupling. As Z increases the approach becomes worse and for large Z the spin–orbit coupling has to be treated from the beginning. Since $(\boldsymbol{L} \cdot \boldsymbol{S})$ does not commute with $\boldsymbol{L}$ and $\boldsymbol{S}$, L and S are no longer good quantum numbers. Instead one uses to the so-called *jj coupling*, where first an individual coupling to a total angular momentum $\boldsymbol{j}_i = \boldsymbol{l}_i + \boldsymbol{s}_i$ is introduced and then to $\boldsymbol{J} = \sum_i \boldsymbol{j}_i$.

11.9 The Zeeman Effect

Although we have already discussed the Zeeman effect in another volume of this lecture series,[18] it is wise to bring up again the most important ingredients in order to complete the discussions. The Zeeman effect describes the splitting of the levels in an atom in the presence of a weak magnetic field $\boldsymbol{B}$. Again we apply perturbation theory in order to treat this problem. The weak magnetic interaction reads

$$\hat{H}_{\text{mag}} = -\frac{e}{2mc}\hat{\boldsymbol{L}}\cdot\boldsymbol{B} - \frac{e}{mc}\hat{\boldsymbol{S}}\cdot\boldsymbol{B}\,, \tag{11.92}$$

with the gyromagnetic factors 1 for orbital angular momentum and 2 for spin. For a magnetic field in the z direction an energy shift results in first-order perturbation theory:

$$\begin{aligned}\Delta E_{QLSJ}(M) &= \langle QLSJM|\hat{H}_{\text{mag}}|QLSJM\rangle \\ &= -\frac{eB}{2mc}\langle QLSJM|(\hat{L}_z + 2\hat{S}_z)|QLSJM\rangle \\ &= -\frac{eB}{2mc}(M\hbar + \langle QLSJM|\hat{S}_z|QLSJM\rangle)\,.\end{aligned} \tag{11.93}$$

Here $J_z = S_z + L_z = M$ was used. In order to determine the matrix element of $\hat{S}_z$ we use the Wigner–Eckart theorem (see Example 11.6). It states that

$$\frac{\langle QLSJM|\hat{S}_z|QLSJM\rangle}{\langle QLSJM|\hat{J}_z|QLSJM\rangle} = \frac{(J1J|M0M)\langle QLSJ\|\hat{S}\|QLSJ\rangle}{(J1J|M0M)\langle QLSJ\|\hat{J}\|QLSJ\rangle}\,. \tag{11.94}$$

The elements on the right-hand side are the reduced matrix elements. The same Clebsch–Gordan coefficients $(J1J|M0M)$ appear as factors in the numerator and denominator and therefore cancel. The Wigner–Eckart theorem states

$$\begin{aligned}&\langle QLSJM|\hat{\boldsymbol{J}}\cdot\hat{\boldsymbol{S}}|QLSJM\rangle \\ &\qquad \sim \langle QLSJ\|\hat{\boldsymbol{J}}\|QLSJ\rangle\langle QLSJ\|\hat{\boldsymbol{S}}\|QLSJ\rangle\end{aligned} \tag{11.95}$$

and analogously

$$\langle QLSJM|\hat{\boldsymbol{J}}^2|QLSJM\rangle \sim \langle QLSJ\|\hat{\boldsymbol{J}}\|QLSJ\rangle^2 \tag{11.96}$$

with the same reduced matrix element $\langle QLSJ\|\hat{\boldsymbol{J}}\|QLSJ\rangle$ as in (11.95). The ratio between (11.95) and (11.96) is equal to (11.94), i.e.

$$\frac{\langle QLSJM|\hat{S}_z|QLSJM\rangle}{\langle QLSJM|\hat{J}_z|QLSJM\rangle} = \frac{\langle QLSJ|\hat{\boldsymbol{J}}\cdot\hat{\boldsymbol{S}}|QLSJ\rangle}{\langle QLSJ\|\hat{\boldsymbol{J}}^2\|QLSJ\rangle}\,, \tag{11.97}$$

or with the replacements $\boldsymbol{J}^2 \to J(J+1)\hbar^2$ and $\hat{J}_z \to +M\hbar$

$$\langle QLSJM|\hat{S}_z|QLSJM\rangle = \frac{M}{J(J+1)\hbar}\langle QLSJM|\hat{\boldsymbol{J}}\cdot\hat{\boldsymbol{S}}|QLSJM\rangle\,. \tag{11.98}$$

[18] W. Greiner: *Quantum Mechanics – An Introduction*, 3rd ed. (Springer, Berlin, Heidelberg 1994).

On the right-hand side we can substitute $\hat{\boldsymbol{J}} \cdot \hat{\boldsymbol{S}}$ by

$$\hat{\boldsymbol{J}} \cdot \hat{\boldsymbol{S}} = \hat{\boldsymbol{S}}^2 + \hat{\boldsymbol{L}} \cdot \hat{\boldsymbol{S}} = \frac{(\hat{\boldsymbol{J}}^2 - \hat{\boldsymbol{L}}^2 + \hat{\boldsymbol{S}}^2)}{2} . \tag{11.99}$$

With this we obtain for (11.98)

$$\langle QLSJM|\hat{S}_z|QLSJM\rangle = \frac{M\hbar}{J(J+1)} \frac{[J(J+1) + S(S+1) - L(L+1)]}{2} . \tag{11.100}$$

The energy splitting (11.93) now becomes

$$\Delta E_{QLSJ}(M) = -\frac{e\hbar B}{2mc} gM \tag{11.101}$$

with the *gyromagnetic factor*

$$g = 1 + \frac{J(J+1) + S(S+1) - L(L+1)}{2J(J+1)} . \tag{11.102}$$

The last equation is also called *Landé's g factor* of the LS coupling. The atom behaves as if it had a magnetic moment $ge\hbar/2mc$. If we have $S = 0$ ($L = J$), then the orbital part only contributes to the magnetic moment and we have $g = 1$. For $L = 0$ there is a contribution from the spin and with $S = J$ we obtain a gyromagnetic factor of $g = 2$. In general g is between 1 and 2. The splitting of the energy levels caused by a weak magnetic field is called *Zeeman's effect.*

A magnetic field that is stronger than the spin–orbit field is responsible mainly for the splitting of the LS multiplet. The state $|QLSM_LM_S\rangle$ then has the energy shift

$$\langle QLSM_LM_S|\hat{H}_{\mathrm{mag}}|QLSM_LM_S\rangle = -\frac{e\hbar B}{2mc}(M_L + 2M_S) . \tag{11.103}$$

States with the same $M_L + 2M_S$ are degenerate in energy. Adding the spin–orbit interaction results in an additional energy shift, which is

$$\langle QLSM_LM_S|\hat{H}_{\mathrm{SO}}|QLSM_LM_S\rangle = \xi(NLS)M_LM_S . \tag{11.104}$$

The total energy shift is then

$$\Delta E_{QLS}(M_L, M_S) = -\frac{e\hbar B}{2mc}(M_L + 2M_S) + \xi(NLS)M_LM_S . \tag{11.105}$$

This combined splitting is called the *Paschen–Back effect.* If, however, the magnetic field becomes comparable to the spin–orbit field, then $\hat{H}_{\mathrm{mag}} + \hat{H}_{\mathrm{SO}}$ should be considered together. If the magnetic field becomes too large, $\boldsymbol{A}^2$ terms ($\boldsymbol{A}$ for vector potential) have to be added and we are confronted with a highly nonlinear problem.

EXERCISE

11.9 The Stark Effect

Problem. Discuss the splitting of levels under the influence of a constant electric field $E_z \boldsymbol{e}_z$ along the z axis (Stark effect). Apply perturbation theory. How do the states split in dependence on E_z?

Solution. The Hamiltonian of an atom in an electric field $\boldsymbol{E} = E_z \boldsymbol{e}_z$ can be written as follows:

$$\hat{H} = \hat{H}_0 - \hat{d}E_z \tag{1}$$

with

$$\hat{d} = e\sum_{i=1}^{N}(\boldsymbol{r}_i \cdot \boldsymbol{e}_z) = \sum_{i=1}^{N} d_i\,. \tag{2}$$

Here d_i is the dipole moment of the ith electron. All electrons feel the same field E_z.

The splitting of the nth energy level in first-order perturbation theory is given by

$$\langle n\alpha'|\hat{d}E_z|n\alpha\rangle = \langle n\alpha'|\hat{d}|N\alpha\rangle E_z\,. \tag{3}$$

However, the dipole operator $\hat{d}$ changes parity, so the matrix element in (3) is different from zero only if $|n\alpha'\rangle$ has the opposite parity to $|n\alpha\rangle$. Equation (3) then contributes only if odd and even parities are mixed in the nth energy level. For larger atoms the levels split because of the electron–electron interaction. For *hydrogen* the energy levels split $\sim E_z$. For larger atoms one has to consider second-order perturbation theory. The energy splitting then is

$$\Delta E_\alpha = E_z^2 \sum_{\alpha'\neq\alpha} \frac{|\langle n\alpha'|\hat{d}|n\alpha\rangle|^2}{E_{n\alpha} - E_{n\alpha'}}\,. \tag{4}$$

The splitting behaves like the square of E_z. The squared term also becomes noticeable in hydrogen for increasing E_z.

Key to the symbols
Element^{Z}
Ground state configuration
Groundstate $^{2S+1}L_J$

	Alkali metals	Alkaline earth metals									Noble metals						Halo-genes	Rare gases
I	H^{1} 1s $^{2}S_{1/2}$																	He^{2} $1s^2$ $^{1}S_0$
II	Li^{3} $1s^22s$ $^{2}S_{1/2}$	Be^{4} $1s^22s^2$ $^{1}S_0$											B^{5} $2s^22p$ $^{2}P_{1/2}$	C^{6} $2s^22p^2$ $^{3}P_0$	N^{7} $2p^3$ $^{4}S_{3/2}$	O^{8} $2p^4$ $^{3}P_2$	F^{9} $2p^5$ $^{2}P_{3/2}$	Ne^{10} $2p^6$ $^{1}S_0$
III	Na^{11} 3s $^{2}S_{1/2}$	Mg^{12} $3s^2$ $^{1}S_0$											Al^{13} $3s^23p$ $^{2}P_{1/2}$	Si^{14} $3s^23p^2$ $^{3}P_0$	P^{15} $3p^3$ $^{4}S_{3/2}$	S^{16} $3p^4$ $^{3}P_2$	Cl^{17} $3p^5$ $^{2}P_{3/2}$	Ar^{18} $3s^23p^6$ $^{1}S_0$
IV	K^{19} 4s $^{2}S_{1/2}$	Ca^{20} $4s^2$ $^{1}S_0$	Sc^{21} $4s^23d$ $^{2}D_{3/2}$	Ti^{22} $4s^23d^2$ $^{3}F_2$	V^{23} $4s^23d^3$ $^{4}F_{3/2}$	Cr^{24} $4s3d^5$ $^{7}S_3$	Mn^{25} $4s^23d^5$ $^{6}S_{5/2}$	Fe^{26} $4s^23d^6$ $^{5}D_4$	Co^{27} $4s^23d^7$ $^{4}F_{9/2}$	Ni^{28} $4s^23d^8$ $^{3}F_4$	Cu^{29} $4s3d^{10}$ $^{2}S_{1/2}$	Zn^{30} $4s^23d^{10}$ $^{1}S_0$	Ga^{31} $4s^23d^{10}4p$ $^{2}P_{1/2}$	Ge^{32} $3d^{10}4p^2$ $^{3}P_0$	As^{33} $3d^{10}4p^3$ $^{4}S_{3/2}$	Se^{34} $3d^{10}4p^4$ $^{3}P_2$	Br^{35} $3d^{10}4p^5$ $^{2}P_{3/2}$	Kr^{36} $4s^24p^6$ $^{1}S_0$
V	Rb^{37} 5s $^{2}S_{1/2}$	Sr^{38} $5s^2$ $^{1}S_0$	Y^{39} $5s^24d$ $^{2}D_{3/2}$	Zr^{40} $5s^24d^2$ $^{3}F_2$	Nb^{41} $5s4d^4$ $^{6}D_{1/2}$	Mo^{42} $5s4d^5$ $^{7}S_3$	Tc^{43} $5s^24d^5$ $^{6}S_{5/2}$	Ru^{44} $5s4d^7$ $^{5}F_5$	Rh^{45} $5s4d^8$ $^{4}F_{9/2}$	Pd^{46} $4d^{10}$ $^{1}S_0$	Ag^{47} $5s4d^{10}$ $^{2}S_{1/2}$	Cd^{48} $5s^24d^{10}$ $^{1}S_0$	In^{49} $5s^24d^{10}5p$ $^{2}P_{1/2}$	Sn^{50} $4d^{10}5p^2$ $^{3}P_0$	Sb^{51} $4d^{10}5p^3$ $^{4}S_{3/2}$	Te^{52} $4d^{10}5p^4$ $^{3}P_2$	I^{53} $4d^{10}5p^5$ $^{2}P_{3/2}$	Xe^{54} $5s^25p^6$ $^{1}S_0$
VI	Cs^{55} 6s $^{2}S_{1/2}$	Ba^{56} $6s^2$ $^{1}S_0$	La^{57} $6s^25d$ $^{2}D_{3/2}$	Hf^{72} $6s^25d^2$ $^{3}F_2$	Ta^{73} $6s^25d^3$ $^{4}F_{3/2}$	W^{74} $6s^25d^4$ $^{5}D_0$	Re^{75} $6s^25d^5$ $^{6}S_{5/2}$	Os^{76} $6s^25d^6$ $^{5}D_4$	Ir^{77} $6s^25d^7$ $^{4}F_{9/2}$	Pt^{78} $6s5d^9$ $^{3}D_3$	Au^{79} $6s5d^{10}$ $^{2}S_{1/2}$	Hg^{80} $6s^25d^{10}$ $^{1}S_0$	Tl^{81} $6s^25d^{10}6p$ $^{2}P_{1/2}$	Pb^{82} $6p^2$ $^{3}P_0$	Bi^{83} $6p^3$ $^{4}S_{3/2}$	Po^{84} $6p^4$ $^{3}P_2$	At^{85} $6p^5$ $^{2}P_{3/2}$	Rn^{86} $6p^6$ $^{1}S_0$
VII	Fr^{87} 7s $^{2}S_{1/2}$	Ra^{88} $7s^2$ $^{1}S_0$	Ac^{89} $7s^26d$ $^{2}D_{3/2}$	Ku^{104} $7s^26d^2$	Ha^{105}	Sg^{106}	Bh^{107}	Hs^{108}	Mt^{109}	110	111	112						
Lanthanides (rare earths)			La^{57} $6s^25d$ $^{2}D_{3/2}$	Ce^{58} $6s^25d4f$ $^{3}H_6$	Pr^{59} $6s^24f^3$ $^{4}I_{9/2}$	Nd^{60} $6s^24f^4$ $^{5}I_4$	Pm^{61} $6s^24f^5$ $^{6}H_{5/2}$	Sm^{62} $6s^24f^6$ $^{7}F_0$	Eu^{63} $6s^24f^7$ $^{8}S_{7/2}$	Gd^{64} $6s^25d4f^7$ $^{9}D_2$	Tb^{65} $6s^25d4f^8$	Dy^{66} $6s^24f^{10}$	Ho^{67} $6s^24f^{11}$	Er^{68} $6s^24f^{12}$	Tm^{69} $6s^24f^{13}$ $^{2}F_{7/2}$	Yb^{70} $6s^24f^{14}$ $^{1}S_0$	Lu^{71} $6s^25d4f^{14}$ $^{2}D_{3/2}$	
Actinides			Ac^{89} $7s^26d$ $^{2}D_{3/2}$	Th^{90} $7s^26d^2$	Pa^{91} $6d5f^2$	U^{92} $6d5f^3$ $^{5}L_6$	Np^{93} $6d5f^4$	Pu^{94} $5f^6$	Am^{95} $5f^7$ $^{8}S_{7/2}$	Cm^{96} $6d5f^7$	Bk^{97} $5f^9$	Cf^{98} $5f^{10}$	Es^{99} $5f^{11}$	Fm^{100} $5f^{12}$	Md^{101} $5f^{13}$	No^{102} $7s^2$	Lr^{103} $7s^26d5f^{14}$	

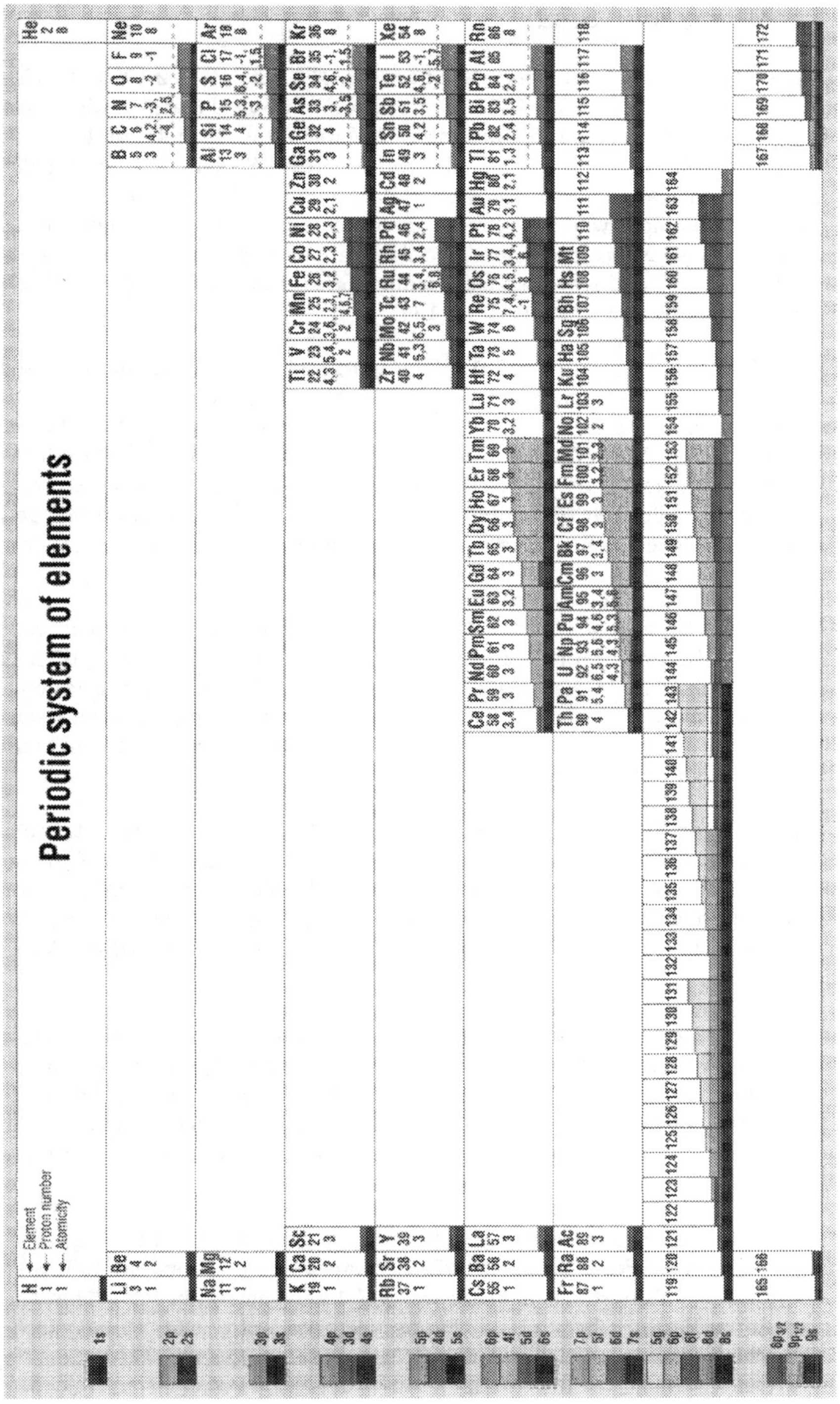

Fig. 11.8. Table of the periodic system of elements. Notice the incorporation of the recently discovered pre-superheavy elements with $Z = 104, \ldots, Z = 112$. They were predicted by theory, as were their chemical properties.

11.10 Biographical Notes

HARTREE, Douglas Rayner, *Cambridge 27.3.1897, †12.2.1958, British physicist and mathematician. 1929–37 professor for applied mathematics, then for theoretical physics at the University of Manchester. 1946–58 professor in Cambridge. 1932 he became a member of the Royal Society. Hartree's most important achievement was his approximate method for the calculation of quantum-mechanical wavefunctions of many-particle systems. Among other things, Hartree has also considered problems about digital calculators, ballistics, and physics of the atmosphere. Main works: numerical analysis (1952), the calculation of atomic structures (1957).

FERMI, Enrico, *Rom 29.9.1901, †Chicago 28.11.1954. Assistent professor at the University of Florence between 1924 and 1926. Between 1927 and 1938 professor of theoretical physics at the university of Chicago. In 1938 he received the Noble Prize of physics for the creation of new radioactive elements by neutron bombardment and for the discovery of chain reactions, which were caused by slow neutrons. He mainly worked in nuclear physics and developed the Fermi statistics named after him. In 1942 he put the first atomic reactor into operation.

THOMAS, Llewellyn Hilleth, *London 21.10.1903, American physicist, British by birth. Between 1929 and 1946 professor in Columbus (Ohio), then at Columbia University in New York. Between 1968 and 1976 he was in Raleigh (NC). Most important works were among, about other things, quantum mechanics of atoms (especially about the spin–orbit interaction of electrons) and about the theory of particle accelerators.

HUND, Friedrich, *Karlsruhe 4.2.1896, †31.03.1997, German physicist. From 1929 to 1946 professor of mathematical physics at the University of Leipzig. Between 1946 and 1951 and between 1951 and 1956 he was professor for theoretical physics at the universities in Jena and Frankfurt/M, respectively, and since 1956 at the University of Göttingen. In 1943 he received the Max-Planck medal.

LANDÉ, Alfred, *Elberfeld, †30.10.1976 (today belonging to Wuppertal) 13.12.1888, American physicist of German origin. 1922–31 professor in Tübingen, then at the Capital University in Columbus (Ohio). In 1921–23 he developed a theory of the mulitplet spectra and of the Zeeman effect (Landé vector model), where he introduced the g factor, now called the Landé factor after him.

PASCHEN, Friedrich Louis Carl Heinrich, *Schwerin (Mecklenburg, Germany) 22.1.1865, †Potsdam 25.2.1947, German physicist. Between 1901 and 1923 professor for physics at the University of Tübingen and in 1924–33 at the University of Berlin, where in addition he was president of the "Physikalisch-Technische Reichsanstalt". His work was in quantum and spectral physics. He discovered the first two lines of the spectral series of hydrogen named after him. The discovery of the splitting of the spectral lines in strong magnetic fields is due to him and Back (Paschen–Back effect).

BACK, Ernst, *21.10.1881, †20.7.1959; German physicist, professor in Hohenheim and Tübingen.

STARK, Johannes, German physicist, *15.4.1874 Schickenhof, in Thansüs, district of Amberg, †21.6.1957 Traunstein. S. became a professor in Hannover; in 1909 he went to Aachen, in 1917 to Greifswald, and in 1920 to Würzburg. He founded the "Jahrbuch der Radioaktivität und Elektronik" in 1904 and discovered in 1905 the (optical) *Doppler effect* in so-called channel rays and in 1913 the *Stark effect*. He was awarded the Nobel Prize in 1919. In 1933 he became the president of the "Notgemeinschaft der Deutschen Wissenschaft". He was a friend of P. Lenard, a supporter of "German physics", which dismissed quantum theory and the theory of relativity as the "product of Jewish thinking".

12. Elementary Structure of Molecules

For molecules a more difficult situation arises, because, in general, we are no longer dealing with a spherical symmetric problem. Generally the electrons do not experience a central potential. In order to find out more about this problem some approximations have to be applied. The atomic nuclei are located at classical equilibrium positions, around which they slowly oscillate (large mass of the nuclei!). On the other hand the electrons are moving fast in the Coulomb field of both nuclei (if we deal with a two-atom molecule). This offers an adiabatic approximation for the motion of the atomic nuclei. This only works because the masses of the atomic nuclei are much larger than the ones of the electrons. The ratio between nucleon (M) and electron (m) mass is

$$\frac{M}{m} \approx 1836\,. \tag{12.1}$$

For heavier atoms this ratio changes to typical values of about $10^{+4} - 10^{+5}$. Because of this mass ratio the electrons move much faster than the atomic nuclei. To a high degree the positions of the nuclei can be regarded as fixed. Vibrations of the nuclei around their equilibrium positions can be treated adiabatically. This shows up as a slow deformation of the molecular orbits.

In addition to electronic excitations, atomic nuclei can *vibrate* against each other and *rotate* as a whole. We estimate the energies of the corresponding excitations. With a being the typical extension of the molecule it follows that $\Delta p \approx \hbar/a$ from Heisenberg's uncertainty principle for a typical electronic excitation and therefore

$$E_a = \frac{(\Delta p)^2}{2m} \approx \frac{\hbar^2}{2ma^2}\,. \tag{12.2}$$

As for atoms the orders of magnitude are some eV. For the relative vibration of the nuclei we assume a harmonic potential $M\omega^2(R-R_0)^2/2$. Here $R-R_0$ corresponds to the distance from the equilibrium position R_0 (see Fig. 12.1). If the relative distance between the two nuclei is $(R-R_0) \approx a$, then the corresponding excitations are approximately equal to the electronic excitations. Hence, we can set

$$\frac{1}{2}M\omega^2a^2 \approx \frac{\hbar^2}{2ma^2}$$

and obtain an estimate for ω:

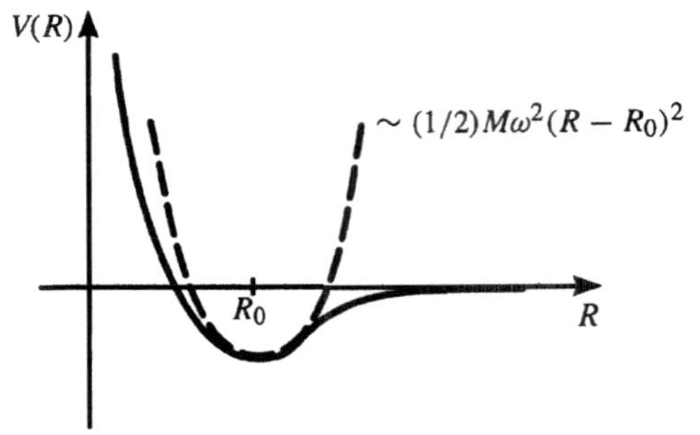

Fig. 12.1. Typical molecular potential for the nuclei in the molecule. R_0 characterizes the relative equilibrium position of the two nuclei. R represents the relative nuclear distance

$$\hbar\omega \approx (m/M)^{1/2} \frac{\hbar^2}{ma^2} . \tag{12.3}$$

The factor $(m/M)^{1/2}$ appears in the vibrational energy and $\hbar^2/ma^2$ is the electronic excitation. We deduce that

$$E_{\text{vib}} \approx \left(\frac{m}{M}\right)^{1/2} E_{\text{el}} . \tag{12.4}$$

It is easier to excite a molecule by rotation than by vibration. For a rotation the nuclei do not have to stretch out of their equilibrium position. In zeroth approximation the equilibrium distance remains the same during the rotation. Only the rotation–vibration interaction (centrifugal force!) is able to change the distance between the rotating nuclei. The rotational energy is given by

$$E_{\text{rot}} = \frac{\hbar^2 l(l+1)}{2\theta} . \tag{12.5}$$

Here l represents the angular momentum of the molecule and θ its moment of inertia ($\theta \approx Ma^2$). Hence, we obtain

$$E_{\text{rot}} \approx \frac{\hbar^2}{ma^2} l(l+1) \left(\frac{m}{M}\right) . \tag{12.6}$$

Here the factor (m/M) appears; it lowers the rotational energy with respect to the electronic excitation:

$$E_{\text{rot}} \approx \left(\frac{m}{M}\right) E_{\text{el}} . \tag{12.7}$$

The excitation of a molecule in general is given by the combination of all excitations mentioned, i.e.

$$E = E_{\text{el}} + E_{\text{vib}} + E_{\text{rot}} . \tag{12.8}$$

The ratio of the energies is

$$E_{\text{el}} : E_{\text{vib}} : E_{\text{rot}} = 1 : \left(\frac{m}{M}\right)^{1/2} : \frac{m}{M} . \tag{12.9}$$

Obviously the rotational energy is lowest: the excitation steps for a rotation are smaller by a factor of $\sqrt{m/M} \approx \sqrt{1/2000} \approx 1/40$ than for those a vibration and these are again smaller by the same factor compared to those for the electrons.

12.1 Born–Oppenheimer Approximation

To describe the various motions of the molecule we begin with the Schrödinger equation. The Hamiltonian is given by

$$\hat{H} = \hat{T}_e + \hat{T}_N + V_{ee} + V_{eN} + V_{NN} , \tag{12.10}$$

where

$$\hat{T}_e = \sum_{i=1}^{N} \frac{\hat{p}_i^2}{2m} \tag{12.11}$$

represents the kinetic energy of the electrons and

$$\hat{T}_N = \sum_{\nu=1}^{2} \frac{\hat{P}_\nu^2}{2M_\nu} \tag{12.12}$$

is the kinetic energy of the nuclei. V_{eN} represents the attractive electron–nuclei potential. V_{ee} describes the repelling electron–electron interaction. V_{NN} indicates the repelling Coulomb interaction between the nuclei. Since the masses of the nuclei are very large, $\hat{T}_N$ can be neglected. This step is called the *Born–Oppenheimer approximation*. In the following, we will explain the approximation in more detail.

If we neglect the kinetic energy $\hat{T}_N$ of the nuclei (static approximation: fixed distance R of the nuclei), the relative distance R between the nuclei only occurs as a parameter. The Schrödinger equation becomes

$$\begin{aligned} &[\hat{T}_e + V_{ee}(\boldsymbol{r}) + V_{eN}(\boldsymbol{r},\boldsymbol{R})]\, \varphi_n(\boldsymbol{r},\boldsymbol{R}) \\ &\quad = [\varepsilon_n(\boldsymbol{R}) - V_{NN}(\boldsymbol{R})]\, \varphi_n(\boldsymbol{r},\boldsymbol{R}) . \end{aligned} \tag{12.13}$$

Here $\boldsymbol{r}$ indicates the position of the electron. The solutions $\varphi_n(\boldsymbol{r},\boldsymbol{R})$ depend parametricaly on the distance between the nuclei. The energy of this state is given by the electronic energy $\varepsilon_n(\boldsymbol{R})$ lowered by $V_{NN}(\boldsymbol{R})$. The solutions $\varphi_n(\boldsymbol{r},\boldsymbol{R})$ represent a complete set of functions. The true wavefunction $\psi(\boldsymbol{r},\boldsymbol{R})$ can be expanded within this set:

$$\psi(\boldsymbol{r},\boldsymbol{R}) = \sum_m \phi_m(\boldsymbol{R})\varphi_m(\boldsymbol{r},\boldsymbol{R}) . \tag{12.14}$$

The coefficients $\phi_m(\boldsymbol{R})$ are to be found and, in general, depend on $\boldsymbol{R}$. $\psi(\boldsymbol{r},\boldsymbol{R})$ is the solution of the full Schrödinger equation, which takes into consideration the kinetic energy $\hat{T}_N$ of the atomic nuclei, i.e.

$$(\hat{T}_e + \hat{T}_N + V_{ee} + V_{eN} + V_{NN})\psi(\boldsymbol{r},\boldsymbol{R}) = E\psi(\boldsymbol{r},\boldsymbol{R}) . \tag{12.15}$$

Inserting (12.14) into (12.15) and using (12.13), we obtain

$$\sum_m (\varepsilon_m(\boldsymbol{R}) + \hat{T}_N)\phi_m(\boldsymbol{R})\varphi_m(\boldsymbol{r},\boldsymbol{R}) = E\sum_m \phi_m(\boldsymbol{R})\varphi_m(\boldsymbol{r},\boldsymbol{R}) . \tag{12.16}$$

Now we multiply from the left-hand side with $\varphi_n^\dagger(\boldsymbol{r},\boldsymbol{R})$, integrate over the full space, and get

$$\sum_m \int \mathrm{d}^3r\, \varphi_n^\dagger(\boldsymbol{r},\boldsymbol{R})\hat{T}_\mathrm{N}\phi_m(\boldsymbol{R})\varphi_m(\boldsymbol{r},\boldsymbol{R}) + \varepsilon_n(\boldsymbol{R})\phi_n(\boldsymbol{R}) \;=\; E\phi_n(\boldsymbol{R})\,. \tag{12.17}$$

Here we have used the orthogonality of the functions $\varphi_n(\boldsymbol{r},\boldsymbol{R})$. $\hat{T}_\mathrm{N}$ is proportional to the Laplace operator Δ_R, which acts on $\phi_m\varphi_m$. It holds that

$$\Delta_R(\phi\varphi) \;=\; (\Delta_R\phi)\varphi + 2\boldsymbol{\nabla}_R\Phi\cdot\boldsymbol{\nabla}_R\varphi + \phi\Delta_R\varphi\,. \tag{12.18}$$

The index R indicates the action of the operators in R space. The first term in (12.18) is proportional to $\hat{T}_\mathrm{N}\phi_n$. The rest is brought to the right-hand side of (12.17). The result reads

$$[\hat{T}_\mathrm{N} + \varepsilon_n(\boldsymbol{R})]\phi_n(\boldsymbol{R}) \;=\; E\phi_n(\boldsymbol{R}) - \sum_m C_{nm}\phi_m(\boldsymbol{R}) \tag{12.19a}$$

with

$$C_{nm}\phi_m(\boldsymbol{R}) \;=\; -\hbar^2\sum_\alpha \frac{1}{2M_\alpha}\int \mathrm{d}^3r\,\varphi_n^\dagger(\boldsymbol{r},\boldsymbol{R}) \tag{12.19b}$$
$$\times\,[2\boldsymbol{\nabla}_{R_\alpha}\phi_m(\boldsymbol{R})\cdot\boldsymbol{\nabla}_{R_\alpha}\varphi_m(\boldsymbol{r},\boldsymbol{R}) + \phi_m(\boldsymbol{R})\Delta_{R_\alpha}\varphi_m(\boldsymbol{r},\boldsymbol{R})]\,.$$

The sum over α comes from $\hat{T}_\mathrm{N}$ [see (12.12)] and $\boldsymbol{\nabla}_{R_\alpha}$ acts only on the coordinate $\boldsymbol{R}_\alpha$ of the nucleus α, which appears in $R = \sqrt{(\boldsymbol{R}_2-\boldsymbol{R}_1)^2}$. Now, the order of magnitude of C_{nm} is $(m/M)^{1/2}$ times smaller than the electronic kinetic energy. This can be seen as follows. The order of magnitude of the term $\sim \hbar^2\Delta_{R_\alpha}\varphi_m/2M_\alpha$ (the kinetic energy of the nucleus) is proportional to $-(m/M_\alpha)\hbar^2(\Delta_r\varphi_m/2m)$; we have simply replaced Δ_{R_α} by Δ_r and introduced the electronic kinetic energy $-\hbar^2\Delta_r\varphi_m/2m$. The factor m/M_α indicates that the contribution of Δ_{R_α} to C_{nm} is smaller by this factor than the kinetic energy of the electron.

The first term in (12.19b) remains to be estimated. For this we approximate ϕ_m by a harmonic oscillator wavefunction: $\phi_m \approx \exp(-(\boldsymbol{R}-\boldsymbol{R}_0)^2M\omega/2\hbar)$, R_0 being the equilibrium position of nucleus α. We have

$$\boldsymbol{\nabla}_{R_\alpha}\phi_m \;\approx\; |\boldsymbol{R}-\boldsymbol{R}_0|\frac{M\omega}{\hbar}\phi_m \;\approx\; \frac{(\delta R)M\omega}{\hbar}\phi_m\,. \tag{12.20}$$

δR indicates the shift from the equilibrium position. The factor M is canceled by $1/M$ in (12.19b) and the contribution is proportional to the vibrational energy $\hbar\omega$. As noted earlier, this goes like $\sim (m/M)^{1/2}$. As a summary, the C_{nm} term can be neglected or treated with the help of perturbation theory. Without the C_{nm} term, (12.19a) reduces to

$$[\hat{T}_\mathrm{N} + \varepsilon_n(\boldsymbol{R})]\phi_n(\boldsymbol{R}) \;=\; E\phi_n(\boldsymbol{R})\,. \tag{12.21}$$

This equation has an interesting interpretation: the energy of the electron states $\varepsilon_n(\boldsymbol{R})$ acts like an effective potential in $\boldsymbol{R}$. We imagine that the electrons build a "medium" in which the atomic nuclei move. This medium acts as an elastic band. If the nuclei try to leave the equilibrium position, they will be drawn back. There is an equilibrium position where $\varepsilon(\boldsymbol{R})$ has a minimum deep enough to generate binding. The elastic band behavior is then nothing other than the expansion up to the order $(\boldsymbol{R}-\boldsymbol{R}_0)^2$.

The C_{nm} produce a mixing between different states φ_n and φ_m. This mixing between the $\varphi_n(\boldsymbol{R})$ states can be neglected in lowest order, because the C_{nm} are small [of order $(m/M)^{1/2}$, as explained previously]. Accordingly the wavefunction is approximately given by

$$\psi_{n\nu}(\boldsymbol{r},\boldsymbol{R}) = \phi_{n\nu}(\boldsymbol{R})\varphi_n(\boldsymbol{r},\boldsymbol{R}) . \tag{12.22}$$

Here ν stands for all quantum numbers of level n. $E_{n\nu}$ indicates the energy of the molecule, which is calculated from (12.21).

In order to describe vibrations and rotations of the molecule $\varepsilon_n(\boldsymbol{R})$ is expanded in coordinates describing vibration and rotation, respectively. The expansion in $\delta R = |R - R_0|$ up to the squared order leads to a harmonic vibrational potential (see Fig. 12.1). $\varepsilon_N(R)$ does not depend on the angles (Euler angles). Hence the rotations of the molecule are free. An excitation of the molecule is a combination of excitations of the harmonic vibrational oscillator and of the rotations.

We summarize: in the Born–Oppenheimer approximation, first the energy levels of the electrons are determined for fixed distances $\boldsymbol{R}$ of the nuclear centers. The electron energy $\varepsilon_n(R)$ plays the role of a potential, in which the nuclei are moving. If this potential has one or several deep enough minima, one or several bound states of the molecule can exist. If the minima are only weak or do not exist at all, then the molecule is not bound.

12.2 The H_2^+ Ion as an Example

The simplest molecule is represented by the H_2^+ ion. We neglect the motion of the nuclei and the spin–orbit coupling, so that the Hamiltonian takes on the form

$$\hat{H} = -\frac{\hbar^2\Delta}{2m} - \frac{e^2}{|\boldsymbol{r}-\boldsymbol{R}_1|} - \frac{e^2}{|\boldsymbol{r}-\boldsymbol{R}_2|} + \frac{e^2}{|\boldsymbol{R}_1-\boldsymbol{R}_2|} . \tag{12.23}$$

The first term is the kinetic energy of the electron, the second and third terms describe the attractive Coulomb interactions between electron and nuclei 1 and 2 with position vectors $\boldsymbol{R}_1$ and $\boldsymbol{R}_2$, respectively. The last term represents the repulsive interaction between the two protons (nuclei). Furthermore, $\boldsymbol{r}$ describes the electron's position and $\boldsymbol{R}_1$ and $\boldsymbol{R}_2$ the positions of the two protons (see Fig. 12.2). It is impossible to determine the eigenfunctions of (12.23) exactly and analytically. Hence, test wavefunctions are constructed and the energy is minimized by variation.

In case of the H_2^+ ion we use a test wavefunction that is a linear combination of hydrogen wavefunctions referring to the first (ψ_1) and second (ψ_2) proton:

$$\psi(\boldsymbol{r}) = \alpha\psi_1(\boldsymbol{r}) + \beta\psi_2(\boldsymbol{r}) . \tag{12.24}$$

This crude method is particularly instructive and transparent. For the lowest state we take the 1s wavefunction, i.e.

$$\psi_1 = \frac{1}{\sqrt{\pi a_0^3}} \mathrm{e}^{-|\boldsymbol{r}-\boldsymbol{R}_1|/a_0} , \tag{12.25a}$$

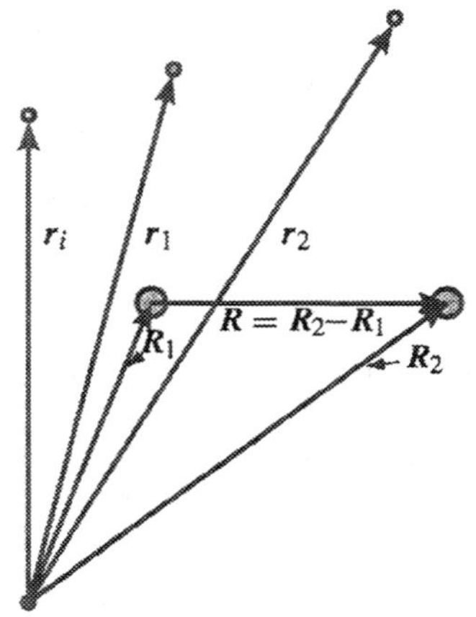

Fig. 12.2. Illustration of the coordinates used: $\boldsymbol{r}_i$ is the position of the ith electron and $\boldsymbol{R}_\alpha$ of the αth nucleus. The distance $R = \sqrt{(\boldsymbol{R}_2 - \boldsymbol{R}_1)^2}$ between the nuclei depends on both $\boldsymbol{R}_1$ and $\boldsymbol{R}_2$

and, correspondingly,

$$\psi_2 = \frac{1}{\sqrt{(\pi a_0^3)}} \mathrm{e}^{-|\boldsymbol{r}-\boldsymbol{R}_2|/a_0} . \tag{12.25b}$$

Here $a_0 = \hbar^2/em = 0.529\,\text{Å}$ represents the Bohr radius. The problem of the H_2^+ ion is reflection symmetric with respect to a plane lying exactly between the two protons. This implies that the wavefunction possesses the same symmetry and that the wavefunction can be classified by its parity. If the parity is positive we have $\alpha = \beta$, and if it is negative $\alpha = -\beta$. So we end up with

$$\psi_\pm(\boldsymbol{r}) = N_\pm[\psi_1(\boldsymbol{r}) \pm \psi_2(\boldsymbol{r})] , \tag{12.26}$$

where $N_\pm$ is an appropriate normalization factor. For the norm $\langle\psi_\pm|\psi_\pm\rangle$ we find

$$1 = N_\pm^2 \left[2 \pm 2 \int \mathrm{d}^3r\, \psi_1^\dagger(\boldsymbol{r})\psi_2(\boldsymbol{r})\right] .$$

With the so-called *overlap integral*

$$\begin{aligned} U(R) &= \int \mathrm{d}^3r\, \psi_1^\dagger(\boldsymbol{r})\psi_2(\boldsymbol{r}) \\ &= \left(1 + \frac{R}{a_0} + \frac{R^2}{3a_0^2}\right) \mathrm{e}^{-R/a_0} \end{aligned}$$

(see Exercise 12.1), in which $R = \sqrt{(\boldsymbol{R}_1 - \boldsymbol{R}_2)^2}$, we have

$$N_\pm = \frac{1}{\sqrt{2[1 \pm U(R)]}} . \tag{12.27}$$

Obviously the overlap integral gives the spatial overlap of the two 1s wavefunctions.

The expectation value of $\hat{H}$ from (12.23) is now given by

$$\begin{aligned}\langle \hat{H} \rangle_\pm &= \varepsilon_\pm(\boldsymbol{R}) \\ &= \frac{\langle \psi_1|\hat{H}|\psi_1\rangle + \langle \psi_2|\hat{H}|\psi_2\rangle \pm 2\langle \psi_1|\hat{H}|\psi_2\rangle}{2(1\pm U)} \,. \end{aligned} \tag{12.28}$$

We now use the abbreviation $|1\rangle$ and $|2\rangle$ for the states ψ_1 and ψ_2, respectively. Since by symmetry we have $\langle 1|\hat{H}|1\rangle = \langle 2|\hat{H}|2\rangle$, it follows that

$$\varepsilon_\pm(\boldsymbol{R}) = \frac{\langle 1|\hat{H}|1\rangle \pm \langle 1|\hat{H}|2\rangle}{(1\pm U)} \,. \tag{12.29}$$

Now, $\langle 1|\hat{H}|1\rangle$ is given by

$$\begin{aligned}\langle 1|\hat{H}|1\rangle &= \int \mathrm{d}^3r\, \psi_1^\dagger(\boldsymbol{r})\hat{H}\psi_1(\boldsymbol{r}) \\ &= \varepsilon_1 + \frac{e^2}{R} - \int \mathrm{d}^3r\, |\psi_1(\boldsymbol{r})|^2 \frac{e^2}{|\boldsymbol{r}-\boldsymbol{R}|} \,. \end{aligned} \tag{12.30}$$

ε_1 is the energy of the 1s state in the hydrogen atom, which is $-1\,\mathrm{Ry}$. This contribution comes from the first two terms in (12.23). The Coulomb energy e^2/R, with $R = |\boldsymbol{R}_1 - \boldsymbol{R}_2|$, results from the repulsive interaction between both protons and the last term in (12.30) comes from the contribution of the interaction between the electron and the second proton. Inserting the wavefunction, we arrive at (see Exercise 12.1)

$$\langle 1|\hat{H}|1\rangle = \varepsilon_1 + \frac{e^2}{R}\left(1+\frac{R}{a_0}\right)\mathrm{e}^{-2R/a_0} \,. \tag{12.31}$$

For the second term in (12.29) we get

$$\begin{aligned}\langle 1|\hat{H}|2\rangle &= \int \mathrm{d}^3r\, \psi_1^\dagger(\boldsymbol{r})\hat{H}\psi_2(\boldsymbol{r}) \\ &= \left(\varepsilon_1 + \frac{e^2}{R}\right)U - \int \mathrm{d}^3r\, \psi_1^\dagger(\boldsymbol{r})\psi_2(\boldsymbol{r})\frac{e^2}{|\boldsymbol{r}-\boldsymbol{R}_2|} \,. \end{aligned} \tag{12.32}$$

The last integral in (12.32) is the so-called *exchange integral*

$$\begin{aligned} A &= \int \mathrm{d}^3r\, \psi_1^\dagger(\boldsymbol{r})\psi_2(\boldsymbol{r})\frac{e^2}{|\boldsymbol{r}-\boldsymbol{R}_2|} \\ &= \frac{e^2}{a_0}\left(1+\frac{R}{a_0}\right)\mathrm{e}^{-R/a_0} \,. \end{aligned} \tag{12.33}$$

This integral will be determined in Exercise 12.1. With

$$U = \left(1+\frac{R}{a_0}+\frac{R^2}{3a_0}\right)\mathrm{e}^{-R/a_0}$$

we can calculate the energy $\varepsilon_\pm(R)$ as a function of R. The behavior of $\varepsilon_\pm(R)$ is depicted in Fig. 12.3. The state with positive parity is bound, since $\varepsilon_+(R)$ has a minimum. On the other hand, the energy of the state with negative parity has no minimum; this state is never bound. Experimentally one finds a binding energy of $-2.8\,\mathrm{eV}$ for $\psi_+(\boldsymbol{r})$ when the distance between the two protons is

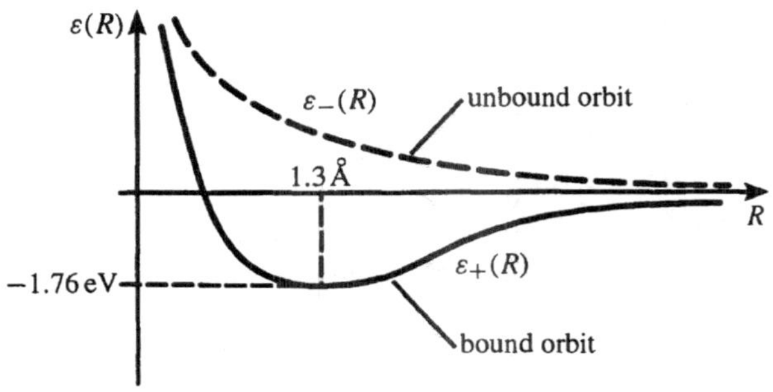

Fig. 12.3. Energy for the bound [$\varepsilon_+(R)$] and unbound [$\varepsilon_-(R)$] orbit of the H_2^+ ion as a function of the distance R between the two protons

1.06 Å. Our calculated energy is still to high, which is to be expected with the crude test wavefunction (12.26).

The wavefunction of an electron, for example, $\psi_\pm(\boldsymbol{r})$, is called a *molecular orbit wavefunction.* The form of the test wavefunction (12.24) is also called a *linear combination of atomic orbits* (LCAO). ψ_+ is called the *bound* and ψ_- the *unbound orbit.* ψ_- is not bound, which can be seen from the following argument. Since ψ_- has negative parity, $\psi_-(\boldsymbol{r}) = -\psi_-(-\boldsymbol{r})$, the wavefunction has a zero at $r = 0$. However, the electron profits especially from the attractive interaction with the protons, which is at a maximum between the protons.

For very large proton distances the approximated function $\psi_+(\boldsymbol{r})$ becomes more realistic. For $R \to 0$ the wavefunction (12.26) becomes the wavefunction for a He^+ atom, because for $R \ll a_0$ the electron practically sees a point nucleus with charge number $Z = 2$; indeed $\psi_+(\boldsymbol{r})$ becomes the 1s wavefunction of the hydrogen atom. This means that for small proton distances the approximation (12.26) becomes inaccurate. This is the reason why the binding energy calculated within this method is too small (see Fig. 12.4).

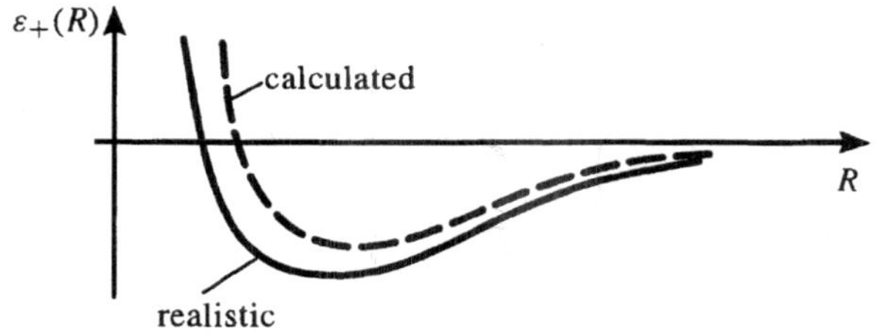

Fig. 12.4. Schematic comparision between calculated and realistic energy $\varepsilon_+(R)$ for the bound orbit

This phenomenon is illustrated in Fig. 12.5. For large proton distances R we have a H atom and an isolated proton. The calculated value approaches the accurate one, which is -1 Ry. But for $R \to 0$ our crude calculation yields -3 Ry, whereas the realistic value is -4 Ry (this is the ground-state energy for a He^+ atom).

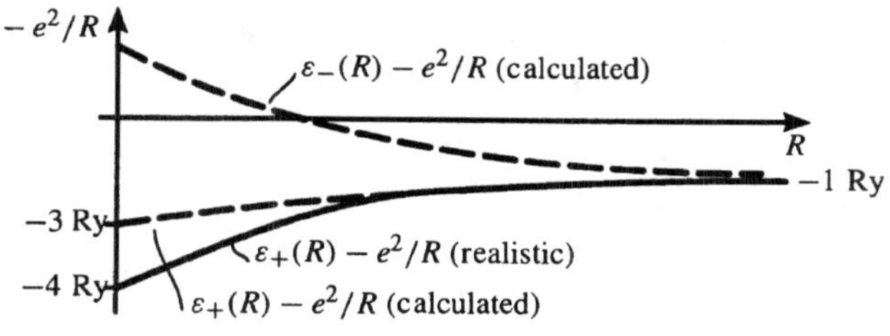

Fig. 12.5. H_2^+ ion: illustration of the calculated and realistic energies (minus the Coulomb repulsion of the protons) for bound $[\varepsilon_+(R)]$ and unbound $[\varepsilon_-(R)]$ orbits. For small R ($R \to 0$) the limit of a He^{2+} ion is approached

EXERCISE

12.1 Calculation of an Overlap Integral and Some Matrix Elements for the H_2^+ Ion

Problem. Calculate

(a) the overlap integral $U(R) = \langle\psi_1(\boldsymbol{R})|\psi_2(\boldsymbol{R})$,
(b) the energy $\langle\psi_1|\hat{H}|\psi_1\rangle$, and
(c) the interaction integral $\langle\psi_1(\boldsymbol{R})|\hat{H}|\psi_2(\boldsymbol{R})\rangle$

for the Hamiltonian (12.23) and the wavefunction (12.25a,b).

Solution.
(a) The overlap integral is

$$U(R) = \frac{1}{\pi a_0^3}\int d^3r \exp\left(-|\boldsymbol{r}-\boldsymbol{R}_1|/a_0 - |\boldsymbol{r}-\boldsymbol{R}_2|/a_0\right) .$$

The introduction of relative $\boldsymbol{R} = \boldsymbol{R}_1 - \boldsymbol{R}_2$ and center-of-mass coordinates $\boldsymbol{R}_s = \frac{1}{2}(\boldsymbol{R}_1 + \boldsymbol{R}_2)$ of the two nucleis gives

$$\begin{aligned} I_1 &= \int d^3r \exp\left(-\left|\boldsymbol{r}-\frac{1}{2}\boldsymbol{R}-\boldsymbol{R}_s\right|\Big/a_0 - \left|\boldsymbol{r}+\frac{1}{2}\boldsymbol{R}-\boldsymbol{R}_s\right|\Big/a_0\right) \\ &= \int d^3r \exp\left(-\left|\boldsymbol{r}-\frac{1}{2}\boldsymbol{R}\right|\Big/a_0 - \left|\boldsymbol{r}+\frac{1}{2}\boldsymbol{R}\right|\Big/a_0\right) \end{aligned}$$

because of the obvious translation invariance of I_1. The azimuthal symmetry of the integrand of I_1 offers the use of *prolate-elliptical coordinates*:[1]

$$\begin{aligned} x &= \frac{R}{2}[(\xi^2-1)(1-\eta^2)]^{1/2}\cos\varphi , \\ y &= \frac{R}{2}[(\xi^2-1)(1-\eta^2)]^{1/2}\sin\varphi , \\ z &= \frac{R}{2}\xi\eta \end{aligned}$$

[1] H. Stöcker: *Taschenbuch mathematischer Formeln und moderner Verfahren* (Harri Deutsch, Thun 1996).

Exercise 12.1.

with $\xi \in [1,\infty)$, $\eta \in [-1,1]$, $\varphi \in [0,2\pi)$ and $R = |\boldsymbol{R}_1 - \boldsymbol{R}_2|$. Using

$$\left|\boldsymbol{r} \mp \frac{1}{2}\boldsymbol{R}\right| = \frac{R}{2}(\xi \mp \eta)\,,$$

it then follows for I_1:

$$\begin{aligned} I_1 &= \left(\frac{R}{2}\right)^3 \int_0^{2\pi} \mathrm{d}\varphi \int_{-1}^{1} \mathrm{d}\eta \int_1^\infty \mathrm{d}\xi\,(\xi^2-\eta^2) \\ &\quad \times \exp\left\{-R\left[(\xi-\eta)+(\xi+\eta)\right]/2a_0\right\} \\ &= 2\pi\left(\frac{R}{2}\right)^3 \int_{-1}^{1} \mathrm{d}\eta \int_1^\infty \mathrm{d}\xi\,(\xi^2-\eta^2)\,\mathrm{e}^{-R\xi/a_0} \\ &= 2\pi\left(\frac{R}{2}\right)^3 \int_1^\infty \mathrm{d}\xi \left(2\xi^2 - \frac{2}{3}\right)\mathrm{e}^{-R\xi/a_0} \\ &= \frac{\pi}{4}R^3\,\mathrm{e}^{-R/a_0}\frac{2a_0}{R}\left(\frac{2}{3} - \frac{2a_0}{R} + \frac{a_0^2}{R^2}\right)\,. \end{aligned}$$

Hence, we have

$$U(R) = \left(\frac{1}{2} + \frac{R}{a_0} + \frac{R^2}{3a_0^2}\right)\mathrm{e}^{-R/a_0}\,.$$

(b) The expectation value of $\hat{H}$ with respect to $\psi_1(\boldsymbol{R})$ is

$$\begin{aligned} \overline{H}_{11} &= \langle\psi_1(\boldsymbol{R})|\hat{H}|\psi_1(\boldsymbol{R})\rangle \\ &= \frac{1}{\pi a_0^3}\int \mathrm{d}^3r \exp\left(-|\boldsymbol{r}-\boldsymbol{R}_1|/a_0\right)\hat{H}\exp\left(-|\boldsymbol{r}-\boldsymbol{R}_1|/a_0\right)\,. \end{aligned}$$

The first two terms of $\hat{H}$ together give the Hamiltonian, which has the eigenfunction ψ_1 as the ground state; the coordinate system has been translated by $\boldsymbol{R}_1$, i.e.

$$\begin{aligned} \hat{H}_0\psi_1 &= \left(-\frac{\hbar^2}{2m}\Delta_r - \frac{e^2}{|\boldsymbol{r}-\boldsymbol{R}_1|}\right)\psi_1 \\ &= \varepsilon_1\psi_1 \end{aligned}$$

with $\varepsilon_1 = -e^2/2a_0$. The expectation value of ψ_1, $\overline{H}_0 = \langle\psi_1|\hat{H}_0|\psi_1\rangle$ is then ε_1. The fourth term of $\hat{H}$ is independent of $\boldsymbol{r}$; hence it follows that

$$\begin{aligned} \overline{H}_1 &= \frac{1}{\pi a_0^3}\int \mathrm{d}^3r \exp\left(-2|\boldsymbol{r}-\boldsymbol{R}_1|/a_0\right)\frac{e^2}{|\boldsymbol{R}_1-\boldsymbol{R}_2|} \\ &= \frac{e^2}{R} \end{aligned}$$

with $R = |\boldsymbol{R}_1 - \boldsymbol{R}_2|$.

The expectation value of the third term of $\hat{H}$ is

$$\overline{H}_2 = \frac{1}{\pi a_0^3}\int \mathrm{d}^3r \exp\left(-2|\boldsymbol{r}-\boldsymbol{R}_1|/a_0\right)\frac{-e^2}{|\boldsymbol{r}-\boldsymbol{R}_2|}\,.$$

The introduction of prolate-elliptical coordinates [see part (a) of this exercise] gives

Exercise 12.1.

$$
\begin{aligned}
\overline{H}_2 &= \frac{1}{\pi a_0^3}\left(\frac{R}{2}\right)^3 \int_0^{2\pi} d\varphi \int_{-1}^{1} d\eta \int_1^{\infty} d\xi\,(\xi^2-\eta^2) \\
&\quad \times \exp\left[-2R(\xi-\eta)/2a_0\right] \frac{-e^2}{R(\xi+\eta)/2} \\
&= \frac{-R^2 e^2}{2a_0^3} \int_{-1}^{1} d\eta \int_1^{\infty} d\xi\,(\xi-\eta) \exp\left[-R(\xi-\eta)/a_0\right] \\
&= \frac{e^2}{a_0} e^{-2R/a_0} .
\end{aligned}
$$

The sum of all terms leads to (12.31):

$$
\begin{aligned}
\overline{H}_{11} &= \langle \psi_1 | \hat{H} | \psi_1 \rangle \\
&= \varepsilon_0 + \overline{H}_1 + \overline{H}_2 \\
&= \varepsilon_0 + \frac{e^2}{R} + \frac{e^2}{a_0} e^{-2R/a_0} \\
&= \varepsilon_0 + \frac{e^2}{R}\left(1 + \frac{R}{a_0} e^{-2R/a_0}\right) .
\end{aligned}
$$

(c) The interaction integral is given by

$$
\begin{aligned}
\overline{H}_{12} &= \langle \psi_1(\boldsymbol{R}) | \hat{H} | \psi_2(\boldsymbol{R}) \rangle \\
&= \frac{1}{\pi a_0^3} \int d^3 r \exp\left(-|\boldsymbol{r}-\boldsymbol{R}_1|/a_0\right) \hat{H} \exp\left(-|\boldsymbol{r}-\boldsymbol{R}_2|/a_0\right) .
\end{aligned}
$$

With the help of parts (a) and (b) of this exercise, the contribution of $\overline{H}_{12}$ can be written down immediately; from the terms 1, 3, and 4 of $\hat{H}$ we get

$$
\begin{aligned}
\overline{H}_{12} &= \varepsilon_0 U(R) + \frac{e^2}{R} U(R) + \frac{1}{\pi a_0^3} \int d^3 r \frac{-e^2}{|\boldsymbol{r}-\boldsymbol{R}_1|} \\
&\quad \times \exp\left(-|\boldsymbol{r}-\boldsymbol{R}_1|/a_0 - |\boldsymbol{r}-\boldsymbol{R}_2|/a_0\right) \\
&= \left(\varepsilon_0 + \frac{e^2}{R}\right) U(R) - A(R) .
\end{aligned}
$$

Here, $A(R)$ corresponds to the already mentioned exchange integral (12.33). In prolate-elliptical coordinates one obtains

$$
\begin{aligned}
A &= \frac{2e^2}{a_0^3}\left(\frac{R}{2}\right)^2 \int_{-1}^{1} d\eta \int_1^{\infty} d\xi\,(\xi^2-\eta^2)\frac{1}{\xi-\eta} e^{-R\xi/a_0} \\
&= \frac{e^2}{2a_0}\left(\frac{R}{a_0}\right)^2 \int_{-1}^{1} d\eta \int_1^{\infty} d\xi\,(\xi+\eta)\, e^{-R\xi/a_0} \\
&= \frac{e^2}{a_0}\left(\frac{R}{a_0}\right)^2 \int_1^{\infty} d\xi\,\xi\, e^{-R\xi/a_0} \\
&= \frac{e^2}{a_0}\left(1 + \frac{R}{a_0}\right) e^{-R/a_0} .
\end{aligned}
$$

12.3 The Hydrogen Molecule

The next, more difficult molecule is the hydrogen molecule, with two electrons. Again we choose a test wavefunction and calculate the expectation value of the Hamiltonian. We construct a symmetric test wavefunction and make the following ansatz:

$$\psi_S(1,2) = \frac{1}{2[1+U(R)]}[\psi_1(\boldsymbol{r}_1)+\psi_2(\boldsymbol{r}_1)][\psi_1(\boldsymbol{r}_2)+\psi_2(\boldsymbol{r}_2)]\,\chi_{\text{singlet}}\,. \tag{12.34}$$

The index "S" stands for "singlet". The subscripts 1 and 2 on ψ indicate the atomic centers with respect to which the one-particle wavefunctions are defined. The arguments $\boldsymbol{r}_1$ and $\boldsymbol{r}_2$ in the wavefunctions describe the positions of the two electrons. We choose again a symmetric wavefunction, since we want to have a state that is energetically as low as possible. Consequently, we set both electrons in the same molecular orbit $\psi_+(\boldsymbol{r}) = \psi_1(\boldsymbol{r}) + \psi_2(\boldsymbol{r})$, which is in accordance with our last section about the H_2^+ ion. Because both electrons are now in the same spatial state, their spins have to be antiparallel due to the Pauli principle: therefore we have a state χ_{singlet} with $S = 0$. If we want a triplett state, i.e. $S = 1$, we have to put one electron into the ψ_+ orbit and the other one into the ψ_- orbit. Since ψ_- is energetically unfavored, the triplett states are higher in energy.

The test wavefunction (12.34) has two disadvantages. First, for very small distances between the two protons the results have to be identical to those for the Helium atom. This is not the case here, because for $R \to 0$ both electrons are in a 1s orbit of a H atom and not in the 1s orbit of the He atom. We see the second disadvantage once we multiply out the wavefunction (12.34):

$$\begin{aligned}\psi_S(1,2) \sim\ & [\psi_1(\boldsymbol{r}_1)\psi_1(\boldsymbol{r}_2)+\psi_2(\boldsymbol{r}_1)\psi_2(\boldsymbol{r}_2)] \\ & +[\psi_1(\boldsymbol{r}_1)\psi_2(\boldsymbol{r}_2)+\psi_1(\boldsymbol{r}_2)\psi_2(\boldsymbol{r}_1)]\,.\end{aligned} \tag{12.35}$$

The expression in the first bracket $[\dots]$ describes two electrons both attached to the vicinity of the first or second proton. The wavefunction of the second bracket stands for one electron attached to the first proton and the other electron to the second proton. There is some problem with the wavefunction of the first bracket because it will be quite improbable that the two electrons are bound to the same proton for large distances R between the atomic centers. This is mainly because the energy of the H^- ion is higher than that of the two H atoms. For large separations R, the true two-electron wavefunction will therefore be better described by the terms in the second bracket in (12.35).

An alternative method for determining the energy of the H_2 molecule is given by the *valence binding* or *Heitler–London* method. In (12.35) only the term for which one electron can be found around each proton is used, i.e. for the singlet state

$$\begin{aligned}\psi_+(\boldsymbol{r}_1,\boldsymbol{r}_2) &\equiv \psi_s(\boldsymbol{r}_1,\boldsymbol{r}_2) \\ &= \frac{1}{\sqrt{2(1+U^2)}}[\psi_1(\boldsymbol{r}_1)\psi_2(\boldsymbol{r}_2)+\psi_1(\boldsymbol{r}_2)\psi_2(\boldsymbol{r}_1)]\,.\end{aligned} \tag{12.36}$$

Analogously we have for the triplet state

$$\begin{aligned}\psi_-(\boldsymbol{r}_1,\boldsymbol{r}_2) &\equiv \psi_t(\boldsymbol{r}_1,\boldsymbol{r}_2) \\ &= \frac{1}{\sqrt{2(1+U^2)}}[\psi_1(\boldsymbol{r}_1)\psi_2(\boldsymbol{r}_2) - \psi_2(\boldsymbol{r}_1)\psi_1(\boldsymbol{r}_2)]\,. \qquad (12.37)\end{aligned}$$

Both functions are already normalized. $U(R)$ is the overlap integral given after (12.26).

We take the expectation value of the Hamiltonian in order to estimate the energy:

$$\begin{aligned}\varepsilon_\pm &= \langle\hat{H}\rangle_\pm \\ &= \frac{\langle 12|\hat{H}|12\rangle + \langle 21|\hat{H}|21\rangle}{2(1\pm U^2)} \\ &\quad \pm\frac{\langle 12|\hat{H}|21\rangle + \langle 21|\hat{H}|12\rangle}{2(1\pm U^2)}\,. \qquad (12.38)\end{aligned}$$

Furthermore we use

$$\langle 12|\hat{H}|12\rangle = \langle 21|\hat{H}|21\rangle$$

and

$$\langle 12|\hat{H}|21\rangle = \langle 21|\hat{H}|12\rangle\,.$$

Hence, (12.38) is simplified to

$$\varepsilon_\pm = \frac{\langle 12|\hat{H}|12\rangle}{1\pm U^2} \pm \frac{\langle 21|\hat{H}|12\rangle}{1\pm U^2}\,. \qquad (12.39)$$

The Hamiltonian has the form

$$\begin{aligned}\hat{H} &= -\frac{\boldsymbol{\nabla}_1^2}{2m} - \frac{\boldsymbol{\nabla}_2^2}{2m} - \frac{e^2}{|\boldsymbol{r}_1-\boldsymbol{R}_1|} - \frac{e^2}{|\boldsymbol{r}_2-\boldsymbol{R}_2|} \\ &\quad - \frac{e^2}{|\boldsymbol{r}_1-\boldsymbol{R}_2|} - \frac{e^2}{|\boldsymbol{r}_2-\boldsymbol{R}_1|} + \frac{e^2}{|\boldsymbol{R}_1-\boldsymbol{R}_2|} + \frac{e^2}{|\boldsymbol{r}_1-\boldsymbol{r}_2|}\,. \qquad (12.40)\end{aligned}$$

The first two terms describe the kinetic energies of the first and second electron, respectively. The next four terms represent the attractive Coulomb interaction between electrons and nuclei. The second from last term represents the repulsive interaction between the nuclei, and the last term that between the two electrons. It holds that

$$\langle 12|\hat{H}|12\rangle = 2\varepsilon_1 + \frac{e^2}{R} + V_c(R)\,, \qquad (12.41)$$

where $R \equiv |\boldsymbol{R}_1 - \boldsymbol{R}_2|$. Here we have taken advantage of

$$\left\langle 1\left| -\frac{\boldsymbol{\nabla}_1^2}{2m} - \frac{e^2}{|\boldsymbol{r}_1-\boldsymbol{R}_1|}\right|1\right\rangle = \varepsilon_1\,;$$

the same holds for $\langle 2|\hat{H}|2\rangle$. Furthermore, we have

$$\begin{aligned}V_c(R) &= \int \mathrm{d}^3r_1\,\mathrm{d}^3r_2\,|\psi_1(\boldsymbol{r}_1)|^2|\psi_2(\boldsymbol{r}_2)|^2 \\ &\quad\times\left(\frac{e^2}{|\boldsymbol{r}_1-\boldsymbol{r}_2|} - \frac{e^2}{|\boldsymbol{r}_2-\boldsymbol{R}_1|} - \frac{e^2}{|\boldsymbol{r}_1-\boldsymbol{R}_2|}\right)\,, \qquad (12.42)\end{aligned}$$

and

$$\langle 12|\hat{H}|21\rangle = U^2\left(2\varepsilon_1 + \frac{e^2}{R}\right) + A(R) \qquad (12.43)$$

with

$$A(R) = \int d^3r_1\, d^3r_2\, \psi_1^\dagger(\boldsymbol{r}_1)\psi_2^\dagger(\boldsymbol{r}_2)\psi_1(\boldsymbol{r}_2)\psi_2(\boldsymbol{r}_1) \times \left(\frac{e^2}{|\boldsymbol{r}_1 - \boldsymbol{r}_2|} - \frac{e^2}{|\boldsymbol{r}_2 - \boldsymbol{R}_1|} - \frac{e^2}{|\boldsymbol{r}_1 - \boldsymbol{R}_2|}\right) . \qquad (12.44)$$

$A(R)$ is the contribution from the exchange interaction. The exchange integral (12.44) is not easy to calculate. As a crude estimate it is proportional to the square of the overlap intergral

$$U^2 = \int d^3r_1\, d^3r_2\, \psi_1^\dagger(\boldsymbol{r}_1)\psi_2^\dagger(\boldsymbol{r}_2)\psi_1(\boldsymbol{r}_2)\psi_2(\boldsymbol{r}_1) = \left(\int d^3r_1\, \psi_1^\dagger(\boldsymbol{r}_1)\psi_2(\boldsymbol{r}_1)\right)\left(\int d^3r_2\, \psi_2^\dagger(\boldsymbol{r}_2)\psi_1(\boldsymbol{r}_2)\right) . \qquad (12.45)$$

Inserting (12.41)–(12.44) into (12.39) we obtain

$$\varepsilon_\pm(R) = 2\varepsilon_1 + \frac{e^2}{R} + \frac{V_c(R) \pm A(R)}{1 \pm U^2} , \qquad (12.46)$$

or equivalently

$$\varepsilon_\pm = 2\varepsilon_1 + \frac{(V_c(R) + e^2/R) \pm (A(R) + U^2e^2/R)}{1 \pm U^2} . \qquad (12.47)$$

Now it turns out, that $V_c(R) + e^2/R$ is always positive and $A(R) + U^2e^2/R$ is mostly negative. For this reason $\varepsilon_+(R)$ is lower in energy than $\varepsilon_-(R)$. If the "binding energy" $\varepsilon_-(R) - 2\varepsilon_1$ or $\varepsilon_+(R) - 2\varepsilon_1$ is a function of R, the first expression shows no and the last one minimum (see Fig. 12.6).

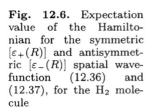

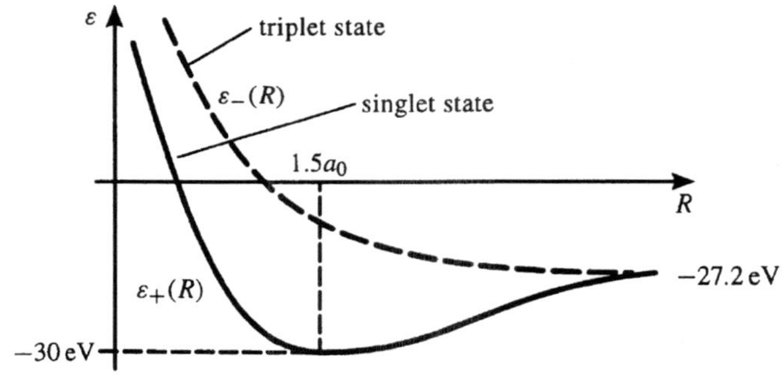

Fig. 12.6. Expectation value of the Hamiltonian for the symmetric $[\varepsilon_+(R)]$ and antisymmetric $[\varepsilon_-(R)]$ spatial wavefunction (12.36) and (12.37), for the H_2 molecule

From this result it is obvious that the hydrogen molecule is bound in the state ψ_+, i.e. the ground state is a spin singlet state. Note that the binding is due to the negative term $A(R) + U^2e^2/R$. It is thus the large overlap of the wavefunctions that causes the binding. The strength of the binding is, as a rough estimate, proportional to the overlap of the two electron wavefunctions. The triplet state is higher in energy because ψ_- disappears in the middle of the two protons. This follows from the antisymmetry of ψ_- [see (12.37)]. Between the protons is the place where the attractive interaction between electron and proton can be used in an optimum way.

For larger distances R between the two protons we have two hydrogen atoms and for $R \to 0$ the He atom has to appear. The energy of the ground state has to be somewhere in between. The ground state of the He atom is a singlet state. Also the H_2 molecule should have such a state. However, the test wavefunction produces an energy that is too high for $R \to 0$; the binding energy is too low. This has to be expected from a test wavefunction.

A real molecule shows a $1/R^6$ behavior for large R, which is known as the *van der Waals interaction*. It is attractive. We cannot obtain such an interaction here since more accurate calculations have to be performed: for the two hydrogen atoms, a second-order perturbation calculation with respect to the electron–proton interaction and to the electron–electron interaction provides the correct $1/R^6$ behavior. The van der Waals interaction makes up approximately 1/1000 of the molecular binding calculated here. This demonstrates that our wavefunction, which also has the wrong limit for $R \to 0$, cannot describe such a small effect. For $R \gg 1$ the energy is described fairly well, but the behavior of the wavefunction is not.

12.4 Electron Pairing

For the H_2 molecule we have seen that a large overlap of the two electron wavefunctions causes binding. This has been the case for the the spin singlet state. In the triplet state the overlap is small, which leads to a repulsion. Now we can imagine why there is no H-He molecule: the two electrons of the He atom in the ground state are in a spin singlet state. Both electrons are coupled antiparallel. If both nuclei are very close to each other, the three electrons (one from H and two from He) see a Li nucleus. Since the 1s orbit is completely occupied, the third electron has to go into an orbit with one main quantum number higher. In the Li^- ion the last electron has an ionization energy of only 0.4 Ry. As a consequence we expect that there is no binding in the H-He molecule. If H and He are separated (see Fig. 12.7), the electron of the H atom has the possibility of interacting both with the electron of the He atom that is parallel to the electron of the H atom and with the electron that is oriented antiparallel. The last case does not work, since the interaction is spin independent and cannot change the spin in the exchange integral (see Fig. 12.7). Therefore it can only interact with the electron of the He atom that is aligned parallel. This is similar to the triplet state of He, where (by definition) two electrons are aligned parallel and interact with each other. The triplet state is not bound: just as it is not bound the H-He molecule. In other words, there is no H-He atom.

The reason for this can also be seen directly from the wavefunctions. For this we build the Slater determinant of the three-electron system:

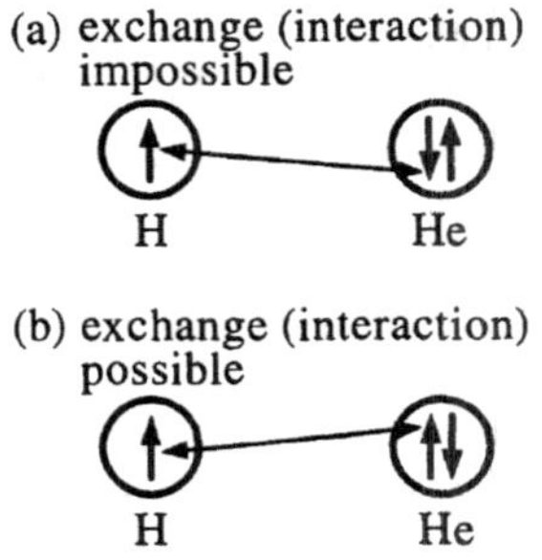

Fig. 12.7. Illustration of the impossible (**a**) and possible (**b**) interactions of the electrons in a H-He molecule. However, the triplet interaction in case (**b**) is repulsive. Therefore no H-He molecule exists

$$\psi(1,2,3) = N \begin{vmatrix} \psi_1(1)\chi_\uparrow(1) & \psi_1(1)\chi_\downarrow(1) & \psi_2(1)\chi_\uparrow(1) \\ \psi_1(2)\chi_\uparrow(2) & \psi_1(2)\chi_\downarrow(2) & \psi_2(2)\chi_\uparrow(2) \\ \psi_1(3)\chi_\uparrow(3) & \psi_1(3)\chi_\downarrow(3) & \psi_2(3)\chi_\uparrow(3) \end{vmatrix} . \qquad (12.48)$$

Here ψ_1 is the 1s helium wavefunction and ψ_2 the 1s hydrogen wavefunction. $\chi_\uparrow$ and $\chi_\downarrow$ are the spin wavefunctions with $m_s = \pm 1/2$, respectively. We calculate the scalar product of (12.48) and obtain, after a direct calculation,

$$\begin{aligned} N^{-2} &= 6 \int d^3r_1\, d^3r_2\, d^3r_3\, \psi_1^\dagger(1)\psi_1^\dagger(2)\psi_2^\dagger(3) \\ &\quad \times [\psi_1(1)\psi_2(3) - \psi_2(1)\psi_1(3)]\psi_1(2) \\ &= 6(1-U^2) . \end{aligned} \qquad (12.49)$$

N is the normalization constant and U denotes the overlap integral between ψ_1 and ψ_2:

$$U = \int d^3r\, \psi_1^\dagger(\boldsymbol{r})\psi_2(\boldsymbol{r}) .$$

The factor 6 in (12.49) appears because we find, for example,

$$\int d^3r_1\, \psi_1^\dagger(1)\psi_2(1) = \int d^3r_2\, \psi_1^\dagger(2)\psi_2(2) .$$

In other words: a permutation was performed at some places in order to show that two terms are equal. Furthermore, $\chi_\uparrow$ and $\chi_\downarrow$ are orthogonal to each other.

A similar calculation leads to the energy:

$$\begin{aligned} \varepsilon(R) &= \frac{1}{1-U^2} \int d^3r_1\, d^3r_2\, d^3r_3\, \psi_1^\dagger(1)\psi_1^\dagger(2)\psi_2^\dagger(3) \\ &\quad \times \hat{H}[\psi_1(1)\psi_2(3) - \psi_2(1)\psi_1(3)]\psi_1(2) . \end{aligned} \qquad (12.50)$$

Here the normalized wavefunction has already been used with the norm of (12.49). The factor 6 cancels, because it also occurs in the numerator of (12.50). From (12.50) we see that only that part appears that describes electrons with the same spin in H and He atoms, i.e.

$$[\psi_1(1)\psi_2(3) - \psi_2(1)\psi_1(3)] .$$

Only electrons with the same spin direction can interact with each other because of the spin-independent Coulomb interaction. Look again at the spin wavefunctions in (12.48)! Two electrons in the same spatial state but with opposite spin are called *paired electrons*. In the example of the H-He molecule we observed that paired electrons interact *repulsively with other electrons*. This explains why rare gases show no chemical activity. In rare gases the shells are all closed, so all electrons are paired. The chemical activity arises as a result of nonpaired electrons. They still have a "free" place, which can be occupied by an electron belonging to another atom. The number of possible bindings is determined by the number of *valence electrons*, i.e. the nonpaired electrons in the outer shell. In addition, there is still the repulsive interaction with the paired electrons belonging to deeper shells. But this can be neglected, because the overlap of the outer-shell electrons with the inner-shell electrons is small.

The chemically active electrons are mostly outer (i.e. from higher shells) s or p electrons. The d and f electrons are often lying too close to the individual nucleus to bind with electrons of other atoms. The binding results in all electrons being paired for a molecule. Hence, such molecules have total spin zero. One exception is given by the molecules of the transition elements (with unpaired d and f electrons). Since the unpaired d and f electrons hardly overlap, they remain unpaired and cause little or no binding.

12.5 Spatially Oriented Orbits

The *alkaline earth metals*, which are found in the second main group of the periodic system, have closed orbits in their ground states: for example, Be has the occupation $1s^22s^2$. In this state they are chemically inactive. However, it does not cost much energy (some eV) to lift an s electron into the next p orbit. The p electron now has three different possibilities for orientation according to the projection onto the z axis ($m_e = 0, \pm 1$). Here, we have placed the z axis along the line connecting the two atomic nuclei. Figure 12.8 shows the state for $m = 0$. For $m = \pm 1$ the wavefunction is perpendicular to the connecting z axis. If the p electron binds with an another p electron (of the other atom), very strong binding results as long as both p electrons are in a $m = 0$ state. Then the two electrons can interact actively at a relatively large two-center distance, and therefore keep the Coulomb repulsion of the nuclei relatively low.

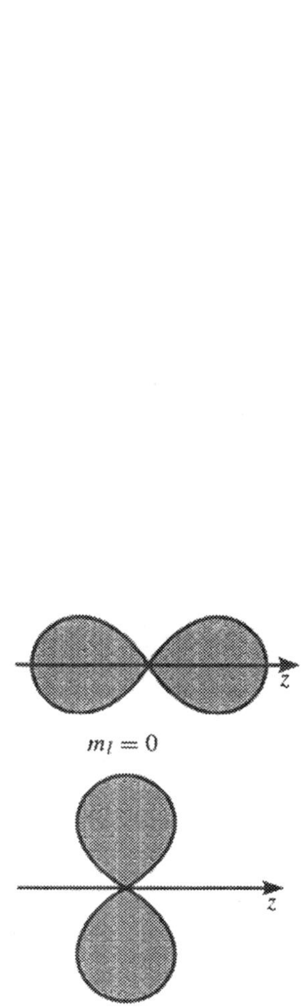

Fig. 12.8. Qualitative picture of the angular distribution of a p elctron with angular-momentum projection $m_l = 0$ and $m_l = \pm 1$

We call states orientated along the x, y, and z axis $|p_x\rangle$, $|p_y\rangle$, and $|p_z\rangle$; p stands for the p orbit. The spatial part of the wavefunction has an angular dependence of the form

$$|p_z\rangle = Y_{10} = \sqrt{\frac{3}{4\pi}}\cos\theta = \sqrt{\frac{3}{4\pi}}\frac{z}{r}, \tag{12.51a}$$

$$|p_y\rangle = \frac{1}{\sqrt{2}}(Y_{11} - Y_{1-1}) = \sqrt{\frac{3}{4\pi}}\frac{y}{r}, \tag{12.51b}$$

$$|p_x\rangle = \frac{1}{\sqrt{2}}(Y_{11} + Y_{1-1}) = \sqrt{\frac{3}{4\pi}}\frac{x}{r} . \tag{12.51c}$$

Here the cartesian coordinates are expressed in terms of the spherical ones.

In order to see how the bindings come about in a p orbit, we look, for example, at the C_2 molecule. In the ground state C has two unpaired 2p electrons, i.e. the *valence number* is 2. The two 2s electrons can be excited and also contribute to binding; we will not take this into account for the present consideration. The question is: Which state (configuration) is lowest in energy? We search for an arrangement with the largest overlap of the electron wavefunctions. This is the case if both electrons are in the $|p_z\rangle$ state (see Fig. 12.9).

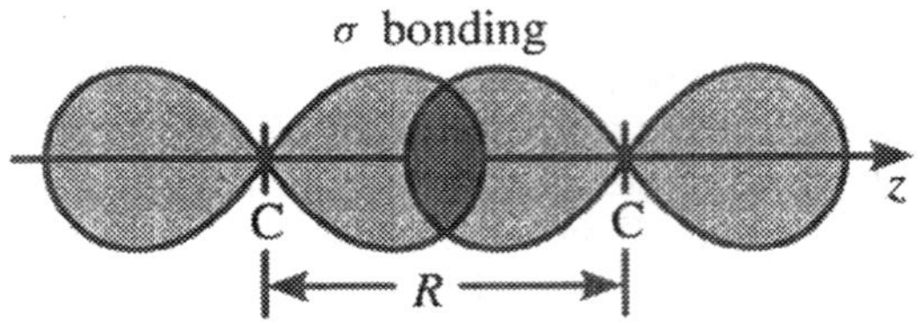

Fig. 12.9. σ bond of two C atoms: the $|p_z\rangle$ orbits of the two carbon nuclei have a large overlap at a relatively large distance R between the C nuclei

Once this binding has occured, there will be no place anymore for another electron because the interaction with it would be repulsive. The binding described above is also called *σ binding* (σ bond), since it possesses $m_e = 0$, in analogy to the notation of the s orbit with $l = 0$. Since the spherical symmetry is violated for a molecule, the molecular orbits cannot be classified by the angular momentum. On the other hand, the projection onto the z axis (L_z) still represents a good quantum number for two-atom molecules.

The next p electrons cannot occupy the $|p_z\rangle$ state anymore as they have to switch over to the $|p_x\rangle$ and $|p_y\rangle$ states. As can be seen in Fig. 12.10 these electrons will have a weaker binding since the overlap is lower: With the reduction in the central distance R the overlap increases but, in addition, the Coulomb repulsion of the nuclei increases. The wavefunction of this binding has the projection $m_e = \pm 1$ onto the z axis. Thus it is also called *π binding* (π bond). So the C_2 molecule has two bindings, a σ and a π binding.

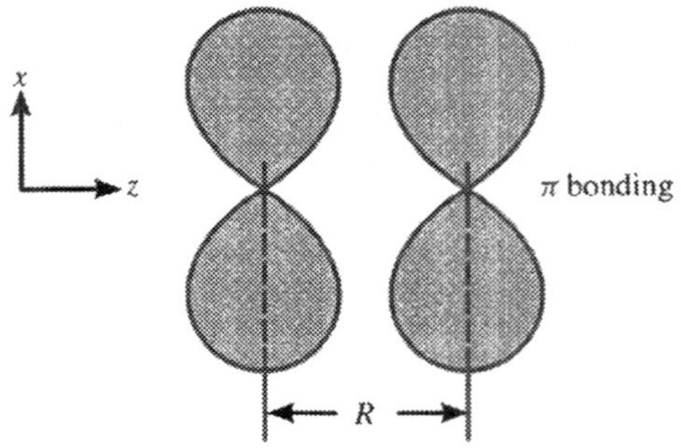

Fig. 12.10. Molecular orbits with $m_l = \pm 1$ only have relatively small overlap for a given two-center distance R. This results in weak binding. Since $m_e = \pm 1$ one speaks of π binding

Nitrogen has three 2p valence electrons. One takes a σ bond and the other two have to switch over to the $|p_x\rangle$ and $|p_y\rangle$ orbits. They experience a weaker π binding.

An interesting exception is the O_2 molecule. The O atom has four 2p valence electrons. The first three take one σ and two π bonds as for the nitrogen molecule. For the fourth electron there is now no binding orbit available. Together with the fourth electron of the other atom it forms a state with an overlap that is as low as possible: one electron in $|p_x\rangle$ and the other in $|p_y\rangle$. Both electrons now form an *antibound state* (as for ψ_- in H_2). A triplet state has spin 1. The O_2 molecule represents an exception from the rule that states that all two-atom molecules for which the atoms are identical and do not belong to transition elements have spin 0 in the ground state.

12.6 Hybridization

One way to amplify the σ binding is to deform the orbit in such a way that the overlap becomes larger (see Fig. 12.11a,b).

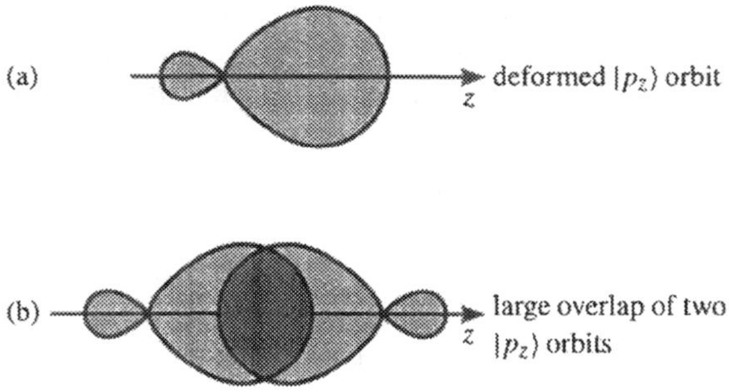

Fig. 12.11a,b. Hybridization of a state: schematic illustration of the amplification of the overlap effect. The σ bond is increased by deformation and, hence, an increased overlap of the $|p_z\rangle$ orbit. **(a)** $|p_z\rangle$ orbit. **(b)** Overlap of two $|p_z\rangle$ orbits

This only works if different angular momenta are mixed with each other. It becomes possible because the rotational symmetry is broken for a two-atom molecule. An example is the Li_2 *molecule*. Li has one 2s electron in the outer shell. But only little energy is needed to shift the 2s electron into a 2p orbit. We denote the s orbit by $|s\rangle$ and the p orbit by $|p\rangle$; then we are able to construct a test wavefunction

$$|s\rangle + \lambda|p_z\rangle\,. \tag{12.52}$$

λ is the variation parameter. The s wavefunction is spherically symmetric (see Fig. 12.12). The p wavefunction has positive sign in the range $z > 0$ and a negative one for $z < 0$ [see Fig. 12.12 and (12.51a)]. We have put the p electron into the $|p_z\rangle$ state.

For $z > 0$ both wavefunctions amplify each other. For $z < 0$ a negative interference (attenuation) appears. For this reason the shape appears on the right-hand side of Fig. 12.12. The overlap with the electron of the other atom is now larger and therefore the binding is stronger. The phenomenon of mixing two atomic orbits described here is called *hybridization*.

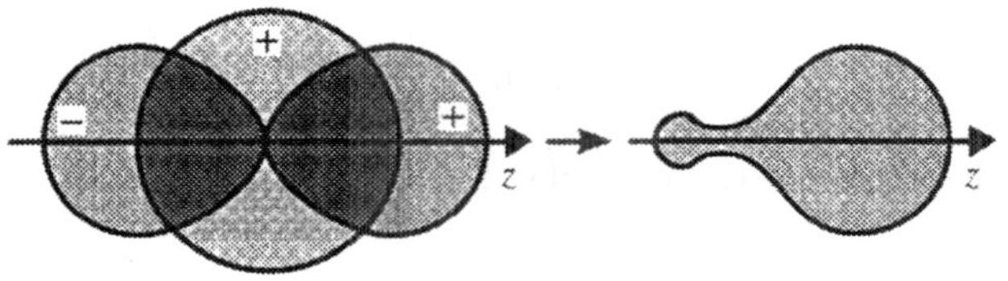

Fig. 12.12. Mixing of the $|s\rangle$ and $|p_z\rangle$ orbits causes a strongly deformed wavefunction. This illustrates the principle of hybridization

Next we consider the water molecule H_2O. The atoms are not lined up, which leads to a larger dipole moment. In the following, we will explain the phenomenon and follow two opposite extremes: no hybridization and maximum hybridization. The truth lies somewhere in between.

Without hybridization we can assume the following: The O atom has four p electrons; if no hybridization occurs, they remain p electrons. We place one electron into the $|p_x\rangle$ and one into the $|p_y\rangle$ state. The last two are paired in the $|p_z\rangle$ state. The electrons in $|p_x\rangle$ and $|p_y\rangle$ bind with the electrons of the H atoms. Since $|p_x\rangle$ and $|p_y\rangle$ are oriented perpendicular to each other, the calculated angle of the binding is 90°. Experimentally one finds 105°. If one takes into account that the two protons of the H atoms repel each other, then the angle becomes larger. Thus, we can imagine how the 105° arises.

The other way of arguing is to assume that the 2s and 2p orbits are energetically degenerate. (In reality the 2p orbit is somewhat higher.) Now mixing can occur. The states with lowest energy, which are obtained by a variation procedure, are

$$|1\rangle = \tfrac{1}{2}(|s\rangle + |p_x\rangle + |p_y\rangle + |p_z\rangle)\,, \tag{12.53a}$$

$$|2\rangle = \tfrac{1}{2}(|s\rangle + |p_x\rangle - |p_y\rangle - |p_z\rangle)\,, \tag{12.53b}$$

$$|3\rangle = \tfrac{1}{2}(|s\rangle - |p_x\rangle + |p_y\rangle - |p_z\rangle)\,, \tag{12.53c}$$

$$|4\rangle = \tfrac{1}{2}(|s\rangle - |p_x\rangle - |p_y\rangle + |p_z\rangle)\,. \tag{12.53d}$$

They have maximum mixing.

Now we consider the spatial orientation of these states (12.53a–d). In (12.53a) the p part has the form [see (12.51a–c)]

$$|p_x\rangle + |p_y\rangle + |p_z\rangle \sim \frac{x+y+z}{r}\,. \tag{12.54}$$

The wavefunction is oriented in direction $(1,1,1)$. For positive x, y, and z the wavefunction is positive and for negative x, y, and z it is negative. If we overlap the $|s\rangle$ state with this state, the wavefunctions add up in the direction $+(1,1,1)$ and substract in the direction $-(1,1,1)$. One proceeds similarly with the remaining states in (12.53b–d). $|2\rangle$ is pointing into the $(1,-1,-1)$ direction, $|3\rangle$ into $(-1,1,-1)$, and $|4\rangle$ into $(-1,-1,-1)$. The four orbits form a tetrahedron (see Fig. 12.13). The oxygen nucleus sits in the middle: from it the four hybridized states $|1\rangle, |2\rangle, |3\rangle, |4\rangle$ point starlike to the corners of the tetrahedron. These legs form an angle of 109.6° to each other which follows from the elementary geometrical relation $\sin(\theta/2) = \sqrt{2/3}$.

The calculated angle is too high. But as already mentioned, complete hybridization is an extreme case, because the s and p orbits in the oxygen atom are not degenerate. The truth lies in between no and maximum hybridization.

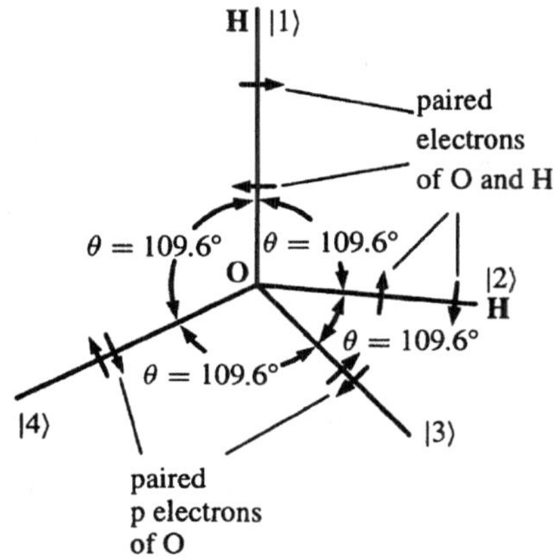

Fig. 12.13. Structure of the water molecule H_2O. The fully hybridized s and p orbits point from oxygen into the direction of the tetrahedron's corners (states $|1\rangle \cdots |4\rangle$). The angle between two neighboring legs is 109.6°, which agrees well with the angle 105° between both H atoms obtained from experiments

The ammonia molecule NH_3 is another interesting example. The molecule forms a tetrahedron with 107° between the axes. The nitrogen nucleus sits in one of the corners, for example, in the upper corner (see Fig. 12.14). If the 2p electrons of the nitrogen atom were only in the $|p_x\rangle$, $|p_y\rangle$, and $|p_z\rangle$ orbits, their angle would be 90°. Via mixing with the 2s state, the angle becomes larger. In the ammonia molecule the ground state splits into two energetically close configurations. This can be explained by the fact that the N atom can be on either side of the pyramid (see Fig. 12.14). The wavefunctions representing the N atom on the upper or lower side may be added or subtracted. The antisymmetric wavefunction belongs to the ground state of NH_3. The symmetrically combined wavefunction slightly above lies. One speaks of a splitting of the ground-state configurations as a result of tunneling of the N atom through the plane spanned by the three hydrogens.

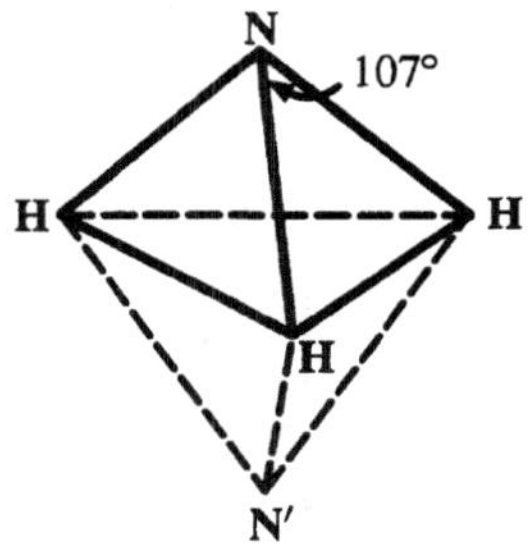

Fig. 12.14. The two, nearly degenerate ammonia configurations (NH_3). N can be located either above or below the plane spanned by the three hydrogens

We can understand some of the properties of water from the electronic properties of the H_2O molecule. The molecule has ten electrons altogether, two from the two H atoms and eight from the oxygen atom. The two 1s electrons in the O atom can be neglected in the following consideration. They belong to a lower-lying closed shell and are, so to speak, inert and therefore inactive. The remaining eight electrons fill the hybridized orbits (12.53a–d). In addition the oxygen attracts the electrons, which implies *polarization of the water molecule.*

Altogether the O atom has four charged legs, two in the directions of the H atoms and two away from the O atom (see Fig. 12.15). If now a second H_2O molecule comes close to the first one, the H atoms, which now appear to be more positively charged, can bind the oxygen atom with a negative leg (see Fig. 12.15). This kind of binding is called a *hydrogen bond.* With only 0.2 eV it is quite weak. The mean distance between a H atom and an O atom of another molecule is about 1.8 Å, whereas the distance between H and O within the same molecule is about 1 Å. This explains the low binding energy! Each water molecule may obviously undergo hydrogen bonding to four neighbors. This mechanism accounts for the accumulation of many water molecules of the form $H_{2n}O_n = (H_2O)_n$ in water (*cluster formation*).

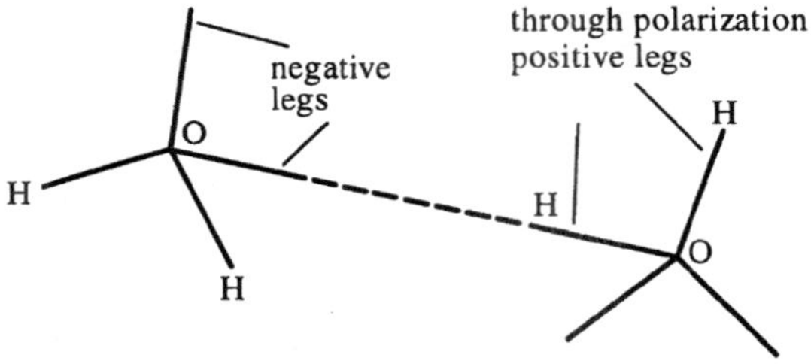

Fig. 12.15. Principle of the hydrogen bond of two H_2O molecules. It leads to an accumulation (building of clusters) of many water molecules, which are connected tetrahedronlike

There are many more molecules that form hydrogen bonds. They belong to the *water-soluble substances* because they untergo hydrogen bonding with the H_2O molecules of the water. Substances that do not form hydrogen bonds, are not water soluble. Oils belong to this category. Heating causes the hydrogen bonds to break and the substance becomes more fluid, i.e. its viscosity decreases.

12.7 Hydrocarbons

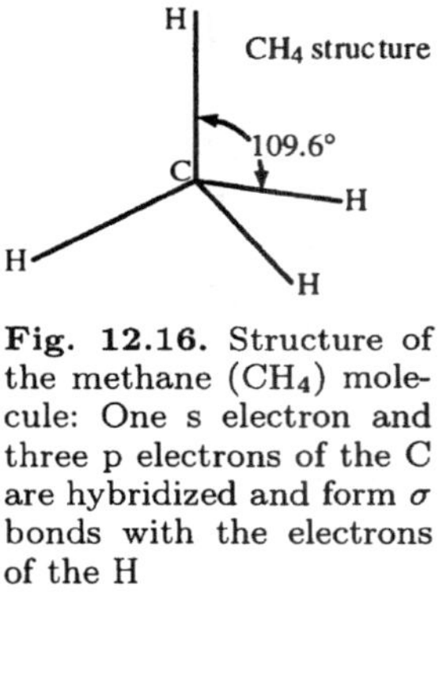

Fig. 12.16. Structure of the methane (CH_4) molecule: One s electron and three p electrons of the C are hybridized and form σ bonds with the electrons of the H

The considerations of the last sections have shown that the geometry of a molecule depends on the structure of the electron orbits and the hybridization. An excellent example of how electron configurations and the geometry of molecules are connected with each other can be found in hydrocarbons. In carbon the ground state is given by the configuration $(1s)^2(2s)^2(2p)^2$. In this state there are two p valence electrons. But the C atom has the most bonds via the $(1s)^2(2s)^1(2p)^3$ configuration, i.e. one 2s electron is shifted into the 2p orbital. Now the C atom has *four* valence electrons, which can bind with electrons of other atoms. We again use the completely hybridized states (12.53a–d) and occupy each of the four states with one electron. These electrons form σ bonds with the electrons of the H atom and give rise to the methane (CH_4) molecule. The angle between the bonds corresponds exactly to the calculated angle of 109.6° (see Fig. 12.16).

In the C_2H_4 molecule, *ethylene*, a planar structure occurs (see Fig. 12.17). This is explained by the following hybridization of the orbits:

$$|1\rangle = |p_z\rangle\,, \tag{12.55a}$$

$$|2\rangle = \sqrt{\frac{1}{3}}|s\rangle + \sqrt{\frac{2}{3}}|p_x\rangle\,, \tag{12.55b}$$

$$|3\rangle = \sqrt{\frac{1}{3}}|s\rangle - \sqrt{\frac{1}{6}}|p_x\rangle + \sqrt{\frac{1}{2}}|p_y\rangle\,, \tag{12.55c}$$

$$|4\rangle = \sqrt{\frac{1}{3}}|s\rangle - \sqrt{\frac{1}{6}}|p_x\rangle - \sqrt{\frac{1}{2}}|p_y\rangle\,. \tag{12.55d}$$

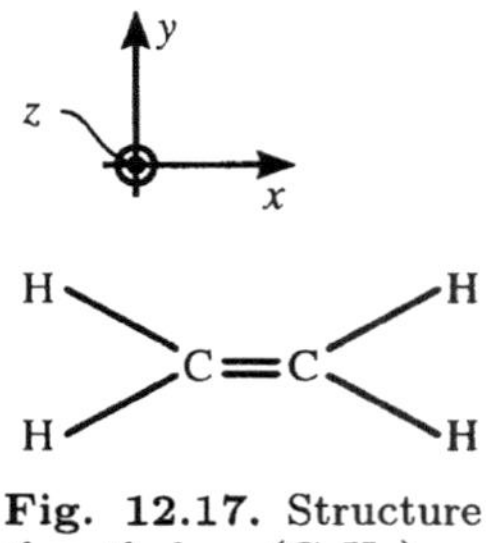

Fig. 12.17. Structure of the ethylene (C_2H_4) molecule. The x, y, z coordinates are indicated

These states follow from a variational calculation. The first state points in $\pm z$ direction, whereas the last three states form an angle of 120° in the x–y plane. The spatial electron-orbit distribution is illustrated in Fig. 12.18. The

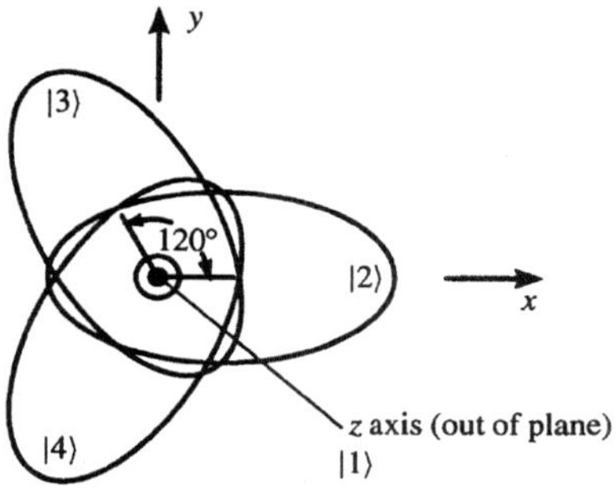

Fig. 12.18. The hybrid orbits $|2\rangle$, $|3\rangle$, and $|4\rangle$ of the C_2H_4 molecule [see (12.55b–d)] form an angle of 120° to each other. The z axis points out of the plane

electrons in states $|3\rangle$ and $|4\rangle$ form a bond with the electrons of the hydrogen atoms.

The electrons in states $|1\rangle$ and $|2\rangle$ form a bond with the electrons of the other C atom. The orbits of the second half (CH_2) of the C_2H_4 molecule are mirror-reflected to the first half. The electrons in $|2\rangle$ form a σ bond and the ones in $|1\rangle$ a π bond, since they are oriented perpendicular to the binding axis. The reason for the planar structure of the C_2H_4 molecule is the π bond. This bound becomes effective only if both CH_2 parts of the molecule are located in the same plane. The orbits for the π bond are oriented perpendicular to it (along the z axis in Fig. 12.18) Rotation around the x axis causes the π bond to break up (see Fig. 12.19). Thus, the π bond enforces the coplanarity of the two CH_2 complexes.

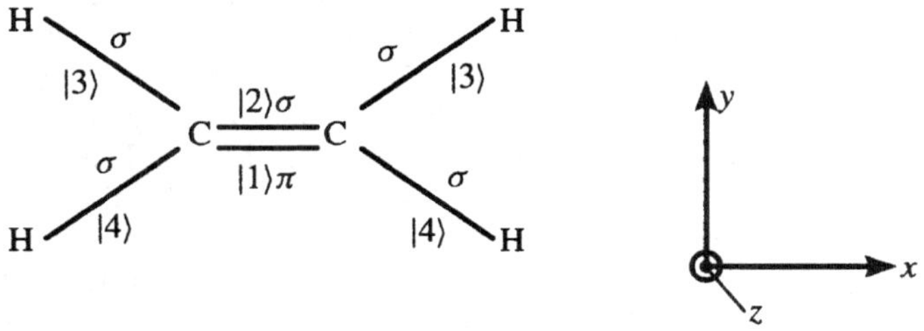

Fig. 12.19. The bond types in ethylene (C_2H_4). The π bond is mediated by the $|1\rangle$ orbit, which originates in the respective carbon atom and points out of the plane of the paper (along the z direction). For a better understanding look also at Figs. 12.10 and 12.18

A linear structure is observed in the *acetylene molecule* C_2H_2. This molecule is of the form H–C $\equiv$ C–H and is oriented along, for example, the x axis. The $|1s\rangle$ state now only mixes with $|p_x\rangle$ in the form

$$\frac{1}{\sqrt{2}}(|s\rangle \pm |p_x\rangle)\,. \tag{12.56}$$

Because of the positive and negative interference of the wavefunctions, we get one state aligned in the $+x$ direction and another one in $-x$ direction. The remaining $|p_y\rangle$ and $|p_z\rangle$ states form π bonds. This is illustrated in Fig. 12.20. The two π bond orbits are oriented in the first case in the y and in the second case in the z direction. It is perhaps useful to have a look again at Fig. 12.10, describing the π bond.

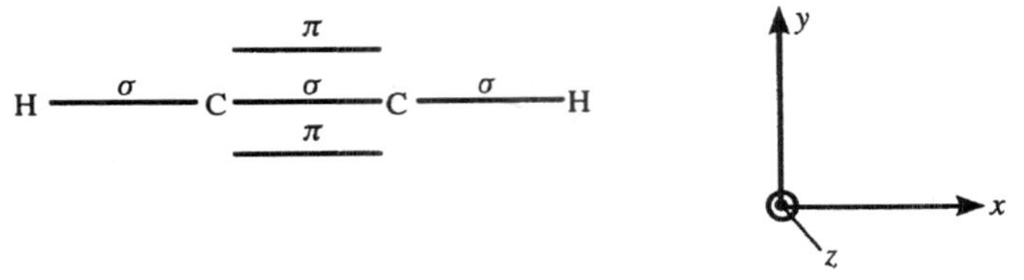

Fig. 12.20. The linear structure of the acetylene molecule (C_2H_2) along the x axis. The two π bonds between the carbon molecules come from the $|p_y\rangle$ and $|p_z\rangle$ orbits, respectively, and are oriented along the y axis and z axis, respectively

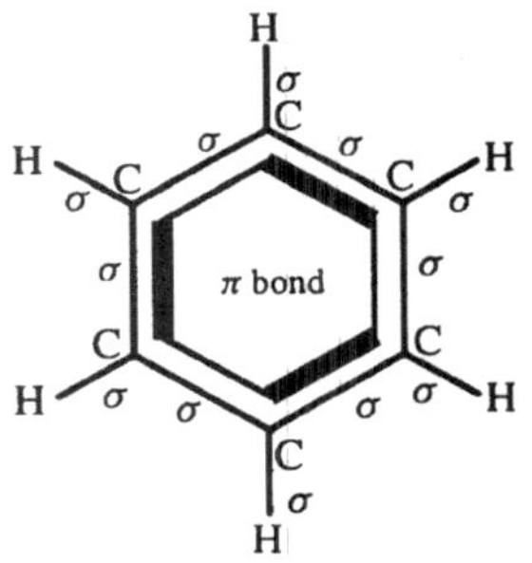

Fig. 12.21. The structure of the benzene molecule C_6H_6. It is ring-shaped and lies in a plane

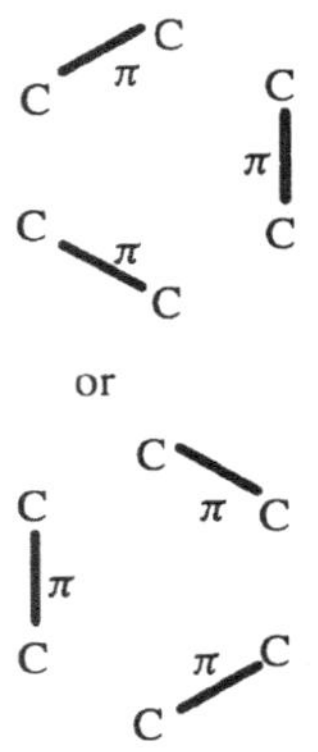

Fig. 12.22. Two equal possibilities for the π bond between the C atoms in the benzene ring. Because of the symmetry of the wavefunction one cannot distinguish between the two cases

In the carbon compounds discussed so far all bonds were localized, i.e. we know which bonds appear between which atoms. A particularly interesting situation appears in the case of the "aromatic" compound *benzene*. It consists of a ring with six carbon atoms, on each of which a hydrogen atom is bound (see Fig. 12.21).

The bonds C–H are all hybridized σ bonds. One bond between the C atoms is also of σ type. Hence, six out of eight valence electrons of two neighboring carbon atoms are involved in σ bonding, the other two will form a π bond. Therfore, between the six C atoms of the benzene-ring only three π bonds will be possible. This is illustrated in Fig. 12.21 by thicker lines. The states of the electrons causing the π bonds are oriented perpendicular to the σ bonds, i.e. out of the plane of the benzene ring. Because of the symmetry it is not possible to say between which C atoms the π bond appears. In Fig. 12.22 two possibilities are depicted.

Indeed, the state with lowest energy is represented by a mixing of these two possibilities. The electrons of the π bonds can be visualized as moving freely in the ring. Due to the larger extension available for these electrons their kinetic energy decreases (uncertainty principle!). With the lower kinetic energy the state is additionally lowered and, hence, stronger bonds result.

Summarizing, we can say that one can already reach good qualitative understanding of the structure of molecules by simple arguments such as the overlap of the electrons in singlet states and hybridization or by increasing the number of valence electrons by exciting an atom (e.g. carbon atom). But this procedure is not sufficient to make accurate, quantitative predictions of the structure of molecules. This can be accomplished only by large-scale computer calculations.

12.8 Biographical Notes

VAN DER WAAL, Johannes D., *Leiden 23.11.1837, †Amsterdam 9.3.1923, Dutch physicist. He worked as a teacher, whose science was largely self-taught. He studied at the University of Leiden and became physics professor in 1877 at the University of Amsterdam. In 1910 v.d.W was awarded the Nobel Prize in physics for his work on the equation of state of gases and liquids.

BORN, Max, *Breslau 11.12.1882, †Göttingen 5.1.1970, German physicist. B. was the son of an anatomy professor at Breslau. He was educated at the universities of Breslau, Heidelberg, Zürich, and Göttingen, where he reveived his PhD in 1907. With the rise of Hitler he moved to Britain, returning in 1953. Born shared with W. Bothe the 1954 Nobel Prize for physics. On his tombstone in Göttingen on finds the fundamental equation of quantum mechanics: $pq - qp = h/2\pi i$.

OPPENHEIMER, J. Robert, *New York City 22.4.1904, †Princeton 18.2.1967, American physicist. O. came from a wealthy family. He studied at Harvard, Cambridge, and Göttingen, where he obtained his PhD in 1927. He worked on the development of the atom bomb in Los Alamos during World War II. In 1963 he received the Fermi award.

LONDON, Fritz, *Breslau 7.3.1900, †Durham 30.3.1954, German physicist. L. was the son of a mathematics professor at Bonn and the brother of the physicist Heinz London. He studied in Frankfurt/Main, Munich, and Bonn where he received his degree in philosophy in 1921. He mainly worked with Heitler on superfluidity, superconductivity, and wave mechanics.

13. Feynman's Path Integral Formulation of Schrödinger's Wave Mechanics

In this final chapter we present an alternative formulation of quantum mechanics operating with new mathematical methods: the so-called path integrals. The first step towards such a description were made by Dirac, but the mathematical foundation and beauty was put forward by Feynman. It contributes essentially to a fundamental understanding of quantum mechanics and allows a derivation of exact equations in complex quantum field theory. It should be noted from the beginning that up to now path integral formulations have not played such an important role for solving certain field-theoretical problems; on the one hand, analytical solutions are only possible in very simple cases, and on the other hand, numerical calculations are extremely computer intense. However, path integrals often allow an approximate solution for physical processes, such as phase transitions, where perturbative methods fail.

In the following, we first remember the connection between classical mechanics and Schrödinger's wave mechanics in order to emphasize the importance of the *action* S. It turns out that the action represents the central quantity that enters into the path integral description of transition amplitudes (Greens functions). Its derivation implicitely assumes Born's probability interpretation of Schrödinger's wave field Ψ as "background knowledge". After we exclusively derive the transition amplitudes with the help of the path integration, we will derive the path integral representation from the conventional description of Schrödinger's Greens function. This shows the equivalence of both formulations. Taking for granted the connection between wavefunction and transition amplitude, we will derive the Schrödinger equation in the framework of the path integral formulation. For further reading see the literature.[1]

[1] R.P. Feynman, A.-R. Hibbs: *Quantum Mechanics and Path Integrals* (McGraw-Hill, New York 1965); C. Itzykson, J.-B. Zuber: *Quantum Field Theory* (McGraw-Hill, New York 1980); W. Greiner, J. Reinhardt: *Field Quantization* (Springer, Berlin, Heidelberg 1996).

13.1 Action Functional in Classical Mechanics and Schrödinger's Wave Mechanics

We begin with a short review of the classical action S, which plays a central role for the path integral formalism. An elegant access to the description of a classical particle is offered by Hamilton's action principle. The starting point is the action functional. L is the Lagrangian of the system and t_a and t_b are two fixed times; then

$$S[b,a] \equiv S[x(t_b),x(t_a)] = \int_{t_a}^{t_b} \mathrm{d}t'\, L(x(t'),\dot{x}(t'),t') \tag{13.1}$$

defines the so-called *action functional*, which assigns a defined value for any path $x(t)$ connecting the fixed end points $x_a = x(t_a)$ and $x_b = x(t_b)$. Within an infinite number of such paths the classical trajectory of a particle is represented by the special path $\bar{x}(t)$, for which the action (13.1) takes on an extremum, more precisely a minimum. The necessary condition for this is that for a small variation δx of the classical path $\bar{x}$ the corresponding action remaines unchanged in first order in δx:

$$\delta S = S[\bar{x}+\delta x] - S[\bar{x}] = 0\,. \tag{13.2}$$

As we know this leads to

$$\delta S = \delta x \frac{\partial L}{\partial \dot{x}}\bigg|_{t_a}^{t_b} + \int_{t_a}^{t_b} \mathrm{d}t' \left(\frac{\partial L}{\partial x} - \frac{\mathrm{d}}{\mathrm{d}t'}\frac{\partial L}{\partial \dot{x}}\right)\delta x\,. \tag{13.3}$$

For fixed end points $\delta x(t_a) = \delta x(t_b) = 0$. Hence the Euler Lagrange equation for the classical trajectory follows:

$$\frac{\partial L}{\partial x} - \frac{\mathrm{d}}{\mathrm{d}t}\frac{\partial L}{\partial \dot{x}} = 0\,. \tag{13.4}$$

In classical mechanics the form of the action functional is most interesting, and not the value of the action itself. We consider the action along a classical trajectory as a function of the upper bound

$$S_{\text{class.}}[\bar{x}(t)] = \int_{t_a}^{t} \mathrm{d}t'\, L(\bar{x}(t'),\dot{\bar{x}}(t'),t')\,. \tag{13.5}$$

From this, the energy and momentum at the end point can be calculated (see Exercise 13.1):

$$p = \frac{\partial S_{\text{class.}}}{\partial \bar{x}}\,, \tag{13.6a}$$

$$E = -\frac{\partial S_{\text{class.}}}{\partial t}\,. \tag{13.6b}$$

We now turn to quantum mechanics. Schrödinger's concept[2] for the derivation of an equation for the wavefunction Ψ followed the analogy between quantum mechanics and classical mechanics, on the one side, and to wave optics and its borderline case, i.e. geometrical optics, on the other side. If we describe the complex field Ψ in the form

[2] E. Schrödinger: Ann. d. Physik **79** (4) (1926) 361; **79** (4) (1926) 489.

$$\Psi(x,t) = a(x,t)\,\mathrm{e}^{\mathrm{i}S[x]/\hbar} \tag{13.7}$$

the wave equation for Ψ has to be fixed in such a way that, first, $|a(x,t)|^2 \equiv \rho(x,t) = |\Psi\Psi^*|$ fulfills the continuity equation, and that, second, the Hamilton Jacobi equation ("eikonal equation") emerges in the limit $\hbar \to 0$ for S. This classical limit of wave mechanics strengthens the importance of the action. Nevertheless Schrödinger's equation, understood as a wave equation, has only a partial connection to quantum mechanics. Only Born's interpretation of the field Ψ as a guiding field for the particles, i.e. of Ψ being the probability amplitude and $|\Psi|^2$ the probability density, established the fundamental significance of the Schrödinger equation

$$\left(\mathrm{i}\hbar\frac{\partial}{\partial t} - \hat{H}\right)|\Psi(t)\rangle = 0\,. \tag{13.8}$$

From this it becomes plausible that the action $S(x)$ enters somehow into the phase of probability amplitudes.

These preliminary remarks suffice and we now introduce the Schrödinger propagator for later use. We define the *Greens operator* by

$$\left(\mathrm{i}\hbar\frac{\partial}{\partial t} - \hat{H}\right)\hat{G}(t,t') = \mathrm{i}\hbar\,\hat{\mathbf{1}}\,\delta(t-t')\,. \tag{13.9}$$

The spatial Greens function

$$G(x,t;x',t') = \langle x|\hat{G}(t,t')|x'\rangle$$

fulfills the equation

$$\left(\mathrm{i}\hbar\frac{\partial}{\partial t} - \hat{H}_x\right)G(x,t;x',t') = \mathrm{i}\hbar\delta(x-x')\delta(t-t')\,. \tag{13.10}$$

Once $\hat{G}$ is known, the time-dependent solution $|\Psi(t)\rangle$ can be deduced according to

$$|\Psi(t)\rangle = \hat{G}(t,t_0)|\Psi(t_0)\rangle\,. \tag{13.11}$$

For the case of a Hamiltonian $\hat{H}$ that does not depend explicitly on time, a formal solution of (13.9) can be given immediately:

$$\hat{G}(t,t') = \theta(t-t')\exp\left(-\frac{\mathrm{i}}{\hbar}\hat{H}(t-t')\right)\,. \tag{13.12}$$

The Greens function in space and time results from this as a matrix element

$$G(x,t;x',t') = \theta(t-t')\langle x|\exp\left(-\frac{\mathrm{i}}{\hbar}\hat{H}(t-t')\right)|x'\rangle\,. \tag{13.13}$$

From here we will derive later the path integral description for the Greens function (13.13).

EXERCISE

13.1 Momentum and Energy at the End Point of a Classical Trajectory

Problem. Determine the momentum and energy at any end point of a classical trajectory.

Solution. The starting point is the classical action along the trajectory $\bar{x}(t)$ as a function of the upper bound:

$$S_{\text{class.}}[\bar{x}(t), x_0] = \int_{t_0}^{t} \mathrm{d}t' \, L[\bar{x}(t'), \dot{\bar{x}}(t'), t'] \,. \tag{1}$$

Generally, the variation of the action is then

$$\delta S = \delta x \frac{\partial L}{\partial \dot{x}}\bigg|_{t_0}^{t} + \int_{t_0}^{t} \mathrm{d}t' \, \delta x \left(\frac{\partial L}{\partial x} - \frac{\mathrm{d}}{\mathrm{d}t'} \frac{\partial L}{\partial \dot{x}} \right) . \tag{2}$$

Along the classical trajectories the integral term vanishes. Assuming, furthermore, $\delta x(t_0) = 0$ and $p = \partial L/\partial \dot{x}$ yields

$$\delta S = p \, \delta x \,. \tag{3}$$

This equation has to be understood in the following way. It compares the variation of the action with respect to paths that differ by the endpoints $\bar{x}(t)$ taken at the same time t. Equation (3) tells us that the partial derivative of the action with respect to the coordinate of the upper end point is equal to the corresponding momentum, i.e.

$$p = \frac{\partial S_{\text{class.}}}{\partial \bar{x}} \,. \tag{4}$$

Analogously, we can take the action as an explicit function of time as we consider paths that end in the given location $\bar{x}$ at different times t.

The definition of the action (1) means the total time derivative of the action along a trajectory is

$$\frac{\mathrm{d}S_{\text{class.}}}{\mathrm{d}t} = L[\bar{x}(t), \dot{\bar{x}}(t), t] \,. \tag{5}$$

On the other hand one has

$$\frac{\mathrm{d}}{\mathrm{d}t} S_{\text{class.}} = \frac{\partial S_{\text{class.}}}{\partial t} + \frac{\partial S_{\text{class.}}}{\partial \bar{x}} \dot{\bar{x}} \,, \tag{6}$$

if we understand $S_{\text{class.}}$ in the above mentioned sense as a function of coordinates and time. From this it follows that

$$\frac{\partial S}{\partial t} = L - p\dot{\bar{x}} = -E \,. \tag{7}$$

13.2 Transition Amplitude as a Path Integral

Consider, for example, the double-slit experiment in quantum mechanics. An electron source Q is mounted in point a. At the screen S there appears an interference pattern, which is described by $|\Psi|^2$. We seek the probability amplitude $K(b,a)$ with which a particle propagates from point a to point b. Now, in quantum mechanics not only the classical trajectories with minimal action contribute to the probability amplitude in b, but also all other possible paths joining the points a and b. We are led to the following postulates.

1) All possible paths contribute equally, i.e. formally in the same way as the amplitude; but different paths contribute with different phases.
2) The phase of the contribution of a given path is determined by the action S along this path (measured in $\hbar$).

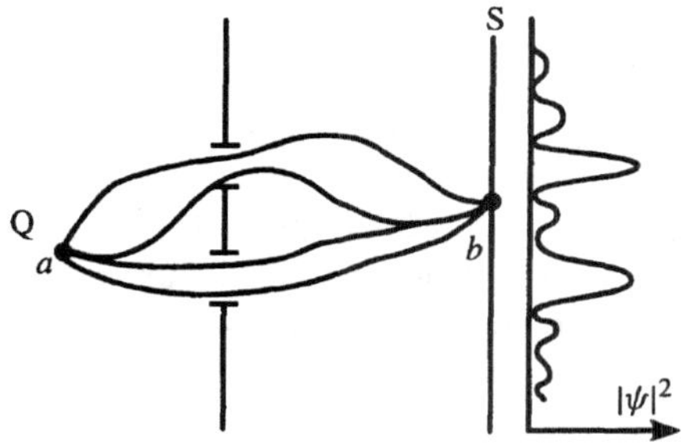

Fig. 13.1. Paths in the double-slit experiment

The probability $P(b,a)$ of reaching point x_b at time t_b from point x_a at time t_a is defined by the absolute square $P(b,a) = |K(b,a)|^2$ of an amplitude $K(b,a)$. Futhermore the amplitude is given by a sum of contributions $\phi[x(t)]$ of all paths joining a and b, i.e.

$$K(b,a) = \sum_{\substack{\text{sum over all paths}\\ \text{from } a \text{ to } b}} \phi[x(t)] \,. \tag{13.14}$$

The contribution of each path has a phase proportional to the action S along the path, i.e.

$$\phi[x(t)] = C \exp\left\{\frac{\mathrm{i}}{\hbar} S[x(t)]\right\} . \tag{13.15}$$

Here C is a normalization constant that has yet to be determined. It is the same for all paths, since all paths contribute with the same weight. Before we go into detail with the mathematical tools and specifications, in particular before we explain the meaning of what a "sum over all paths" is in (13.14), we shall clarify why in the classical limit one particular path, i.e. the classical trajectory, is most important: the classical limit means $S \gg \hbar$. Now, every path contributes in an equal manner, but for the classical path $S/\hbar$ is very large. The classical path $\bar{x}(t)$ is the one for which the phase in (13.15) becomes

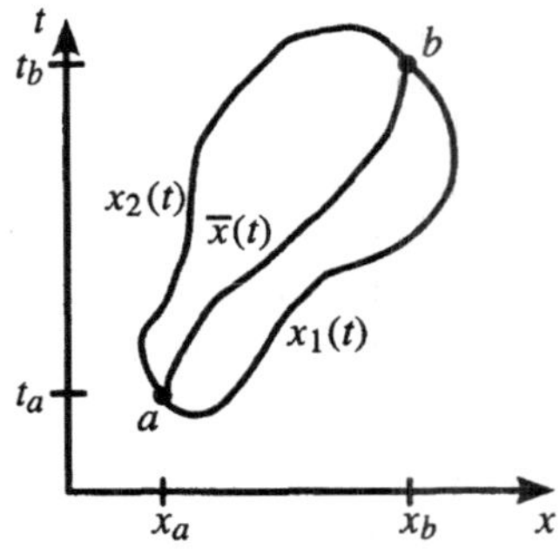

Fig. 13.2. The classical path $\bar{x}(t)$ and neighboring paths $x_\nu(t)$

stationary. Hence, infinitesimal neighboring paths have the same phase in first order. Compared to the classical one other paths lead to strongly oscillating phases (see Fig. 13.2). The corresponding contributions in (13.14) interfere destructively and cancel each other. Therefore the amplitude (13.14) essentially reduces to

$$K(b,a) \approx f(b,a)\exp\left(\frac{\mathrm{i}}{\hbar}S_{\mathrm{class.}}\right) , \tag{13.16}$$

where $f(b,a)$ will be a weakly changing function.

We will now discuss the sum contained in (13.14) in more detail and we will be guided by the analogy to Riemann's definition for an integral. For this we discretize the time interval $T = t_b - t_a$ into N equal partial intervals of length ϵ:

$$\begin{aligned} N\epsilon &= t_b - t_a , & t_b &> t_a , \\ \epsilon &= t_{i+1} - t_i , & i &= 1,\dots,N-1 , \\ t_0 &= t_a , & t_N &= t_b , \\ x_0 &= x_a , & x_N &= x_b . \end{aligned} \tag{13.17}$$

Let x_a, t_a and x_b, t_b be the two fixed end points. We now construct a certain path by choosing special points x_i for all intermediate time points t_i and connect the selected points by straight lines (see Fig. 13.3). By refining this

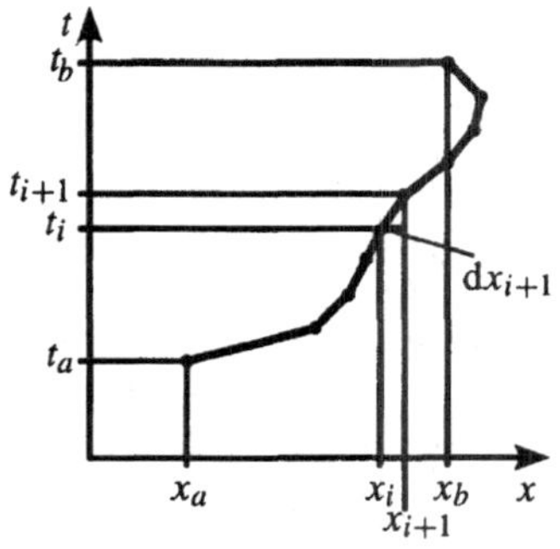

Fig. 13.3. A discretized path from (x_a, t_a) to (x_b, t_b)

lattice, we can approximate every path with any desired quality. Hence, it is natural to define the sum over all paths as a multiple integral over all values of x_i ($i = 1, \dots, N-1$; x_0, x_N fixed):

$$K(b,a) \sim \int \mathrm{d}x_1 \dots \mathrm{d}x_{N-1}\, \phi[x(t)] \,.$$

Let us now define the sum implied in (13.14) more precisely. In Riemann's integral ϵ can be made smaller and smaller. Here one cannot directly generate a well-defined limiting value in this way. In order to force convergence, a normalization factor $A(\epsilon)$, depending on ϵ, has to be introduced. Summation leads to the *path integral*:

$$K(b,a) = \lim_{\substack{\epsilon \to 0 \\ N \to \infty}} \frac{1}{A} \int \frac{\mathrm{d}x_1}{A} \cdots \frac{\mathrm{d}x_{N-1}}{A} \exp\left\{ \frac{\mathrm{i}}{\hbar} S[b,a] \right\} , \tag{13.18}$$

with the action

$$S[b,a] = \int_{t_a}^{t_b} \mathrm{d}t'\, L(x, \dot{x}, t') \,.$$

S is a line integral along a route $\mathrm{d}x_1 \dots \mathrm{d}x_{N-1}$ through the points x_i. The above construction of the path integral is not unequivocal. Instead of connecting the x_i by straight lines it is more useful in general to choose sections of classical paths. Such a construction is possible for any Lagrange function.

We introduce the abbreviated notation:

$$K(b,a) = \int_a^b \mathcal{D}x(t) \exp\left\{ \frac{\mathrm{i}}{\hbar} S[b,a] \right\} . \tag{13.19}$$

For the amplitude (13.18) and (13.19), an important property can be derived. Let us consider the path integral for two successive events (see Fig. 13.4) with $t_a < t_c < t_b$. The action fulfills the obvious property:

$$S[b,a] = S[b,c] + S[c,a] \,.$$

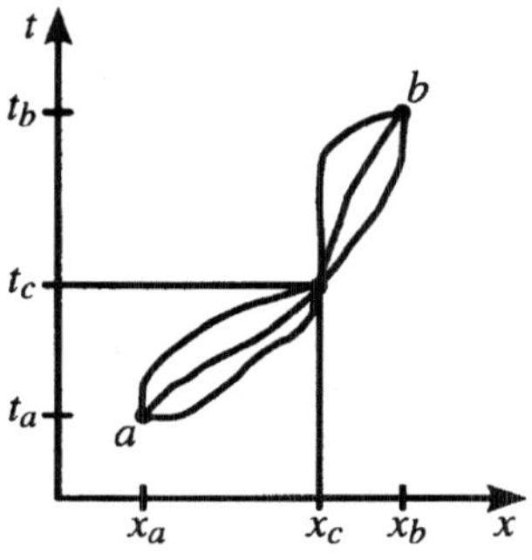

Fig. 13.4. Two successive events (x_c, t_c) and (x_b, t_b) and the corresponding paths

Now we consider (13.18):

$$\begin{aligned} K(b,a) &= \lim_{\substack{\epsilon\to 0\\ N\to\infty}} \frac{1}{A}\int \frac{\mathrm{d}x_1}{A}\cdots\frac{\mathrm{d}x_{M-1}}{A}\frac{\mathrm{d}x_c}{A}\frac{\mathrm{d}x_{M+1}}{A}\cdots \\ &\quad \times \frac{\mathrm{d}x_{N-1}}{A}\exp\left(\frac{\mathrm{i}}{\hbar}(S[b,c]+S[c,a])\right), \end{aligned} \tag{13.20}$$

with $t_a < t_1 \dots t_{M-1} < t_M \equiv t_c < t_{M+1} \dots < t_b$.

The integrations can be performed in any order, in particular as follows. First, for a fixed x_c, $S[a,c]$ will be a constant with respect to the integrations over x_{M+1} up to x_{N-1}. The same is valid for $S[c,b]$ with respect to the integrations over x_1 up to x_{M-1}. The integration over the point in between, x_c, is performed at the end. Rewriting (13.20) accordingly yields

$$\begin{aligned} K(b,a) &= \int \mathrm{d}x_c \lim_{\epsilon\to 0}\left(\frac{1}{A}\int\frac{\mathrm{d}x_1}{A}\cdots\frac{\mathrm{d}x_{M-1}}{A}\exp\left\{\frac{\mathrm{i}}{\hbar}S[c,a]\right\}\right) \\ &\quad \times\left(\frac{1}{A}\int\frac{\mathrm{d}x_{M+1}}{A}\cdots\frac{\mathrm{d}x_{N-1}}{A}\exp\left\{\frac{\mathrm{i}}{\hbar}S[b,c]\right\}\right), \end{aligned}$$

which can also be denoted as

$$K(b,a) = \int \mathrm{d}x_c\, K(b,c)K(c,a)\,. \tag{13.21}$$

Therefore the amplitude $K(b,a)$ can be calculated in such a way that first the path integration is performed from a to a fixed intermediate point c and afterwards from c to b and then over all possible intermediate points c. Equation (13.21) expresses therefore, how amplitudes of successive time events are multiplied. We can immediately generalize this to

$$\begin{aligned} K(b,a) &= \int \mathrm{d}x_1\dots\mathrm{d}x_{N-1}\dots\mathrm{d}x_i\dots K(b,N-1)K(N-1,N-2)\dots \\ &\quad \times K(i+1,i)\dots K(1,a)\,. \end{aligned} \tag{13.22}$$

For an infinitesimal time interval $\epsilon = t_{i+1} - t_i$ between points x_{i+1} and x_i the following is valid up to first order in ϵ:

$$\begin{aligned} K(i+1,i) &= \frac{1}{A}\exp\left(\frac{\mathrm{i}}{\hbar}S[i+1,i]\right) \\ &= \frac{1}{A}\exp\left[\frac{\mathrm{i}}{\hbar}\int_{t_i}^{t_{i+1}}\mathrm{d}t'\, L[x(t'),\dot{x}(t'),t']\right] \\ &\approx \frac{1}{A}\exp\left[\frac{\mathrm{i}\epsilon}{\hbar}L\left(\frac{x_{i+1}+x_i}{2},\frac{x_{i+1}-x_i}{\epsilon},\frac{\epsilon}{2}\right)\right]. \end{aligned} \tag{13.23}$$

Thus the integrand from (13.22), i.e. the amplitude for any complete path, can be written as

$$\phi[x(t)] = \lim_{\substack{\epsilon\to 0\\ N\to\infty}} \prod_{i=0}^{N-1} K(i+1,i)\,. \tag{13.24}$$

Let us specify (13.22) for the case of a particle's motion in a potential $V(x)$. According to the discretized form implied by (13.23) the corresponding Lagrangian reads

$$L = \frac{m}{2}\left(\frac{x_{i+1}-x_i}{\epsilon}\right)^2 - V(x_i)\,. \tag{13.25}$$

Insertion into (13.22) yields according to (13.23)

$$\begin{aligned} K(b,a) &= \lim_{\substack{\epsilon\to 0\\ N\to\infty}} \int \mathrm{d}x_1 \ldots \mathrm{d}x_{N-1} \\ &\quad \times \frac{1}{A^N} \exp\left[\frac{\mathrm{i}\epsilon}{\hbar}\sum_{i=0}^{N-1}\left(\frac{m}{2}\left(\frac{x_{i+1}-x_i}{\epsilon}\right)^2 - V(x_i)\right)\right]\ ; \end{aligned} \tag{13.26}$$

this is the probability amplitude of finding a particle at b coming from a. The sum in the exponent of (13.26) has to be understood as a Riemann sum of the action integral along a certain path. This will be further illuminated in the following exercise.

EXERCISE

13.2 The Transition Amplitude for a Free Particle

Problem. Determine the transition amplitude $K(b,a)$ for a free particle with the help of (13.26).

Solution. The starting point is (13.26) with $V = 0$ (free particle):

$$K(b,a) = \lim_{\substack{\epsilon\to 0\\ N\to\infty}} \int \mathrm{d}x_1 \ldots \mathrm{d}x_{N-1}\frac{1}{A^N}\exp\left[\frac{\mathrm{i}m}{2\hbar\epsilon}\sum_{j=0}^{N-1}(x_{j+1}-x_j)^2\right]\,, \tag{1}$$

with $x_0 = x_a$, $x_N = x_b$, and $N\epsilon = (t_b - t_a)$. The various integrations appearing in (1) are reducible to simple Gaussian integrals by means of the quadratic supplement in the exponent. The successive performance of the $N-1$ integrations leads to a set of Gaussian integrations:

$$\int_{-\infty}^{\infty} \mathrm{d}y\, \mathrm{e}^{\alpha(x-y)^2+\beta(z-y)^2} = \left(\frac{-\pi}{\alpha+\beta}\right)^{1/2}\exp\left[\frac{\alpha\beta}{\alpha+\beta}(x-z)^2\right]\,. \tag{2}$$

Let us begin with the integration over x_1 ($\mu = \mathrm{i}m/2\hbar\epsilon$):

$$\int_{-\infty}^{\infty} \mathrm{d}x_1\, \mathrm{e}^{\mu(x_2-x_1)^2+\mu(x_1-x_0)^2} = \left(\frac{-\pi}{2\mu}\right)^{1/2}\mathrm{e}^{\mu(x_2-x_0)^2/2}\,.$$

Integration over x_2 gives:

$$\begin{aligned} &\int_{-\infty}^{\infty} \mathrm{d}x_1\,\mathrm{d}x_2\, \mathrm{e}^{\mu[(x_3-x_2)^2+(x_2-x_1)^2+(x_1-x_0)^2]} \\ &= \left(\frac{-\pi}{2\mu}\right)^{1/2}\int_{-\infty}^{\infty}\mathrm{d}x_2\,\mathrm{e}^{\mu(x_2-x_0)^2/2+\mu(x_3-x_2)^2} \\ &= \left(\frac{-\pi}{2\mu}\right)^{1/2}\left(\frac{-\pi}{3\mu/2}\right)^{1/2}\mathrm{e}^{\mu(x_3-x_0)^2/3} \\ &= \left(\frac{(-\pi)^2}{3\mu^2}\right)^{1/2}\mathrm{e}^{\mu(x_3-x_0)^2/3}\,. \end{aligned}$$

Exercise 13.2.

Performing the $N-1$ integrations one after the other yields the expression

$$\int_{-\infty}^{\infty} \mathrm{d}x_1 \ldots \mathrm{d}x_{N-1} \exp\left\{\mu\left[(x_N - x_{N-1})^2 + \cdots + (x_1 - x_o)^2\right]\right\}$$
$$= \frac{1}{\sqrt{N}}\left(\frac{-\pi}{\mu}\right)^{(N-1)/2} \mathrm{e}^{\mu(x_N - x_o)^2/N} . \tag{3}$$

The amplitude (1) then takes the following form:

$$K(b,a) = \lim_{\substack{\epsilon\to 0 \\ N\to\infty}} \left(\frac{1}{A(\epsilon)}\right)^N \left(\frac{2\pi \mathrm{i}\hbar\epsilon}{m}\right)^{N/2}$$
$$\times \left(\frac{m}{2\pi\mathrm{i}\hbar(t_b - t_a)}\right)^{1/2} \exp\left[\frac{\mathrm{i}m}{2\hbar}\frac{(x_b - x_a)^2}{(t_b - t_a)}\right] . \tag{4}$$

But now the limits $\epsilon \to 0$, $N \to \infty$ should exist. The only way to guarantee the convergence is to fix the normalization factor:

$$A(\epsilon) = \left(\frac{2\pi\mathrm{i}\hbar\epsilon}{m}\right)^{1/2} . \tag{5}$$

We will derive this factor in another way in the next subsection.

Then the result is

$$K(b,a) = \left(\frac{m}{2\pi\mathrm{i}\hbar(t_b - t_a)}\right)^{1/2} \exp\left[\frac{\mathrm{i}m}{2\hbar}\frac{(x_b - x_a)^2}{(t_b - t_a)}\right] . \tag{6}$$

It is remarkable that this expression is identical with the spatial representation of the Greens function of the one-dimensional free Schrödinger equation.[3]

13.3 Path Integral Representation of the Schrödinger Propagator

Now we will represent the one-dimensional Greens function (13.13) as a path integral. It is given by

$$G(x_a t_a; x_b t_b) = \langle x_b | \exp\left(-\frac{\mathrm{i}}{\hbar}\hat{H}\tau\right) | x_a \rangle , \qquad \tau = t_b - t_a > 0 , \tag{13.27}$$

with $\hat{H} = \hat{T} + \hat{V}$. The following identity holds for the operator function:

$$\mathrm{e}^{\hat{O}} = \left(\mathrm{e}^{\hat{O}/N}\right)^N . \tag{13.28}$$

If we substitute $\lambda = \mathrm{i}(t_b - t_a)/\hbar$ into (13.27), the Greens function becomes

$$G(x_a t_a; x_b t_b) = \lim_{N\to\infty} \langle x_b | \,\mathrm{e}^{-\lambda(\hat{T}+\hat{V})/N}\, \mathrm{e}^{-\lambda(\hat{T}+\hat{V})/N} \ldots \mathrm{e}^{-\lambda(\hat{T}+\hat{V})/N} | x_a \rangle .$$

We now apply the product formula (see Exercise 13.3):

[3] See e.g. W. Greiner, J. Reinhardt: *Quantum Electrodynamics*, 2nd ed. (Springer, Berlin, Heidelberg 1994).

$$\lim_{N\to\infty}\left[\left(e^{-\lambda(\hat{T}+\hat{V})/N}\right)^N - \left(e^{-\lambda\hat{T}/N}\,e^{-\lambda\hat{V}/N}\right)^N\right] = 0\,, \qquad (13.29)$$

and obtain:

$$G(x_a t_a; x_b t_b) = \lim_{N\to\infty}\langle x_b|\left(e^{-\lambda\hat{T}/N}\,e^{-\lambda\hat{V}/N}\right)^N|x_a\rangle\,. \qquad (13.30)$$

From here only a few more steps lead to the path integral. Inserting a complete set of spatial states

$$\hat{1} = \int dx_i\,|x_i\rangle\langle x_i|\,, \qquad i = 1,\ldots,N-1\,,$$

yields

$$G(x_a t_a; x_b t_b) = \lim_{N\to\infty}\int dx_1\ldots dx_{N-1} \times \prod_{i=0}^{N-1}\langle x_{i+1}|\,e^{-\lambda\hat{T}/N}\,e^{-\lambda\hat{V}/N}|x_i\rangle\,, \qquad (13.31)$$

where $x_0 = x_a$ and $x_N = x_b$. Now the matrix elements appearing in (13.31) have to be determined. Since the operator for the potential energy $\hat{V}$ is diagonal in space, we have

$$\langle x_{i+1}|\,e^{-\lambda\hat{T}/N}\,e^{-\lambda\hat{V}/N}|x_i\rangle = \langle x_{i+1}|\,e^{-\lambda\hat{T}/N}|x_i\rangle\,e^{-\lambda V(x_i)/N}\,. \qquad (13.32)$$

In order to calculate the spatial matrix element of the operator $e^{-\lambda\hat{T}/N}$ we insert a complete set of momentum eigenstates:

$$\begin{aligned}\langle x_{i+1}|\,e^{-\lambda\hat{T}/N}|x_i\rangle &= \int dp\,\langle x_{i+1}|\,e^{-\lambda\hat{T}/N}|p\rangle\langle p|x_i\rangle \\ &= \int dp\,\langle x_{i+1}|p\rangle\langle p|x_i\rangle\,e^{-\lambda p^2/2mN} \\ &= \frac{1}{2\pi\hbar}\int_{-\infty}^{\infty} dp\,e^{-\lambda p^2/2mN}\,e^{ip(x_{i+1}-x_i)}\,.\end{aligned}$$

Such Gaussian integrals can be calculated, resulting in

$$\int_{-\infty}^{\infty} dx\,e^{-\alpha x^2+\beta x} = \sqrt{\frac{\pi}{\alpha}}\,e^{\beta^2/4\alpha}\,.$$

Thus, we obtain

$$\langle x_{i+1}|\,e^{-\lambda\hat{T}/N}|x_i\rangle = \left(\frac{mN}{2\pi\lambda\hbar}\right)^{1/2}\exp\left[\frac{-mN(x_{i+1}-x_i)^2}{2\lambda\hbar^2}\right]\,. \qquad (13.33)$$

Inserting (13.32) into (13.31) results in

$$G(x_a t_a; x_b t_b) = \lim_{N\to\infty}\int dx_1\ldots dx_{N-1}\left(\frac{mN}{2\pi\lambda\hbar}\right)^{N/2} \times \prod_{i=0}^{N-1}\exp\left[-\frac{mN}{2\lambda\hbar^2}(x_{i+1}-x_i)^2 - \frac{\lambda}{N}V(x_i)\right]\,.$$

We insert $\epsilon = (t_b - t_a)/N = \hbar\lambda/iN$ and sum the exponential expressions:

$$G(x_a t_a; x_b t_b) = \lim_{\substack{\epsilon\to 0\\ N\to\infty}} \int \mathrm{d}x_1 \ldots \mathrm{d}x_{N-1} \left(\frac{m}{2\pi\mathrm{i}\hbar\epsilon}\right)^{N/2} \times \exp\left\{\frac{\mathrm{i}\epsilon}{\hbar}\sum_{i=0}^{N-1}\left[\frac{m}{2}(x_{i+1}-x_i)^2 - V(x_i)\right]\right\} . \quad (13.34)$$

With this we have deduced the path integral expression for the Schrödinger propagator. It is identical in form with the expression for the transition amplitude $K(b,a)$ in (13.26). Now the undetermined normalization factor appearing in (13.26) can be identified as

$$A(\epsilon) = \left(\frac{2\pi\mathrm{i}\hbar\epsilon}{m}\right)^{1/2} . \quad (13.35)$$

Note that we were led to the Greens function $K(b,a)$ of the Schrödinger equation exclusively by the path integration method.

EXERCISE

13.3 Trotter's Product Rule

Problem. Prove Trotter's product rule:

$$\lim_{N\to\infty}\left\{\left(\exp\left[-\frac{\lambda}{N}(\hat{T}+\hat{V})\right]\right)^N - \left(\exp\left[-\frac{\lambda}{N}\hat{T}\right]\exp\left[-\frac{\lambda}{N}\hat{V}\right]\right)^N\right\} = 0 . \quad (1)$$

Solution. We first show that the two operator functions

$$\hat{F}(\alpha) = \mathrm{e}^{-\alpha(\hat{T}+\hat{V})} \quad \text{and} \quad \hat{G}(\alpha) = \mathrm{e}^{-\alpha\hat{T}}\,\mathrm{e}^{-\alpha\hat{V}} \quad \text{with } \alpha = \frac{\lambda}{N}$$

differ only by commutation terms, which vanish in the limit $N \to \infty$.

An operator function is defined by its Taylor series, e.g.

$$\hat{G}(\alpha) = \sum_{n=0}^{\infty}\frac{(\alpha)^n}{n!}\left(\frac{\mathrm{d}^n\hat{G}}{\mathrm{d}\alpha^n}\right)\Bigg|_{\alpha=0} . \quad (2)$$

In the following, a useful operator identity will be applied:

$$\hat{K}(\alpha) = \mathrm{e}^{\alpha\hat{A}}\hat{B}\,\mathrm{e}^{-\alpha\hat{A}} = \sum_{n=0}^{\infty}\frac{(\alpha)^n}{n!}[\hat{A},\hat{B}]_{(n)} , \quad (3)$$

with $[\hat{A},\hat{B}]_{(0)} = \hat{B}$, $[\hat{A},\hat{B}]_{(1)} = [\hat{A},\hat{B}]$, $[\hat{A},\hat{B}]_{(2)} = [\hat{A},[\hat{A},\hat{B}]]$, ... For the proof of (3) the coefficients $(\mathrm{d}^n\hat{K}/\mathrm{d}\alpha^n)|_{\alpha=0}$ of the Taylor series have to be calculated:

$$\hat{K}(\alpha) = \sum_{n=0}^{\infty}\frac{(\alpha)^n}{n!}\left(\frac{\mathrm{d}^n\hat{K}}{\mathrm{d}\alpha^n}\right)\Bigg|_{\alpha=0} .$$

Thus: *Exercise 13.3.*

$n = 0$

$$\hat{K}(0) = [\hat{A},\hat{B}]_{(0)} = \hat{B}\,;$$

$n = 1$

$$\frac{\mathrm{d}\hat{K}}{\mathrm{d}\alpha} = \hat{A}\,\mathrm{e}^{+\alpha\hat{A}}\hat{B}\,\mathrm{e}^{-\alpha\hat{A}} - \mathrm{e}^{\alpha\hat{A}}\hat{B}\hat{A}\,\mathrm{e}^{-\alpha\hat{A}} \tag{4}$$

$$= \mathrm{e}^{\alpha\hat{A}}[\hat{A},\hat{B}]\,\mathrm{e}^{-\alpha\hat{A}}\,, \tag{5}$$

$$\left.\frac{\mathrm{d}\hat{K}}{\mathrm{d}\alpha}\right|_{\alpha=0} = [\hat{A},\hat{B}]_{(1)} = [\hat{A},\hat{B}]\,; \tag{6}$$

$n = 2$

$$\frac{\mathrm{d}^2\hat{K}}{\mathrm{d}\alpha^2} = \hat{A}\,\mathrm{e}^{\alpha\hat{A}}[\hat{A},\hat{B}]\,\mathrm{e}^{-\alpha\hat{A}} - \mathrm{e}^{\alpha\hat{A}}[\hat{A},\hat{B}]\hat{A}\,\mathrm{e}^{-\alpha\hat{A}} \tag{7}$$

$$= \mathrm{e}^{\alpha\hat{A}}[\hat{A},[\hat{A},\hat{B}]]\,\mathrm{e}^{-\alpha\hat{A}}\,, \tag{8}$$

$$\left.\frac{\mathrm{d}^2\hat{K}}{\mathrm{d}\alpha^2}\right|_{\alpha=0} = [\hat{A},[\hat{A},\hat{B}]] = [\hat{A},\hat{B}]_{(2)}\,. \tag{9}$$

For any n one has

$$\left.\frac{\mathrm{d}^n\hat{K}}{\mathrm{d}\alpha^n}\right|_{\alpha=0} = \underbrace{[\hat{A},\cdots[\hat{A},\hat{B}]]}_{n\ \text{times}} = [\hat{A},\hat{B}]_{(n)}\,. \tag{10}$$

Inserting (10) into the Taylor series for $\hat{K}(\alpha)$ yields the identity (3).

Turning to the operator function $\hat{G}(\alpha) = \mathrm{e}^{-\alpha\hat{T}}\,\mathrm{e}^{-\alpha\hat{V}}$ and calculating explicitly the first terms of its Taylor series, we obtain:

$n = 0$

$$\hat{G}(\alpha)|_{\alpha=0} = \hat{\mathbf{1}}\,;$$

$n = 1$

$$\frac{\mathrm{d}\hat{G}}{\mathrm{d}\alpha} = (-)\hat{T}\hat{G}(\alpha) + (-)\,\mathrm{e}^{-\alpha\hat{T}}V\,\mathrm{e}^{-\alpha\hat{V}} \tag{11}$$

$$= (-)\hat{T}\hat{G}(\alpha) + (-)\,\mathrm{e}^{-\alpha\hat{T}}V\,\mathrm{e}^{\alpha\hat{T}}\,\mathrm{e}^{-\alpha\hat{T}}\,\mathrm{e}^{-\alpha\hat{V}} \tag{12}$$

$$= (-)\hat{T}\hat{G}(\alpha) + (-)\left(\hat{V} + \sum_{m=1}^{\infty}\frac{(-\alpha)^m}{m!}[\hat{T},\hat{V}]_{(m)}\right)\hat{G}(\alpha)\,, \tag{13}$$

$$= (-)(\hat{T}+\hat{V})\hat{G}(\alpha) + (-)\sum_{m=1}^{\infty}\frac{(-\alpha)^m}{m!}[\hat{T},\hat{V}]_{(m)}\hat{G}(\alpha)\,, \tag{14}$$

$$\left.\frac{\mathrm{d}\hat{G}}{\mathrm{d}\alpha}\right|_{\alpha=0} = (-)(\hat{T}+\hat{V})\,; \tag{15}$$

Exercise 13.3.

$n = 2$

$$\frac{\mathrm{d}^2\hat{G}}{\mathrm{d}\alpha^2} = \left((-1)(\hat{T}+\hat{V}) + (-1)\sum_{m=1}^{\infty}\frac{(-\alpha)^m}{m!}[\hat{T},\hat{V}]_{(m)}\right)\frac{\mathrm{d}\hat{G}}{\mathrm{d}\alpha} \tag{16}$$

$$+(-1)^2\sum_{m=1}^{\infty}\frac{(-\alpha)^{m-1}}{(m-1)!}[\hat{T},\hat{V}]_{(m)}\hat{G}(\alpha)\,, \tag{17}$$

$$\left.\frac{\mathrm{d}^2\hat{G}}{\mathrm{d}\alpha^2}\right|_{\alpha=0} = (-1)^2(\hat{T}+\hat{V})^2 + (-1)^2[\hat{T},\hat{V}]\,. \tag{18}$$

In the way indicated, all higher derivatives can be determined. Then one gets

$$\left.\frac{\mathrm{d}^n\hat{G}}{\mathrm{d}\alpha^n}\right|_{\alpha=0} = (-1)^n(\hat{T}+\hat{V})^n + \{\text{commutator terms}\}\,.$$

Inserting this into the Taylor expansion (2) and performing the summation one obtains

$$\hat{G}(\alpha) = \hat{F}(\alpha) + \frac{\alpha^2}{2}[\hat{T},\hat{V}] + O(\alpha^3) = \frac{\lambda}{N}\,. \tag{19}$$

Hence we find

$$\left[\hat{F}(\alpha)\right]^N - \left(\hat{G}(\alpha)\right)^N = O(\alpha^2)\,, \tag{20}$$

i.e. the above difference is at least proportional to $\alpha^2 = \lambda^2/N^2$. In the limit $N \to \infty$ the right-hand side of (20) vanishes, which proves the validity of Trotter's formula (1).

13.4 Alternative Derivation of the Schrödinger Equation

In the preceding subsection we convinced ourselves that the path integral for the transition amplitude is identical with the Greens function for the one-dimensional Schrödinger equation

$$\left(-\frac{\hbar^2}{2m}\frac{\partial^2}{\partial x^2} + V(x,t)\right)\Psi(x,t) = \mathrm{i}\hbar\frac{\partial}{\partial t}\Psi(x,t)\,. \tag{13.36}$$

The Greens function $K(b,a)$, or if we specify the arguments in more detail, $K(xt, x't')$, represents the probability amplitude for the propagation of a particle from place x' at time t' to x at a later time t. The probability amplitude for finding a particle at time t at position x is described by the wavefunction $\Psi(x,t)$. If we are not interested in the particle's past, we can define a wavefunction by

$$\Psi(x,t) = \int \mathrm{d}x'\,K(xt, x't')\Psi(x',t')\,. \tag{13.37}$$

The integral core $K(xt, x't')$ propagates the wavefunction from time t' to time t. We now consider the special situation in which t and t' differ only by an infinitesimal time interval ϵ. Applying the corresponding expression (13.23) for the transition amplitude results in

$$\Psi(x,t+\epsilon) = \frac{1}{A}\int_{-\infty}^{\infty} \mathrm{d}x' \exp\left[\frac{\mathrm{i}\epsilon}{\hbar}L\left(\frac{x+x'}{2},\frac{x-x'}{\epsilon},t+\frac{\epsilon}{2}\right)\right]\Psi(x,t)\,. \tag{13.38}$$

The Lagrangian is given by

$$L = \frac{m}{2}\dot{x}^2 - V(x,t)\,.$$

Hence, we obtain explicitly

$$\Psi(x,t+\epsilon) = \int_{-\infty}^{\infty} \mathrm{d}x'\frac{1}{A}\exp\left[\frac{\mathrm{i}m}{2\hbar\epsilon}(x-x')^2 - \frac{\mathrm{i}\epsilon}{\hbar}V\left(\frac{x+x'}{2},t+\frac{\epsilon}{2}\right)\right]\Psi(x',t)\,. \tag{13.39}$$

The exponent includes all phases $(x-x')^2/\epsilon$. It leads to strong oscillations in the case of large deviations $x'-x=\xi$. Consequently, the main contribution to the integral is expected for close points x and x'. After substitution of $x'=x+\xi$,

$$\Psi(x,t+\epsilon) = \int_{-\infty}^{\infty} \mathrm{d}\xi\frac{1}{A}\exp\left(\frac{\mathrm{i}m}{2\hbar\epsilon}\xi^2\right)\exp\left[-\frac{\mathrm{i}\epsilon}{\hbar}V\left(x+\frac{\xi}{2},t+\frac{\epsilon}{2}\right)\right]\Psi(x+\xi,t)\,,$$

we expand the integrand and the left-hand side of the above equation, keeping only terms up to linear in ϵ and quadratic in ξ. This leads to

$$\Psi(x,t)+\epsilon\frac{\partial}{\partial t}\Psi(x,t) = \int_{-\infty}^{\infty}\mathrm{d}\xi\frac{1}{A}\exp\left(\frac{\mathrm{i}m}{2\hbar\epsilon}\xi^2\right) \times\left(1-\frac{\mathrm{i}\epsilon}{\hbar}V(x,t)\right)\left(\Psi(x,t)+\xi\frac{\partial}{\partial x}\Psi(x,t)+\frac{1}{2}\xi^2\frac{\partial^2}{\partial x^2}\Psi(x,t)\right)\,.$$

The integrals are given by

$$\int_{-\infty}^{\infty}\mathrm{d}\xi\exp\left(\frac{\mathrm{i}m}{2\hbar\epsilon}\xi^2\right) = \left(\frac{2\pi\mathrm{i}\hbar\epsilon}{m}\right)^{1/2}\,, \tag{13.40}$$

$$\int_{-\infty}^{\infty}\mathrm{d}\xi\exp\left(\frac{\mathrm{i}m}{2\hbar\epsilon}\xi^2\right)\xi = 0\,, \tag{13.41}$$

$$\int_{-\infty}^{\infty}\mathrm{d}\xi\exp\left(\frac{\mathrm{i}m}{2\hbar\epsilon}\xi^2\right)\xi^2 = \frac{\mathrm{i}\hbar\epsilon}{m}\left(\frac{2\pi\mathrm{i}\hbar\epsilon}{m}\right)^{1/2}\,. \tag{13.42}$$

Thus we obtain, up to $O(\epsilon^2)$,

$$\Psi(x,t)+\epsilon\frac{\partial}{\partial t}\Psi(x,t) = \frac{1}{A}\left[\left(\frac{2\pi\mathrm{i}\hbar\epsilon}{m}\right)^{1/2}\left(\Psi(x,t)-\frac{\mathrm{i}\epsilon}{\hbar}V(x,t)\Psi(x,t)+\frac{\mathrm{i}\hbar\epsilon}{2m}\frac{\partial^2}{\partial x^2}\Psi(x,t)\right)\right]\,. \tag{13.43}$$

The normalization factor $A(\epsilon)$ has to be choosen in such a way that (13.43) is fulfilled in the limit $\epsilon \to 0$. Therefore we have to choose $A(\epsilon) = (2\pi\hbar\epsilon/m)^{-1/2}$, which we have seen already [see (13.35) and Exercise 13.2]. It follows that

$$\Psi + \epsilon\frac{\partial}{\partial t}\Psi = \Psi - \frac{\mathrm{i}\epsilon}{\hbar}V\Psi - \frac{\hbar\epsilon}{2\mathrm{i}\,m}\frac{\partial^2}{\partial x^2}\Psi .$$

In first order in ϵ this equation is fulfilled only if

$$\mathrm{i}\hbar\frac{\partial}{\partial t}\Psi = -\frac{\hbar^2}{2m}\frac{\partial^2}{\partial x^2}\Psi + V\Psi$$

is valid. This is the Schrödinger equation.

13.5 Biographical Notes

DIRAC, Paul Adrien Maurice, English physicist, *Bristol 8.8.1902, †Florida 20.10.1984. Since 1927 D. was a member of St. Johns College at the University of Cambridge and from 1932 he was professor of mathematics and physics. From 1953 he was professor of mathematics and physics at the University of Oxford. Together with E. Schrödinger, Dirac obtained the Noble Prize for physics in 1933. His main work was quantum mechanics and nuclear physics. He set up the relativistic wave equation with spin 1/2 (Dirac equation).

FEYNMAN, Richard Philips, American physicist, *New York 11.5.1918, †Los Angeles 15.12.1988. Between 1950 and 1988 Feynman was professor at the California Institute of Technology in Pasadena. He formulated quantum electrodynamics and developed a graphical description for calculating complicated field-theoretical processes (Feynman diagrams). In 1965 he received the Noble Prize for physics.

Subject Index

图书在版编目（CIP）数据

量子力学专论 = Quantum Mechanics：Special Chapters：英文 /（德）W. 格雷纳著 . —影印本 . —北京：世界图书出版有限公司北京分公司，2019.3
ISBN 978-7-5192-6107-8

Ⅰ . ①量… Ⅱ . ① W… Ⅲ . ①量子力学—英文 Ⅳ .O413.1

中国版本图书馆 CIP 数据核字（2019）第 055842 号

中文书名 量子力学专论
英文书名 Quantum Mechanics: Special Chapters
著　　者 W. Greiner
责任编辑 刘　慧　高　蓉

出版发行 世界图书出版有限公司北京分公司
地　　址 北京市东城区朝内大街 137 号
邮　　编 100010
电　　话 010-64038355（发行）　64033507（总编室）
网　　址 http://www.wpcbj.com.cn
邮　　箱 wpcbjst@vip.163.com
销　　售 新华书店
印　　刷 北京建宏印刷有限公司
开　　本 787 mm × 1092 mm　1/16
印　　张 25
字　　数 476 千字
版　　次 2019 年 5 月第 1 版
印　　次 2020 年 5 月第 2 次印刷
版权登记 01-2019-1310
国际书号 ISBN 978-7-5192-6107-8
定　　价 129.00 元
